# Heuristics, Probability and Causality
## A Tribute to Judea Pearl

Titles published in the "Tributes" series

Volume 1
We Will Show Them! Essays in Honour of Dov Gabbay, Volume 1
S. Artemov, H. Barringer, A. d'Avila Garcez, L. Lamb and J. Woods, eds.

Volume 2
We Will Show Them! Essays in Honour of Dov Gabbay, Volume 2
S. Artemov, H. Barringer, A. d'Avila Garcez, L. Lamb and J. Woods, eds.

Volume 3
Probability and Inference: Essays in Honour of Henry E. Kyburg
Bill Harper and Greg Wheeler, eds.

Volume 4
The Way Through Science and Philosophy:
Essays in Honour of Stig Andur Pedersen
H. B. Andersen, F. V. Christiansen, K. F. Jørgensen, and V. F. Hendricks, eds.

Volume 5
Approaching Truth: Essays in Honour of Ilkka Niiniluoto
Sami Pihlström, Panu Raatikainen and Matti Sintonen, eds.

Volume 6
Linguistics, Computer Science and Language Processing.
Festschrift for Franz Guenthner on the Occasion of his $60^{th}$ Birthday
Gaston Gross and Klaus U. Schulz, eds.

Volume 7
Dialogues, Logics and Other Strange Things.
Essays in Honour of Shahid Rahman.
Cédric Dégremont, Laurent Keiff and Helge Rückert, eds.

Volume 8
Logos and Language.
Essays in Honour of Julius Moravcsik.
Dagfinn Follesdal and John Woods, eds.

Volume 9
Acts of Knowledge: History, Philosophy and Logic.
Essays dedicated to Göran Sundholm
Giuseppe Primiero and Shahid Rahman, eds.

Volume 10
Witnessed Years. Essays in Honor of Petr Hájek
Petr Cintula, Zuzana Haniková and Vítězslav Švejdar, eds.

Volume 11
Heuristics, Probability and Causality. A Tribute to Judea Pearl
Rina Dechter, Hector Geffner and Joseph Y. Halpern, eds.


Tributes Series Editor
Dov Gabbay                                            dov.gabbay@kcl.ac.uk

# Heuristics, Probability and Causality
## A Tribute to Judea Pearl

edited by
Rina Dechter
Hector Geffner
and
Joseph Y. Halpern

© Individual author and College Publications 2010. All rights reserved.

ISBN 978-1-904987-66-6 (Hardback)
ISBN 978-1-904987-65-9 (Paperback)

College Publications
Scientific Director: Dov Gabbay
Managing Director: Jane Spurr
Department of Computer Science
King's College London, Strand, London WC2R 2LS, UK

http://www.collegepublications.co.uk

Original cover design by orchid creative       www.orchidcreative.co.uk
Printed by Lightning Source, Milton Keynes, UK

---

All rights reserved. No part of this publication may be reproduced, stored in a retrieval system or transmitted in any form, or by any means, electronic, mechanical, photocopying, recording or otherwise without prior permission, in writing, from the publisher.

# Table of Contents

*List of Contributors* .................................................................... ix
*Preface* ........................................................................................ xi

## I. Heuristics ............................................................................. 1

1. **Heuristic Search for Planning Under Uncertainty**
   Blai Bonet and Eric A. Hansen ............................................. 3
2. **Heuristics, Planning, and Cognition**
   Hector Geffner ....................................................................... 23
3. **Mechanical Generation of Admissible Heuristics**
   Robert Holte, Jonathan Schaeffer, and Ariel Felner ............. 43
4. **Space Complexity of Combinatorial Search**
   Richard E. Korf ..................................................................... 53
5. **Paranoia Versus Overconfidence in Imperfect-Information Games**
   Austin Parker, Dana Nau and V.S. Subrahmanian ............... 63
6. **Heuristic Search: Pearl's Significance From a Personal Perspective**
   Ira Pohl .................................................................................. 89

## II. Probability ......................................................................... 103

7. **Inference in Bayesian Networks: A Historical Perspective**
   Adnan Darwiche .................................................................... 105
8. **Graphical Models of the Visual Cortex**
   Thomas Dean ......................................................................... 121
9. **On the Power of Belief Propagation: A Constraint Propagation Perspective**
   Rina Dechter, Bozhena Bidyuk, Robert Mateescu, and Emma Rollon ..... 143
10. **Bayesian Nonparametric Learning: Expressive Priors for Intelligent Systems**
    Michael I. Jordan .................................................................. 167

11. Judea Pearl and Graphical Models for Economics
    Michael Kearns ........................................................... 187

12. Belief Propagation in Loopy Graphs
    Daphne Koller ............................................................ 191

13. Extending Bayesian Networks to the Open-Universe Case
    Brian Milch and Stuart Russell ........................................... 217

14. A Heuristic Procedure for Finding Hidden Variables
    Azaria Paz ............................................................... 239

15. Probabilistic Programming Languages: Independent
    Choices and Deterministic Systems
    David Poole .............................................................. 253

16. Arguing with a Bayesian Intelligence
    Ingrid Zukerman .......................................................... 271

## III. Causality ........................................................... 293

17. Instrumental Sets
    Carlos Brito ............................................................. 295

18. Seeing and Doing: The Pearlian Synthesis
    Philip Dawid ............................................................. 309

19. Effect Heterogeneity and Bias in Main-Effects-Only
    Regression Models
    Felix Elwert and Christopher Winship ..................................... 327

20. Causal and Probabilistic Reasoning in P-Log
    Michael Gelfond and Nelson Rushton ....................................... 337

21. On Computers Diagnosing Computers
    Moises Goldszmidt ........................................................ 359

22. Overthrowing the Tyranny of Null Hypotheses
    Hidden in Causal Diagrams
    Sander Greenland ......................................................... 365

23. Actual Causation and the Art of Modeling
    Joseph Y. Halpern and Christopher Hitchcock .............................. 383

24. From C-Believed Propositions to Causal Calculator
    Vladimir Lifschitz ....................................................... 407

25. Analysis of the Binary Instrumental Variable Model
    Thomas S. Richardson and James M. Robins ................................. 415

26. **Pearl Causality and the Value of Control**
    Ross Shachter and David Heckerman .................................... 445

27. **Cause for Celebration, Cause for Concern**
    Yoav Shoham .......................................................... 463

28. **Automated Search for Causal Relations: Theory and Practice**
    Peter Spirtes, Clark Glymour, Richard Scheines, and Robert Tillman .... 467

29. **The Structural Model and the Ranking-Theoretic Approach to Causation: A Comparison**
    Wolfgang Spohn ....................................................... 507

30. **On Identifying Causal Effects**
    Jin Tian and Ilya Shpitser ........................................... 523

# IV. Reminiscences ..................................................... 545

31. **Questions and Answers**
    Nils J. Nilsson ...................................................... 547

32. **Fond Memories From an Old Student**
    Edward T. Purcell .................................................... 553

33. **Reverend Bayes and Inference Engines**
    David Spiegelhalter .................................................. 559

34. **An Old-Fashioned Scientist Shaping a New Discipline**
    Hector Geffner ....................................................... 563

35. **Sticking with the Crowd of Four**
    Rina Dechter ......................................................... 565

# List of Contributors

Bozhena Bidyuk, Google, USA

Blai Bonet, Departamento de Computación, Universidad Simón Bolívar, Venezuela

Carlos Brito, Departamento de Computacao, Universidade Federal do Ceara, Brazil

Adnan Darwiche, Computer Science Department, UCLA, USA

Philip Dawid, Statistical Laboratory, University of Cambridge, UK

Thomas Dean, Google, USA

Rina Dechter, School of Information and Computer Sciences, UC Irvine, USA

Felix Elwert, Department of Sociology, University of Wisconsin-Madison, USA

Ariel Felner, Department of Information Systems Engineering, Ben-Gurion University, Israel

Hector Geffner, ICREA and Universitat Pompeu Fabra, Spain

Michael Gelfond, Department of Computer Science, Texas Tech University, USA

Clark Glymour, Department of Philosophy, CMU, USA

Moises Goldszmidt, Microsoft Research, Silicon Valley, USA

Sander Greenland, Departments of Epidemiology and Statistics, UCLA, USA

Joseph Y. Halpern, Computer Science Department, Cornell University, USA

Eric A. Hansen, Department of Computer Science and Engineering, Mississippi State University, USA

David Heckerman, Microsoft Research, USA

Christopher Hitchcock, Division of the Humanities and Social Sciences, Caltech, USA

Robert Holte, Department of Computing Science, University of Alberta, Canada

Michael I. Jordan, Departments of EECS & Statistics, UC Berkeley, USA

Michael Kearns, Computer and Information Science Department, University of Pennsylvania, USA

Daphne Koller, Computer Science Department, Stanford University, USA

Richard E. Korf, Computer Science Department, UCLA, USA

Vladimir Lifschitz, Department of Computer Science, The University of Texas at Austin, USA

Robert Mateescu, Microsoft Research, Cambridge, UK

Brian Milch, Google, USA

Dana Nau, Computer Science Department, University of Maryland USA

Nils J. Nilsson, Computer Science Department, Stanford University, USA

Austin Parker, Computer Science Department, University of Maryland, USA

Azaria Paz, Computer Science Department, Technion – Israel Institute of Technology, Israel

Ira Pohl , Computer Science Department, UC Santa Cruz, USA

David Poole, Department of Computer Science, University of British Columbia, Canada

Edward T. Purcell, Los Angeles, USA

Thomas S. Richardson, Department of Statistics, University of Washington, USA

James M. Robins, Departments of Epidemiology and Biostatistics, Harvard University, USA

Emma Rollon, School of Information and Computer Sciences, UC Irvine, USA

Nelson Rushton, Department of Computer Science, Texas Tech University, USA

Stuart Russell, Department of EECS, UC Berkeley, USA

Jonathan Schaeffer, Department of Computing Science, University of Alberta, Canada

Richard Scheines, Department of Philosophy, CMU, USA

Ross Shachter, Department of Management Science and Engineering, Stanford University, USA

Yoav Shoham, Computer Science Department, Stanford University, USA

Ilya Shpitser, Department of Epidemiology, Harvard University, USA

David Spiegelhalter, Statistics Laboratory, University of Cambridge, UK

Peter Spirtes, Department of Philosophy, CMU, USA

Wolfgang Spohn, Department of Philosophy, University of Konstanz, Germany

V. S. Subrahmanian, Computer Science Department, University of Maryland, USA

Jin Tian, Department of Computer Science, Iowa State University, USA

Robert Tillman, Department of Philosophy and Machine Learning Department, CMU, USA

Christopher Winship, Department of Sociology, Harvard University, USA

Ingrid Zukerman, Faculty of Information Technology, Monash University, Australia

# Preface

This book is a collection of articles in honor of Judea Pearl written by close colleagues and former students. Its three main parts, heuristics, probabilistic reasoning, and causality, correspond to the titles of the three ground-breaking books authored by Judea, and are followed by a section of short reminiscences.

Judea Pearl was born in Tel Aviv and is a graduate of the Technion - Israel Institute of Technology. He came to the United States for postgraduate work in 1960. He received his Master's degree in physics from Rutgers University and his Ph.D. degree in electrical engineering from the Brooklyn Polytechnic Institute, both in 1965. Until 1969, he held research positions at RCA David Sarnoff Research Laboratories in Princeton, New Jersey and at Electronic Memories, Inc. at Hawthorne, California. In 1969 Pearl joined the UCLA faculty where he is currently an emeritus professor of computer science and director of the cognitive systems laboratory.

Judea started his research work in artificial intelligence (AI) in the mid-1970s, not long after joining UCLA. In the eyes of a hard scientist, AI must have been a fascinating but slippery scientific discipline then; a lot of AI was done through introspection and programming, building systems that could display some form of intelligence.

Since then, AI has changed a great deal. Arguably no one has played a larger role in that change than Judea. Judea Pearl's work made probability the prevailing language of modern AI and, perhaps more significantly, it placed the elaboration of crisp and meaningful models, and of effective computational mechanisms, at the center of AI research. This work is conveyed in the more than 300 scientific papers, and in his three landmark books *Heuristics* (1984), *Probabilistic Reasoning* (1988), and *Causality* (2000), where he deals with the basic questions concerning the acquisition, representation, and effective use of heuristic, probabilistic, and causal knowledge. He tackled these issues not as a philosopher or mathematician, but as an engineer and a cognitive scientist. His "burning question" was (and still is) how does the human mind "do it", and he set out to answer this question with an unusual combination of intuition, passion, intellectual honesty, and technical skill.

Judea is the recipient of numerous scientific awards. In 1996 he was selected by the UCLA Academic Senate as the 81st Faculty Research Lecturer to deliver an annual research lecture which presents the university's most distinguished scholars to the public. He received the 1999 IJCAI Research Excellence Award in Artificial Intelligence for "his fundamental work on heuristic search, reasoning under uncertainty, and causality", the 2001 London School of Economics Lakatos Award for the "best book in the philosophy of science", the 2004 ACM Allen Newell Award for "seminal contributions that extend to philosophy, psychology, medicine, statistics, econometrics, epidemiology and social science", and the 2008 Benjamin Franklin

Medal for "creating the first general algorithms for computing and reasoning with uncertain evidence".

Judea has had more than 20 PhD students at UCLA, many of whom have become successful AI researchers on their own and many have contributed to this volume. Chronologically, they are: Antonio Leal (1976), Alan Chrolotte (1977), Ed Purcell (1978), Joseph Saleh (1980), Jin Kim (1983), Gerard Michon (1983), Rina Dechter (1985), Ingrid Zukerman (1986), Hector Geffner (1989), Dan Geiger (1990), Moises Goldszmidt (1992), Tom Verma (1990), Itay Meiri (1992), Rachel Ben-Eliyahu (1993), Sek-Wah Tan (1995), Alexander Balke (1995), Max Chickering (1996), Jin Tian (2002), Carlos Brito (2004), Blai Bonet (2004), Mark Hopkins (2005), Chen Avin (2006), and Ilya Shpitser (2008).

On a sadder note, Judea is the father of slain Wall Street Journal reporter Daniel Pearl and president of the Daniel Pearl Foundation, which he co-founded with his wife Ruth in April 2002 to continue Daniel's life-work of dialogue and understanding and to address the root causes of his tragic death.

This book will be presented to Judea on March 12, 2010 at a special event at UCLA honoring his life and work, where many of the contributing authors to this book will speak. Two of the editors of this volume, Rina and Hector, are former students of Judea, and the third, Joe, is a close colleague and collaborator. The three of us would like to thank all the authors whose articles are included in this volume. Special thanks go to Adnan Darwiche and Rich Korf of the UCLA Computer Science Department, who helped to organize this event, and to Avi Dechter, Randy Hess, Nir Lipovetzky, Felix Elwert, and Jane Spurr, who helped in the production of the book.

Judea, on behalf of those present in the book, and the many of your students and colleagues who are not, we would like to express our most profound gratitude and admiration to you, as an advisor, a scientist, and a great human being. It has been a real privilege to know you, to benefit from your (truly enjoyable!) company, to watch you, and to learn from you. As students, we couldn't have hoped for a better role model. As colleagues, we couldn't have benefited more from your collaboration and leadership. We know that you don't like compliments, but you are certainly the light in our candle!

Thank you Judea!!!

Rina, Hector, and Joe

# Part I: Heuristics

# 1
# Heuristic Search for Planning under Uncertainty

BLAI BONET AND ERIC A. HANSEN

## 1 Introduction

The artificial intelligence (AI) subfields of heuristic search and automated planning are closely related, with planning problems often providing a stimulus for developing and testing search algorithms. Classical approaches to heuristic search and planning assume a deterministic model of sequential decision making in which a solution takes the form of a sequence of actions that transforms a start state into a goal state. The effectiveness of heuristic search for classical planning is illustrated by the results of the planning competitions organized by the AI planning community, where optimal planners based on A*, and satisficing planners based on variations of best-first search and enforced hill climbing, have performed as well or better than many other planners in the deterministic track of the competition [Edelkamp, Hoffmann, and Littman 2004; Gerevini, Bonet, and Givan 2006].

Beginning in the 1990's, AI researchers became increasingly interested in the problem of planning under uncertainty and adopted Markov decision theory as a framework for formulating and solving such problems [Boutilier, Dean, and Hanks 1999]. The traditional dynamic programming approach to solving Markov decision problems (MDPs) [Bertsekas 1995; Puterman 1994] can be viewed as a form of "blind" or uninformed search. Accordingly, several AI researchers considered how to generalize well-known heuristic-search techniques in order to develop more efficient planning algorithms for MDPs. The advantage of heuristic search over traditional, blind dynamic programming is that it uses an admissible heuristic and intelligent search control to focus computation on solving the problem for relevant states, given a start state and goal states, without considering irrelevant or unreachable parts of the state space.

In this article, we present an overview of research on heuristic search for problems of sequential decision making where state transitions are stochastic instead of deterministic, an important class of planning problems that corresponds to the most basic kind of Markov decision process, called a fully-observable Markov decision process. For this special case of the problem of planning under uncertainty, a fairly mature theory of heuristic search has emerged over the past decade and a half. In reviewing this work, we focus on two key issues: how to generalize classic heuristic search algorithms in order to solve planning problems with stochastic state transitions, and how to compute admissible heuristics for these search problems.

Judea Pearl's classic book, *Heuristics*, provides a comprehensive overview of heuristic search theory as of its publication date in 1984. One of our goals in this article is to show that the twin themes of that book, admissible heuristics and intelligent search control, have been central issues in the subsequent development of a class of algorithms for problems of planning under uncertainty. In this short survey, we rely on references to the literature for many of the details of the algorithms we review, including proofs of their properties and experimental results. Our objective is to provide a high-level overview that identifies the key ideas and contributions in the field and to show how the new search algorithms for MDPs relate to the classical search algorithms covered in Pearl's book.

## 2 Planning with uncertain state transitions

Many planning problems can be modeled by a set of states, $S$, that includes an initial state $s_{init} \in S$ and a set of goal states, $G \subseteq S$, and a finite set of applicable actions, $A(s) \subseteq A$, for each non-goal state $s \in S \backslash G$, where each action incurs a positive cost $c(s,a)$. In a classical, deterministic planning problem, an action $a \in A(s)$ causes a *deterministic* transition, where $f(s,a)$ is the next state after applying action $a$ in state $s$. The objective of a planner is to find a sequence of actions, $\langle a_0, a_1, \ldots, a_n \rangle$, that when applied to the initial state results in a trajectory, $\langle s_0 = s_{init}, a_0, s_1, a_1, \ldots, a_n, s_{n+1} \rangle$, that ends in a goal state, $s_{n+1} \in G$, where $a_i \in A(s_i)$ and $s_{i+1} = f(s_i, a_i)$. Such a plan is optimal if its cost, $\sum_{i=0}^{n} c(s_i, a_i)$, is minimum among all possible plans that achieve a goal.

To model the uncertain effects of actions, we consider a generalization of this model in which the deterministic transition function is replaced by a *stochastic* transition function, $p(\cdot|s,a)$, where $p(s'|s,a)$ is the probability of making a transition to state $s'$ after taking action $a$ in state $s$. In general, the cost of an action depends on the successor state; but usually, it is sufficient to consider the expected cost of an action, denoted $c(s,a)$.

With this simple change of the transition function, the planning problem is changed from a deterministic shortest-path problem to a *stochastic shortest-path problem*. As defined by Bertsekas and Tsitsiklis [Bertsekas and Tsitsiklis 1991], a stochastic shortest-path problem can have actions that incur positive or negative costs. But several subsequent researchers, including Barto et al. [Barto, Bradtke, and Singh 1995], assume that a stochastic shortest-path problem only has actions that incur positive costs. The latter assumption is in keeping with the model of planning problems we sketched above, as well as classical models of heuristic search, and so we ordinarily assume that the actions of a stochastic shortest-path problem incur positive costs only. In case where we allow actions to have both positive and negative costs, we make this clear.

Defined in either way, a stochastic shortest-path problem is a special case of a fully-observable infinite-horizon Markov decision process (MDP). There are several

MDP models with different optimization criteria, and almost all of the algorithms and results we review in this article apply to other MDPs. The most widely-used model in the AI community is the discounted infinite-horizon MDP. In this model, there are rewards instead of costs, $r(s, a)$ denotes the expected reward for taking action $a$ in state $s$, which can be positive or negative, $\gamma \in (0, 1)$ denotes a discount factor, and the objective is to maximize expected total discounted reward over an infinite horizon. Interestingly, any discounted infinite-horizon MDP can be reduced to an equivalent stochastic shortest-path problem [Bertsekas 1995; Bonet and Geffner 2009]. Thus, we do not sacrifice any generality by focusing our attention on stochastic shortest-path problems.

Adoption of a stochastic transition model has important consequences for the structure of a plan. A plan no longer takes the simple form of a sequence of actions. Instead, it is typically represented by a mapping from states to actions, $\pi : S \to A$, called a *policy* in the literature on MDPs. (For the class of problems we consider, where the horizon is infinite, a planner only needs to consider *stationary* policies, which are policies that are not indexed by time.) Note that this representation of a plan assumes closed-loop plan execution instead of open-loop plan execution. It also assumes that an agent always knows the current state of the system; this is what is meant by saying the MDP is *fully observable*.

A stochastic shortest-path problem is solved by finding a policy that reaches a goal state with probability one after a finite number of steps, beginning from any other state. Such a policy is called a *proper policy*. Given a stochastic transition model, it is not possible to bound the number of steps of plan execution it takes to achieve a goal, even for proper policies. Thus, a stochastic shortest-path problem is an *infinite-horizon* MDP. In the infinite-horizon framework, the termination of a plan upon reaching a goal state is modeled by specifying that goal states are zero-cost absorbing states, which means that for all $s \in G$ and $a \in A$, $c(s, a) = 0$ and $p(s|a, s) = 1$. Equivalently, we can assume that no actions are applicable in a goal state. To reflect the fact that plan execution terminates after a finite, but uncertain and unbounded, number of steps, this kind of infinite-horizon MDP is also called an *indefinite-horizon* MDP. Note that when the state set is finite and the number of steps of plan execution is unbounded, the same state can be visited more than once during execution of a policy. Thus, a policy specifies not only conditional behavior, but cyclic behavior too.

For a process that is controlled by a fixed policy $\pi$, stochastic trajectories beginning from state $s_0$, of the form $\langle s_0, \pi_0(s_0), s_1, \pi_1(s_1), \ldots \rangle$, are generated with probability $\prod_{i=0}^{\infty} p(s_{i+1}|s_i, \pi(s_i))$. These probabilities uniquely define a probability measure $P_\pi$ on the set of trajectories from which the costs incurred by $\pi$ can be calculated. Indeed, the cost (or value) of $\pi$ for state $s$ is the expected cost of these trajectories when $s_0 = s$, defined as

$$V_\pi(s) = E_\pi\left[\sum_{k=0}^\infty c(X_k, \pi(X_k)) \,\bigg|\, X_0 = s\right],$$

where the $X_k$'s are *random variables* that denote states of the system at different time points, distributed according to $P_\pi$, and where $E_\pi$ is the expectation with respect to $P_\pi$. The function $V_\pi$ is called the state evaluation function, or simply the value function, for policy $\pi$. For a stochastic shortest-path problem, it is well-defined as long as $\pi$ is a proper policy, and $V_\pi(s)$ equals the expected cost to reach a goal state from state $s$ when using policy $\pi$.

A policy $\pi$ for a stochastic shortest-path problem is optimal if its value function satisfies the Bellman optimality equation:

$$V^*(s) = \begin{cases} 0 & \text{if } s \in G, \\ \min_{a \in A(s)}\{c(s,a) + \sum_{s' \in S} p(s'|s,a) V^*(s')\} & \text{otherwise.} \end{cases} \quad (1)$$

The unique solution of this functional equation, denoted $V^*$, is the optimal value function; hence, all optimal policies have the same value function. Given the optimal value function, one can recover an optimal policy by acting greedily with respect to the value function. A *greedy policy* with respect to a value function $V$ is defined as follows:

$$\pi_V(s) = \operatorname*{argmin}_{a \in A(s)}\left\{c(s,a) + \sum_{s' \in S} p(s'|s,a) V(s')\right\}.$$

Thus, the problem of finding an optimal policy for an MDP is reduced to the problem of solving the optimality equation.

There are two basic dynamic programming approaches for solving Equation (1): value iteration and policy iteration. The value iteration approach is used by all of the heuristic search algorithms we consider, and so we review it here. Starting with an initial value function $V_0$, satisfying $V_0(s) = 0$ for $s \in G$, value iteration computes a sequence of updated value functions by performing, at each iteration, the following *backup* for all states $s \in S$:

$$V_{n+1}(s) := \min_{a \in A(s)}\left\{c(s,a) + \sum_{s' \in S} p(s'|s,a) V_n(s')\right\}. \quad (2)$$

For a stochastic shortest-path problem, the sequence of value functions computed by value iteration is guaranteed to converge to an optimal value function if the following conditions are satisfied: (i) a proper policy exists, and (ii) any policy that is not proper has infinite cost for some state. (Note that if all action costs are positive, any policy that is not proper has infinite cost for some state.) The algorithm described by Equation (2) is called *synchronous value iteration* since all state values are updated in parallel. A variation of this algorithm, called *asynchronous value iteration*, updates only a subset of states at each iteration. As long as every state is guaranteed to be updated infinitely often over time, convergence is still guaranteed.

The convergence of value iteration is asymptotic. In practice, value iteration is stopped when the residuals, $|V_{n+1}(s) - V_n(s)|$, for all states are sufficiently small. The *Bellman residual*, $\max_{s \in S} |V_{n+1}(s) - V_n(s)|$, can be used to bound the suboptimality of a policy or value function for discounted MDPs. For stochastic shortest-path problems, however, suboptimality bounds are not generally possible, as shown by Bertsekas and Tsitsiklis [Bertsekas and Tsitsiklis 1991], yet there is always a sufficiently small (positive) Bellman residual that yields an optimal solution.

## 3 Heuristic search algorithms

Traditional dynamic programming algorithms for MDPs, such as value iteration and policy iteration, solve the optimization problem for the entire state space. By contrast, heuristic search algorithms focus on finding a solution for just the states that are reachable from the start state by following an optimal policy, and use an admissible heuristic to "prune" large parts of the remaining state space. For deterministic shortest-path problems, the effectiveness of heuristic search is well-understood, especially in the AI community. For example, dynamic programming algorithms such as Dijkstra's algorithm and the Bellman-Ford algorithm compute *all* single-source shortest paths, solving the problem for every possible starting state, whereas heuristic search algorithms such as A* and IDA* compute a shortest path from a particular start state to a goal state, usually considering just a fraction of the entire state space. This is the method used to optimally solve problems such as the Rubik's Cube from arbitrary initial configurations, when the enormous size of the state space, which is $4.3 \times 10^{19}$ states for Rubik's Cube [Korf 1997], renders exhaustive methods inapplicable.

In the following, we show that the strategy of heuristic search can also be effective for stochastic shortest-path problems, and, in general, MDPs. The strategy is to solve the problem only for states that are reachable from the start state by following an optimal policy. This means that a policy found by heuristic search is a partial function from the state space to the action space, sometimes called a *partial policy*. A policy $\pi$ is said to be *closed* with respect to state $s$ if it is defined over all states that can be reached from $s$ by following policy $\pi$, and it is said to be closed with respect to the initial state (or just closed) if it is closed with respect to $s_{init}$. Thus, the objective of a heuristic search algorithm for MDPs is to find a partial policy that is closed with respect to the initial state and optimal. The states that are reachable from the start state by following an optimal policy are sometimes called the *relevant states* of the problem. In solving a stochastic shortest-path problem for a given initial state, it is not necessarily the case that the set of relevant states is much smaller than the entire state space, nor is it always easy to estimate its size as a fraction of the state space. But when the set of relevant states *is* much smaller than the entire state set, the heuristic search approach can have a substantial advantage, similar to the advantage heuristic search has over traditional dynamic programming algorithms in solving deterministic shortest-path problems.

**Algorithm 1** RTDP with admissible heuristic $h$.

Let $V$ be the empty hash table whose entries $V(s)$ are initialized to $h(s)$ as needed.
**repeat**
  $s := s_{init}$.
  **while** $s$ is not a goal state **do**
    For each action $a$, set $Q(s,a) := c(s,a) + \sum_{s' \in S} p(s'|s,a)V(s')$.
    Select a best action $\mathbf{a} := \mathrm{argmin}_{a \in A} Q(s,a)$.
    Update value $V(s) := Q(s, \mathbf{a})$.
    Sample the next state $s'$ with probability $p(s'|s, \mathbf{a})$ and set $s := s'$.
  **end while**
**until** some termination condition is met.

## 3.1 Real-Time Dynamic Programming

The first algorithm to apply a heuristic search approach to solving MDPs is called *Real-Time Dynamic Programming (RTDP)* [Barto, Bradtke, and Singh 1995]. RTDP generalizes a heuristic search algorithm developed by Korf [Korf 1990], called *Learning Real-Time A\* (LRTA\*)*, by allowing state transitions to be stochastic instead of deterministic.

Except for the fact that RTDP solves a more general class of problems, it is very similar to LRTA*. Both algorithms interleave planning with execution of actions in a real or simulated environment. They perform a series of *trials*, where each trial begins with an "agent" at the start state $s_{init}$. The agent takes a sequence of actions where each action is selected greedily based on the current state evaluation function. The trial ends when the agent reaches a goal state. The algorithms are called "real-time" because they perform a limited amount of search in the time interval between each action. At minimum, they perform a *backup* for the current state, as defined by Equation (2), which corresponds to a one-step lookahead search; but more extensive search and backups can be performed if there is enough time. They are called "learning" algorithms because they cache state values computed in the course of the search. In an efficient implementation, a hash table is used to store the updated state values and only values for states visited during a trial are stored in the hash table. For all other states, state values are given by an admissible heuristic function $h$. Algorithm 1 shows pseudocode for a trial of RTDP.

The properties of RTDP generalize the properties of Korf's LRTA* algorithm, and can be summarized as follows. First, if all state values are initialized with an admissible heuristic function $h$, then updated state values are always admissible. Second, if there is a proper policy, a trial of RTDP cannot get trapped in a loop and must terminate in a goal state after a finite number of steps. Finally, for the set of states that is reachable from the start state by following an optimal policy, which Barto et al. call the set of *relevant states*, RTDP converges asymptotically to optimal state values and an optimal policy. These results depend on the assumptions that

(i) all immediate costs incurred by transitions from non-goal states are positive, and (ii) the initial state evaluation function is admissible, with all goal states having an initial value of zero.[1]

Although we classify RTDP as a heuristic search algorithm, it is also a dynamic programming algorithm. We consider an algorithm to be a form of dynamic programming if it solves a dynamic programming recursion such as Equation (1) and caches results for subproblems in a table, so that they can be reused without needing to be recomputed. We consider it to be a form of heuristic search if it uses an admissible heuristic and reachability analysis, beginning from a start state, to prune parts of the state space. By these definitions, LRTA* and RTDP are both dynamic programming algorithms and heuristic search algorithms, and so is A*. We still contrast heuristic search to *simple* dynamic programming, which solves the problem for the entire state space. Value iteration and policy iteration are simple dynamic programming algorithms, as are Dijkstra's algorithm and Bellman-Ford. But heuristic search algorithms can often be viewed as a form of enhanced or focused dynamic programming, and that is how we view the algorithms we consider in the rest of this survey.[2] The relationship between heuristic search and dynamic programming comes into clearer focus when we consider LAO*, another heuristic search algorithm for solving MDPs.

## 3.2 LAO*

Whereas RTDP generalizes LRTA*, an online heuristic search algorithm, the next algorithm we consider, LAO* [Hansen and Zilberstein 2001], generalizes the classic AO* search algorithm, which is an offline heuristic search algorithm. The 'L' in LAO* indicates that it can find solutions with *loops*, unlike AO*. Table 1 shows how various dynamic programming and heuristic search algorithms are related, based on the structure of the solutions they find. As we will see, the branching *and cyclic* behavior specified by a policy for an indefinite-horizon MDP can be represented explicitly in the form of a cyclic graph.

Both AO* and LAO* represent the search space of a planning problem as an AND/OR graph. In an AND/OR graph, an OR node represents the choice of an action and an AND node represents a set of outcomes. AND/OR graph search was

---

[1] Although the convergence proof given by Barto et al. depends on the assumption that all action costs are positive, Bertsekas and Tsitsiklis [Bertsekas and Tsitsiklis 1996] prove that RTDP also converges for stochastic shortest-path problems with both positive and negative action costs, given the additional assumption that all improper policies have infinite cost. If action costs are positive and negative, however, the assumption that all improper policies have infinite cost is difficult to verify. In practice, it is often more convenient to assume that all action costs are positive.

[2] Not every heuristic search algorithm is a dynamic programming algorithm. Tree-search heuristic search algorithms, in particular, do not cache the results of subproblems and thus do not qualify as dynamic programming algorithms. For example, IDA*, which explores the tree expansion of a graph, does not cache the results of subproblems and thus does not qualify as a dynamic programming algorithm. On the other hand, IDA* extended with a transposition table caches the results of subproblems and thus is a dynamic programming algorithm.

|  | Solution form | | |
| --- | --- | --- | --- |
|  | simple path | acyclic graph | cyclic graph |
| Dynamic programming | Dijkstra's | backwards induction | value iteration |
| Offline heuristic search | A* | AO* | LAO* |
| Online heuristic search | LRTA* | RTDP | RTDP |

Table 1. Classification of dynamic programming and heuristic search algorithms.

originally developed to model problem-reduction search problems, where a problem is solved by recursively dividing it into subproblems. But it can also be used to model conditional planning problems where the state transition caused by an action is stochastic, and each possible successor state must be considered by the planner.

In AND/OR graph search, a solution is a subgraph of an AND/OR graph that is defined as follows: (i) the root node (corresponding to the start state) belongs to the solution graph, (ii) for every OR node in the solution graph, exactly one of its branches (typically, the one with the lowest cost) belongs to the solution graph, and (iii) for every AND node in the solution graph, all of its branches belong to the solution graph. A solution graph is *complete* if every directed path that begins at the root node ends at a goal node. It is a *partial solution graph* if any directed path ends at an open (i.e., unexpanded) node.

The heuristic search algorithm AO* finds an acyclic solution graph by iteratively expanding nodes on the fringe of the best partial solution graph (beginning from a partial solution graph that consists only of the root node), until the best solution graph is complete. At each step, the best partial solution graph (corresponding to a partial policy) is determined by "greedily" choosing, for each OR node, the branch (or action) with the best expected value. For conditional planning problems with stochastic state transitions, AO* solves the dynamic programming recursion of Equation (1). It does so by repeatedly alternating two steps until convergence. In the forward or *expansion step*, it expands one or more nodes on the fringe of the current best partial solution graph. In the backward or *cost-revision step*, it propagates any change in the heuristic state estimates for the states in the fringe backwards through the graph. The first step is a form of forward reachability analysis, beginning from the start state. The second step is a form of dynamic programming, using backwards induction since the graph is assumed to be acyclic. Thus, AND/OR graph heuristic search is a form of dynamic programming that is enhanced by forward reachability analysis guided by an admissible heuristic.

The classic AO* algorithm only works for problems with acyclic spaces. But stochastic planning problems, such as MDPs, often contain cycles in space and their solutions may include cycles too. To generalize AO* on these models, the key idea is to use a more general dynamic programming algorithm in the cost-revision step, such as value iteration. This simple generalization is the key difference between AO*

---
**Algorithm 2** Improved LAO* with admissible heuristic $h$.
---
The explicit graph initially consists of the start state $s_{init}$.
**repeat**
    Depth-first traversal of states in the current best (partial) solution graph.
    **for** each visited state $s$ in postorder traversal **do**
        If state $s$ is not expanded, expand it by generating each successor state $s'$ and initializing its value $V(s')$ to $h(s')$.
        Set $V(s) := \min_{a \in A(s)} c(s,a) + \sum_{s'} p(s'|s,a)V(s')$ and mark the best action.
    **end for**
**until** the best solution graph has no unexpanded tip state and residual $< \epsilon$.
**return** An $\epsilon$-optimal solution graph.
---

and LAO*. However, allowing a solution to contain loops substantially increases the complexity of the cost-revision step. For AO*, the cost-revision step requires at most one update per node. For LAO*, many updates per node may be required before convergence to exact values. As a result, a naive implementation of LAO* that expands a single fringe node at a time and performs value iteration in the cost-revision step until convergence to exact values can be extremely slow.

However, a couple of simple changes create a much more efficient version of LAO*. Although Hansen and Zilberstein did not give the modified algorithm a distinct name, it has been referred to in the literature as *Improved LAO\**. Recall that in its expansion step, LAO* does a depth-first traversal of the current best partial solution graph in order to identify the open nodes on its fringe, and expands one or more of the open nodes. To improve efficiency, Improved LAO* expands all open nodes on the fringe of the best current partial solution graph (yet it is easily modified to expand less or more nodes), and then, during the cost-revision step, it performs only *one* backup for each node in the current solution graph. Conveniently, both the expansion and cost-revision steps can be performed in the same depth-first traversal of the best partial solution graph, since node expansions and backups can be performed when backtracking during a depth-first traversal. Thus, the complexity of a single iteration of the expansion and cost-revision steps is bounded by the number of nodes in the current best (partial) solution graph. Algorithm 2 shows the pseudocode.

RTDP and the more efficient version of LAO* have many similarities. Principal among them, both perform backups only for states that are reachable from the start state by choosing actions greedily based on the current value function. The key difference is how they choose the order in which to visit states and perform backups. RTDP relies on stochastic exploration based on real or simulated trials (an online strategy), whereas LAO* relies on systematic depth-first traversals (an offline strategy). In fact, all of the other heuristic search algorithms we review in the rest of this article rely on one of the other of these two general strategies for traversing the reachable state space and updating the value function.

Experiments show that Improved LAO* finds a good solution as quickly as RTDP and converges to an optimal solution much faster; faster convergence is due to its use of systematic search instead of stochastic simulation to explore the state space. The test for convergence to an optimal solution generalizes the convergence test for AO*: the best solution graph is optimal if it is complete (i.e., it does not contain any unexpanded nodes), and if state values have converged to exact values for all nodes in the best solution graph. If the state values are not exact, it is possible to bound the suboptimality of the solution by adapting the error bounds developed for value iteration.

## 3.3 Bounds and faster convergence

In comparing the performance of Improved LAO* and RTDP, Hansen and Zilberstein made a couple of observations that inspired subsequent improvements of RTDP. One observation was that the convergence test used by LAO* could be adapted for use by RTDP. As formulated by Barto et al., RTDP is guaranteed to converge asymptotically but does not have an explicit convergence test or a way of bounding the suboptimality of a solution. A second observation was that RTDP's slow convergence relative to Improved LAO* is due to its reliance on stochastic exploration of the state space, instead of systematic search, and its rate of convergence could be improved by exploring the state space more systematically. We next consider several improved methods for testing for convergence and increasing the rate of convergence.

**Labeling solved states.** Bonet and Geffner [Bonet and Geffner 2003a; Bonet and Geffner 2003b] developed a pair of related algorithms, called Labeled RTDP (LRTDP) and Heuristic Dynamic Programming (HDP), that combine both of these ideas with a third idea adopted from the original AO* algorithm: the idea of labeling 'solved' states. In the classic AO* algorithm, a state $s$ is labeled as 'solved' if it is a goal state or if every state that is reachable from $s$ by taking the best action at each OR node is labeled 'solved'. Labeling speeds up the search because it is unnecessary to expend search effort in parts of the solution that have already converged; AO* terminates when the start node is labeled 'solved'.

When a solution graph contains loops, however, labeling states as 'solved' cannot be done in the traditional way. It is not even guaranteed to be useful; if the start state is reachable from every other state, for example, it is not possible to label any state as 'solved' before the start state itself is labeled as 'solved'. But in many cases, a solution graph with loops has a "partly acyclic" structure. Stated precisely, the solution graph can often be decomposed into strongly-connected components, using Tarjan's well-known algorithm. In this case, the states in one strongly-connected component can be labeled as 'solved' before the states in other, predecessor components are labeled.

Tarjan's algorithm decomposes a graph into strongly-connected components in the course of a depth-first traversal of the graph. Since Improved LAO* expands and updates the states in the current best solution graph in the course of a depth-first traversal of the graph, the two algorithms are easily combined. In fact, Bonet and Geffner [Bonet and Geffner 2003a] present their HDP algorithm as a synthesis of Tarjan's algorithm and a depth-first search algorithm, similar to the one used in Improved LAO*.

The same idea of labeling states as 'solved' can also be combined with RTDP. In Labeled RTDP (LRTDP), trials are very much like RTDP trials except that they terminate when a solved stated is reached. (Initially only the goal states are solved.) At the end of a trial, a labeling procedure is invoked for each unsolved state visited in the trial, in reverse order from the last unsolved state to the start state. For each state $s$, the procedure performs a depth-first traversal of the states that are reachable from $s$ by selecting actions greedily based on the current value function. If the residuals of these states are less than a threshold $\epsilon$, then *all* of them are labeled as 'solved'. Like AO*, Labeled RTDP terminates when the initial state is labeled as 'solved'. The labeling procedure used by LRTDP is similar to the traversal procedures used in HDP and Improved LAO*. However, the innovation of LRTDP is that instead of always traversing the solution graph from the start state, it begins the traversal at each state visited in a trial, in backwards order from the last unsolved state, which allows the convergence of states near the goal to be recognized before states near the initial state have converged.

Experiments show that LRTDP converges much faster than RTDP, and somewhat faster than Improved LAO*, in solving benchmark "racetrack" problems. In general, the amount of improvement is problem-dependent since it depends on the extent to which the solution graph decomposes into strongly-connected components. In the racetrack domain, the improvement over Improved LAO* is due to labeling states as 'solved'; the more substantial improvement over RTDP is partly due to labeling, but also due to the more systematic traversal of the state space.

**Lower and upper bounds.** Both LRTDP and HDP gradually reduce the Bellman residual until it falls below a threshold $\epsilon$. If the threshold is sufficiently small, the policy is optimal. But the residual, by itself, does not bound the suboptimality of the solution. To bound its suboptimality, we need an upper bound on the value of the starting state in addition to the lower-bound values computed by heuristic search. Once a closed policy is found, an obvious way to bound its suboptimality is to evaluate the policy; its value for the start state is an upper bound that can be compared to the admissible lower-bound value computed by heuristic search. But this approach does not allow the suboptimality of an incomplete solution (one for which the start state is not yet labeled 'solved') to be bounded.

McMahan et al. [McMahan, Likhachev, and Gordon 2005] and Smith and Simmons [Smith and Simmons 2006] describe two algorithms, called *Bounded RTDP (BRTDP)* and *Focused RTDP (FRTDP)* respectively, that compute upper bounds in order to bound the suboptimality of a solution, including incomplete solutions, and use the difference between the upper and lower bounds on state values to focus search effort. The key assumption of both algorithms is that in addition to an admissible heuristic function that returns lower bounds for any state, there is a function that returns upper bounds for any state. Every time BRTDP or FRTDP visit a state, they perform two backups: a standard RTDP backup to compute a lower-bound value and another backup to compute an upper-bound value. In simulated trials, action outcomes are determined based on their probability *and* the largest difference between the upper and lower bound values of the possible successor states, which has the effect of biasing state exploration to where it is most likely to improve the value function.

This approach has a lot of attractive properties. In particular, being able to bound the suboptimality of an incomplete solution is useful when it is computationally prohibitive to compute a policy that is closed with respect to the start state. However, the approach is based on the assumption that an upper-bound value function is available and easily computed, and this assumption may not be realistic for many stochastic shortest-path problems. For discounted MDPs, on the other hand, such bounds are easily computed, as we show in Section 4.[3]

### 3.4 Learning Depth-First Search

AND/OR graphs can represent the search space of problem-reduction problems and MDPs, by appropriately defining the cost of complete solution graphs, and they can also be used to represent the search space of adversarial game-playing problems, non-deterministic planning problems, and even deterministic planning problems. Bonet and Geffner [Bonet and Geffner 2005a; Bonet and Geffner 2006] describe a *Learning Depth-First Search (LDFS)* algorithm that provides a unified framework for solving search problems in these different AI models. LDFS performs iterated depth-first searches over the current best partial solution graph, enhanced with backups and labeling of 'solved' states. Bonet and Geffner show that LDFS generalizes well-known algorithms in some cases and points to novel algorithms in other cases. For deterministic planning problems, for example, they show that LDFS instantiates to IDA* with transposition tables. For game-search problems, they show that LDFS corresponds to an Alpha-Beta search algorithm with null windows called MTD [Plaat, Schaeffer, Pijls, and de Bruin 1996], which is reminiscent of Pearl's SCOUT algorithm [Pearl 1983]. For MDPs, LDFS corresponds to a version of Improved LAO* enhanced with labeling of 'solved' states. For max AND/OR search problems, LDFS instantiates to a novel algorithm that experiments show is more efficient than existing algorithms [Bonet and Geffner 2005a].

---

[3]Before developing FRTDP, Smith and Simmons [Smith and Simmons 2005] developed a very similar heuristic search algorithm for partially observable Markov decision processes (POMDPs) that backs up both lower-bound and upper-bound state values in AND/OR graph search. A similar AND/OR graph-search algorithm for POMDPs was described earlier by Hansen [Hansen 1998]. Since both algorithms solve discounted POMDPs, both upper and lower bounds are easily available.

## 3.5 Symbolic heuristic search

The algorithms we have considered so far assume a "flat" state space and enumerate states, actions, and transitions individually. For very large state spaces, it is often more convenient to adopt a structured or symbolic representation that exploits regularities to represent the same information more compactly and manipulate it more efficiently, in terms of sets of states and sets of transitions. As an example, Hoey et al. [Hoey, St-Aubin, Hu, and Boutilier 1999] show how to perform symbolic value iteration for *factored MDPs*, which are represented in a propositional language, using algebraic decision diagrams as a compact data structure. Based on their approach, Feng and Hansen [Feng and Hansen 2002; Feng, Hansen, and Zilberstein 2003] describe a symbolic LAO* algorithm and a symbolic version of RTDP for factored MDPs. Boutilier et al. [Boutilier, Reiter, and Price 2001] show how to perform symbolic dynamic programming for MDPs represented in a first-order language, and Karabaev and Skvortsova [Karabaev and Skvortsova 2005] show that symbolic heuristic search can also be performed over such MDPs.

## 4 Admissible heuristics

Heuristic search algorithms require admissible heuristics to prune large state spaces effectively. As advocated by Pearl, an effective and domain-independent strategy for obtaining admissible heuristics consists in optimally solving a relaxation of the problem, an MDP in our case. In this section, we review some relaxation-based heuristics for MDPs. However, we first consider admissible heuristics that are not based on relaxations. Although such heuristics are not informative, they are useful when informative heuristics cannot be easily computed.

### 4.1 Non-informative heuristics

For stochastic shortest-path problems where all actions incur positive costs, a simple admissible heuristic assigns the value of zero to every state, $h(s) = 0, \forall s \in S$, since zero is a lower bound on the cost of an optimal solution. Note that this heuristic is equivalent to using a zero-constant admissible heuristic for A* when solving deterministic shortest-path problems. In problems with uniform costs this is equivalent to a breadth-first search.

For the more general model of stochastic shortest-path problems that allows both negative and positive action costs, it is not possible to bound the optimal value function in such a simple way, and simple, non-informative heuristics are not readily available. But for discounted infinite-horizon MDPs, the optimal value function is easily bounded both above and below. Note that for this class of MDPs, we adopt the reward-maximization framework. Let $R^U = \max_{s \in S, a \in A(s)} r(s, a)$ denote the maximum immediate reward for an MDP and let $R^L = \min_{s \in S, a \in A(s)} r(s, a)$ denote the minimum immediate reward. For an MDP with discount factor $\gamma \in (0, 1)$, the function $h(s) = R^U/1 - \gamma$ is an upper bound on the optimal value function and provides admissible heuristic estimates, and the function $l(s) = R^L/1 - \gamma$ is a lower bound on the optimal value function. The time required to compute these bounds is linear in the number of states and actions, but the bounds need to be computed just once as their value does not depend on the state $s$.

## 4.2 Relaxation-based heuristics

The relaxations that are used for obtaining admissible heuristics in deterministic planning can be used for MDPs as well, as we will see. But first, we consider a relaxation that applies only to search problems with uncertain transitions. It assumes the agent can control the transition by choosing the best outcome among the set of possible outcomes of an action.

Recall from Equation (1) that the equation that characterizes the optimal value function of a stochastic shortest-path problem has the form $V^*(s) = 0$ for goal states $s \in G$, and

$$V^*(s) = \min_{a \in A(s)} \left\{ c(s,a) + \sum_{s' \in S} p(s'|s,a) V^*(s') \right\},$$

for non-goal states $s \in S \backslash G$. A lower bound on $V^*$ is immediately obtained if the expectation in the equation is replaced by a minimization over the values of the successor states, as follows,

$$V_{min}(s) = \min_{a \in A(s)} \left\{ c(s,a) + \min_{s' \in S(s,a)} V_{min}(s') \right\},$$

where $S(s,a) = \{s' : p(s'|s,a) > 0\}$ is the subset of successor states of $s$ through the action $a$. Interestingly, this equation is the optimality equation for a deterministic shortest-path problem over the graph $G_{min} = (V, E)$ where $V = S$, and there is an edge $(s, s')$ with cost $c(s, s') = \min\{c(s,a) : p(s'|s,a) > 0, a \in A(s)\}$ for $s' \in S(s,a)$. The graph $G_{min}$ is a *relaxation of the MDP* on which the non-deterministic outcomes of an action are separated along different deterministic actions, in a way that the agent has the ability to choose the most convenient one. If this relaxation is solved optimally, the state values $V_{min}(s)$ provide an admissible heuristic for the MDP. This relaxation is called the *min-min relaxation* of the MDP [Bonet and Geffner 2005b]; its optimal value at state $s$ is denoted by $V_{min}(s)$.

When the number of states is relatively small and can fit in memory, the state values $V_{min}(s)$ can be obtained using Dijkstra's algorithm in time polynomial in the number of states and actions. Otherwise, the values can be obtained, as needed, using a search algorithm such as A* or IDA* on the graph $G_{min}$. Indeed, the state value $V_{min}(s)$ is the cost of a minimum-cost path from $s$ to any goal state. A* and IDA* require an admissible heuristic function $h(s)$ for searching $G_{min}$; if nothing better is available, the non-informative heuristic $h(s) = 0$ can be used.

Given a deterministic relaxation of an MDP, such as this, another approach to computing admissible heuristics for the original MDP is based on the recognition that any admissible heuristic for the deterministic relaxation is also admissible for the original MDP. That is, if an estimate $h(s)$ is a lower bound on the value $V_{min}(s)$, it is also a lower bound on the value $V^*(s)$ for the MDP. Therefore, we can use any method for computing admissible heuristics for deterministic shortest-path problems in order to compute admissible heuristics for the corresponding stochastic shortest-path problems. Since such methods often rely on state abstraction, the heuristics can be stored in memory even when the state space of the original problem is much too large to fit in memory.

Instead of applying relaxation methods for deterministic shortest-path problems to a deterministic relaxation of an MDP, another approach is to apply similar relaxation methods directly to the MDP. This strategy was explored by Dearden and Boutilier [Dearden and Boutilier 1997], who describe an approach to state abstraction for factored MDPs that can be used to compute admissible heuristics. Their approach ignores certain state variables of the original MDP in order to create an exponentially smaller abstract MDP that can be solved more easily. Such a relaxation can be useful when it is not desirable to abstract away all stochastic aspects of a problem.

## 4.3 Planning languages and heuristics

Most approaches to state abstraction for MDPs, including that of Dearden and Boutilier, assume the MDP has a factored or otherwise structured representation, instead of a "flat" representation that explicitly enumerates individual states, actions, and transitions. To allow scalability, the representation languages used by most planners are high-level languages based on propositional logic or a fragment of first-order logic, that permits the description of large problems in a succinct way; often, a problem with $n$ states and $m$ actions can be described with $O(\log nm)$ bits.

As an example, the PPDDL language [Younes and Littman 2004] has been used in the International Planning Competition to describe MDPs [Bryce and Buffet 2008; Gerevini, Bonet, and Givan 2006]. PPDDL is an extension of PDDL [McDermott, Ghallab, Howe, Knoblock, Ram, Veloso, Weld, and Wilkins 1998] that handles actions with non-deterministic effects and multiple initial situations. Like PDDL, it is a STRIPS language extended with types, conditional effects, and disjunctive goals and conditions.

The fifth International Planning Competition used a fragment of PPDDL, consisting of STRIPS extended with negative conditions, conditional effects and simple probabilistic effects. The fragment disallows the use of existential quantification, disjunction of conditions, nested conditional effects, and probabilistic effects inside conditional effects. What remains, nevertheless, is a simple representation language for probabilistic planning in which a large collection of challenging problems can be modeled. For our purposes, it is a particularly interesting fragment because it allows standard admissible heuristics for classical STRIPS planning to be easily adapted and thus "lifted" for probabilistic planning. In the rest of this section, we briefly present a STRIPS language extended with conditional effects, some of its variants for probabilistic planning, and how to compute admissible heuristics for it.

A STRIPS planning problem with conditional effects (simply STRIPS) is a tuple $\langle F, I, G, O \rangle$ where $F$ is a set of fluent symbols, $I \subseteq F$ is the initial state, $G \subseteq F$ denotes the set of goal states, and $O$ is a set of operators. A state is a valuation of fluent symbols that is denoted by the subset of fluents are true in the state. An operator $a \in O$ consists of a precondition $Pre \subseteq F$, and a collection $CE$ of conditional effects of the form $C \rightarrow L$, where $C$ and $L$ are sets of literals that denote the condition and effect of the conditional effect.

A simple probabilistic STRIPS problem (simply sp-STRIPS) is a STRIPS problem in which each operator $a$ has a precondition $Pre$ and a list of probabilistic outcomes of the form $\langle (p_1, CE_1), \ldots, (p_n, CE_n) \rangle$ where $p_i > 0$, $\sum_i p_i \leq 1$, and each $CE_i$ is a set of conditional effects. In sp-STRIPS, the state that results after applying action $a$ on state $s$ is equal to the state that result after applying the conditional effects in $CE_i$ on $s$ with probability $p_i$, or the same state $s$ with probability $1 - \sum_{i=1}^{n} p_i$.

In PPDDL, probabilistic effects are expressed using statements of the form

$$(\texttt{probabilistic } \{\texttt{<rational> <det-effect>}\}^+)$$

where <rational> is a rational number and <det-effect> is a deterministic, possibly compound, effect. The intuition is that the deterministic effect occurs with the given probability and that no effect occurs with the remaining probability. A specification without probabilistic effects can be converted in polynomial time to STRIPS. However, when there are probabilistic effects involved, it is necessary to consider all possible simultaneous executions. For example, an action that simultaneously tosses three coins can be specified as follows:

```
(:action toss-three-coins
  :parameters (c1 c2 c3 - coin)
  :precondition (and (not (tossed c1)) (not (tossed c2)) (not (tossed c3)))
  :effect (and (tossed c1)
               (tossed c2)
               (tossed c3)
               (probabilistic 1/2 (heads c1) 1/2 (tails c1))
               (probabilistic 1/2 (heads c2) 1/2 (tails c2))
               (probabilistic 1/2 (heads c3) 1/2 (tails c3))))
```

This action is not an sp-STRIPS action since its outcomes are factored along multiple probabilistic effects. An equivalent sp-STRIPS action has as precondition the same precondition but effects of the form $\langle (1/8, CE_1), (1/8, CE_2), \ldots, (1/8, CE_8) \rangle$ where each $CE_i$ stands for a deterministic outcome of the action; e.g., $CE_1 =$ (and (heads c1) (heads c2) (heads c3)).

Under the assumptions that there are no probabilistic effects inside conditional effects and that there are no nested conditional effects, a probabilistic planning problem described with PPDDL can be transformed into an equivalent sp-STRIPS problem by taking the cross products of the probabilistic effects within each action; a translation that takes exponential time in the maximum number of probabilistic effects per action. However, once in sp-STRIPS, the problem can be further relaxed into (deterministic) STRIPS by converting each action of form $\langle Pre, \langle (p_1, CE_1), \ldots, (p_n, CE_n) \rangle \rangle$ into $n$ deterministic actions of the form $\langle Pre, CE_i \rangle$. This relaxation is the min-min relaxation now implemented at the level of the representation language, without the need to explicitly generate the state and action spaces of the MDP.

The min-min relaxation of a PPDDL problem is a deterministic planning problem whose optimal solution provides an admissible heuristic for the probabilistic planning problem. Thus, any admissible heuristic for the deterministic problem provides an admissible heuristic for the probabilistic problem. (This is the approach used in the mGPT planner for probabilistic planning [Bonet and Geffner 2005b].)

Above relaxation gives an interesting and fruitful connection with the field of (deterministic) automated planning in which the computation of domain-independent and admissible heuristics is an important area of research. Over the last decade, the field has witnessed important progresses in the development of novel and powerful heuristics that can be used for probabilistic planning.

## 5 Conclusions

We have shown that increased interest in the problem of planning under uncertainty has led to the development of a new class of heuristic search algorithms for these planning problems. The effectiveness of these algorithms illustrates the wide applicability of the heuristic search approach. This approach is influenced by ideas that can be traced back to some of the fundamental contributions in the field of heuristic search laid down by Pearl.

In this brief survey, we only reviewed search algorithms for the special case of the problem of planning under uncertainty in which state transitions are uncertain. Many other forms of uncertainty may need to be considered by a planner. For example, planning problems with imperfect state information are often modeled as partially observable Markov decision processes for which there are also algorithms based on heuristic search [Bonet and Geffner 2000; Bonet and Geffner 2009; Hansen 1998; Smith and Simmons 2005]. For some planning problems, there is uncertainty about the parameters of the model. For other planning problems, there is uncertainty due to the presence of multiple agents. The development of effective heuristic search algorithms for these more complex planning problems remains an important and active area of research.

## References

Barto, A., S. Bradtke, and S. Singh (1995). Learning to act using real-time dynamic programming. *Artificial Intelligence* 72(1), 81–138.

Bertsekas, D. (1995). *Dynamic Programming and Optimal Control, (2 Vols)*. Athena Scientific.

Bertsekas, D. and J. Tsitsiklis (1991). Analysis of stochastic shortest path problems. *Mathematics of Operations Research 16*(3), 580–595.

Bertsekas, D. and J. Tsitsiklis (1996). *Neuro-Dynamic Programming*. Belmont, Massachusetts: Athena Scientific.

Bonet, B. and H. Geffner (2000). Planning with incomplete information as heuristic search in belief space. In S. Chien, S. Kambhampati, and C. Knoblock (Eds.), *Proc. 6th Int. Conf. on Artificial Intelligence Planning and Scheduling (AIPS-00)*, Breckenridge, CO, pp. 52–61. AAAI Press.

Bonet, B. and H. Geffner (2003a). Faster heuristic search algorithms for planning with uncertainty and full feedback. In G. Gottlob and T. Walsh (Eds.), *Proc. 18th Int. Joint Conf. on Artificial Intelligence (IJCAI-03)*, Acapulco, Mexico, pp. 1233–1238. Morgan Kaufmann.

Bonet, B. and H. Geffner (2003b). Labeled RTDP: Improving the convergence of real-time dynamic programming. In E. Giunchiglia, N. Muscettola, and D. S. Nau (Eds.), *Proc. 13th Int. Conf. on Automated Planning and Scheduling (ICAPS-03)*, Trento, Italy, pp. 12–21. AAAI Press.

Bonet, B. and H. Geffner (2005a). An algorithm better than AO*? In M. M. Veloso and S. Kambhampati (Eds.), *Proc. 20th National Conf. on Artificial Intelligence (AAAI-05)*, Pittsburgh, USA, pp. 1343–1348. AAAI Press.

Bonet, B. and H. Geffner (2005b). mGPT: A probabilistic planner based on heuristic search. *Journal of Artificial Intelligence Research 24*, 933–944.

Bonet, B. and H. Geffner (2006). Learning depth-first search: A unified approach to heuristic search in deterministic and non-deterministic settings, and its application to MDPs. In D. Long, S. F. Smith, D. Borrajo, and L. McCluskey (Eds.), *Proc. 16th Int. Conf. on Automated Planning and Scheduling (ICAPS-06)*, Cumbria, UK, pp. 142–151. AAAI Press.

Bonet, B. and H. Geffner (2009). Solving POMDPs: RTDP-Bel vs. point-based algorithms. In C. Boutilier (Ed.), *Proc. 21st Int. Joint Conf. on Artificial Intelligence (IJCAI-09)*, Pasadena, California, pp. 1641–1646. AAAI Press.

Boutilier, C., T. Dean, and S. Hanks (1999). Decision-theoretic planning: Structural assumptions and computational leverage. *Journal of Artificial Intelligence Research 11*, 1–94.

Boutilier, C., R. Reiter, and B. Price (2001). Symbolic dynamic programming for first-order MDPs. In B. Nebel (Ed.), *Proc. 17th Int. Joint Conf. on Artificial Intelligence (IJCAI-01)*, Seattle, WA, pp. 690–697. Morgan Kaufmann.

Bryce, D. and O. Buffet (Eds.) (2008). *6th International Planning Competition: Uncertainty Part*, Sydney, Australia.

Dearden, R. and C. Boutilier (1997). Abstraction and approximate decision-theoretic planning. *Artificial Intelligence 89*, 219–283.

Edelkamp, S., J. Hoffmann, and M. Littman (Eds.) (2004). *4th International Planning Competition*, Whistler, Canada.

Feng, Z. and E. Hansen (2002). Symbolic heuristic search for factored Markov decision processes. In R. Dechter, M. Kearns, and R. S. Sutton (Eds.), *Proc. 18th National Conf. on Artificial Intelligence (AAAI-02)*, Edmonton, Canada, pp. 455–460. AAAI Press.

Feng, Z., E. Hansen, and S. Zilberstein (2003). Symbolic generalization for on-line planning. In C. Meek and U. Kjaerulff (Eds.), *Proc. 19th Conf. on Uncertainty in Artificial Intelligence (UAI-03)*, Acapulco, Mexico, pp. 209–216. Morgan Kaufmann.

Gerevini, A., B. Bonet, and R. Givan (Eds.) (2006). *5th International Planning Competition*, Cumbria, UK.

Hansen, E. (1998). Solving POMDPs by searching in policy space. In *Proc 14th Conf. on Uncertainty in Artificial Intelligence (UAI-98)*, Madison, WI, pp. 211–219.

Hansen, E. and S. Zilberstein (2001). LAO*: A heuristic search algorithm that finds solutions with loops. *Artificial Intelligence 129*(1–2), 139–157.

Hoey, J., R. St-Aubin, A. Hu, and C. Boutilier (1999). SPUDD: Stochastic planning using decision diagrams. In *Proc. 15th Conf. on Uncertainty in Artificial Intelligence (UAI-99)*, Stockholm, Sweden, pp. 279–288. Morgan Kaufmann.

Karabaev, E. and O. Skvortsova (2005). A heuristic search algorithm for solving first-order MDPs. In F. Bacchus and T. Jaakkola (Eds.), *Proc 21st Conf. on Uncertainty in Artificial Intelligence (UAI-05)*, Edinburgh, Scotland, pp. 292–299. AUAI Press.

Korf, R. (1990). Real-time heuristic search. *Artificial Intelligence 42*, 189–211.

Korf, R. (1997). Finding optimal solutions to rubik's cube using pattern databases. In B. Kuipers and B. Webber (Eds.), *Proc. 14th National Conf. on Artificial Intelligence (AAAI-97)*, Providence, RI, pp. 700–705. AAAI Press / MIT Press.

McDermott, D., M. Ghallab, A. Howe, C. Knoblock, A. Ram, M. M. Veloso, D. Weld, and D. Wilkins (1998). PDDL – The Planning Domain Definition Language. Technical Report CVC TR-98-003/DCS TR-1165, Yale Center for Computational Vision and Control, New Haven, USA.

McMahan, H. B., M. Likhachev, and G. Gordon (2005). Bounded real-time dynamic programming: RTDP with monotone upper bounds and performance guarantees. In L. D. Raedt and S. Wrobel (Eds.), *Proc. 22nd Int. Conf. on Machine Learning (ICML-05)*, Bonn, Germany, pp. 569–576. ACM.

Pearl, J. (1983). *Heuristics*. Morgan Kaufmann.

Plaat, A., J. Schaeffer, W. Pijls, and A. de Bruin (1996). Best-first fixed-depth minimax algorithms. *Artificial Intelligence 87*(1-2), 255–293.

Puterman, M. (1994). *Markov Decision Processes – Discrete Stochastic Dynamic Programming*. John Wiley and Sons, Inc.

Smith, T. and R. Simmons (2005). Point-based POMDP algorithms: Improved analysis and implementation. In F. Bacchus and T. Jaakkola (Eds.), *Proc. 21st Conf. on Uncertainty in Artificial Intelligence (UAI-05)*, Edinburgh, Scotland, pp. 542–547. AUAI Press.

Smith, T. and R. G. Simmons (2006). Focused real-time dynamic programming for MDPs: Squeezing more out of a heuristic. In Y. Gil and R. J. Mooney (Eds.), *Proc. 21st National Conf. on Artificial Intelligence (AAAI-06)*, Boston, USA, pp. 1227–1232. AAAI Press.

Younes, H. and M. Littman (2004). PPDDL1.0: An extension to PDDL for expressing planning domains with probabilistic effects. `http://www.cs.cmu.edu/~lorens/papers/ppddl.pdf`.

# 2

# Heuristics, Planning and Cognition

Hector Geffner

## 1 Introduction

In the book *Heuristics,* Pearl studies the strategies for the control of problem solving processes in human beings and machines, pondering how people manage to solve an extremely broad range of problems with so little effort, and how machines could do the same [Pearl 1983, pp. vii]. The central concept in the book, as captured in the title, are the *heuristics*: the "criteria, methods, or principles for deciding which among several alternative courses of action promises to be the most effective in order to achieve some goal" [Pearl 1983, pp. 3]. Pearl places special emphasis on heuristics that take the form of *evaluation functions* and which provide quick but approximate estimates of the distance or cost-to-go from a given state to the goal. These heuristic evaluation functions provide the search with a sense of direction with actions resulting in states that are closer to the goal being preferred. An informative heuristic $h(s)$ in the 15-puzzle, for example, is the well known 'sum of Manhattan distances', that adds up the Manhattan distance of each tile, from its location in the state $s$ to its goal location.

The book *Heuristics* laid the foundations for the work in automated problem solving in Artificial Intelligence (AI) and is still a basic reference in the field. On the other hand, as an account of human problem solving, the book has not been as influential. A reason for this is that while the book devotes one chapter to discuss the derivation of *heuristics,* most of the book is devoted to the formulation and analysis of heuristic search *algorithms*. Most of these algorithms, such as A* and AO*, are complete and optimal, meaning that they will find a solution if there is one, and that the solution found will have minimal cost (provided that the heuristic does not overestimate the true costs). Yet, while people excel at solving a wide variety of problems almost effortlessly, it's only in puzzle-like problems where they need to restore to search, and then, they are not particularly good at it and are even worse when solutions must be optimal.

Thus, the account of problem solving in the book exhibits a gap that has been characteristic of AI systems, that result in programs that rival the best human experts in specialized domains but are no match to children in their general problem solving abilities.

In this article, I aim to present recent work in AI Planning, a form of *domain-independent problem solving,* that builds on Pearl's work and bears on this gap.

Planners are general problem solvers aimed at solving an infinite collection of problems automatically. The problems are instances of various classes of models all of which are intractable in the worst case. In order to solve these problems effectively thus, a planner must automatically recognize and exploit their structure. This is the key challenge in planning and, more generally, in domain-independent problem solving. In planning, this challenge has been addressed *by deriving the heuristic evaluations functions automatically* from the problems, an idea explored by Pearl and developed more fully in recent planning research. The resulting domain-independent planners are not as efficient as specialized solvers but are more general, and thus, behave in a way that is closer to people. Moreover, the resulting evaluation functions often enable the solution of problems with almost no search, and appear to play the role of the 'intuitions' and 'feelings' that guide human problem solving and have been difficult to capture explicitly by means of rules. We will see indeed how such heuristic evaluation functions are defined and computed in a domain-independent fashion, and why they can be regarded as relevant from a cognitive point of view.

The organization of the article is the following. We consider in order, planning models, languages, and algorithms (Section 2), the automatic extraction of heuristic evaluation functions and other developments in planning (Sections 3 and 4), the cognitive interpretation of these heuristics (Section 5), and then, more generally, the relation between AI and Cognitive Science (Section 6).

## 2 Planning

Planning is an area of AI concerned with the selection of actions for achieving goals. The first AI planner and one of the first AI programs was the General Problem Solver (GPS) developed by Newell, Shaw, and Simon in the late 50's [Newell, Shaw, and Simon 1958; Newell and Simon 1963]. Since then, planning has remained a central topic in AI while changing in significant ways: on the one hand, it has become *more mathematical*, with a variety of planning problems defined and studied; on the other, it has become *more empirical*, with planning algorithms evaluated experimentally and planning competitions held periodically.

Planning can be understood as representing one of the three main approaches for *selecting the action to do next*; a problem that is central in the design of autonomous systems, called often the *control problem* in AI.

In the *programming-based approach*, the programmer solves the control problem in its head and makes the solution explicit in the program. For example, for a robot moving in an office environment, the program may say to back up when too close to a wall, to search for a door if the robot has to move to another room, etc. [Brooks 1987; Mataric 2007].

In the *learning-based approach*, the control knowledge is not provided explicitly by a programmer but is learned by trial and error, as in reinforcement learning [Sutton and Barto 1998], or by generalization from examples, as in supervised learning [Mitchell 1997].

Figure 1. Planning is the model-based approach to autonomous behavior: a planner is a solver that accepts a compact model of the actions, sensors, and goals, and outputs a plan or controller that determines the action to do next given the observations.

Finally, in the *model-based approach*, the control knowledge is derived automatically from a model of the actions, sensors, and goals.

Planning is the model-based approach to autonomous behavior. A planner is a solver that accepts a model of the actions, sensors, and goals, and outputs a plan or controller that determines the action to do next given the observations gathered (Fig. 1). Planners come in a wide variety, depending on the type of model that they target [Ghallab, Nau, and Traverso 2004]. *Classical planners* address deterministic state models with full information about the initial situation, while *conformant planners* address state models with non-deterministic actions and incomplete information about the initial state. In both cases, the resulting plans are open-loop controllers that do not take observations into account. On the other hand, contingent and POMDP planners address scenarios with both uncertainty and feedback, and output genuine closed-loop controllers where the selection of actions depends on the observations gathered.

In all cases, the models are intractable in the worst case, meaning that brute force methods do not scale up to problems involving many actions and variables. Domain-independent approaches aimed at solving these models effectively must thus automatically *recognize* and *exploit* the structure of the individual problems that are given. Like in other AI models such as Constraint Satisfaction Problems and Bayesian Networks [Dechter 2003; Pearl 1988], the key to exploiting the structure of problems in planning models, is *inference*. The most common form of inference in planning is the automatic derivation of heuristic evaluation functions to guide the search. Before considering such domain-independent heuristics, however, we will make precise some of the models used in planning and the languages used for representing them.

## 2.1 Planning Models

*Classical planning* is concerned with the selection of actions in environments that are *deterministic* and whose initial state is *fully known*. The model underlying classical planning can thus be described as a state space featuring:

- a finite and discrete set of states $S$,
- a *known initial state* $s_0 \in S$,
- a set $S_G \subseteq S$ of goal states,

- actions $A(s) \subseteq A$ applicable in each state $s \in S$,
- a *deterministic state transition function* $f(a, s)$ for $a \in A(s)$ and $s \in S$, and
- positive *action costs* $c(a, s)$ that may depend on the action and the state.

A solution or *plan* is a sequence of actions $a_0, \ldots, a_n$ that generates a state sequence $s_0, s_1, \ldots, s_{n+1}$ such that $a_i$ is applicable in the state $s_i$ and results in the state $s_{i+1} = f(a_i, s_i)$, the last of which is a goal state.

The cost of a plan is the sum of the action costs, and a plan is optimal if it has minimum cost. The cost of a problem is the cost of its optimal solutions. When action costs are all 1, a situation that is common in classical planning, plan cost reduces to plan length, and the optimal plans are simply the shortest ones.

The computation of a classical plan can be cast as a *path-finding* problem in a directed graph whose nodes are the states, and whose source and target nodes are the initial state $s_0$ and the goal states $S_G$. Algorithms for solving such problems are polynomial in the number of nodes (states), which is exponential in the number of problem variables (see below). The use of heuristics for guiding the search for plans in large graphs is aimed at improving such worst case behavior.

The model underlying classical planning does not account for either uncertainty or sensing and thus gives rise to plans that represent open-loop controllers where observations play no role. Other planning models in AI take these aspects into account and give rise to different types of controllers.

*Conformant planning* is planning in the presence of uncertainty in the initial situation and action effects. In the resulting model, the initial state $s_0$ is replaced by a *set* $S_0$ of possible initial states, and the deterministic transition function $f(a, s)$ that maps the state $s$ into the unique successor state $s' = f(a, s)$, is replaced by a non-deterministic transition function $F(a, s)$ that maps $s$ into a *set* of possible successor states $s' \in F(a, s)$. A solution to such model, called a conformant plan, is an action sequence that achieves the goal with certainty for *any* possible initial state and *any* possible state transition [Goldman and Boddy 1996]. The search for conformant plans can also be cast as a path-finding problem but over a different, exponentially larger graph whose nodes represent *belief states*. In this formulation, a belief state $b$ stands for the set of states deemed possible, the initial belief state is $b_0 = S_0$, and actions $a$, whether deterministic or not, map a belief state $b$ into a unique successor belief state $b_a$, where $s' \in b_a$ if there is a state $s$ in $b$ such that $s' \in F(a, s)$ [Bonet and Geffner 2000].

*Planning with sensing,* often called contingent planning in AI, refers to planning in the face of both uncertainty and feedback. The model extends the one for conformant planning with a characterization of sensing. A *sensor model* expresses the relation between the observations and the true but possibly hidden states, and can be codified through a set $o \in O$ of observation tokens and a function $o(s)$ that maps states $s$ into observation tokens. An environment is fully observable if different states give rise to different observations, i.e., $o(s) \neq o(s')$ if $s \neq s'$, and partially observable otherwise. While the model for planning with sensing is a slight variation of the model for conformant planning, the resulting solution or plan forms are quite different as observations can and must be taken into account in the selection of actions. Indeed, solution to planning with sensing problems can be expressed equivalently as either *trees* [Weld, Anderson, and Smith 1998], *policies* mapping beliefs into actions [Bonet and Geffner 2000], or *finite-state controllers* [Bonet, Palacios, and Geffner 2009]. A finite-state controller is an automata defined by a collection of tuples of the form $\langle q, o, a, q' \rangle$ that prescribe to do action $a$ and move to the controller state $q'$ after getting the observation $o$ in the controller state $q$.

The probabilistic versions of these models are also used in planning. The models that result when the actions have stochastic effects and the states are fully observable are the familiar Markov Decision Processes (MDPs) used in Operations Research and Control Theory [Bertsekas 1995], while the models that result when action and sensors are stochastic, are the Partial Observable MDPs (POMDPs) [Kaelbling, Littman, and Cassandra 1998].

## 2.2 Planning Languages

A domain-independent planner is a general solver over a class of models: classical planners are solvers over the class of basic state models where actions are deterministic and the initial state is fully known, conformant planners are solvers over the class of models where actions are non-deterministic and the initial state is partially known, and so on. In all cases, the corresponding state model that characterizes a given planning problem is not given explicitly but in a compact form, with the states associated with the values of a given set of variables.

One of the most common languages for representing classical problems is Strips, a planning language that can be traced back to the early 70's [Fikes and Nilsson 1971]. A planning problem in Strips is a tuple $P = \langle F, O, I, G \rangle$ where

- $F$ stands for the set of relevant *variables* or *fluents*,
- $O$ stands for the set of relevant *operators* or *actions*,
- $I \subseteq F$ stands for the *initial situation*, and
- $G \subseteq F$ stands for the *goal situation*.

In Strips, the actions $o \in O$ are represented by three sets of atoms from $F$ called the Add, Delete, and Precondition lists, denoted as $Add(o)$, $Del(o)$, $Pre(o)$. The first, describes the atoms that the action $o$ makes true, the second, the atoms that $o$ makes false, and the third, the atoms that must be true in order for the action to be applicable. A Strips problem $P = \langle F, O, I, G \rangle$ encodes in compact form the state model $\mathcal{S}(P)$ where

- the states $s \in S$ are the possible *collections of atoms* from $F$,
- the initial state $s_0$ is $I$,

- the goal states $s$ are those for which $G \subseteq s$,
- the actions $a$ in $A(s)$ are the ones in $O$ with $Prec(a) \subseteq s$,
- the state transition function is $f(a,s) = (s \setminus Del(a)) \cup Add(a)$, and
- the action costs $c(a)$ are equal to 1 by default.

The states in $S(P)$ represent the possible valuations over the boolean variables in $F$. Thus, if the set of variables $F$ has cardinality $|F| = n$, the number of states in $S(P)$ is $2^n$. A state $s$ represents the valuation where the variables appearing in $s$ are taken to be true, while the variables not appearing in $s$ are false.

As an example, a planning domain that involves three locations $l_1$, $l_2$, and $l_3$, and three tasks $t_1$, $t_2$, and $t_3$, where $t_i$ can be performed only at $l_i$, can be modeled with a set $F$ of fluents $at(l_i)$ and $done(t_i)$, and a set $O$ of actions $go(l_i, l_j)$ and $do(t_i)$, $i, j = 1, \ldots, 3$, with precondition, add, and delete lists

$$Pre(a) = \{at(l_i)\} \ , \ Add(a) = \{at(l_j)\} \ , \ Del(a) = \{at(l_i)\}$$

for $a = go(l_i, l_j)$, and

$$Pre(a) = \{at(l_i)\} \ , \ Add(a) = \{done(t_i)\} \ , \ Del(a) = \{\}$$

for $a = do(t_i)$. The problem of doing tasks $t_1$ and $t_2$ starting at location $l_3$ can then be modeled by the tuple $P = \langle F, I, O, G \rangle$ where

$$I = \{at(l_3)\} \ \text{ and } \ G = \{done(t_1), done(t_2)\} \ .$$

A solution to $P$ is an applicable action sequence that maps the state $s_0 = I$ into a state where the goals in $G$ are all true. In this case one such plan is the sequence

$$\pi = \{go(l_3, l_1), do(t_1), go(l_1, l_2), do(t_2)\} \ .$$

The number of states in the problem is $2^6$ as there are 6 boolean variables. Still, it can be shown that many of these states are not reachable from the initial state. Indeed, the atoms $at(l_i)$ for $i = 1, 2, 3$ are mutually exclusive and exhaustive, meaning that every state reachable from $s_0$ makes one and only one of these atoms true. These boolean variables encode indeed the possible values of the multi-valued variable that represents the agent location.

Strips is a planning language based on variables that are boolean, yet planning languages featuring primitive multi-valued variables and richer syntactic constructs are commonly used for describing both classical and non-classical planning models [McDermott 1998; Younes, Littman, Weissman, and Asmuth 2005].

## 2.3 Planning Algorithms

We have presented some of the models used in domain-independent planning, and one of the languages used for describing them in compact form. We focus now on the algorithms developed for solving them.

GPS, the first AI planner introduced by Newell, Shaw, and Simon, used a technique called means-ends analysis where differences between the current state and the goal situation were identified and mapped into operators that could decrease those differences [Newell and Simon 1963]. Since then, the idea of means-ends analysis has been refined and extended in many ways, seeking planning algorithms that are *sound* (only produce plans), *complete* (produce a plan if one exists), and *effective* (scale up to large problems). By the early 90's, the state-of-the-art method was UCPOP [Penberthy and Weld 1992], an elegant algorithm based on partial-order causal link planning [Sacerdoti 1975; Tate 1977; McAllester and Rosenblitt 1991], a planning method that is sound and complete, but which doesn't scale up too well.

The situation in planning changed drastically in the middle 90's with the introduction of Graphplan [Blum and Furst 1995], a planning algorithm based on the Strips representation but which otherwise had little in common with previous approaches, and scaled up better. Graphplan works iteratively in two phases. In the first phase, Graphplan builds a *plan graph* in polynomial time, made up of a sequence of layers $F_0, A_0, \ldots, F_{n-1}, A_{n-1}, F_n$ where $F_i$ and $A_i$ denote sets of fluents and actions respectively. $F_0$ is the set of fluents true in the initial situation and $n$ is a planning horizon, initially the index of the first layer $F_i$ where all the goals appear. In this construction, certain pairs of actions and certain pairs of fluents are marked as mutually exclusive or mutex. The meaning of these layers and mutexes is roughly the following: if a fluent $p$ is not in layer $F_i$, then no plan can achieve $p$ in $i$ steps or less, while if the pair $p$ and $q$ is in $F_i$ but marked as mutex, then no plan can achieve $p$ and $q$ *jointly* in $i$ steps or less. Graphplan makes then an attempt to extract a plan from the graph, a computation that is exponential in the worst case. If the plan extraction fails, the planning horizon $n$ is increased by 1, the plan graph is extended one level, and the plan extraction procedure is tried again. Blum and Furst showed that the planning algorithm is sound, complete, and *optimal*, meaning that the plan obtained minimizes the number of time steps provided that certain sets of actions can be done in parallel. More importantly, they showed *experimentally* that this planning approach scaled up much better than previous approaches.

Due to the new ideas and the emphasis on the empirical evaluation of planning algorithms, Graphplan had a great influence in planning research that has seen two new approaches in recent years that scale up better than Graphplan using methods that are not specific to planning.

In the SAT approach to planning [Kautz and Selman 1996], Strips problems are converted into *satisfiability* problems expressed as a set of clauses (a formula in Conjunctive Normal Form) that are fed into state-of-the-art SAT solvers. If for some horizon $n$, the clauses are satisfiable, a parallel plan that solves the problem can be read from the model returned by the solver. If not, like in Graphplan, the plan horizon is increased by 1 and the process is repeated until a plan is found. The approach works well when the required horizon is not large and optimal parallel plans are sought.

In the *heuristic search* approach [McDermott 1996; Bonet, Loerincs, and Geffner 1997], the planning problem is solved by heuristic search algorithms with heuristic evaluation functions extracted automatically from the problem encoding. In forward or progression-based planning, the state space $S(P)$ for a problem $P$ is searched for a path connecting the initial state with a goal state. In backward or regression-based planning, plans are searched backwards from the goal. Heuristic search planners have been shown to scale up to very large problems when solutions are not required to be optimal.

The heuristic search approach has actually not only delivered performance but also an explanation for why Graphplan scaled up better than its predecessors. While not described in this form, Graphplan is a heuristic search planner using a heuristic evaluation function encoded implicitly in the planning graph, and a well known admissible search algorithm [Bonet and Geffner 2001]. The difference in performance between recent and older planning algorithms is thus the result of *inference:* while planners searched for plans blindly until Graphplan, they all search with automatically derived heuristics now, or with unit resolution and clause learning when based on the SAT formulation. Domain-independent solvers whose search is not informed by inference of some sort, do not scale up, as there are too many alternatives to choose from, with a few of them leading to the goal.

## 3 Domain-Independent Planning Heuristics

The main novelty in state-of-the-art planners is the use of automatically derived heuristics to guide the search for plans. In *Heuristics*, Pearl showed how heuristics such as the sum of Manhattan distances for the 15-puzzle, the Euclidian distance for Road Map finding, and the Minimum Spanning Tree for the Travelling Saleman Problem, can all be understood as optimal cost functions of suitable problem *relaxations*. Moreover, for the 15-puzzle, Pearl explicitly considered relaxations obtained mechanically from a Strips representation, showing that both the number of misplaced tiles and the sum of Manhattan distances heuristics are optimal cost functions of relaxations where some *preconditions* of the actions for moving tiles are dropped.

Pearl focused then on the conditions under which a problem relaxation is 'simple enough' so that its optimal cost can be computed in polynomial time. This research problem attracted his attention at the time, and explains his interest on the *graphical structures* underlying various types of problems, including problems of combinatorial optimization, constraint satisfaction, and probabilistic inference. One kind of structure that appeared to result in 'easy' problems in all these contexts was *trees*. Pearl and his students showed indeed that inference on probabilistic Bayesian Trees and Constraint Satisfaction Trees was *polynomial* [Pearl 1982; Dechter and Pearl 1985], even if the general problems are NP-hard (see also [Mackworth and Freuder 1985]). The notion of graphical structures underlying inference problems and the conditions under which they render inference polynomial have been generalized since then in the notion of *treewidth,* a parameter that measures how tree-like is a graph structure [Pearl 1988; Dechter 2003].

Research on the automatic derivation of heuristics in planning builds on Pearl's intuition but takes a different path. The relaxation $P^+$ that underlies most current heuristics in domain-independent planning is obtained from a Strips problem $P$ by dropping, not the preconditions, but the *delete lists*. This relaxation is quite informative but is not 'easy'; indeed finding an optimal solution to a delete-free problem $P^+$ is not easier from a complexity point of view than finding an optimal solution to the original problem $P$. On the other hand, finding *one solution* to $P^+$, not necessarily optimal, can be done easily, in low polynomial time. The result is that heuristics obtained from $P^+$ are informative but not *admissible* (they may overestimate the true cost), and hence, they can be used effectively for finding plans but not for finding *optimal* plans.

If $P(s)$ refers to a planning problem that is like $P = \langle F, I, O, G \rangle$ but with $I = s$, and $\pi(s)$ is the solution found for the delete-relaxation $P^+(s)$, the heuristic $h(s)$ that estimates the cost of the problem $P(s)$ is defined as

$$h(s) = Cost(\pi(s)) = \sum_{a \in \pi(s)} cost(a) \ .$$

The plans $\pi(s)$ for the relaxation $P^+(s)$ are called *relaxed plans*, and there have been many proposals for defining and computing them. We explain below one such method that corresponds to running Graphplan on the delete-relaxation $P^+(s)$ [Hoffmann and Nebel 2001]. In delete-free problems, Graphplan runs in polynomial time and its plan graph construction is simplified as there are no mutex relations to keep track of.

The layers $F_0$, $A_0$, $F_1$, ..., $F_{n-1}$, $A_{n-1}$, $F_n$ in the plan graph for $P^+(s)$ are computed starting with $F_0 = s$, by placing in $A_i$, $i = 1, \ldots, n-1$, all the actions $a$ in $P$ whose preconditions $Pre(a)$ are in $F_i$, and placing in $F_{i+1}$, the add effects of those actions along with the fluents in $F_i$. This construction is terminated when the goals $G$ are all in $F_n$, or when $F_n = F_{n+1}$. Then if $G \not\subseteq F_n$, $h(s) = \infty$, as it can be shown then that the relaxed problem $P^+(s)$ and the original problem $P(s)$ have no solution. Otherwise, a (relaxed) parallel plan $\pi(s)$ for $P^+(s)$ can be obtained backwards from the layer $F_n$ by collecting the actions that add the goals, and recursively, the actions that add the preconditions of those actions that are not true in the state $s$.

More precisely, for $G_n = G$ and $i$ from $n-1$ to $0$, $B_i$ is set to a minimal collection of actions in $A_i$ that add all the atoms in $G_{i+1} \setminus F_i$, and $G_i$ is set to $Pre(B_i) \cup (G_{i+1} \cap F_i)$ where $Pre(B_i)$ is the collection of fluents that are preconditions of actions in $B_i$. It can be shown then that $\pi(s) = B_0, \ldots, B_{n-1}$ is a parallel plan for the relaxation $P^+(s)$; the plan being parallel because the actions in each set $B_i$ are assumed to be done in parallel. The heuristic $h(s)$ is then just $Cost(\pi(s))$. This

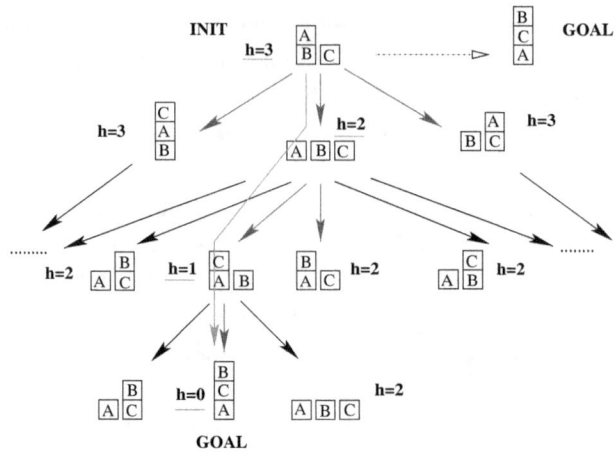

Figure 2. A simple planning problem involving three blocks with initial and goal situations $I$ and $G$ as shown. The actions allow to move a clear block on top of another clear block or to the table. A plan for the problem is a path that connects $I$ with $G$ in the directed graph partially shown. In this example, the plan can be found greedily by taking in each state $s$, starting with $s = I$, the action that results in a state $s'$ that is closer to the goal according to the heuristic. The heuristic values (shown) are derived automatically from the problem as described in the text.

is indeed the heuristic introduced in the FF planner [Hoffmann and Nebel 2001], which is suitable when action costs are uniform. For non-uniform action costs, other heuristics are more convenient [Keyder and Geffner 2008].

## 4 Meaning of Domain-Independent Heuristics

In order to illustrate the meaning and derivation of domain-independent heuristics, let us consider the example shown in Fig. 2, where blocks $a$, $b$, and $c$ initially arranged so that $a$ is on $b$, and $b$ and $c$ are on the table, must be rearranged so that $b$ is on $c$, and $c$ is on $a$. The actions allow to move a clear block (a block with no block on top) on top of another clear block or to the table. The problem can be expressed as a Strips problem $P = \langle F, I, O, G \rangle$ with a set of atoms $F$ given by $on(x, y)$, $ontable(x)$, and $clear(x)$, where $x$ and $y$ range over the block labels $a$, $b$, and $c$. In the heuristic search approach to planning, the solution to $P$ becomes a path-finding problem in the directed graph associated with the state model $S(P)$, where the nodes stand for the states in $S(P)$, and the actions $a \in O$ are mapped into edges connecting a state $s$ with a state $s'$ when $a$ is applicable in $s$ and maps $s$ into $s'$.

The Blocks World is simple for people, but until recently, not so simple for *domain-independent* planners. Indeed, the size of the graph to search is exponential in the number of blocks $n$, with $n!$ possible towers of $n$ blocks, and additional combinations of shorter towers.

Figure 2 shows the search that results from a planner using the heuristic described above, whose value $h(s)$ for each of the states in the graph is shown. All action costs are assumed to be 1. With the heuristic shown, the solution to the problem can be found with no search at all by just selecting in each state $s$ the action that leads to the state $s'$ that is closest to the goal (lowest heuristic value). In the initial state, this action is the one that places block $a$ on the table, in the following state, the action that places $c$ on $a$, and so on.

In order to understand the numbers shown in the figure, let us see how the value $h(s) = 3$ for the initial state $s$ is derived. The heuristic $h(s)$ is $|\pi(s)|$ where $\pi(s)$ is the plan found for the relaxation $P^+(s)$. The relaxed plan $\pi(s)$ is obtained by constructing first the layered graph $F_0$, $A_0$, ..., $F_{n-1}$, $A_{n-1}$, $F_n$, where $n > 0$ as none of the goals $on(b,c)$ and $on(c,a)$ are in $F_0 = s$. The actions in $A_0$ are the actions applicable given the atoms in $F_0$, i.e., the actions $a$ with $Pre(a) \subseteq F_0$. This set includes the actions of moving $c$ to $a$, $a$ to $c$, and $a$ to the table, but does not include actions that move $b$ as the precondition $clear(b)$ is not part of $F_0$. The set $F_1$ extends $F_0$ with all the atoms added by the actions in $A_0$, and includes $on(c,a)$, $on(a,c)$, $ontable(a)$, and $clear(b)$, but not the goal $on(b,c)$. Yet with $clear(b)$ and $clear(c)$ in $F_1$, the action for moving $b$ to $c$ appears in layer $A_1$, and therefore, the other goal atom $on(b,c)$ appears in $F_2$. By collecting the actions that first add the goal atoms $on(c,a)$ and $on(b,c)$, and recursively, the preconditions of those actions that are not in $s$, a relaxed plan $\pi(s)$ with 3 actions is obtained so that $h(s) = 3$. There are several choices for the actions in $\pi(s)$ that result from the way ties in the plan extraction procedure are broken. One possible relaxed plan involves moving $a$ to the table and $c$ to $a$ in the first step, and $b$ to $c$ in the second step. Another involves moving $a$ to $c$ and $c$ to $a$ first, and then $b$ to $c$.

It is important to notice that fluent layers such as $F_1$ in the plan graph do not represent any 'real' states in the original problem $P$ as they include atoms pairs like $on(a,c)$ and $on(c,a)$ that cannot be achieved jointly in any state $s'$ reachable from the initial state. The layer $F_1$ is instead an *abstraction* that *approximates* the set of all states reachable in one step from the initial state by taking their union. This approximation implies that finding an atom $p$ in a layer $F_n$ with $n > 1$ is no guarantee that there is a real plan for $p$ in $P(s)$ that achieves $p$ in $n$ time steps, rather than one such parallel plan exists in the relaxation. Similarly, the relaxed plans $\pi(s)$ obtained above are quite 'meaningless'; they move $a$ to the table or to $c$ at the same time that they move $c$ to $a$. Yet, these 'meaningless' relaxed plans $\pi(s)$ yield the heuristic values $h(s)$ that provide the search with a very meaningful and effective sense of direction.

Let us finally point out that the computation of the domain-independent heuristic $h(s)$ results in valuable information that goes beyond the *numbers* $h(s)$. Indeed, from the computation of the heuristic value $h(s)$, it is possible to determine the actions applicable in the state $s$ that are most relevant to the goal, and then focus on the evaluation of the states that result from those actions only. This type of *action pruning* has been shown to be quite effective [Hoffmann and Nebel 2001], and in slightly different form is part of state-of-the-art planners [Richter, Helmert, and Westphal 2008].

## 5 Other Developments in Planning

Domain-independent planning is concerned with non-classical models also where information about the initial situation is incomplete, actions may have non-deterministic effects, and states may be fully or partially observable. A number of *native solvers* for such models, that include Markov Decision Processes (MDPs) and Partially Observable MDPs have been developed, and progress in the area has been considerable too. Moreover, many of these solvers are also based on *heuristic search* methods (see the article by Bonet and Hansen in this volume). I will not review this literature here but focus instead on two ways in which the results obtained for *classical planning* are relevant to such richer settings too.

First, it's often possible to plan under *uncertainty* without having to model the uncertainty explicitly. This is well known by control engineers that normally design closed-loop controllers for stochastic systems ignoring the 'noise'. Indeed, the error in the model is compensated by the feedback loop. In planning, where non-linear models are considered, the same simplification works too. For instance, in a Blocks World where the action of moving a block may fail, an effective closed-loop policy can be obtained by replanning from the current state when things didn't progress as predicted by the simplified model. Indeed, the planner that did best in the first probabilistic planning competition [Younes, Littman, Weissman, and Asmuth 2005], was not an MDP planner, but a classical replanner of this type. Of course, this approach is not suitable when it may be hard or impossible to recover from failures, or when the system state is not fully observable. In everyday planning, however, such cases may be the exception.

Second, it has been recently shown that it's often possible to efficiently transform problems featuring uncertainty and sensing into classical planning problems that do not. For example, problems $P$ involving uncertainty in the initial situation and no sensing, namely conformant planning problems, can be compiled into classical problems $K(P)$ by adding new actions and fluents that express conditionals [Palacios and Geffner 2007]. The translation from the conformant problem $P$ into the classical problem $K(P)$ is sound and complete, and provided that a *width* parameter defined over $P$ is bounded, it is polynomial too. The result is that the conformant plans for $P$ can be read from the plans for $K(P)$ that can be computed using a classical planner. Moreover, this technique has been recently used for deriving *finite-state controllers* that solve problems featuring both incomplete information and sensing [Bonet, Palacios, and Geffner 2009]. A finite-state controller is an automata that given the current (controller) state and the current observation selects an action and updates the controller state, and so on, until reaching the goal. Figure 3 shows one such problem (left) and the resulting controller (right). The problem, motivated by the work on deictic representations in the selection of actions [Chapman 1989; Ballard, Hayhoe, Pook, and Rao 1997], is about placing a visual marker on top of a green block in a blocks-world scene where the location of the green blocks is not known. The visual marker, initially at the lower left corner of the scene (shown as an eye), can be moved in the four directions, one cell at a time. The observations are whether the cell beneath the marker is empty ('C'), a non-green block ('B'), or a green block ('G'), and whether it is on the table ('T') or not ('-'). The controller shown on the right has been derived by running a classical planner over a classical problem obtained by an automatic translation from the original problem that involves both uncertainty and sensing. In the figure, the controller states $q_i$ are shown in circles while the label $o/a$ on an edge connecting two states $q$ to $q'$ means to do action $a$ when observing $o$ in $q$ and then switching to $q'$. In the classical planning problem obtained from the translation, the actions are tuples $(f_q, f_o, a, f_{q'})$ whose effects are those of the action $a$ but conditional on the fluents $f_q$ and $f_o$ representing the controller state $q$ and observation $o$ being true. In such a case, the fluent $f_{q'}$ representing the controller state $q'$ is made true and $f_q$ is made false. The two appealing features of this formulation is that the resulting classical plans encode very succint closed-loop controllers, and that these controllers are quite general. Indeed, the controller shown in the figure not only solves the problem for the configuration of blocks shown, but for *any configuration* involving *any number of blocks*. The controller prescribes to move the 'eye' up until there are no blocks, then to move it down until reaching the table and right, and to repeat this process until a green block is found ('G'). Likewise, the 'eye' must move right when there are no blocks in a given spot (both 'T' and 'C' observed). See [Bonet, Palacios, and Geffner 2009] for details.

## 6 Heuristics and Cognition

Heuristic evaluation functions are used also in other settings such as Chess playing programs [Pearl 1983] and reinforcement learning [Sutton and Barto 1998]. The difference between evaluations functions in Chess, reinforcement learning, and domain-independent planning mimic actually quite closely the relation among the three approaches to action selection mentioned in the introduction: programming-based, learning-based, and model-based. Indeed, the evaluation functions are programmed by hand in Chess, are learned by trial-and-error in reinforcement learning, and are derived from a (relaxed) model in domain-independent planning.

Heuristic evaluation functions in reinforcement learning, called simply valuation functions, are computed by stochastic sampling and dynamic programming updates. This is a *model-free method* that has been shown to be effective in low-level tasks that do not involve large state spaces, and which provides an accurate account of learning in the brain [Schultz, Dayan, and Montague 1997].

Heuristic evaluation functions as used in domain-independent planning are com-

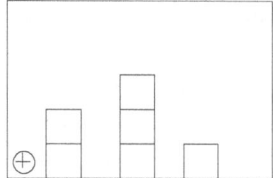 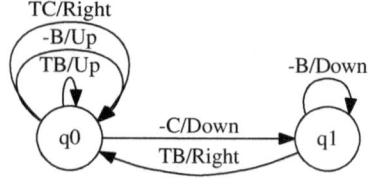

Figure 3. *Left:* The visual marker shown as an 'eye' must be placed on a green block in the blocks-world scene shown, where the locations of the green blocks are not known. The visual marker can be moved in the four directions, one cell at a time. The observations are whether the cell beneath the marker is empty ('C'), a non-green block ('B'), or a green block ('G'), and whether the marker is on the table ('T') or not ('-'). *Right:* The controller derived for this problem using a classical planner over a suitable automatic transformation. The controller states $q_i$ are shown in circles while the label $o/a$ on an edge connecting $q$ to $q'$ means to do $a$ when observing $o$ in $q$ switching then to $q'$. The controller works not only for the problem instance shown on the left, but for *any* instance resulting from changes in the configuration or in the number of blocks.

puted by *model-based methods* where suitable relaxations are solved from scratch. The technique has been shown to work over large problems involving hundred of actions and fluents. Here I want to argue these methods also have features that make them interesting from a cognitive point of view as a plausible basis for an account of 'feelings', 'emotions', or 'appraisals' in high-level human problem solving. I focus on three of these features.

First, domain-independent heuristics are fast (low-polynomial time) and effective, as the 'fast and frugal' heuristics advocated by Gigerenzer and others [Gigerenzer and Todd 1999; Gigerenzer 2007], and yet, they are general too: they apply indeed to all the problems that fit the (classical planning) model and to problems that can be cast in that form (like the visual-marker problem above).

Second, the derivation of these heuristics sheds light on why appraisals may be opaque from a cognitive point of view, and thus not conscious. This is because the heuristic values are obtained from a relaxed model where the meaning of the symbols is different than the meaning of the symbols in the 'true' model. For example, the action of moving an object from one place to another, deletes the old place in the true model but not in the delete-relaxation where an object can thus appear in multiple places at the same time. Thus, if the agent selecting the actions with the resulting heuristic does not have access to the relaxation, it won't be able to explain how the heuristic evaluations are produced nor what they stand for. The importance of the unconscious in everyday cognition is a topic that has been receiving increased attention in recent years, with conscious, deliberate reasoning, appearing to rely heavily on unconscious processing and representing just the tip of the 'cognitive iceberg' [Wilson 2002; Hassin, Uleman, and Bargh 2005; Evans

2008]. While this is evident in vision and natural language processing, where it is clear that one does not have access to how one 'sees' or 'understands', this is likely to be true in most cognitive tasks, including apparently simple problems such as the Blocks World where our ability to find reasons for the actions selected, does not explain how such actions are selected in the first place. In this sense, the focus of cognitive psychology on puzzles such as the Tower of Hanoi may be misplaced: simple problems, such as the Blocks World, are not simple for *domain-independent solvers,* and there is no question that people are capable of solving domains that they have never seen where the combinatorics would defy a naive, blind solver.

Third, the heuristics provide the agent with a sense of direction or 'gut feelings' that guide the action selection in the presence of many alternatives, while avoiding an infinite regress in the decision process. Indeed, emotions long held to interfere with the decision process and rationality, are now widely perceived as a requisite in contexts where it is not possible to consider all alternatives. Emotions and gut feelings are thus perceived as the 'invisible hand' that successfully guides us out of these mental labyrinths [Ketelaar and Todd 2001; Evans 2002].[1] The 'rationality of the emotions' have been defended on theoretical grounds by philosophers [De Sousa 1990; Elster 1999], and on empirical grounds by neuroscientists that have studied the impairments in the decision process that result from lesions in the frontal lobes [Damasio 1995]. The link between emotions and evaluation functions, point to their computational role as well.

While emotions are currently thought as providing the appraisals that are necessary for navigating in a complex world, there are actually very few accounts of *how such appraisals may be computed.* Reinforcement learning methods provide one such account that works well in low level tasks without requiring a model. Heuristic planning methods provide another account that works well in high-level tasks where the model is known. Moreover, as discussed above, heuristic planning methods do not only provide an account of the appraisals, but also of the actions that are worth evaluating. These are the actions $a$ in the state $s$ that are deemed relevant to the goal in the computation of the heuristic $h(s)$; the so-called helpful actions [Hoffmann and Nebel 2001]. This form of *action pruning* may account for a key difference between programs and humans in games such as Chess: while the former consider all possible moves and responses (up to a certain depth), the latter focus on the analysis and evaluation of a few moves and countermoves. Domain-independent heuristics can account in principle for both the focus and the evaluation, the latter in the *value* of the heuristic function $h(s)$, the former in its *structure*.

---

[1] Some philosophers and cognitive scientists refer to this combinatorial problem as the 'frame problem' in AI. This terminology, however, is not accurate. The frame problem in AI [McCarthy and Hayes 1969] refers to the problem that arises in logical accounts of actions and change where the description of the action effects does not suffice to capture what does not change. E.g., the number of chairs in the room does not change if the bell rings. The frame problem is the problem of capturing what does not change from a concise logical description of what changes [Ford and Pylyshyn 1996].

## 7 AI and Cognitive Science: Past and Future

Pearl's ideas on the mechanical discovery of heuristics has received renewed attention in the area of domain-independent planning where heuristic evaluation functions, derived automatically from the problem encoding, are used to guide the search for plans in large spaces. Heuristic search planners are powerful domain-independent solvers that have been *empirically tested* over many large domains involving hundred of actions and variables.

The developments in planning parallel those in other areas of AI and bear on the relevance of Artificial Intelligence to the understanding of the human mind. AI and Cognitive Science were twin disciplines until the 80's, with AI looking to the human mind for inspiration, and Cognitive Science looking to computation as a language for modeling. The relationship between AI and Cognitive Science has changed, however, and the two disciplines do not appear to be that close now. Below, I go over some of the relevant changes that explain this divorce, and explain why, in spite to them, AI remains and will likely remain critically relevant for understanding the human mind, a premise that underlies and motivates the work of Judea Pearl and others AI scientists.

A lot of work in AI until the 80's was about writing programs capable of displaying intelligence over ill-defined problems, either by appealing to introspection or by interviewing an expert. Many good ideas came out from this work, yet few have had a lasting scientific value. The methodological problem with the 'knowledge-based' approach in AI was that the resulting programs were not robust and they always appeared to be missing critical knowledge; either declarative (e.g., that men don't get pregnant), procedural (e.g., which rule or action to apply next), or both. This situation led to an impasse in the 80's, and to many debates and criticisms, like that 'good old fashioned AI' is 'rule application' but human intelligence is not [Haugeland 1993], that representation is not needed for intelligent behavior and gets in the way [Brooks 1991], that subsymbolic neural networks and genetic algorithms are the way to go [Rumelhart and McClelland 1986; Holland 1992], etc.

In part due to the perceived limitations of the knowledge-based approach and the criticisms, and in part due to its own evolution, mainstream AI has changed substantially since the 80's. One of the key methodological changes is that many researchers have moved from the early paradigm of *writing programs for ill-defined problems* to *writing solvers for well-defined mathematical models*. These models include Constraint Satisfaction Problems, Strips Planning, Bayesian Networks and Partially Observable Markov Decision Processes, among others. Solvers are programs that take a compact description of a particular model instance (a planning problem, a CSP instance, and so on) and automatically compute its solution. Unlike the early AI programs, solvers are *general* as they must deal with any problem that fits the model (any instance). Moreover, some of these models, like POMDPs, are extremely expressive. The challenge in this research agenda is mainly *com-*

*putational:* how to make these domain-independent solvers scale up to large and interesting problems given that all these models are intractable in the worst case. Work in these areas has uncovered techniques that accomplish this by automatically recognizing and exploiting the structure of the problems at hand. In planning, these techniques have to do with the automatic derivation and use of heuristic evaluation functions; in SAT and CSPs, with constraint propagation and learning, while in CSPs and Bayesian Networks, with the use of the underlying graphical structure.

The relevance of the early work in AI to Cognitive Science was based on *intuition:* programs provided a way for specifying intuitions precisely and for trying them out. The more recent work on *domain-independent solvers* is more technical and experimental, and is focused not on reproducing intuitions but on *scalability*. This may give the impression, confirmed by the current literature, that recent work in AI is less relevant to Cognitive Science than work in the past. This impression, however, may prove wrong on at least two grounds. First, intuition is not what it used to be, and it is now regarded as the tip of an iceberg whose bulk is made of massive amounts of shallow, fast, but unconscious inference mechanisms that cannot be rendered explicit in the form of rules [Wilson 2002; Hassin, Uleman, and Bargh 2005; Gigerenzer 2007]. Second, whatever these mechanisms are, they appear to work pretty well and to scale up. This is no small feat, given that most methods, whether intuitive or not, do not. Indeed, if the techniques that really scale up are not that many, a plausible conjecture at this point, it may well be the case that *the twin goals of accounting reliably for the intuitions and of scaling up* have a large overlap. By focusing then on the study of meaningful models and the computational methods for dealing with them *effectively*, AI may prove its relevance to human cognition in ways that may go well beyond the rules, cognitive architectures, and knowledge structures of the 80's. Human Cognition, indeed, still provides the inspiration and motivation for a lot of research in AI. The use of Bayesian Networks in Development Psychology for understanding how children acquire and use causal relations [Gopnik, Glymour, Sobel, Schulz, , Kushnir, and Danks 2004], and the use of Reinforcement Learning algorithms in Neuroscience for interpreting the activity of dopamine cells in the brain [Schultz, Dayan, and Montague 1997], are two examples of general AI techniques that have made it recently into Cognitive Science. As AI focuses on models and solvers able to scale up, more techniques are likely to follow. One such candidate is the automatic derivation of heuristic functions as used in planning, which like the research on Bayesian Networks, owes a lot to the work of Judea Pearl.

## References

Ballard, D., M. Hayhoe, P. Pook, and R. Rao (1997). Deictic codes for the embodiment of cognition. *Behavioral and Brain Sciences 20*(4), 723–742.

Bertsekas, D. (1995). *Dynamic Programming and Optimal Control, Vols 1 and 2.*

Athena Scientific.

Blum, A. and M. Furst (1995). Fast planning through planning graph analysis. In *Proceedings of IJCAI-95*, pp. 1636–1642. Morgan Kaufmann.

Bonet, B. and H. Geffner (2000). Planning with incomplete information as heuristic search in belief space. In *Proc. of AIPS-2000*, pp. 52–61. AAAI Press.

Bonet, B. and H. Geffner (2001). Planning as heuristic search. *Artificial Intelligence 129*(1–2), 5–33.

Bonet, B., G. Loerincs, and H. Geffner (1997). A robust and fast action selection mechanism for planning. In *Proceedings of AAAI-97*, pp. 714–719. MIT Press.

Bonet, B., H. Palacios, and H. Geffner (2009). Automatic derivation of memoryless policies and finite-state controllers using classical planners. In *Proc. Int. Conf. on Automated Planning and Scheduling (ICAPS-09)*.

Brooks, R. (1987). A robust layered control system for a mobile robot. *IEEE J. of Robotics and Automation 2*, 14–27.

Brooks, R. (1991). Intelligence without representation. *Artificial Intelligence 47*(1–2), 139–159.

Chapman, D. (1989). Penguins can make cake. *AI magazine 10*(4), 45–50.

Damasio, A. (1995). *Descartes' Error: Emotion, Reason, and the Human Brain.* Quill.

De Sousa, R. (1990). *The rationality of emotion.* MIT Press.

Dechter, R. (2003). *Constraint Processing.* Morgan Kaufmann.

Dechter, R. and J. Pearl (1985). The anatomy of easy problems: a constraint-satisfaction formulation. In *Proc. International Joint Conference on Artificial Intelligence (IJCAI-85)*, pp. 1066–1072.

Elster, J. (1999). *Alchemies of the Mind: Rationality and the Emotions.* Cambridge University Press.

Evans, D. (2002). The search hypothesis of emotion. *British J. Phil. Science 53*, 497–509.

Evans, J. (2008). Dual-processing accounts of reasoning, judgment, and social cognition. *Annual Review of Pschycology 59*, 255–258.

Fikes, R. and N. Nilsson (1971). STRIPS: A new approach to the application of theorem proving to problem solving. *Artificial Intelligence 1*, 27–120.

Ford, K. and Z. Pylyshyn (1996). *The robot's dilemma revisited: the frame problem in artificial intelligence.* Ablex Publishing.

Ghallab, M., D. Nau, and P. Traverso (2004). *Automated Planning: theory and practice.* Morgan Kaufmann.

Gigerenzer, G. (2007). *Gut feelings: The intelligence of the unconscious.* Viking Books.

Gigerenzer, G. and P. Todd (1999). *Simple Heuristics that Make Us Smart.* Oxford University Press.

Goldman, R. P. and M. S. Boddy (1996). Expressive planning and explicit knowledge. In *Proc. AIPS-1996*.

Gopnik, A., C. Glymour, D. Sobel, L. Schulz, , T. Kushnir, and D. Danks (2004). A theory of causal learning in children: Causal maps and Bayes nets. *Psychological Review 111*(1), 3–31.

Hassin, R., J. Uleman, and J. Bargh (2005). *The new unconscious.* Oxford University Press, USA.

Haugeland, J. (1993). *Artificial intelligence: The very idea.* MIT press.

Hoffmann, J. and B. Nebel (2001). The FF planning system: Fast plan generation through heuristic search. *Journal of Artificial Intelligence Research 14*, 253–302.

Holland, J. (1992). *Adaptation in natural and artificial systems.* MIT Press.

Kaelbling, L., M. Littman, and T. Cassandra (1998). Planning and acting in partially observable stochastic domains. *Artificial Intelligence 101*(1–2), 99–134.

Kautz, H. and B. Selman (1996). Pushing the envelope: Planning, propositional logic, and stochastic search. In *Proc. AAAI*, pp. 1194–1201.

Ketelaar, T. and P. M. Todd (2001). Framing our thoughts: Evolutionary psychology's answer to the computational mind's dilemma. In H. Holcomb (Ed.), *Conceptual Challenges in Evolutionary Psychology*. Kluwer.

Keyder, E. and H. Geffner (2008). Heuristics for planning with action costs revisited. In *Proc. ECAI-08*, pp. 588–592.

Mackworth, A. and E. C. Freuder (1985). The complexity of some polynomial network consistency algorithms for constraint satisfaction problems. *Artificial Intelligence 25*(1), 65–74.

Mataric, M. J. (2007). *The Robotics Primer*. MIT Press.

McAllester, D. and D. Rosenblitt (1991). Systematic nonlinear planning. In *Proceedings of AAAI-91*, Anaheim, CA, pp. 634–639. AAAI Press.

McCarthy, J. and P. Hayes (1969). Some philosophical problems from the standpoint of artificial intelligence. In *Machine Intelligence 4*, pp. 463–502. Edinburgh University Press.

McDermott, D. (1996). A heuristic estimator for means-ends analysis in planning. In *Proc. Third Int. Conf. on AI Planning Systems (AIPS-96)*.

McDermott, D. (1998). PDDL – the planning domain definition language. At http://ftp.cs.yale.edu/pub/mcdermott.

Mitchell, T. (1997). *Machine Learning*. McGraw-Hill.

Newell, A., J. Shaw, and H. Simon (1958). Elements of a theory of human problem solving. *Psychology Review 23*, 342–343.

Newell, A. and H. Simon (1963). GPS: a program that simulates human thought. In E. Feigenbaum and J. Feldman (Eds.), *Computers and Thought*, pp. 279–293. McGraw Hill.

Palacios, H. and H. Geffner (2007). From conformant into classical planning: Efficient translations that may be complete too. In *Proc. 17th Int. Conf. on Planning and Scheduling (ICAPS-07)*.

Pearl, J. (1982). Reverend Bayes on inference engines: A distributed hierarchical approach. In *Proceedings of the National Conference on Artificial Intelligence (AAAI-82)*, pp. 133–136.

Pearl, J. (1983). *Heuristics*. Addison Wesley.

Pearl, J. (1988). *Probabilistic Reasoning in Intelligent Systems*. Morgan Kaufmann.

Penberthy, J. and D. Weld (1992). UCPOP: A sound, complete, partially order planner for ADL. In *Proceedings KR'92*.

Richter, S., M. Helmert, and M. Westphal (2008). Landmarks revisited. In *Proc. AAAI*, pp. 975–982.

Rumelhart, D. and J. McClelland (Eds.) (1986). *Parallel distributed processing: explorations in the microstructure of cognition. Vol. 1*. MIT Press.

Sacerdoti, E. (1975). The nonlinear nature of plans. In *Proceedings of IJCAI-75*, Tbilisi, Georgia, pp. 206–214.

Schultz, W., P. Dayan, and P. Montague (1997). A neural substrate of prediction and reward. *Science 275*(5306), 1593–1599.

Sutton, R. and A. Barto (1998). *Introduction to Reinforcement Learning*. MIT Press.

Tate, A. (1977). Generating project networks. In *Proc. IJCAI*, pp. 888–893.

Weld, D., C. Anderson, and D. Smith (1998). Extending Graphplan to handle uncertainty and sensing actions. In *Proc. AAAI-98*, pp. 897–904. AAAI Press.

Wilson, T. (2002). *Strangers to ourselves*. Belknap Press.

Younes, H., M. Littman, D. Weissman, and J. Asmuth (2005). The first probabilistic track of the international planning competition. *Journal of Artificial Intelligence Research 24*, 851–887.

# 3
# Mechanical Generation of Admissible Heuristics

ROBERT HOLTE, JONATHAN SCHAEFFER, AND ARIEL FELNER

## 1 Introduction

This chapter takes its title from Section 4.2 of Judea Pearl's landmark book *Heuristics* [Pearl 1984], and explores how the vision outlined there has unfolded in the quarter-century since its appearance. As the book's title suggests, it is an in-depth summary of classical artificial intelligence (AI) heuristic search, a subject to which Pearl and his colleagues contributed substantially in the early 1980s.

The purpose of heuristic search is to find a least-cost path in a state space from a given start state to a goal state. In principle, such problems can be solved by classical shortest path algorithms, such as Dijkstra's algorithm [Dijkstra 1959], but in practice the state spaces of interest in AI are far too large to be solved in this way. One of the seminal insights in AI was recognizing that even extremely large search problems can be solved quickly if the search algorithm is provided with additional information in the form of a heuristic function $h(s)$ that estimates the distance from any given state $s$ to the nearest goal state [Doran and Michie 1966; Hart, Nilsson, and Raphael 1968]. A heuristic function $h(s)$ is said to be *admissible* if, for every state $s$, $h(s)$ is a lower bound on the true cost of reaching the nearest goal from state $s$. Admissibility is desirable because it guarantees the optimality of the solution found by the most widely-used heuristic search algorithms.

Most of the chapters in *Heuristics* contain mathematically rigorous definitions and analysis. In contrast, Chapter 4 offers a conceptual account of where heuristic functions come from, and a vision of how one might create algorithms for automatically generating effective heuristics from a problem description. An early version of the chapter had been published previously in the widely circulated *AI Magazine* [Pearl 1983].

Chapter 4's key idea is that distances in the given state space can be estimated by computing exact distances in a "simplified" version of the state space. There are many different ways a state space can be simplified. Pearl focused almost exclusively on *relaxation*, which is done by weakening or eliminating one or more of the conditions that restrict how one is allowed to move from one state to another. For example, to estimate the driving distance between two cities, one can ignore the constraint that driving must be done on roads. In this relaxed version of the problem, the distance between two cities is simply the straight-line distance. It is

| | 1 | 2 | 3 |
|---|---|---|---|
| 4 | 5 | 6 | 7 |
| 8 | 9 | 10 | 11 |
| 12 | 13 | 14 | 15 |

| 5 | 9 | 7 | 14 |
|---|---|---|---|
| 3 | 1 | 10 | 15 |
| 4 | 11 | | 8 |
| 2 | 13 | 12 | 6 |

Figure 1. 15-puzzle

easy to see, in general, that distances in a relaxed space cannot exceed distances in the given state space, and therefore the heuristic functions defined in this way are guaranteed to be admissible. An alternate way of looking at this is to view the elimination of conditions as equivalent to adding new edges to the search graph. Therefore, optimal solutions to the relaxed graph (with the additional edges) must be a lower bound on the solution to the original problem.

As a second example of relaxation, consider the 15-puzzle shown in Figure 1, which consists of a set of tiles numbered 1-15 placed in a 4 × 4 grid, leaving one square in the grid unoccupied (called the "blank" and shown as a black square). The only moves that are permitted are to slide a tile that is adjacent to the blank into the blank position, effectively exchanging the tile with the blank. For example, four moves are possible in the right-hand side of Figure 1: tile 10 can be moved down, tile 11 can be moved right, tile 8 can be moved left, and tile 12 can be moved up. To solve the puzzle is to find a sequence of moves that transforms a given scrambled state (right side of Figure 1) into a goal state (such as the one on the left). One possible relaxation of the 15-puzzle state space can be defined by removing the restriction that a tile must be adjacent to the blank to be moveable. In this relaxation any tile can move from its current position to any adjacent position at any time, regardless of whether the adjacent position is occupied or not. The number of moves required to solve this relaxed version (called the Manhattan Distance) is clearly less than or equal to the number of moves required to solve the 15-puzzle itself. Note that in this case the relaxed state space has many more states than the original 15-puzzle (many tiles can now occupy a single location) but it is easier to solve, at least for humans (tiles move entirely independently of one another).

Pearl observes that in AI a state space is almost always defined implicitly by a set of operators that describe a successor relation between states. Each operator has a precondition defining the states to which it can be applied and a postcondition describing how the operator changes the values of the variables used to describe a state. This implies that relaxing a state space description by eliminating one or more preconditions is a simple syntactic operation, and the set of all possible relaxations of a state space description (by eliminating combinations of preconditions) is well-defined and, in fact, easy to enumerate. Hence it is entirely feasible for a mechanical system to generate heuristic functions and, indeed, to search through the space of heuristic functions defined by eliminating preconditions in all possible ways.

The mechanical search through a space of heuristic functions has as its goal, in Pearl's view, a heuristic function with two properties. First, the heuristic function should return values that are as close to the true distances as possible (Chapter 6 in *Heuristics* justifies this). Second, the heuristic function must be efficiently computable, otherwise the reduction in search effort that the heuristic function produces might be outweighed by the increase in computation time caused by the calculation of the heuristic function. Pearl saw the second requirement as the more difficult to detect automatically and proposed that mechanically-recognizable forms of decomposability of the relaxed state space would be the key to mechanically generating efficiently-computable heuristic functions. Pearl recognized that the search for a good heuristic function might itself be quite time-consuming, but argued that this cost was justified because it could be amortized over an arbitrarily large number of problem instances that could all be solved much more efficiently using the same heuristic function.

The preceding paragraphs summarize Pearl's vision for how effective heuristics might be generated automatically from a state space description. The remainder of our chapter contains a brief look at the research efforts directed towards realizing Pearl's vision. We conclude that Pearl correctly anticipated a fundamental breakthrough in heuristic search in the general terms he set out in Chapter 4 of *Heuristics* although not in all of its specifics. Our discussion is informal and the ideas presented and their references are illustrative, not exhaustive.

## 2 The Vision Emerges

The idea of using a solution in a simplified state space to guide the search for a solution in the given state space dates to the early days of AI [Minsky 1963] and was first implemented and shown to be effective in the ABSTRIPS system [Sacerdoti 1974]. However, these early methods did not use the cost of the solution in the simplified space as a heuristic function; they used the solution itself as a skeleton which was to be *refined* into a solution in the given state space by inserting additional operators.

The idea of using distances in a simplified space as heuristic estimates of distances in the given state space came later. It did not originate with Judea Pearl (in fact, he credits Stan Rosenschein for drawing the idea to his attention). However, by devoting a chapter of his otherwise technical book to the speculative idea that admissible heuristic functions could be created automatically, he became an important early promoter of it.

The idea was first developed in the Milan Polytechnic Artificial Intelligence Project in the period 1973-1979. In a series of papers (*e.g.* [Sangiovanni-Vincentelli and Somalvico 1973; Guida and Somalvico 1979]) the Milan group developed the core elements of Pearl's vision. They proposed defining a heuristic function as the exact distance in a relaxed state space and proved that such heuristic functions would be both admissible and *consistent*.[1] To make the computation of such heuristic functions efficient the Milan group envisaged a hierarchy of relaxed spaces, with search at one level being guided by a heuristic function defined by distances in the level above. The Milan group also foresaw the possibility of algorithms for searching through the space of possible simplified state spaces, although the first detailed articulation of this idea, albeit in a somewhat different context, was by Richard Korf [1980].

John Gaschnig [1979] picked up on the Milan work. He made the key observation that if a heuristic function is calculated by searching in a relaxed space, the total time required to solve the problem using the heuristic function could exceed the time required to solve the problem directly with breadth-first search (*i.e.* without using the heuristic function). This was formally proven shortly afterwards by Marco Valtorta [1981, 1984]. This observation led to a focus on the efficiency with which distances in the simplified space could be computed. The favorite approach to doing this (as exemplified in *Heuristics*) was to search for simplified spaces that could be decomposed.

## 3  The Vision Becomes a Reality

Directly inspired by Pearl's vision, Jack Mostow and Armand Prieditis set themselves the task of automating what had hitherto been paper-and-pencil speculation. The result was their ABSOLVER system [Mostow and Prieditis 1989; Prieditis 1993], which fully vindicated Pearl's enthusiasm for the idea of mechanically generating effective, admissible heuristics.

The input to ABSOLVER was a state space description in the standard STRIPS notation [Fikes and Nilsson 1971]. ABSOLVER had a library containing two types of transformations, each of which would take as input a STRIPS representation of a state space and produce as output one or more other STRIPS representations. The first type of transformation were *abstracting* transformations. Their purpose was to create a simplification (or "abstraction") of the given state space. One of these was *drop precondition*, exactly as Pearl had proposed. Their other abstracting transformations were a type of simplification that Pearl had not anticipated—they were *homomorphisms*, which are many-to-one mappings of states in the given space to states in the abstract space. Homomorphic state space abstractions for the purpose of defining heuristic functions were first described by Dennis Kibler in an unpublished report [1982], but their importance was not appreciated until ABSOLVER and the parallel work done by Keki Irani and Suk Yoo [1988].

An example of a homomorphic abstraction of the 15-puzzle is shown in Figure 2. Here tiles 9-15 and the blank are just as in the original puzzle (Figure 1) but tiles 1-8 have had their numbers erased so that they are not distinguishable from each other. Hence for any particular placement of tiles 9-15 and the blank, all the different ways

---

[1] Heuristic function $h(s)$ is consistent if, for any two states $s_1$ and $s_2$, $h(s_1) \leq dist(s_1, s_2) + h(s_2)$, where $dist(s_1, s_2)$ is the distance from $s_1$ to $s_2$.

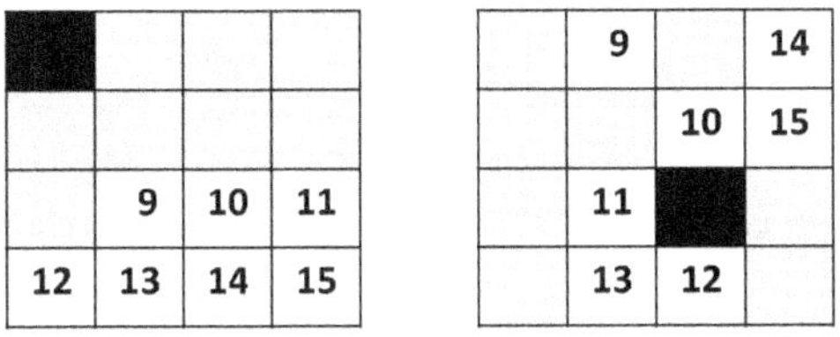

Figure 2. Homomorphic abstraction of the 15-puzzle

of permuting tiles 1-8 among the remaining positions produce 15-puzzle states that map to the same abstract state, even though they would all be distinct states in the original state space. For example, the abstract state in the left part of Figure 2 is the abstraction of the goal state in the original 15-puzzle (left part of Figure 1), but it is also the abstraction of all the non-goal states in the original puzzle in which tiles 9-15 and the blank are in their goal positions but some or all of tiles 1-8 are not. Using this abstraction, the distance from the 15-puzzle state in the right part of Figure 1 to the 15-puzzle goal state would be estimated by calculating the true distance, in the abstract space, from the abstract state in the right part of Figure 2 to the state in the left part of Figure 2.

In addition to abstracting transformations, ABSOLVER's library contained "optimizing" transformations, which would create an equivalent description of a given STRIPS representation in which search could be completed more quickly. This included the "factor" transformation that would, if possible, decompose the state space into independent subproblems, one of the methods Pearl had suggested.

ABSOLVER was applied to thirteen state spaces and found effective heuristic functions in six of them. Five of the functions it discovered were novel, including a simple, effective heuristic for Rubik's Cube that had been overlooked by experts:

> after extensive study, Korf was unable to find a single good heuristic evaluation function for Rubik's Cube [Korf 1985]. He concluded that "if there does exist a heuristic, its form is probably quite complex."

([Mostow and Prieditis 1989], page 701)

## 4 Dawn of the Modern Era

Despite ABSOLVER's success, it did not launch the modern era of abstraction-based heuristic functions. That would not happen until 1994, when Joe Culberson and Jonathan Schaeffer's work on *pattern databases* (PDBs) first appeared [Culberson and Schaeffer 1994]. They used homomorphic abstractions of the kind illustrated in Figure 2 and, as explained above, defined the heuristic function, $h(s)$, of state $s$ to be the actual distance in the abstract space between the abstract state corresponding to $s$ and the abstract goal. The key idea behind PDBs is to store the heuristic function as a lookup table so that its calculation during a search is extremely fast.

To do this, it is necessary to precompute all the distances to the goal state in the abstract space. This is typically done by a backwards breadth-first search starting at the abstract goal state. Each abstract state reached in this way is associated with a specific storage location in the PDB, and the state's distance from the abstract goal is stored in this location as the value in the PDB.

Precomputing abstract distances to create a lookup-table heuristic function was actually one of the optimizing transformations in ABSOLVER, but Culberson and Schaeffer had independently come up with the idea. Unlike the ABSOLVER work, they validated it by producing a two orders of magnitude reduction in the search effort (measured in nodes expanded) needed to solve instances of the 15-puzzle, as compared to the then state-of-the-art search algorithms using an enhanced Manhattan Distance heuristic. To achieve this they used two PDBs totaling almost one gigabyte of memory, a very large amount in 1994 when the experiments were performed [Culberson and Schaeffer 1994]. The paper's referees were sharply critical of the exorbitant memory usage, rejecting the paper three times before it finally was accepted [Culberson and Schaeffer 1996].

Such impressive results on the 15-puzzle could not go unnoticed. The fundamental importance of PDBs was established beyond doubt in 1997 when Richard Korf used PDBs to enable standard heuristic search techniques to find optimal solutions to instances of Rubik's Cube for the first time [Korf 1997].

Since then, PDBs have been used to build effective heuristic functions in numerous applications, including various combinatorial puzzles [Felner, Korf, and Hanan 2004; Felner, Korf, Meshulam, and Holte 2007; Korf and Felner 2002], multiple sequence alignment [McNaughton, Lu, Schaeffer, and Szafron 2002; Zhou and Hansen 2004], pathfinding [Anderson, Holte, and Schaeffer 2007], model checking [Edelkamp 2007], planning [Edelkamp 2001; Edelkamp 2002; Haslum, Botea, Helmert, Bonet, and Koenig 2007], and vertex cover [Felner, Korf, and Hanan 2004].

## 5 Current Status

The use of abstraction to create heuristic functions has profoundly advanced the fields of planning and heuristic search. But the current state of the art is not entirely as Pearl envisaged. Although he recognized that there were other types of state space abstraction, Pearl emphasized relaxation. In this detail, he was too narrowly focused. Researchers have largely abandoned relaxation in favor of homomorphic abstractions, of which many types have been developed and shown useful for defining heuristic functions, such as domain abstraction [Hernádvölgyi and Holte 2000], $h$-abstraction [Haslum and Geffner 2000], projection [Edelkamp 2001], constrained abstraction [Haslum, Bonet, and Geffner 2005], and synchronized products [Helmert, Haslum, and Hoffmann 2007].

Pearl argued for the automatic creation of effective heuristic functions by searching through a space of abstractions. There has been some research in this direction [Prieditis 1993; Hernádvölgyi 2003; Edelkamp 2007; Haslum, Botea, Helmert,

Bonet, and Koenig 2007; Helmert, Haslum, and Hoffmann 2007], but more is needed. However, important progress has been made on the subproblem of evaluating the effectiveness of a heuristic function, with the development of a generic, practical method for accurately predicting how many nodes IDA* (a standard heuristic search algorithm) will expand for any given heuristic function [Korf and Reid 1998; Korf, Reid, and Edelkamp 2001; Zahavi, Felner, Burch, and Holte 2008].

Finally, Pearl anticipated that efficiency in calculating the heuristic function would be achieved by finding abstract state spaces that were decomposable in some way. This has not come to pass, although there is now a general theory of when it is admissible to add the values returned by two or more different abstractions [Yang, Culberson, Holte, Zahavi, and Felner 2008]. Instead, the efficiency of the heuristic calculation has been achieved either by precomputing the heuristic function's values and storing them in a lookup table, as PDBs do, or by creating a hierarchy of abstractions and using distances at one level as a heuristic function to guide the calculation of distances at the level below [Holte, Perez, Zimmer, and MacDonald 1996; Holte, Grajkowski, and Tanner 2005], as anticipated by the Milan group.

# 6 Conclusion

Judea Pearl has received numerous accolades for his prodigious research and its impact. Amidst this impressive body of work are his often-overlooked contributions to the idea of the automatic discovery of heuristic functions. Even though *Heuristics* is over 25 years old (ancient by Computing Science standards), Pearl's ideas still resonate today.

**Acknowledgments:** The authors gratefully acknowledge the support they have received over the years for research in this area from Canada's Natural Sciences and Engineering Research Council (NSERC), Alberta's Informatics Circle of Research Excellence (iCORE), and the Israeli Science Foundation (ISF).

# References

Anderson, K., R. Holte, and J. Schaeffer (2007). Partial pattern databases. In *Symposium on Abstraction, Reformulation and Approximation*, pp. 20–34. Springer-Verlag LNAI #4612.

Culberson, J. and J. Schaeffer (1994). Efficiently searching the 15-puzzle. Technical Report 94-08, Department of Computing Science, University of Alberta.

Culberson, J. and J. Schaeffer (1996). Searching with pattern databases. In G. McCalla (Ed.), *AI'96: Advances in Artificial Intelligence*, pp. 402–416. Springer-Verlag LNAI #1081.

Dijkstra, E. (1959). A note on two problems in connexion with graphs. *Numerische Mathematik 1*, 269–271.

Doran, J. and D. Michie (1966). Experiments with the graph traverser program. In *Proceedings of the Royal Society A*, Volume 294, pp. 235–259.

Edelkamp, S. (2001). Planning with pattern databases. In *European Conference on Planning*, pp. 13–24.

Edelkamp, S. (2002). Symbolic pattern databases in heuristic search planning. In *Artificial Intelligence Planning and Scheduling (AIPS)*, pp. 274–283.

Edelkamp, S. (2007). Automated creation of pattern database search heuristics. In *Model Checking and Artificial Intelligence*, pp. 35–50. Springer-Verlag LNAI #4428.

Felner, A., R. Korf, and S. Hanan (2004). Additive pattern database heuristics. *Journal of Artificial Intelligence Research (JAIR) 22*, 279–318.

Felner, A., R. Korf, R. Meshulam, and R. Holte (2007). Compressed pattern databases. *Journal of Artificial Intelligence Research (JAIR) 30*, 213–247.

Fikes, R. and N. Nilsson (1971). STRIPS: A new approach to the application of theorem proving to problem solving. *Artificial Intelligence 2(3/4)*, 189–208.

Gaschnig, J. (1979). A problem similarity approach to devising heuristics: First results. In *International Joint Conference on Artificial Intelligence (IJCAI)*, pp. 301–307.

Guida, G. and M. Somalvico (1979). A method for computing heuristics in problem solving. *Information Sciences 19*, 251–259.

Hart, P., N. Nilsson, and B. Raphael (1968). A formal basis for the heuristic determination of minimum cost paths. *IEEE Transactions on Systems Science and Cybernetics SCC-4(2)*, 100–107.

Haslum, P., B. Bonet, and H. Geffner (2005). New admissible heuristics for domain-independent planning. In *National Conference on Artificial Intelligence (AAAI)*, pp. 1163–1168.

Haslum, P., A. Botea, M. Helmert, B. Bonet, and S. Koenig (2007). Domain-independent construction of pattern database heuristics for cost-optimal planning. In *National Conference on Artificial Intelligence (AAAI)*, pp. 1007–1012.

Haslum, P. and H. Geffner (2000). Admissible heuristics for optimal planning. In *Artificial Intelligence Planning and Scheduling (AIPS)*, pp. 140–149.

Helmert, M., P. Haslum, and J. Hoffmann (2007). Flexible abstraction heuristics for optimal sequential planning. In *Automated Planning and Scheduling*, pp. 176–183.

Hernádvölgyi, I. (2003). Solving the sequential ordering problem with automatically generated lower bounds. In *Operations Research 2003 (Heidelberg, Germany)*, pp. 355–362.

Hernádvölgyi, I. and R. Holte (2000). Experiments with automatically created memory-based heuristics. In *Symposium on Abstraction, Reformulation and Approximation*, pp. 281–290. Springer-Verlag LNAI #1864.

Holte, R., J. Grajkowski, and B. Tanner (2005). Hierarchical heuristic search revisited. In *Symposium on Abstraction, Reformulation and Approximation*, pp. 121–133. Springer-Verlag LNAI #3607.

Holte, R., M. Perez, R. Zimmer, and A. MacDonald (1996). Hierarchical A*: Searching abstraction hierarchies efficiently. In *National Conference on Artificial Intelligence (AAAI)*, pp. 530–535.

Irani, K. and S. Yoo (1988). A methodology for solving problems: Problem modeling and heuristic generation. *IEEE Transactions on Pattern Analysis and Machine Intelligence 10*(5), 676–686.

Kibler, D. (1982). Natural generation of admissible heuristics. Technical Report TR-188, University of California at Irvine.

Korf, R. (1980). Towards a model of representation changes. *Artificial Intelligence 14*(1), 41–78.

Korf, R. (1985). *Learning to solve problems by searching for macro-operators*. Marshfield, MA, USA: Pitman Publishing, Inc.

Korf, R. (1997). Finding optimal solutions to Rubik's Cube using pattern databases. In *National Conference on Artificial Intelligence (AAAI)*, pp. 700–705.

Korf, R. and A. Felner (2002). Disjoint pattern database heuristics. *Artificial Intelligence 134*(1-2), 9–22.

Korf, R. and M. Reid (1998). Complexity analysis of admissible heuristic search. In *National Conference on Artificial Intelligence (AAAI)*, pp. 305–310.

Korf, R., M. Reid, and S. Edelkamp (2001). Time complexity of iterative-deepening-A*. *Artificial Intelligence 129*(1-2), 199–218.

McNaughton, M., P. Lu, J. Schaeffer, and D. Szafron (2002). Memory efficient A* heuristics for multiple sequence alignment. In *National Conference on Artificial Intelligence (AAAI)*, pp. 737–743.

Minsky, M. (1963). Steps toward artificial intelligence. In E. Feigenbaum and J. Feldman (Eds.), *Computers and Thought*, pp. 406–452. McGraw-Hill.

Mostow, J. and A. Prieditis (1989). Discovering admissible heuristics by abstracting and optimizing: A transformational approach. In *International Joint Conference on Artificial Intelligence (IJCAI)*, pp. 701–707.

Pearl, J. (1983). On the discovery and generation of certain heuristics. *AI Magazine 4*(1), 23–33.

Pearl, J. (1984). *Heuristics – Intelligent Search Strategies for Computer Problem Solving*. Addison-Wesley.

Prieditis, A. (1993). Machine discovery of effective admissible heuristics. *Machine Learning 12*, 117–141.

Sacerdoti, E. (1974). Planning in a hierarchy of abstraction spaces. *Artificial Intelligence 5*(2), 115–135.

Sangiovanni-Vincentelli, A. and M. Somalvico (1973). Theoretical aspects of state space approach to problem solving. In *International Congress on Cybernetics*, Namur, Belgium.

Valtorta, M. (1981). *A Result on the Computational Complexity of Heuristic Estimates for the A* Algorithm*. Ph.D. thesis, Department of Computer Science, Duke University.

Valtorta, M. (1984). A result on the computational complexity of heuristic estimates for the A* algorithm. *Information Sciences 55*, 47–59.

Yang, F., J. Culberson, R. Holte, U. Zahavi, and A. Felner (2008). A general theory of additive state space abstractions. *Journal of Artificial Intelligence Research (JAIR) 32*, 631–662.

Zahavi, U., A. Felner, N. Burch, and R. Holte (2008). Predicting the performance of IDA* with conditional probabilities. In *National Conference on Artificial Intelligence (AAAI)*, pp. 381–386.

Zhou, R. and E. Hansen (2004). Space-efficient memory-based heuristics. In *National Conference on Artificial Intelligence (AAAI)*, pp. 677–682.

# 4
# Space Complexity of Combinatorial Search

RICHARD E. KORF

## 1 Introduction: The Problem

It is well-known that most complete search algorithms take exponential time to run on most combinatorial problems. The reason for this is that many combinatorial problems are NP-hard, and most complete search algorithms guarantee an optimal solution, so unless P=NP, the time complexity of these algorithms must be exponential in the problem size.

What is not so often appreciated is that the limiting resource of many search algorithms is not time, but the amount of memory they require. For example, a simple brute-force breadth-first search (BFS) of an implicit problem space stores every node it generates in memory. If we assume we can generate ten million nodes per second, can store a node in four bytes of memory, and have four gigabytes of memory, we will exhaust our memory in a hundred seconds, or less than two minutes. If our problem is too large to be solved in this amount of time, the memory limitation becomes the bottleneck.

This problem has existed since the first computers were built. While memory capacities have increased by many orders of magnitude over that time, processors have gotten faster at roughly the same pace, and the problem persists. We describe here the major approaches to this problem over the past 25 years. While most of the algorithms discussed have both brute-force and heuristic versions, we focus primarily on the brute-force algorithms, since they are simpler, and the issues are largely the same in both cases.

## 2 Depth-First Search

One solution to this problem in some settings is depth-first search (DFS), which requires memory that is only linear in the maximum search depth. The reason is that at any point in time, it saves only the path from the start node to the current node being expanded, either on an explicit node stack, or on the call stack of a recursive implementation. As a result, the memory requirement of DFS is almost never a limitation. For finite tree-structured problem space graphs, where all solutions are equally desirable, DFS solves the memory problem. For example, chronological backtracking is a DFS for constraint satisfaction problems, and does not suffer any memory limitations in practice.

With an infinite search tree, or when we want a shortest solution path, however, DFS has significant drawbacks. In an infinite search tree, which can result from a depth-first search of a finite graph with multiple paths to the same state, DFS is not complete, but can traverse a single path until it exhausts memory. For example, the problem space graphs of the well-known sliding-tile puzzles are finite, but a depth-first search of these spaces explores a tree-expansion of the graph, which is infinite. Even with a finite search tree, the first solution found by DFS will not be a shortest solution in general.

## 3  Iterative Deepening

### 3.1  Depth-First Iterative Deepening

One solution to these limitations of DFS is depth-first iterative-deepening (DFID) [Korf 1985a]. DFID performs a series of depth-first searches, each to a successively greater depth. DFID simulates BFS, but using memory that is only linear in the maximum search depth. It is guaranteed to find a solution if one exists, even on an infinite tree, and the first solution it finds will be a shortest one.

DFID is essentially the same as iterative-deepening searches used in two-player game programs [Slate and Atkin 1977], but is used to solve a completely different problem. In a two-player game, iterative deepening is used to determine the search depth, because moves must be made within a given time period, and it is difficult to predict how long it will take to search to a given depth. In contrast, DFID is applied to single-agent problems where a shortest solution is required, in order to avoid the memory limitation of BFS.

DFID first appeared in a Columbia University technical report [Korf 1984]. It was independently published in two different papers in IJCAI-85 [Korf 1985b; Stickel and Tyson 1985], and called "consecutively bounded depth-first search" in the latter.

### 3.2  Iterative-Deepening-A*

While discussing DFID with Judea Pearl on a trip to UCLA in 1984, he suggested its extension to heuristic search that became Iterative-Deepening-A* (IDA*) [Korf 1985a]. IDA* overcomes the memory limitation of the A* algorithm [Hart, Nilsson, and Raphael 1968] for heuristic searches the same way that DFID overcomes the memory limitation of BFS for brute-force searches. In particular, it uses the A* cost function of $f(n) = g(n) + h(n)$, where $g(n)$ is the cost of the current path from the root to node $n$, and $h(n)$ is a heuristic estimate of the lowest cost of any path from node $n$ to a goal node. IDA* performs a series of depth-first search iterations, where each branch of the search is terminated when the cost of the last node on that branch exceeds a cost threshold for that iteration. The cost threshold of the first iteration is set to the heuristic estimate of the start state, and the cost threshold of each successive iteration is set to the minimum cost of all nodes generated but not expanded on the previous iteration. Like A*, IDA* guarantees an optimal solution if the heuristic function is *admissible*, or never overestimates actual cost. IDA* was the first algorithm to find optimal solutions to the Fifteen Puzzle, the famous four-by-four sliding-tile puzzle [Korf 1985a]. It was also the first algorithm to find optimal solutions to Rubik's Cube [Korf 1997].

I invited Judea to be a co-author on both the IJCAI paper [Korf 1985b], and a subsequent AI journal paper [Korf 1985a] that described both DFID and IDA*. At the time, I was a young assistant professor, and he declined the co-authorship with typical generosity, saying that he didn't need another paper, and that the paper would be much more important to me at that stage of my career. In retrospect, I regret that I didn't insist on joint authorship.

## 4  Other Limited-Memory Search Algorithms

A number of researchers noted that linear-space search algorithms such as DFID and IDA* use very little space in practice, and explored whether better performance could be achieved by using all the memory available on a machine.

Perhaps the simplest of these algorithms is MREC [Sen and Bagchi 1989]. MREC is a hybrid of A* and IDA*. It runs A* until memory is almost full, and then runs successive iterations of IDA* from each node generated but not expanded by A*.

Perhaps the most elegant of these algorithms is MA* [Chakrabarti, Ghose, Acharya, and de Sarkar 1989]. MA* also runs A* until memory is almost full. Then, in order to get enough memory to expand the best node, it finds a group of sibling leaf nodes with the worst cost values, and deletes them, leaving behind only their parent node, with a stored cost equal to the minimum of its children's values. The algorithm alternates between expanding the best nodes, and contracting the worst nodes, until a solution is chosen for expansion.

Unfortunately, the overhead of this algorithm makes it impractical compared to IDA*. There have been at least two attempts to make this basic algorithm more efficient, namely SMA* for simplified MA* [Russell 1992] and ITS for Iterative Threshold Search [Ghosh, Mahanti, and Nau 1994], but none of these algorithms significantly outperform IDA*.

## 5  Recursive Best-First Search

Best-first search is a general class of search algorithms that maintains both a Closed list of expanded nodes, and an Open list of nodes that have been generated but not yet expanded. Initially, the Open list contains just the start node, and the Closed list is empty. Each node $n$ on Open has an associated cost $f(n)$. At each step of the algorithm, an Open node of lowest cost is chosen for expansion, moved to the Closed list, and its children are placed on Open along with their associated costs. The algorithm continues until a goal node is chosen for expansion. Different best-first search algorithms differ only in their cost functions. For example, if the cost of a node is simply its depth, then best-first search becomes breadth-first search. Alternatively, if the cost of a node $n$ is $f(n) = g(n) + h(n)$, then best-first search becomes the A* algorithm.

A cost function is *monotonic* if the cost of a child node is always greater than or equal to the cost of its parent. The cost function $g(n)$ is monotonic if all edges have non-negative cost. The A* cost function $f(n) = g(n) + h(n)$ is monotonic if the heuristic function $h(n)$ is *consistent*, meaning that $h(n) \leq k(n,n') + h(n')$, where $n'$ is a child of node $n$, and $k(n,n')$ is the cost of the edge from $n$ to $n'$. Many heuristic functions are both admissible and consistent. If the cost function is monotonic, then the order in which nodes are first expanded by IDA* is the same as for a best-first search with the same cost function.

Not all useful cost functions are monotonic, however. For example, Weighted A* (WA*) is a best-first search with the cost function $f(n) = g(n) + w * h(n)$. If $w$ is greater than one, then WA* usually finds solutions much faster than A*, but at a small cost in solution quality. With $w$ greater than one, however, $f(n) = g(n)+w*h(n)$ is not monotonic, even with a consistent $h(n)$. With a non-monotonic cost-function, IDA* does not expand nodes in best-first order. In particular, in parts of the search tree where the cost of nodes is lower than the cost threshold for the current iteration, IDA* behaves as a brute-force search, expanding nodes in the order in which they are generated.

Can any linear-space search algorithm simulate a best-first search with a non-monotonic cost function? Surprisingly, the answer is yes. *Recursive best-first search* [Korf 1993] (RBFS) maintains a path from the start node to the last node generated, along with the siblings of nodes on that path. Stored with each node is a cost value. If the node has never been expanded before, its stored cost is its original cost. If it has been expanded before, its stored cost is the minimum cost of all its descendents that have been generated but not yet expanded, which are not stored in memory. The sibling node off the current path of lowest cost is the ancestor of the next leaf node that would be expanded by a best-first search. By propagating these values up the tree, and inheriting these values down the tree as a previously explored path is regenerated, RBFS always finds the next leaf node expanded by a best-first search. Thus, it simulates a best-first search even with a non-monotonic cost function. Furthermore, with a monotonic cost function it can outperform IDA* if there are many different unique cost values in the search tree. For details on RBFS, see [Korf 1993].

## 6 Drawback of Linear-Space Search Algorithms

The advantage of linear-space searches, such as DFS, DFID, IDA* and RBFS, is that they use very little memory, and hence can run for weeks or months on large problems. They all share a significant liability relative to best-first searches such as BFS, A* or WA*, however. In particular, on search graphs with multiple paths to the same node, or cycles in the graph, linear-space algorithms can generate exponentially more nodes than best-first search.

For example, consider a rectangular grid problem-space graph. From any node in such a graph, moving North and then East generates the same state as moving East and then North. These are referred to as duplicate nodes, since they represent the same state arrived at via two different paths. When a best-first search is run on such a graph, as each node is generated it is checked to see if the same state already appears on the Open or Closed lists. If so, only the node reached by a shortest path is stored, and the duplicate node is eliminated. Thus, by detecting and rejecting such duplicate nodes, a breadth-first search to a radius of $r$ on a grid graph would expand $O(r^2)$ nodes.

A linear-space algorithm doesn't store most of the nodes it generates however, and hence cannot detect most duplicate nodes. In a grid graph, each node has four neighbors. A linear-space search will not normally regenerate its immediate parent as one of its children, reducing the number of children to three for all but the start node. Thus, a depth-first search of a grid graph to a radius $r$ will generate $O(3^r)$ nodes, compared to $O(r^2)$ nodes for a best-first search. This is an enormous overhead on graphs with many paths to the same state, rendering linear-space algorithms completely impractical in such problem spaces.

## 7 Frontier Search

Fortunately, there is another technique that can significantly reduce the memory required by a search algorithm on problem spaces with many duplicate nodes. The basic idea is to save only the Open list and not the Closed list. This algorithm schema is called *frontier search*, since the Open list represents the frontier of nodes that have been generated but not yet expanded [Korf, Zhang, Thayer, and Hohwald 2005]. When a node is expanded in frontier search, it is simply deleted rather than being moved to a Closed list.

The advantage of this technique is that the Open list can be much smaller than the Closed list. In the grid graph, for example, the Closed list grows as $O(r^2)$, whereas the Open list grows only as $O(r)$, where $r$ is the radius of the search.

For ease of explanation, we'll assume a problem space with reversible operators, but the method also applies to some directed problem-space graphs as well. There are two reasons to save the Closed list. One is to detect duplicate nodes, and the other is to return the solution path. We first consider duplicate detection.

### 7.1 Detecting Duplicate Nodes

Imagine the search frontier as a continuous boundary of Open nodes containing a region of Closed nodes. To minimize the memory needed, we need to prevent Closed nodes from being regenerated. There are two ways that this might happen. One is by a child node regenerating its parent node. This is prevented in frontier search by storing with each Open node a *used-operator bit* for each operator that could generate that node. This bit is set to one whenever the corresponding operator is used to generate the Open node. When an Open node is expanded, the inverses of those operators whose used bits are set to one are not applied, thus preventing a node from regenerating its parent.

The other way a Closed node could be regenerated is by the frontier looping back on itself, like a wave breaking through the surface of the water below. If the frontier is unbroken, when this happens the Open node being expanded would first have to generate another Open node on the frontier of the search before generating a Closed node on the interior. When this happens, the duplicate Open node is detected, and the the union of the used operator bits set in each of the two copies is stored with the single copy retained. In other words, one part of the frontier cannot invade another part of the interior without passing through another part of the frontier first, where the intrusion is detected. By storing and managing such used-operator bits, frontier search detects all duplicate node generations and prevents a node from being expanded more than once.

An alternative to used-operator bits is to save several levels of the search at a time [Zhou and Hansen 2003]. In particular, Closed nodes are stored until all their children are expanded, and then deleted.

## 7.2 Reconstructing the Solution Path

The other reason to store the Closed list is to reconstruct the solution path at the end of a search. In a best-first search, this is done by storing with each node a pointer to its parent node. Once a goal state is reached, these pointers are followed back from the goal to the initial state, generating the solution path in reverse order. This can't be done with frontier search directly, since the Closed list is not saved.

Frontier search can reconstruct the solution path using *divide-and-conquer bidirectional frontier search*. We search simultaneously both forward from the initial state and backward from the goal state. When the two search frontiers meet and a "middle" state on a optimal path has been found, we then use the same algorithm recursively to search from the initial state to the middle state, and from the middle node to the goal node.

The solution path can also be constructed using unidirectional search, as long as one can identify nodes that are approximately half way along an optimal path. For example, the problem of two-way sequence alignment in computational biology can be mapped to finding a shortest path in a two-dimensional grid [Needleman and Wunsch 1970]. In such a path, a state on the midline of the grid will be about half way along an optimal solution. In such a problem, we search forward from the initial state to the goal state. For every node on the Open list past the midpoint of the grid, we store a pointer to its ancestor on the midpoint. Once we reach the goal state, its ancestor on the midline is approximately in the middle of the optimal solution path. We then recursively apply the same algorithm to find a path from the initial state to the middle state, and from the middle state to a goal state.

Tree-structured search spaces and densely connected graphs such as grids represent two ends of the connectivity spectrum. On a tree, linear-space search algorithms perform very well, since they generate no duplicate nodes. Frontier search doesn't save much memory on a tree, however, since the number of leaf nodes dominates the number of interior nodes. Conversely, on a grid graph, linear-space search performs very poorly because undetected duplicate nodes dwarf the number

## Space Complexity

of unique nodes, while frontier search performs very well, reducing the memory required from quadratic to linear space.

## 8 Disk-Based Search

Even on problems where frontier search is effective, memory is still the resource that limits its applicability. An additional approach to this memory limitation is to use magnetic disk to store nodes rather than semiconductor memory. While semiconductor memory has gotten much larger and cheaper over time, it still costs about $30 per gigabyte. In contrast, magnetic disk storage costs about $100 per terabyte, which is 300 times cheaper. The problem with simply replacing semiconductor memory with magnetic disks, however, is that random access of a byte on disk can take up to ten milliseconds, which is five orders of magnitude slower than for memory. Thus, disk access must be sequential for efficiency.

Consider a simple breadth-first search (BFS), which is usually implemented with a first-in first-out queue. Nodes are read from the head of the queue, expanded, and their children are written to the tail of the queue. Such a queue can efficiently be stored on disk, since all accesses are sequential.

In order to detect duplicate nodes efficiently, however, the nodes are also stored in a hash table. Nodes are looked up in the hash table as they are generated, and duplicate nodes are discarded. Such a hash table cannot be directly implemented on magnetic disk, however, due to the long latency of random access.

A solution to this problem is called *delayed duplicate detection* [Korf 2008] or DDD for short. The BFS queue is stored on disk, but nodes are not checked for duplicates as they are generated. Rather, duplicate nodes are appended to the queue, and are only eliminated periodically, such as at the end of each depth iteration. There are several ways to eliminate duplicate nodes from a large file stored on disk.

The simplest way is to sort the nodes based on their state representation. This will bring duplicate nodes to adjacent positions. Then, a simple linear scan of the file can be used to detect and merge duplicate nodes. The drawback of this approach is that the sorting takes $O(n \log n)$ time, where $n$ is the number of nodes. With a terabyte of storage, and four bytes per state, $n$ can be as large as 250 billion, and hence $\log n$ as large as 38.

An alternative is to use hash-based DDD. This scheme relies on two orthogonal hash functions defined on the state representation. In the first phase, the input file is read, and nodes are output to separate files based on the value of the first hash function. Thus, any sets of duplicate node will be confined to the same file. In the second phase, the nodes in each individual file are hashed into memory using the second hash function, and duplicates are detected and merged in memory. The advantage of this approach is that the time complexity is only linear in the number of nodes, rather than $O(n \log n)$ time for sorting-based DDD.

The overall DDD algorithm proceeds in alternating phases of node expansion followed by merging duplicate nodes. Combined with frontier search, DDD has been used to perform complete breadth-first searches of sliding-tile puzzles as large as the Fifteen Puzzle, with over $10^{13}$ nodes [Korf and Schultze 2005]. It has also been used for large heuristic searches of the four-peg Towers of Hanoi problem with up to 31 disks, generating over $2.5 \times 10^{13}$ nodes [Korf and Felner 2007]. These searches take weeks to run, and time is the limiting resource, not storage capacity.

An alternative to DDD for disk-based search is called *structured duplicate detection* (SDD) [Zhou and Hansen 2004]. In this approach, the problem space is partitioned into subsets, so that when expanding nodes in one subset, the children only belong to a small number of other subsets, referred to as its *duplicate detection scope*. The subset of nodes currently being expanded is kept in memory, along with the subsets in its duplicate detection scope, detecting any duplicates generated immediately, while storing other subsets on disk. As different subsets of nodes are expanded, currently resident duplicate detection scopes are swapped out to disk to make room for the duplicate detection scopes of the new nodes being expanded.

## 9 Summary and Conclusions

We have presented a number of different algorithms, designed over the past 25 years, to deal with the space complexity of brute-force and heuristic search. They fall into two general categories. The linear-space search algorithms, including depth-first search, depth-first iterative-deepening, iterative-deepening-A*, and recursive best-first search, use very little memory but cannot detect most duplicate nodes. They perform very well on trees, but poorly on highly-connected graphs, such as a grid. The best-first search algorithms, including breadth-first search, A*, weighted A*, frontier search, and disk-based algorithms, detect all duplicate nodes, and hence perform well on highly-connected graphs. The best-first algorithms are limited by the amount of memory available, except for the disk-based techniques, which are limited by time in practice.

**Acknowledgments:** This research was supported continuously by the National Science Foundation, most recently under grant No. IIS-0713178.

## References

Chakrabarti, P., S. Ghose, A. Acharya, and S. de Sarkar (1989, December). Heuristic search in restricted memory. *Artificial Intelligence 41*(2), 197–221.

Ghosh, R., A. Mahanti, and D. Nau (1994, August). An efficient limited-memory heuristic tree search algorithm. In *Proceedings of the Twelfth National Conference on Artificial Intelligence (AAAI-94)*, Seattle, WA, pp. 1353–1358.

Hart, P., N. Nilsson, and B. Raphael (1968, July). A formal basis for the heuristic determination of minimum cost paths. *IEEE Transactions on Systems Science and Cybernetics SSC-4*(2), 100–107.

Korf, R. (1984). The complexity of brute-force search. technical report, Computer Science Department, Columbia University, New York, NY.

Korf, R. (1985a). Depth-first iterative-deepening: An optimal admissible tree search. *Artificial Intelligence 27*(1), 97–109.

Korf, R. (1985b, August). Iterative-deepening-a*: An optimal admissible tree search. In *Proceedings of the Ninth International Joint Conference on Artificial Intelligence (IJCAI-85)*, Los Angeles, CA, pp. 1034–1036.

Korf, R. (1993, July). Linear-space best-first search. *Artificial Intelligence 62*(1), 41–78.

Korf, R. (1997, July). Finding optimal solutions to rubik's cube using pattern databases. In *Proceedings of the Fourteenth National Conference on Artificial Intelligence (AAAI-97)*, Providence, RI, pp. 700–705.

Korf, R. (2008, December). Linear-time disk-based implicit graph search. *Journal of the Association for Computing Machinery 55*(6), 26:1 to 26:40.

Korf, R. and A. Felner (2007, January). Recent progress in heuristic search: A case study of the four-peg towers of hanoi problem. In *Proceedings of the Twentieth International Joint Conference on Artificial Intelligence (IJCAI-07)*, Hyderabad, India, pp. 2334–2329.

Korf, R. and P. Schultze (2005, July). Large-scale, parallel breadth-first search. In *Proceedings of the Twentieth National Conference on Artificial Intelligence (AAAI-05)*, Pittsburgh, PA, pp. 1380–1385.

Korf, R., W. Zhang, I. Thayer, and H. Hohwald (2005, September). Frontier search. *Journal of the Association for Computing Machinery 52*(5), 715–748.

Needleman, S. and C. Wunsch (1970). A general method applicable to the search for similarities in the amino acid sequences of two proteins. *Journal of Molecular Biology 48*, 443–453.

Russell, S. (1992, August). Efficient memory-bounded search methods. In *Proceedings of the Tenth European Conference on Artificial Intelligence (ECAI-92)*, Vienna, Austria.

Sen, A. and A. Bagchi (1989, August). Fast recursive formulations for best-first search that allow controlled use of memory. In *Proceedings of the Eleventh International Joint Conference on Artificial Intelligence (IJCAI-89)*, Detroit, MI, pp. 297–302.

Slate, D. and L. Atkin (1977). Chess 4.5 - the northwestern university chess program. In P. Frey (Ed.), *Chess Skill in Man and Machine*, pp. 82–118. New York, NY: Springer-Verlag.

Stickel, M. and W. Tyson (1985, August). An analysis of consecutively bounded depth-first search with applications in automated deduction. In *Proceedings of the Ninth International Joint Conference on Artificial Intelligence (IJCAI-85)*, Los Angeles, CA, pp. 1073–1075.

Zhou, R. and E. Hansen (2003, August). Sparse-memory graph search. In *Proceedings of the Eighteenth International Joint Conference on Artificial Intelligence (IJCAI-03)*, Acapulco, Mexico, pp. 1259–1266.

Zhou, R. and E. Hansen (2004, July). Structured duplicate detection in external-memory graph search. In *Proceedings of the Nineteenth National Conference on Artificial Intelligence (AAAI-04)*, San Jose, CA, pp. 683–688.

# 5
# Paranoia versus Overconfidence in Imperfect-Information Games

AUSTIN PARKER, DANA NAU, AND V.S. SUBRAHMANIAN

> *Only the paranoid survive.*
> –Andrew Grove, Intel CEO

> *Play with supreme confidence, or else you'll lose.*
> –Joe Paterno, college football coach

## 1 Introduction

In minimax game-tree search, the *min* part of the minimax backup rule derives from what we will call the *paranoid assumption*: the assumption that the opponent will always choose a move that minimizes our payoff and maximizes his/her payoff (or our estimate of the payoff, if we cut off the search before reaching the end of the game). A potential criticism of this assumption is that the opponent may not have the ability to decide accurately what move this is. But in several decades of experience with game-tree search in chess, checkers, and other zero-sum perfect-information games, the paranoid assumption has worked so well that such criticisms are generally ignored.

In game-tree search algorithms for imperfect-information games, the backup rules are more complicated. Many of them (see Section 6) involve computing a weighted average over the opponent's possible moves (or a Monte Carlo sample of them), where each move's weight is an estimate of the probability that this is the opponent's best possible move. Although such backup rules do not take a *min* at the opponent's move, they still tacitly encode the paranoid assumption, by assuming that the opponent will choose optimally from the set of moves he/she is actually capable of making.

Intuitively, one might expect the paranoid assumption to be less reliable in imperfect-information games than in perfect-information games; for without perfect information, it may be more difficult for the opponent to judge which move is best. The purpose of this paper is to examine whether it is better to err on the side of paranoia or on the side of overconfidence. Our contributions are as follows:

1. **Expected utility.** We provide a recursive formula for the expected utility of a move in an imperfect-information game, that explicitly includes the opponent's strategy $\sigma$. We prove the formula's correctness.

2. **Information-set search.** We describe a game-tree search algorithm called *information-set search* that implements the above formula. We show analytically that with an accurate opponent model, information-set search produces optimal results.

3. **Approximation algorithm.** Information-set search is, of course, intractable for any game of interest as the decision problem in an imperfect-information game is complete in double exponential time [Reif 1984]. To address this intractability problem, we provide a modified version of information-set search that computes an approximation of a move's expected utility by combining Monte Carlo sampling of the belief state with a limited-depth search and a static evaluation function.

4. **Paranoia and overconfidence.** We present two special cases of the expected-utility formula (and hence of the algorithm) that derive from two different opponent models: the *paranoid* model, which assumes the opponent will always make his/her best possible move, and the *overconfident* model, which assumes the opponent will make moves at random.

5. **Experimental results.** We provide experimental evaluations of information-set search in several different imperfect-information games. These include imperfect-information versions of P-games [Pearl 1981; Nau 1982a; Pearl 1984], N-games [Nau 1982a], and kalah [Murray 1952]; and an imperfect-information version of chess called kriegspiel [Li 1994; Li 1995]. Our main experimental results are:

   - Information-set search outperformed HS, the best of our algorithms for kriegspiel in [Parker, Nau, and Subrahmanian 2005].
   - In all of the games, the overconfident opponent model outperformed the paranoid model. The difference in performance became more marked when we decreased the amount of information available to each player.

This work was influenced by Judea Pearl's invention of P-games [Pearl 1981; Pearl 1984], and his suggestion of investigating backup rules other than minimax [Pearl 1984]. We also are grateful for his encouragement of the second author's early work on game-tree search (e.g., [Nau 1982a; Nau 1983]).

## 2 Basics

Our definitions and notation are based on [Osborne and Rubinstein 1994]. We consider games having the following characteristics: two players, finitely many moves and states, determinism, turn taking, zero-sum utilities, imperfect information expressed via *information sets* (explained in Section 2.1), and *perfect recall* (explained in Section 2.3). We will let $G$ be any such game, and $a_1$ and $a_2$ be the two players.

Our techniques are generalizable to stochastic multi-player non-zero-sum games,[1] but that is left for future work.

At each state $s$, let $a(s)$ be the player to move at $s$, with $a(s) = \emptyset$ if the game is over in $s$. Let $M(s)$ be the set of available moves at $s$, and $m(s)$ be the state produced by making move $m$ in state $s$. A *history* is a sequence of moves $h = \langle m_1, m_2, \ldots, m_j \rangle$. We let $s(h)$ be the state produced by history $h$, and when clear from context, will abuse notation and use $h$ to represent $s(h)$ (e.g., $m(h) = m(s(h))$). Histories in which the game has ended are called *terminal*. We let $H$ be the set of all possible histories for game $G$.

## 2.1 Information Sets

Intuitively, an information set is a set of histories that are indistinguishable to a player $a_i$, in the sense that each history $h$ provides $a_i$ with the same sequence of observations. For example, suppose $a_1$ knows the entire sequence of moves that have been played so far, except for $a_2$'s last move. If there are two possibilities for $a_2$'s last move, then $a_1$'s information set includes two histories, one for each of the two moves.

In formalizing the above notion, we will not bother to give a full formal definition of an "observation." The only properties we need for an observation are the following:[2]

- We assume that each player $a_i$'s sequence of observations is a function $O_i(h)$ of the current history $h$. The rationale is that if $a_1$ and $a_2$ play some game a second time, and if they both make the same moves that they made the first time, then they should be able to observe the same things that they observed the first time.

- We assume that when two histories $h, h'$ produce the same sequence of observations, they also produce the same set of available moves, i.e., if $O_i(h) = O_i(h')$, then $M(s(h)) = M(s(h'))$. The rationale for this is that if the current history is $h$, $a_i$'s observations won't tell $a_i$ whether the history is $h$ or $h'$, so $a_i$ may attempt to make a move $m$ that is applicable in $s(h')$ but not in $s(h)$. If $a_i$ does so, then $m$ will produce *some* kind of outcome, even if the outcome is just an announcement that $a_i$ must try a different move. Consequently, we can easily make $m$ applicable in $s(h)$, by defining a new state $m(s(h))$ in which this outcome occurs.

---

[1] Nondeterministic initial states, outcomes, and observations can be modeled by introducing an additional player $a_0$ who makes a nondeterministic move at the start of the game and after each of the other players' moves. To avoid affecting the other players' payoffs, $a_0$'s payoff in terminal states is always 0.

[2] Some game-theory textbooks define information sets without even using the notion of an "observation." They simply let a player's information sets be the equivalence classes of a partition over the set of possible histories.

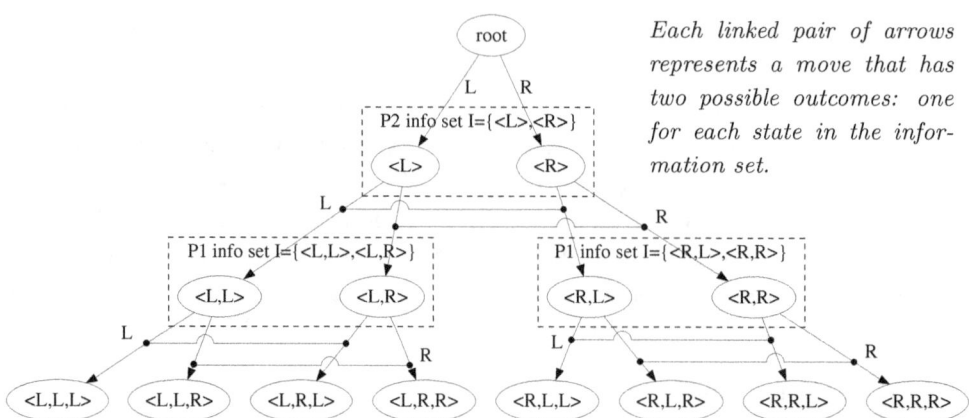

Each linked pair of arrows represents a move that has two possible outcomes: one for each state in the information set.

Figure 1. A game tree for a two-player imperfect-information game between two players P1 and P2 who move in alternation. The players may move either left ($L$) or right ($R$), and their moves are hidden from each other (e.g., after P1's first move, P2 knows that P1 has moved, but not whether the move was $L$ or $R$). Each node is labeled with its associated history (e.g., $\langle L \rangle$ and $\langle R \rangle$ for the two children of the root node). The information set of the player to move is indicated by a dotted box (e.g., after P1's first move, P2's information set is $\{\langle L \rangle, \langle R \rangle\}$).

- We assume that terminal histories with distinct utilities always provide distinct observations, i.e., for terminal histories $h, h' \in T$, if $U_i(h) \neq U_i(h')$ then $O_i(h) \neq O_i(h')$.

We define $a_i$'s information set for $h$ to be the set of all histories that give $a_i$ the same observations that $h$ gives, i.e., $[h]_i = \{h' \in H : O_i(h') = O_i(h)\}$. The set of all possible information sets for $a_i$ is $\mathcal{I}_i = \{[h]_i : h \in H\}$. It is easy to show that $\mathcal{I}_i$ is a partition of $H$.

Figure 1 shows an example game tree illustrating the correspondence between information sets and histories. In that game, player $a_1$ makes the first move, which is hidden to player $a_2$. Thus player $a_2$ knows that the history is either $\langle L \rangle$ or $\langle R \rangle$, which is denoted by putting a dotted box around the nodes for those histories.

### 2.2 Strategies

In a perfect-information game, a player $a_i$'s *strategy* is a function $\sigma_i(m|s)$ that returns the probability $p$ that $a_i$ will make move $m$ in state $s$. For imperfect-information games, where $a_i$ will not always know the exact state he/she is in, $\sigma_i$ is a function of an information set rather than a state; hence $\sigma_i(m|I)$ is the probability that $a_i$ will make move $m$ when their information set is $I$. We let $M(I)$ be the set of moves available in information set $I$.

If $\sigma_i$ is a mixed strategy, then for every information set $I \in \mathcal{I}_i$ where it is $a_i$'s move, there may be more than one move $m \in M(I)$ for which $\sigma_i(m|I) > 0$. But if $\sigma_i$ is a pure strategy, then there will be a unique move $m_I \in M(I)$ such that

$\sigma_i(m|I) = 0 \ \forall m \neq m_I$ and $\sigma_i(m_I|I) = 1$; and in this case we will use the notation $\sigma_i(I)$ to refer to $m_I$.

If $h = \langle m_1, m_2, \ldots, m_n \rangle$ is a history, then its probability $P(h)$ can be calculated from the players' strategies. Suppose $a_1$'s and $a_2$'s strategies are $\sigma_1$ and $\sigma_2$. In the special case where $a_1$ has the first move and the players move in strict alternation,

$$P(h|\sigma_1, \sigma_2) = \sigma_1(m_1|h_0)\sigma_2(m_2|h_1) \ldots \sigma_1(m_j|h_{j-1}), \sigma_2(m_{j+1}|h_j), \ldots, \quad (1)$$

where $h_j = \langle m_1, \ldots, m_j \rangle$ ($h_0 = \langle \rangle$). More generally,

$$P(h|\sigma_1, \sigma_2) = \prod_{j=0}^{n-1} \sigma_{a(h_j)}(m_{j+1}|h_j). \quad (2)$$

Given $\sigma_1$, $\sigma_2$, and any information set $I$, the conditional probability of any $h \in I$ is the normalized probability

$$P(h|I, \sigma_1, \sigma_2) = \frac{P(h|\sigma_1, \sigma_2)}{\sum_{h' \in I} P(h'|\sigma_1, \sigma_2)}. \quad (3)$$

## 2.3 Perfect Recall

*Perfect recall* means that every player always remembers all the moves they've made – we can't have two histories in player $a_i$'s information set which disagree on what player $a_i$ did at some point in the past. One can get a more detailed explanation of perfect and imperfect recall in perfect information games in [Osborne and Rubinstein 1994].

In a game of perfect recall, it is easy to show that if $I \in \mathcal{I}_1$, then all histories in $I$ have the same sequence of moves for $a_1$, whence the probability of $h$ given $I$ is conditionally independent of $\sigma_1$. If $h = \langle m_1, m_2, \ldots, m_n \rangle$, then

$$P(h|I, \sigma_1, \sigma_2) = P(h|I, \sigma_2) = \frac{\prod_{a(h_j)=a_2} \sigma_2(m_{j+1}|h_j)}{\sum_{h' \in I} \prod_{a(h'_j)=a_2} \sigma_2(m_{j+1}|h'_j)}. \quad (4)$$

An analogous result, with the subscripts 1 and 2 interchanged, holds when $I \in \mathcal{I}_2$.

## 2.4 Utility and Expected Utility

If a history $h$ takes us to the game's end, then $h$ is *terminal*, and we let $U(h)$ be the *utility* of $h$ for player $a_1$. Since the game is zero-sum, it follows that $a_2$'s utility is $-U(h)$.

If $a_1$ and $a_2$ have strategies $\sigma_1$ and $\sigma_2$, then the expected utility for $a_i$ is

$$EU(\sigma_1, \sigma_2) = \sum_{h \in T} P(h|\sigma_1, \sigma_2) U(h), \quad (5)$$

where $T$ is the set of all terminal histories, and $P(h|\sigma_1, \sigma_2)$ is as in Eq. (2). Since the game is zero-sum, it follows that $a_2$'s expected utility is $-EU(\sigma_1, \sigma_2)$.

For the expected utility of an individual history $h$, there are two cases:

**Case 1:** History $h$ is terminal. Then $h$'s expected utility is just its actual utility, i.e.,

$$EU(h|\sigma_1, \sigma_2) = EU(h) = U(h). \tag{6}$$

**Case 2:** History $h$ ends at a state where it is $a_i$'s move. Then $h$'s expected utility is a weighted sum of the expected utilities for each of $a_i$'s possible moves, weighted by the probabilities of $a_i$ making those moves:

$$\begin{aligned} EU(h|\sigma_1,\sigma_2) &= \sum_{m \in M(h)} \sigma_i(m|h) \cdot EU(h \circ m|\sigma_1,\sigma_2) \\ &= \sum_{m \in M(h)} \sigma_i(m|[h]_i) \cdot EU(h \circ m|\sigma_1,\sigma_2), \end{aligned} \tag{7}$$

where $\circ$ denotes concatenation.

The following lemma shows that the recursive formulation in Eqs. (6–7) matches the notion of expected utility given in Eq. 5.

LEMMA 1. *For any strategies $\sigma_1$ and $\sigma_2$, $EU(\langle\rangle|\sigma_1,\sigma_2)$ (the expected utility of the empty initial history as computed via the recursive Equations 6 and 7) equals $EU(\sigma_1, \sigma_2)$.*

**Sketch of proof.** This is shown by showing, by induction on the length of $h$, the more general statement that

$$EU(h|\sigma_1,\sigma_2) = \sum_{h' \in T, h'=h \circ m_k, \circ \cdots \circ, m_n} P(h'|\sigma_1,\sigma_2)U(h')/P(h|\sigma_1,\sigma_2), \tag{8}$$

where $k$ is one greater than the size of $h$ and $n$ is the size of each $h'$ as appropriate. The base case occurs when $h$ is terminal, and the inductive case assumes Eq. 8 holds for histories of length $m + 1$ to show algebraically that Eq. 8 holds for histories of length $m$. □

The expected utility of an information set $I \in H$ is the weighted sum of the expected utilities of its histories:

$$EU(I|\sigma_1,\sigma_2) = \sum_{h \in I} P(h|I,\sigma_1,\sigma_2)EU(h|\sigma_1,\sigma_2). \tag{9}$$

COROLLARY 2. *For any strategies $\sigma_1$ and $\sigma_2$, and player $a_i$, $EU([\langle\rangle]_i|\sigma_1,\sigma_2)$ (the expected utility of the initial information set for player $a_i$) equals $EU(\sigma_1,\sigma_2)$.*

## 3 Finding a Strategy

We now develop the theory for a game-tree search technique that exploits an opponent model.

## 3.1 Optimal Strategy

Suppose $a_1$'s and $a_2$'s strategies are $\sigma_1$ and $\sigma_2$, and let $I$ be any information set for $a_1$. Let $M^*(I|\sigma_1,\sigma_2)$ be the set of all moves in $M(I)$ that maximize $a_1$'s expected utility at $I$, i.e.,

$$M^*(I|\sigma_1,\sigma_2) = \underset{m \in M(I)}{\operatorname{argmax}} EU(I \circ m|\sigma_1,\sigma_2)$$

$$= \left\{ m^* \in M(I) \;\middle|\; \begin{array}{l} \forall m \in M(I),\; \sum_{h \in I} P(h|I,\sigma_1,\sigma_2) EU(h \circ m^*|\sigma_1,\sigma_2) \\ \qquad \geq \sum_{h \in I} P(h|I,\sigma_1,\sigma_2) EU(h \circ m|\sigma_1,\sigma_2) \end{array} \right\}. \quad (10)$$

Since we are considering only finite games, every history has finite length. Thus by starting at the terminal states and going backwards up the game tree, applying Eqs. (7) and (9) at each move, one can compute a strategy $\sigma_1^*$ such that:

$$\sigma_1^*(m|I) = \begin{cases} 1/|M^*(I,\sigma_1^*,\sigma_2)|, & \text{if } m \in M^*(I|\sigma_1^*,\sigma_2), \\ 0, & \text{otherwise.} \end{cases} \quad (11)$$

THEOREM 3. *Let $\sigma_2$ be a strategy for $a_2$, and $\sigma_1^*$ be as in Eq. (11). Then $\sigma_1^*$ is $\sigma_2$-optimal.*

**Sketch of proof.** Let $\bar{\sigma}_1$ be any $\sigma_2$-optimal strategy. The basic idea is to show, by induction on the lengths of histories in an information set $I$, that $EU(I|\sigma_1^*,\sigma_2) \geq EU(I|\bar{\sigma}_1,\sigma_2)$.

The induction goes backwards from the end of the game: the base case is where $I$ contains histories of maximal length, while the inductive case assumes the inequality holds when $I$ contains histories of length $k+1$, and shows it holds when $I$ contains histories of length $k$. The induction suffices to show that $EU([\langle\rangle]_1|\sigma_1^*,\sigma_2) \geq EU([\langle\rangle]_1|\bar{\sigma}_1,\sigma_2)$, whence from Lemma 1, $EU(\sigma_1^*,\sigma_2) \geq EU(\bar{\sigma}_1,\sigma_2)$. □

Computing $\sigma_1^*$ is more difficult than computing an optimal strategy in a perfect-information game. Reif [Reif 1984] has shown that the problem of finding a strategy with a guaranteed win is doubly exponential for imperfect-information games (this corresponds to finding $\sigma_1$ such that for all $\sigma_2$, $\sigma_1$ wins).

In the minimax game-tree search algorithms used in perfect-information games, one way of dealing with the problem of intractability is to approximate the utility value of a state by searching to some limited depth $d$, using a *static evaluation function* $\mathcal{E}(\cdot)$ that returns approximations of the expected utilities of the nodes at that depth, and pretending that the values returned by $\mathcal{E}$ are the nodes' actual utility values. In imperfect-information games we can compute approximate values for $EU$ in a similar fashion:

$$EU_d(h|\sigma_1^*, \sigma_2) = \begin{cases} \mathcal{E}(h), & \text{if } d = 0, \\ U(h), & \text{if } h \text{ is terminal,} \\ \sum_{m \in M(h)} \sigma_2(m|[h]_2) \cdot EU_{d-1}(h \circ m|\sigma_1^*, \sigma_2), & \text{if it's } a_2\text{'s move,} \\ EU_{d-1}(h \circ \operatorname{argmax}_{m \in M(h)}(EU_d([h \circ m]_1|\sigma_1^*, \sigma_2))), & \text{if it's } a_1\text{'s move,} \end{cases} \quad (12)$$

$$EU_d(I|\sigma_1^*, \sigma_2) = \sum_{h \in I} P(h|I, \sigma_1^*, \sigma_2) \cdot EU_d(h|I, \sigma_1^*, \sigma_2). \quad (13)$$

### 3.2 Opponent Models

Eqs. (11–12) assume that $a_1$ knows $a_2$'s strategy $\sigma_2$, an assumption that is quite unrealistic in practice. A more realistic assumption is that $a_1$ has a *model* of $a_2$ that provides an approximation of $\sigma_2$. For example, in perfect-information games, the well-known minimax formula corresponds to an opponent model in which the opponent always chooses the move whose utility value is lowest. We now consider two opponent models for imperfect-information games: the *overconfident* and *paranoid* models.

**Overconfidence.** The overconfident model assumes $a_2$ is just choosing moves at random from a uniform distribution; i.e., it assumes $a_2$'s strategy is $\sigma_2(m|I) = 1/|M(I)|$ for every $m \in M(I)$, and second, that $a_1$'s strategy is $\sigma_2$-optimal. If we let $OU_d(h) = EU_d(h|\sigma_1^*, \sigma_2)$ and $OU_d(I) = EU_d(I|\sigma_1^*, \sigma_2)$ be the expected utilities for histories and information sets under these assumptions, then it follows from Eqs. (12–13) that:

$$OU_d(h) = \begin{cases} \mathcal{E}(h), & \text{if } d = 0, \\ U(h), & \text{if } h \text{ is terminal,} \\ \sum_{m \in M(h)} \frac{OU_{d-1}(h \circ m)}{|M(h)|}, & \text{if it's } a_2\text{'s move,} \\ OU_{d-1}(h \circ \operatorname{argmax}_{m \in M(h)} OU_d([h \circ m]_1)), & \text{if it's } a_1\text{'s move,} \end{cases} \quad (14)$$

$$OU_d(I) = \sum_{h \in I} (1/|I|) \cdot OU_d(h). \quad (15)$$

If the algorithm searches to a limited depth (Eq. 12 with $d < \max_{h \in H} |h|$), we will refer to the resulting strategy as *limited-depth overconfident*. If the algorithm searches to the end of the game (i.e., $d \geq \max_{h \in H} |h|$), we will refer to the resulting strategy as *full-depth overconfident*; and in this case we will usually write $OU(h)$ rather than $OU_d(h)$.

**Paranoia.** The paranoid model assumes that $a_2$ will always make the worst possible move for $a_1$, i.e., the move that will produce the minimum expected utility over all of the histories in $a_1$'s information set. This model replaces the summation in the third line of Eq. (12) with a minimization:

$$PU_d(h) = \begin{cases} \mathcal{E}(I), & \text{if } d = 0, \\ U(h), & \text{if } h \text{ is terminal}, \\ PU_{d-1}(h \circ \text{argmin}_{m \in M(h)}(\min_{h' \in [h]_1} PU_d([h \circ m]))), & \text{if it's } a_2\text{'s move}, \\ PU_{d-1}(h \circ \text{argmax}_{m \in M(h)}(\min_{h' \in [h]_1} PU_d([h \circ m]))), & \text{if it's } a_1\text{'s move}, \end{cases} \quad (16)$$

$$PU_d(I) = \min_{h \in I} PU_d(h). \quad (17)$$

Like we did for overconfident search, we will use the terms *limited-depth* and *full-depth* to refer to the cases where $d < \max_{h \in H} |h|$ and $d \geq \max_{h \in H} |h|$, respectively; and for a full-depth paranoid search, we will usually write $PU(h)$ rather than $PU_d(h)$.

In perfect-information games, $PU(h)$ equals $h$'s minimax value. But in imperfect-information games, $h$'s minimax value is the minimum Eq. (11) over all possible values of $\sigma_2$; and consequently $PU(h)$ may be less than $h$'s minimax value.

### 3.3 Comparison with the Minimax Theorem

The best known kinds of strategies for zero-sum games are the strategies based on the famous Minimax Theorem [von Neumann and Morgenstern 1944]. These *minimax strategies* tacitly incorporate an opponent model that we will call the *minimax model*. The minimax model, overconfident model, and paranoid model each correspond to differing assumptions about $a_2$'s knowledge and competence, as we will now discuss.

Let $\Sigma_1$ and $\Sigma_2$ be the sets of all possible pure strategies for $a_1$ and $a_2$, respectively. If $a_1$ and $a_2$ use mixed strategies, then these are probability distributions $P_1$ and $P_2$ over $\Sigma_1$ and $\Sigma_2$. During game play, $a_1$ and $a_2$ will randomly choose pure strategies $\sigma_1$ and $\sigma_2$ from $P_1$ and $P_2$. Generally they will do this piecemeal by choosing moves as the game progresses, but game-theoretically this is equivalent to choosing the entire strategy all at once.

**Paranoia:** If $a_1$ uses a paranoid opponent model, this is equivalent to assuming that $a_2$ knows in advance the pure strategy $\sigma_1$ that $a_1$ will choose from $P_1$ during the course of the game, and that $a_2$ can choose the optimal counter-strategy, i.e., a strategy $P_2^{\sigma_1}$ that minimizes $\sigma_1$'s expected utility. Thus $a_1$ will want to choose a $\sigma_1$ that has the highest possible expected utility given $P_2^{\sigma_1}$. If there is more than one such $\sigma_1$, then $a_1$'s strategy can be any one of them or can be an arbitrary probability distribution over all of them.

**Minimax:** If $a_2$ uses a minimax opponent model, this is equivalent to assuming that $a_2$ will know in advance what $a_1$'s mixed strategy $P_1$ is, and that $a_2$ will be competent enough to choose the optimal counter-strategy, i.e., a mixed strategy $P_2^{P_1}$ that minimizes $P_1$'s expected utility. Thus $a_1$ will want to use a mixed strategy $P_1$

that has the highest possible expected utility given $P_2^{P_1}$.

In perfect-information games, the minimax model is equivalent to the paranoid model. But in imperfect-information games, the minimax model assumes $a_2$ has less information than the paranoid model does: the minimax model assumes that $a_2$ knows the probability distribution $P_1$ over $a_1$'s possible strategies, and the paranoid model assumes that $a_2$ knows which strategy $a_1$ will choose from $P_1$.

**Overconfidence:** If $a_1$ uses an overconfident opponent model, this equivalent to assuming that $a_2$ knows nothing about (or is not competent enough to figure out) how good or bad each move is, whence $a_2$ will use a strategy $P_2^=$ in which all moves are equally likely. In this case, $a_1$ will want to choose a strategy $\sigma_1$ that has the highest expected utility given $P_2^=$. If there is more than one such $\sigma_1$, then $a_1$'s strategy can be any one of them or can be an arbitrary probability distribution over all of them.

In both perfect- and imperfect-information games, the overconfident model assumes $a_2$ has much less information (and/or competence) than in the minimax and paranoid models.

### 3.4 Handling Large Information Sets

Information sets can be quite large. When they are too large for techniques like the above to run in a reasonable amount of time, there are several options.

**Game simplification** reduces the size of the information set by creating an analogous game with smaller information sets. This technique has worked particularly well in poker [Billings, Burch, Davidson, Holte, Schaeffer, Schauenberg, and Szafron 2003; Gilpin and Sandholm 2006a; Gilpin and Sandholm 2006b], as it is possible to create a "simpler" game which preserves win probabilities (within some $\epsilon$). However, these approaches apply only to variants of poker, and the technique is not easily generalizable. Given an arbitrary game $G$ other than poker, we know of no general-purpose way of producing a simpler game whose expected utilities accurately reflect expected utilities in $G$.

**State aggregation** was first used in the game of sprouts [Applegate, Jacobson, and Sleator 1991], and subsequently has been used in computer programs for games such as bridge (e.g., [Ginsberg 1999]), in which many of the histories in an information set are similar, and hence can be reasoned about as a group rather than individually. For example, if one of our opponents has an ace of hearts and a low heart, it usually does not matter *which* low heart the opponent has: generally all low hearts will lead to an identical outcome, so we need not consider them separately. The aggregation reduces the computational complexity by handling whole sets of game histories in the information set at the same time. However, just as with game simplification, such aggregation techniques are highly game dependent. Given an arbitrary game $G$, we do not know of a general-purpose way to aggregate states of $G$ in a way that is useful for computing expected utility values in $G$.

Unlike the previous two techniques, **statistical sampling** [Corlett and Todd

1985] is general enough to fit any imperfect-information game. It works by selecting a manageable subset of the given, large, information set, and doing our computations based on that.

Since we are examining game playing across several imperfect-information games we use the third technique. Let us suppose $\Gamma$ is an expected utility function such as $OU_d$ or $PU_d$. In statistical sampling we pick $I' \subset I$ and compute the value of $\Gamma(I')$ in place of $\Gamma(I)$. There are two basic algorithms for doing the sampling:

1. **Batch:** Pick a random set of histories $I' \subset I$, and compute $\Gamma_s(I')$ using the equations given earlier.

2. **Iterative:** Until the available time runs out, repeatedly pick a random $h \in I$, compute $\Gamma(\{h\})$ and aggregate that result with all previous picks.

The iterative method is preferable because it is a true anytime algorithm: it continues to produce increasingly accurate estimates of $\Gamma(I)$ until no more time is available. In contrast, the batch method requires guessing how many histories we will be able to compute in that time, picking a subset $I'$ of the appropriate size, and hoping that the computation finishes before time is up. For more on the relative advantages of iterative and batch sampling, see [Russell and Wolfe 2005].

Statistical sampling, unlike game simplification and state aggregation, can be used for arbitrary imperfect-information games rather than just on games that satisfy special properties. Consequently, it is what we use in our experiments in Section 5.

## 4 Analysis

Since paranoid and overconfident play both depend on opponent models that may be unrealistic, which of them is better in practice? The answer is not completely obvious. Even in games where each player's moves are completely hidden from the other player, it is not hard to create games in which the paranoid strategy outplays the overconfident strategy and vice-versa. We now give examples of games with these properties.

Figures 2 and 3, respectively, are examples of situations in which paranoid play outperforms overconfident play and vice versa. As in Figure 1, the games are shown in tree form in which each dotted box represents an information set. At each leaf node, $U$ is the payoff for player 1. Based on these values of $U$, the table gives, the probabilities of moving left (L) and right (R) at each information set in the tree, for both the overconfident and paranoid strategies. At each leaf node, pr1 is the probability of reaching that node when player 1 is overconfident and player 2 is paranoid, and pr2 is the probability of reaching that node when player 2 is overconfident and player 1 is paranoid.

In Figure 2, the paranoid strategy outperforms the overconfident strategy, because of the differing choices the strategies will make at the information set I2:

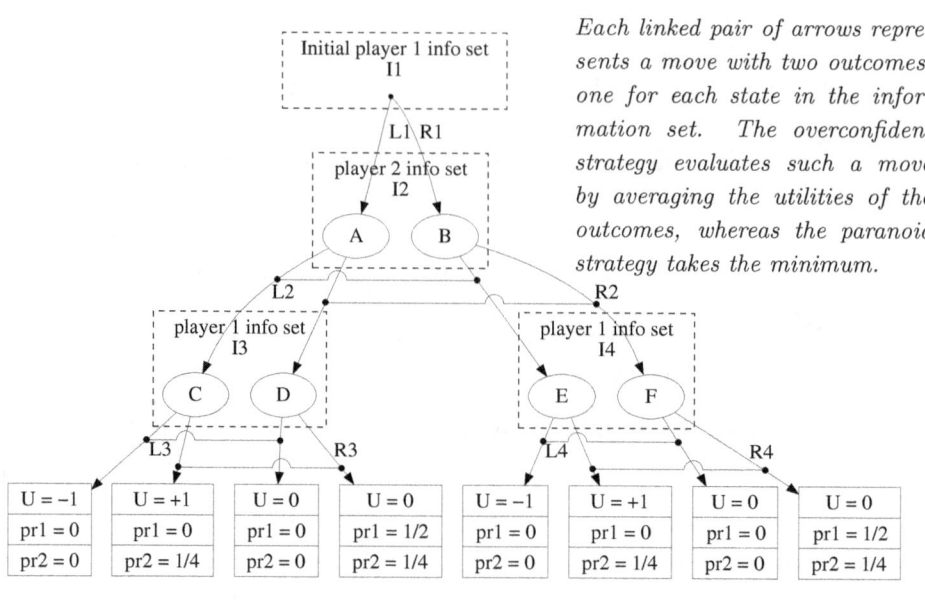

*Each linked pair of arrows represents a move with two outcomes: one for each state in the information set. The overconfident strategy evaluates such a move by averaging the utilities of the outcomes, whereas the paranoid strategy takes the minimum.*

| Info set | Overconfident strategy | | Paranoid strategy | |
|---|---|---|---|---|
| I1 | $P(L1) = 1/2$ | $P(R1) = 1/2$ | $P(L1) = 1/2$ | $P(R1) = 1/2$ |
| I2 | $P(L2) = 1/2$ | $P(R2) = 1/2$ | $P(L2) = 0$ | $P(R2) = 1$ |
| I3 | $P(L3) = 0$ | $P(R3) = 1$ | $P(L3) = 0$ | $P(R3) = 1$ |
| I4 | $P(L4) = 0$ | $P(R4) = 1$ | $P(L4) = 0$ | $P(R4) = 1$ |

Figure 2. An imperfect-information game in which paranoid play beats overconfident play. If an overconfident player plays against a paranoid player and each player has an equal chance of moving first, the expected utilities are $-0.25$ for the overconfident player and $0.25$ for the paranoid player.

- Suppose player 1 is overconfident and player 2 is paranoid. Then at information set I2, player 2 assumes its opponent will always choose the worst possible response. Hence when choosing a move at I2, player 2 thinks it will lose if it chooses L2 and will tie if it chooses R2, so it chooses R2 to avoid the anticipated loss.

- Suppose player 1 is paranoid and player 2 is overconfident. Then at information set I2, player 2 assumes its opponent is equally likely to move left or right. Hence when choosing a move at I2, player 2 thinks that both moves have the same expected utility, so it will choose between them at random—which is a mistake, because its paranoid opponent will win the game by moving right in both information sets I3 and I4.

Figure 3 shows a game in which the overconfident strategy outperforms the paranoid strategy. Again, the pertinent information set is I2:

- Suppose overconfident play is player 1 and paranoid play is player 2. Then

Paranoia versus Overconfidence in Imperfect-Information Games

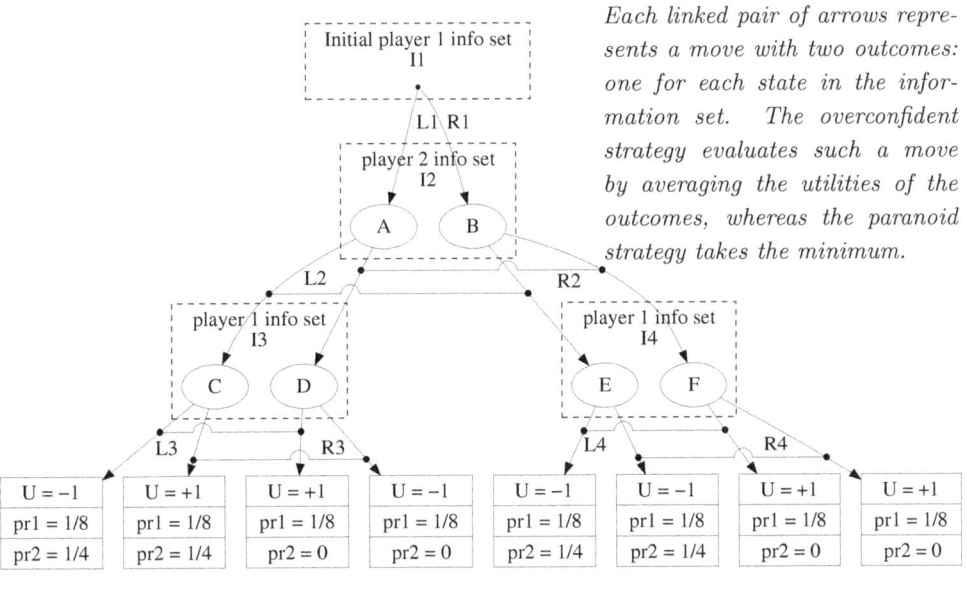

Figure 3. An imperfect-information game where overconfident play beats paranoid play. If an overconfident player plays against a paranoid player and each player has an equal chance of moving first, the expected utilities are 0.25 for the overconfident player and −0.25 for the paranoid player.

paranoid play, assuming the worst, believes both move L2 and R2 are losses. R2 is a loss because the opponent may have made move R1 resulting in a forced loss for player 2 at node F, and L2 is a loss because the opponent may have made move L1 and then may make move R4 resulting in a loss for player 2. Since there is a potential loss in all cases, paranoid play chooses both cases with equal probability.

- When overconfident play is player 2, it makes move L2 at I2, on the theory that the opponent was equally likely to make moves L1 and R1 and therefore giving it a 50% probability of ending up in node E, which is a forced win for player 2. Against paranoid play as player 1, this is a good move, since paranoid play actually does make moves L1 and R1 with 50% probability.

These two examples show that neither strategy is guaranteed to be better in all cases: sometimes paranoid play outperforms overconfident play, and sometimes vice versa. So to determine their relative worth, deeper analysis is necessary.

## 4.1 Analysis of Overconfidence Performance in Perfect Information Games

Let $s$ be a state in a perfect-information zero-sum game. We will say that a child $s'$ of $s$ is *minimax-optimal* if $\mu(s') \geq \mu(s'')$ for every child $s''$ of $s$, where $\mu(s)$ is the minimax value for the player to move at $s$. A *minimax strategy* is any strategy that will always move to a minimax-optimal node. In the game-tree search literature, minimax strategies have often been called "perfect play" because they produce the highest possible value against an opponent who is also using a minimax strategy.

In perfect-information zero-sum games, $PU(s) = \mu(s)$ at every state $s$, hence full-depth paranoid play is a minimax strategy. Surprisingly, if the only outcomes are wins and losses (or equivalently, utility values of 1 and $-1$), full-depth *overconfident* play is also a minimax strategy. To prove this result, we first need a lemma:

LEMMA 4. *Let $G$ be any finite two-player perfect-information game whose outcomes all have utility 1 or $-1$. At every state $s$, if $\mu(s) = 1$ then $OC(s) = 1$, and if $\mu(s) = -1$ then $OC(s) \in [-1, 1)$.*

**Sketch of proof.** This is proven by induction on the height of the state $s$ under consideration. The base case occurs for with terminal nodes of height 0 for which the lemma follows trivially. The inductive case supposes the lemma holds for all states of height $k$ and shows algebraically for states $s$ of height $k+1$ in each of four possible cases: (1) if it is $a_1$'s move and $\mu(s) = -1$ then $OC(s) \in [-1, 1)$, (2) if it is $a_1$'s move and $\mu(s) = 1$ then $OC(s) = 1$, (3) if it is $a_2$'s move and $\mu(s) = -1$ then $OC(s) \in [-1, 1)$, and (4) if it is $a_2$'s move and $\mu(s) = 1$ then $OC(s) = 1$. Since the game allows only wins and losses (so that $\mu(s)$ is 1 or $-1$), these are all the possibilities. □

THEOREM 5. *Let $G$ be any finite two-player perfect-information game whose outcomes all have utility 1 or $-1$. At every nonterminal state $s$, the overconfident strategy, $\sigma_O$, will move to a state $s'$ that is minimax-optimal.*

**Proof.** Immediate from the lemma. □

This theorem says that in head-to-head play in perfect-information games allowing only wins or losses, the full-depth overconfident and full-depth paranoid strategies will be evenly matched. In the experimental section, we will see this to hold in practice.

## 4.2 Discussion

**Paranoid play.** When using paranoid play $a_1$ assumes that $a_2$ has always and will always make the worst move possible for $a_1$, but $a_1$ does this *given only $a_1$'s information set*. This means that for any given information set, the paranoid player will find the history in the information set that is least advantageous to itself and make moves as though that were the game's actual history *even when the game's actual* *history is any other member of the information set.* There is a certain intuitively appealing protectionism occurring here: an opponent that happens to have made the perfect moves cannot trap the paranoid player. However, it really is not clear exactly how well a paranoid player will do in an imperfect-information game, for the following reasons:

- There is no reason to necessarily believe that the opponent has made those "perfect" moves. In imperfect-information games, the opponent has different information than the paranoid player, which may not give the opponent enough information to make the perfect moves paranoid play expects.

- Against non-perfect players, the paranoid player may lose a lot of potentially winnable games. The information set could contain thousands of histories in which a particular move $m$ is a win; if that move is a loss on just one history, and there is another move $m'$ which admits no losses (and no wins), then $m$ will not be chosen.[3]

- In games such as kriegspiel, in which there are large and diverse information sets, usually every information set will contain histories that are losses, hence paranoid play will evaluate all of the information sets as losses. In this case, all moves will look equally terrible to the paranoid player, and paranoid play becomes equivalent to random play.[4]

We should also note the relationship paranoid play has to the "imperfection" of the information in the game. A game with large amounts of information and small information sets should see better play from a paranoid player than a game with large information sets. The reason for this is that as we get more information about the actual game state, we can be more confident that the move the paranoid player designates as "worst" is a move the opponent can discover and make in the actual game. The extreme of this is a perfect information game, where paranoid play has proven quite effective: it is minimax search. But without some experimentation, it is not clear to what extent smaller amounts of information degrade paranoid play.

**Overconfident play.** Overconfident play assumes that $a_2$ will, with equal probability, make all available moves regardless of what the available information tells $a_2$ about each move's expected utility. The effect this has on game play depends on the extent to which $a_2$'s moves diverge from random play. Unfortunately for overconfidence, many interesting imperfect-information games implicitly encourage

---

[3]This argument assumes that the paranoid player examines the entire information set rather than a statistical sample as discussed in Section 3.4. If the paranoid player examines a statistical sample of the information set, there is a good chance that the statistical sample will not contain the history for which $m$ is a loss. Hence in this case, statistical sampling would actually *improve* the paranoid player's play.

[4]We have verified this experimentally in several of the games in the following section, but omit these experiments due to lack of space.

non-random play. In these games the overconfident player will not adequately consider the risks of its moves. The overconfident player, acting under the theory that the opponent is unlikely to make a particular move, will many times not protect itself from a potential loss.

However, depending on the amount of information in the imperfect-information game, the above problem may not be as bad as it seems. For example, consider a situation where $a_1$, playing overconfidently, assumes the opponent is equally likely to make each of the ten moves available in $a_1$'s current information set. Suppose that each move is clearly the best move in exactly one tenth of the available histories. Then, despite the fact that the opponent is playing a deterministic strategy, random play is a good opponent model given the information set. This sort of situation, where the model of random play is reasonable despite it being not at all related to the opponent's actual mixed strategy, is more likely to occur in games where there is less information. The larger the information set, the more likely it is that every move is best in enough histories to make that move as likely to occur as any other. Thus in games where players have little information, there may be a slight advantage to overconfidence.

**Comparative performance.** The above discussion suggests that (1) paranoid play should do better in games with "large" amounts of information, and (2) overconfident play might do better in games with "small" amounts of information. But will overconfident play do better than paranoid play? Suppose we choose a game with small amounts of information and play a paranoid player against an overconfident player: what should the outcome be? Overconfident play has the advantage of probably not diverging as drastically from the theoretically correct expected utility of a move, while paranoid play has the advantage of actually detecting and avoiding bad situations – situations to which the overconfident player will not give adequate weight.

Overall, it is not at all clear from our analysis how well a paranoid player and an overconfident player will do relative to each other in a real imperfect-information game. Instead, experimentation is needed.

## 5 Experiments

In this section we report on our experimental comparisons of overconfident versus paranoid play in several imperfect-information games.

One of the games we used was kriegspiel, an imperfect-information version of chess [Li 1994; Li 1995; Ciancarini, DallaLibera, and Maran 1997; Sakuta and Iida 2000; Parker, Nau, and Subrahmanian 2005; Russell and Wolfe 2005]. In kriegspiel, neither player can observe anything about the other player's moves, except in cases where the players directly interact with each other. For example, if $a_1$ captures one of $a_2$'s pieces, $a_2$ now knows that $a_1$ has a piece where $a_2$'s piece used to be. For more detail, see Section 5.2.

In addition, we created imperfect-information versions of three perfect-information games: P-games [Pearl 1984], N-games [Nau 1982a], and a simplified version of kalah [Murray 1952]. We did this by hiding some fraction $0 \leq h \leq 1$ of each player's moves from the other player. We will call $h$ the *hidden factor*, because it is the fraction of information that we hide from each player: when $h = 0$, each player can see all of the other player's moves; when $h = 1$, neither player can see any of the other player's moves; when $h = 0.2$, each player can see 20% of the other player's moves; and so forth.

In each experiment, we played two players head-to-head for some number of trials, and averaged the results. Each player went first on half of the trials.

### 5.1 Experiments with Move-Hiding

We did experiments in move-hiding variants of simple perfect information games. These experiments were run on 3.4 GHz Xeon processors with at least 2 GB of RAM per core. The programs were written in OCaml. All games were 10-ply long, and each player searched all the way to the end of the game.

**Hidden-move P-game experiments.** P-games were invented by Judea Pearl [Pearl 1981], and have been used in many studies of game-tree search (e.g., [Nau 1982a; Pearl 1984]). They are two-player zero-sum games in which the game tree has a constant branching factor $b$, fixed game length $d$, and fixed probability $P_0$ that the first player wins at any given leaf node.[5] One creates a P-game by randomly assigning "win" and "loss" values to the $b^d$ leaf nodes.

We did a set of experiments with P-games with $P_0 = 0.38$, which is the value of $P_0$ most likely to produce a nontrivial P-game [Nau 1982b]. We used depth $d = 10$, and varied the branching factor $b$. We varied the hidden factor $h$ from 0 to 1 by increments of 0.2, so that the number of hidden moves varied from 0 to 10. In particular, we hid a player's $m^{th}$ move if $\lfloor m \cdot h \rfloor > \lfloor (m-1) \cdot h \rfloor$. For instance, in a game where each player makes 5 moves and the hidden factor is 0.6, then the $2^{nd}$, $4^{th}$, and $5^{th}$ moves of both players are hidden.

For each combination of parameters, we played 2000 games: 1000 in which one of the players moved first, and 1000 in which the other player moved first. Thus in each of our figures, each data point is the average of 2000 runs.

Figure 4(a) shows the results of head-to-head play between the overconfident and paranoid strategies. These results show that in hidden-move P-games, paranoid play does indeed perform worse than overconfident play with hidden factors greater than 0. The results also confirm theorem 5, since overconfident play and paranoid play did equally well with hidden factor 0. From these experiments, it seems that paranoid play may not be as effective in imperfect-information games as it is in perfect information games.

**Hidden-move N-game experiments.** P-games are known to have a property called *game-tree pathology* that does not occur in "natural" games such as chess [Nau

---
[5] Hence [Pearl 1984] calls P-games $(d, b, P_0)$-games.

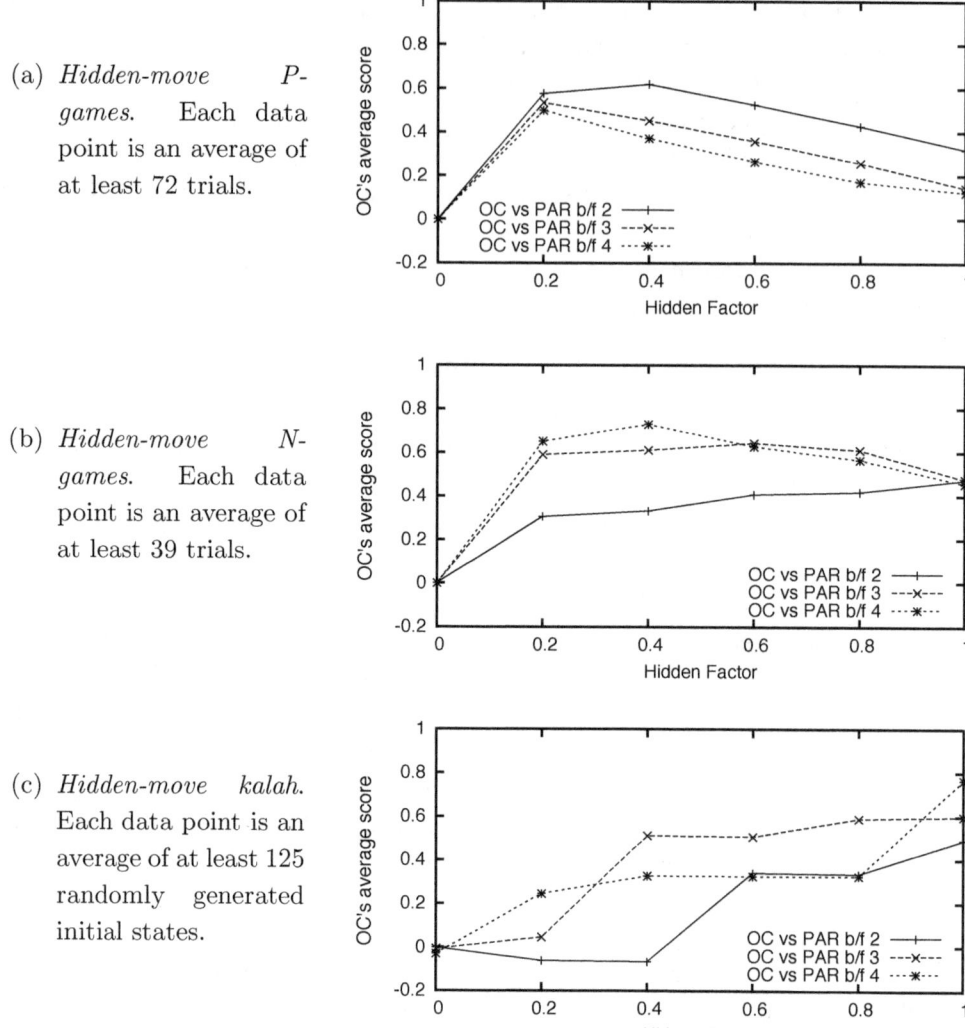

(a) *Hidden-move P-games.* Each data point is an average of at least 72 trials.

(b) *Hidden-move N-games.* Each data point is an average of at least 39 trials.

(c) *Hidden-move kalah.* Each data point is an average of at least 125 randomly generated initial states.

Figure 4. Average scores for overconfident (OC) play against paranoid (PAR) play.

1982a], and we wanted to ascertain whether this property might have influenced our experimental results on hidden-move P-games. N-games are similar to P-games but do not exhibit game-tree pathology, so we did a similar set of experiments on hidden-move N-games.

An N-game is specified by a triple $(d, b, P_0)$, where $d$ is the game length, $b$ is the branching factor, and $P_0$ is a probability. An N-game specified by this triple has a game tree of height $d$ and branching factor $b$, and each arc in the game tree is randomly assigned a value of $+1$ with probability $P_0$, or $-1$ otherwise. A leaf node is a win for player 1 (and a loss for player 2) if the sum of the values on the arcs between the root and the leaf node is greater than zero; otherwise the leaf node is a loss for player 1 (and a win for player 2).

Figure 4(b) shows our experimental results for hidden-move N-games. Just as before, overconfident and paranoid play did equally well with hidden factor 0, and overconfident play outperformed paranoid play with hidden factors greater than 0.

**Kalah experiments.** Kalah [Murray 1952] is also called mankalah, mancala, warri, and other names. It is an ancient African game played on a board with a number of pits that contain seeds, in which the objective is to acquire more seeds than the opponent, either by moving them to a special pit (called a kalah) or by capturing them from the opponent's pits.

In kalah, there are two rows of 6 pits. Flanking the rows of pits on both sides are the larger kalahs. Players sit on opposite sides of the board with one of the rows of pits nearer to each player. Each player owns the kalah on their left. The game starts with 6 stones in each of the pits except the kalahs. The player moves by picking up all the stones from one of the pits in the near row and placing one stone in each pit clockwise around the board including their kalah but excluding the opponent's kalah. If the last stone is placed in their kalah, the player moves again. If the last stone is placed in an empty pit, the player moves all stones from the opposite pit to their kalah. The game ends when the player to move has no moves because all pits on their side are empty. At that point, all stones in pits on the other player's side are placed in the player to move's kalah and the player with the most stones wins; ties occur when both plays own the same number of stones.

Because of the computation requirements of playing a full game of kalah, our experiments were on a simplified version of kalah that we call randomized kalah. The game differs from kalah in several ways:

- We vary the number of pits on the board. This varies the branching factor.

- To ensure a constant branching factor, we allow players to "move" from a pit that contains no stones. These are null moves that have no effect on the board.

- We end the game after 10 ply, to ensure that the algorithms can search the entire tree.

- We eliminate the move-again rule, to ensure alternating moves by the players.

- We start with a random number of stones in each pit to ensure that at each branching factor there will be games with non-trivial decisions.

Since randomized kalah is directly motivated by a very old game that people still play, its game trees are arguably much less "artificial" than those of P-games or N-games.

The results of playing overconfidence versus paranoia in hidden-move versions of randomized kalah are shown in Figure 4(c). The results are roughly similar to the

P-game and N-game results, in the sense that overconfidence generally outperforms paranoia; but the results also differ from the P-game and N-game results in several ways. First, overconfidence generally does better at high hidden factors than at low ones. Second, paranoia does slightly better than overconfidence at hidden factor 0 (which does not conflict with Theorem 5, since kalah allows ties). Third, paranoia does better than overconfidence when the branching factor is 2 and the hidden factor is 0.2 or 0.4. These are the only results we saw where paranoia outperformed overconfidence.

The fact that with the same branching factor, overconfidence outperforms paranoia with hidden factor 0.6, supports the hypothesis that as the amount of information in the game decreases, paranoid play performs worse with respect to overconfident play. The rest of the results support that hypothesis as well: overconfidence generally increases in performance against paranoia as the hidden factor increases.

## 5.2 Kriegspiel Experiments

For experimental tests in an imperfect-information game people actually play, we used kriegspiel, an imperfect-information version of chess in which the players cannot see their opponent's pieces. Kriegspiel is useful for this study because (i) it is clearly a game where each player has only a small amount of information about the current state, and (ii) due to its relationship to chess, it is complicated enough strategically to allow for all sorts of subtle and interesting play. A further advantage to kriegspiel is that it is played competitively by humans even today [Li 1994; Li 1995; Ciancarini, DallaLibera, and Maran 1997].

Kriegspiel is a chess variant played with a chess board. When played in person, it requires three chess kits: one for each player and one for the referee. All boards are set up as in normal chess, but neither player is allowed to see their opponent's or the referee's board. The players then move in alternation as in standard chess, keeping their moves hidden from the other player. All player's moves are also played by the referee on the referee's board. Since neither player can see the referee's board, the referee acts as a mediator, telling the players if the move they made is legal or illegal, and giving them various other observations about the move made. We use the ICC's kriegspiel observations, described at http://www.chessclub.com/help/Kriegspiel. Observations define the information sets. Any two histories that have the same observations at each move and all the same moves for one of the players are in the same information set.

When played on the internet, the referee's job can be automated by a computer program. For instance, on the Internet Chess Club one can play kriegspiel, and there have been thousands of kriegspiel games played on that server.

We ran our experiments on a cluster of computers runing linux, with between 900 MB and 1.5 GB RAM available to each process. The processors were Xeons, Athlons, and Pentiums, ranging in clockspeed from 2 GHz to 3.2 GHz. We used

Table 1. Average scores for overconfident play against paranoid play, in 500 kriegspiel games using the ICC ruleset. $d$ is the search depth.

| Over-confident | Paranoid | | |
|---|---|---|---|
| | $d=1$ | $d=2$ | $d=3$ |
| $d=1$ | +0.084 | +0.186 | +0.19 |
| $d=2$ | +0.140 | +0.120 | +0.156 |
| $d=3$ | +0.170 | +0.278 | +0.154 |

Table 2. Average scores for overconfident and paranoid play against HS, with 95% confidence intervals. $d$ is the search depth.

| $d$ | Paranoid | Overconfident |
|---|---|---|
| 1 | $-0.066 \pm 0.02$ | $+0.194 \pm 0.038$ |
| 2 | $+0.032 \pm 0.035$ | $+0.122 \pm 0.04$ |
| 3 | $+0.024 \pm 0.038$ | $+0.012 \pm 0.042$ |

time controls and always forced players in the same game to ensure the results were not biased by different hardware. The algorithms were written in C++. The code used for overconfident and paranoid play is the same, with the exception of the opponent model. We used a static evaluation function that was developed to reward conservative kriegspiel play, as our experience suggests such play is generally better. It uses position, material, protection and threats as features.

The algorithms used for kriegspiel are depth-limited versions of the paranoid and overconfident players. To handle the immense information-set sizes in kriegspiel, we used iterative statistical sampling (see Section 3.4). To get a good sample with time control requires limiting the search depth to at most three ply. Because time controls remain constant, the lower search depths are able to sample many more histories than the higher search depths.

**Head-to-head overconfident vs. paranoid play.** We did experiments comparing overconfident play to paranoid play by playing the two against each other. We gave the algorithms 30 seconds per move and played each of depths one, two, and three searches against each other. The results are in Table 1. In these results, we notice that overconfident play consistently beats paranoid play, regardless of the depth of either search. This is consistent with our earlier results for hidden-move games (Section 5.1); and, in addition, it shows overconfident play doing better than paranoid play in a game that people actually play.

**HS versus overconfidence and paranoia.** We also compared overconfident and paranoid play to the hybrid sampling (HS) algorithm from our previous work [Parker, Nau, and Subrahmanian 2005]. Table 2 presents the results of the experiments, which show overconfidence playing better than paranoia except in depth three search, where the results are inconclusive. The inconclusive results at depth three (which are an average over 500 games) may be due to the sample sizes achieved via iterative sampling. We measured an average of 67 histories in each sample at depth three, which might be compared to an average of 321 histories in each sample at depth two and an average of 1683 histories at depth one. Since both algorithms use iterative sampling, it could be that at depth three, both algorithms examine insufficient samples to do much better than play randomly.

In every case, overconfidence does better than paranoia against HS. Further, overconfidence outperforms HS in every case (though sometimes without statistical significance), suggesting that information-set search is an improvement over the techniques used in HS.

## 6 Related Work

There are several imperfect-information game-playing algorithms that work by treating an imperfect-information game as if it were a collection of perfect-information games [Smith, Nau, and Throop 1998; Ginsberg 1999; Parker, Nau, and Subrahmanian 2005]. This approach is useful in imperfect-information games such as bridge, where it is not the players' moves that are hidden, but instead some information about the initial state of the game. The basic idea is to choose at random a collection of states from the current information set, do conventional minimax searches on those states as if they were the real state, then aggregate the minimax values returned by those searches to get an approximation of the utility of the current information set. This approach has some basic theoretical flaws [Frank and Basin 1998; Frank and Basin 2001], but has worked well in games such as bridge.

Poker-playing computer programs can be divided into two major classes. The first are programs which attempt to approximate a Nash equilibrium. The best examples of these are PsOpti [Billings, Burch, Davidson, Holte, Schaeffer, Schauenberg, and Szafron 2003] and GS1 [Gilpin and Sandholm 2006b]. The algorithms use an intuitive approximation technique to create a simplified version of the poker game that is small enough to make it feasible to find a Nash equilibrium. The equilibrium can then be translated back into the original game, to get an approximate Nash equilibrium for that game. These algorithms have had much success but differ from the approach in this paper: unlike any attempt to find a Nash equilibrium, information-set search simply tries to find the optimal strategy against a given opponent model. The second class of poker-playing programs includes Poki [Billings, Davidson, Schaeffer, and Szafron 2002] which uses expected value approximations and opponent modeling to estimate the value of a given move and Vexbot [Billings, Davidson, Schauenberg, Burch, Bowling, Holte, Schaeffer, and Szafron 2004] which uses search and adaptive opponent modeling.

The above works have focused specifically on creating successful programs for card games (bridge and poker) in which the opponents' moves (card plays, bets) are observable. In these games, the hidden information is which cards went to which players when the cards were dealt. Consequently, the search techniques are less general than information-set search, and are not directly applicable to hidden-move games such as kriegspiel and the other games we have considered in this paper.

# 7 Conclusion

We have introduced a recursive formulation of the expected value of an information set in an imperfect information game. We have provided analytical results showing that this expected utility formulation plays optimally against any opponent if we have an accurate model of the opponent's strategy.

Since it is generally not the case that the opponent's strategy is known, the question then arises as to what the recursive search should assume about an opponent. We have studied two opponent models, a "paranoid" model that assumes the opponent will choose the moves that are best for them, hence worst for us; and an "overconfident" model that assumes the opponent is making moves purely at random.

We have compared the overconfident and paranoid models in kriegspiel, in an imperfect-information version of kalah, and in imperfect-information versions of P-games [Pearl 1984] and N-games [Nau 1982a]. In each of these games, the overconfident strategy consistently outperformed the paranoid strategy. The overconfident strategy even outperformed the best of the kriegspiel algorithms in [Parker, Nau, and Subrahmanian 2005].

These results suggest that the usual assumption in perfect-information game tree search—that the opponent will choose the best move possible—is not as effective in imperfect-information games.

**Acknowledgments:** This work was supported in part by AFOSR grant FA95500610405, NAVAIR contract N6133906C0149, DARPA IPTO grant FA8650-06-C-7606, and NSF grant IIS0412812. The opinions in this paper are those of the authors and do not necessarily reflect the opinions of the funders.

# References

Applegate, D., G. Jacobson, and D. Sleator (1991). Computer analysis of sprouts. Technical report, Carnegie Mellon University.

Billings, D., N. Burch, A. Davidson, R. Holte, J. Schaeffer, T. Schauenberg, and D. Szafron (2003). Approximating game-theoretic optimal strategies for full-scale poker. In *IJCAI*, pp. 661–668.

Billings, D., A. Davidson, J. Schaeffer, and D. Szafron (2002). The challenge of poker. *Artif. Intell. 134*, 201–240.

Billings, D., A. Davidson, T. Schauenberg, N. Burch, M. Bowling, R. Holte, J. Schaeffer, and D. Szafron (2004). Game tree search with adaptation in stochastic imperfect information games. *Computers and Games 1*, 21–34.

Ciancarini, P., F. DallaLibera, and F. Maran (1997). Decision Making under Uncertainty: A Rational Approach to Kriegspiel. In J. van den Herik and J. Uiterwijk (Eds.), *Advances in Computer Chess 8*, pp. 277–298.

Corlett, R. A. and S. J. Todd (1985). A monte-carlo approach to uncertain inference. In *AISB-85*, pp. 28–34.

Frank, I. and D. Basin (2001). A theoretical and empirical investigation of search in imperfect information games. *Theoretical Comp. Sci. 252*, 217–256.

Frank, I. and D. A. Basin (1998). Search in games with incomplete information: A case study using bridge card play. *Artif. Intell. 100*(1-2), 87–123.

Gilpin, A. and T. Sandholm (2006a). Finding equilibria in large sequential games of imperfect information. In *EC '06*, pp. 160–169.

Gilpin, A. and T. Sandholm (2006b). A texas hold'em poker player based on automated abstraction and real-time equilibrium computation. In *AAMAS '06*, pp. 1453–1454.

Ginsberg, M. L. (1999). GIB: Steps toward an expert-level bridge-playing program. In *IJCAI-99*, pp. 584–589.

Li, D. (1994). *Kriegspiel: Chess Under Uncertainty*. Premier.

Li, D. (1995). *Chess Detective: Kriegspiel Strategies, Endgames and Problems*. Premier.

Murray, H. J. R. (1952). *A History of Board Games other than Chess*. London, UK: Oxford at the Clarendon Press.

Nau, D. S. (1982a). An investigation of the causes of pathology in games. *Artif. Intell. 19*(3), 257–278.

Nau, D. S. (1982b). The last player theorem. *Artif. Intell. 18*(1), 53–65.

Nau, D. S. (1983). Decision quality as a function of search depth on game trees. *JACM 30*(4), 687–708.

Osborne, M. J. and A. Rubinstein (1994). *A Course In Game Theory*. MIT Press.

Parker, A., D. Nau, and V. Subrahmanian (2005, August). Game-tree search with combinatorially large belief states. In *IJCAI*, pp. 254–259.

Pearl, J. (1981, August). Heuristic search theory: Survey of recent results. In *Proc. Seventh Internat. Joint Conf. Artif. Intel.*, Vancouver, Canada, pp. 554–562.

Pearl, J. (1984). *Heuristics: intelligent search strategies for computer problem solving*. Boston, MA, USA: Addison-Wesley Longman Publishing Co., Inc.

Reif, J. (1984). The complexity of two-player games of incomplete information. *Jour. Computer and Systems Sciences 29*, 274–301.

Russell, S. and J. Wolfe (2005, August). Efficient belief-state and-or search, with application to kriegspiel. In *IJCAI*, pp. 278–285.

Sakuta, M. and H. Iida (2000). Solving kriegspiel-like problems: Exploiting a transposition table. *ICCA Journal 23*(4), 218–229.

Smith, S. J. J., D. S. Nau, and T. Throop (1998). Computer bridge: A big win for AI planning. *AI Magazine 19*(2), 93–105.

von Neumann, J. and O. Morgenstern (1944). *Theory of Games and Economic Behavior*. Princeton University Press.

# 6
# Heuristic Search: Pearl's Significance from a Personal Perspective

IRA POHL

## 1 Introduction

This paper is about heuristics, and the significance of Judea Pearl's work to the field. The impact of Pearl's monograph was transformative. It heralded a third wave in the practice and theory of heuristic search. First there were the pioneering search algorithms without a theoretical basis, such as GPS or GT. Second came the Nilsson [1980] work that formulated a basic theory of A*. Third, in 1984, was Judea Pearl adding depth and breadth to this theory and adding a more sophisticated probabilistic context. Judea Pearl's book, *Heuristics: Intelligent Search Strategies for Computer Problem Solving*, was a tour de force summarizing the work of three decades.

Heuristic Search is a holy grail of Artificial Intelligence. It attempts to be a universal methodology to achieve AI, demonstrating early success in an era when AI was largely experimental. Pre-1970 programs that were notable include the Doran-Michie Graph Traverser [Doran and Michie 1966], the Art Samuel checker program [Samuel 1959], and the Newell, Simon, and Shaw General Problem Solver [Newell and Simon 1972]. These programs showed what could be called *intelligence* across a gamut of puzzles, games, and logic problems. Having no theoretical basis for predicting success, they were tested and compared to human performance.

The lack of theory for heuristic search changed in 1968 with the publication by Hart, Nilsson, and Raphael [1968] of their A* algorithm and its analysis. A* was provably optimum under some theoretical assumptions. The outcome of this work at the SRI robotics group fueled a series of primary results including my own, and later Pearl's. Pearl's book [Pearl 1984] captured and synthesized much of the A* work, including my work from the late 1960's [Pohl 1967; Pohl 1969] through 1977 [Pohl 1970a; Pohl 1970b; Pohl 1971; Pohl 1973; Pohl 1977]. It built a solid theoretical structure for heuristic search, and inspired much of my own and others subsequent work [Ratner and Pohl 1986; Ratner and Warmuth 1986; Kaindl and Kaintz 1997; Politowski 1984].

## 2 Early Experimentation

In the 1960's there were three premiere AI labs at US universities: CMU, MIT and Stanford; there was one such lab in England: the Machine Intelligence group at

Edinburgh; and there was the AI group at SRI. Each had particular strengths and visions. The CMU group led by Allen Newell and Herb Simon [Newell and Simon 1972] took a cognitive simulation approach. Their primary algorithmic framework was GPS, the General Problem Solver. This algorithm could be viewed as an elementary divide and conquer strategy. Guidance was based on detecting differences between partial solution and goal states. To make progress this algorithm attempted to apply operators to partial solutions, and reduce the difference with a goal state. It demonstrated that a heuristic search strategy could be applied to a wide array of problems that were associated with human intelligence. These included combinatorial puzzles such as crypt-arithmetic and towers of Hanoi.

The Graph Traverser [Doran and Michie 1966] reduced AI questions to the task of heuristic search. It was deployed on the 8 puzzle, a classic combinatorial puzzle typical of mildly challenging human amusements. "The objective of the 8-Puzzle is to rearrange a given initial configuration of eight numbered tiles arranged on a 3 x 3 board into a given final configuration called the goal state [Pearl 1984, p. 6].". Michie and Doran tried to obtain efficient search by discovering useful and computationally simple heuristics that measured perceived effort toward a solution. An example was "how many tiles are out of their goal space." The Graph Traverser demonstrated that a graph representation was a useful general perspective in problem solving, and that computationally knowledgeable heuristics could efficiently guide search.

The MIT AI lab pioneered many projects in both robotics and advanced problem solving. Marvin Minsky and his collaborators solved relatively difficult mathematical problems such as word algebra problems and calculus problems [Slagle 1963]. They used context, mathematical models and search to solve these problems. Ultimately this work led to programs such as Mathematica. They gave a plausible argument for what could be described as $knowledge + deduction = intelligence$.

The Stanford AI lab was led by John McCarthy, who was important in two regards: computational environment and logical representations. McCarthy pioneered with LISP and time sharing: tools and schemes that would profoundly impact the entire computational community. He championed predicate logic as a uniform and complete representation of what was needed to express and reason about the world.

The SRI lab, originally affiliated with Stanford, but later independent, was engaged in robotics. A concern was how an autonomous mobile robot could efficiently navigate harsh terrain, such as the moon. Here the emphasis was more engineering and algorithmic. Here the question was how something could be made to work efficiently, not whether it simulated human intelligence or generated a comprehensive theory of inference.

## 3 Early Theory

The Graph Traverser was a heuristic path finding algorithm attempting to minimize search steps to a goal node. This approach was experimental and explored how to

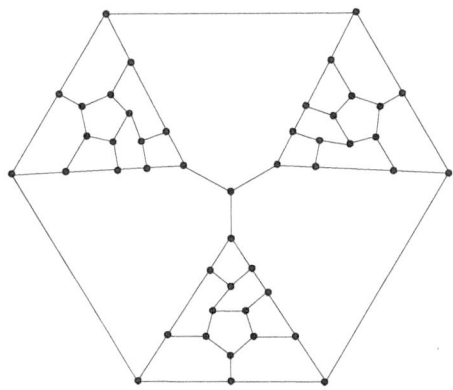

Figure 1. Tutte's graph

formulate and test for useful heuristics. The Dijkstra shortest path algorithm [Dijkstra 1959] was a combinatorial algorithm. It was an improvement on earlier graph theoretic and operations research algorithms for the pure shortest path optimization problem. A* developed in Hart, Nilsson, and Raphael [1968], is the adaptation of Dijkstra's shortest path algorithm to incorporate admissible heuristics. It preserved the need to find an optimal path, while using heuristics that attempted to minimize search.

In the period 1966-1969, I worked on graph theory algorithms, such as the Dijkstra algorithm. The Stanford Computer Science department was dominated by numerical analysts with a strong algorithmic approach. The AI group was metamathematical and inference oriented. Don Knuth had yet to come to Stanford, so there was not yet any systematic work or courses offered on combinatorial algorithms. I was asked to evaluate PL1 for use at Stanford and SLAC and decided to test it out by building a graph algorithm library.

I enjoyed puzzle-like problems and had an idea for improving Warnsdorf's rule for finding a Knight's tour. A Knight's tour is a Hamiltonian path on the 64 square chessboard whose edge connectivity is determined by Knight moves. In graph theory terms the rule was equivalent to going to an unvisited node of minimum out-degree. Warnsdorf's rule is an example of a classic greedy heuristic. I modified it to find a Hamiltonian in the 46 node Tutte's graph [Pohl 1967; Tutte 1946], as well as showed that it worked well on Knight's Tours.

At that time, the elegance of Dijkstra's work had impressed me, and I sought to improve his shortest path algorithm by implementing it bidirectionally. An early attempt by Nicholson [1966] had proved incorrect because of an error in the stopping criteria. These ideas conspired to lead me to test my methods further as variations on A* [Pohl 1971].

The first half of Pearl [1984], Chapter 3 is a sophisticated and succinct summation of search results through 1978. Three theorems principally due to Hart, Nilsson, and Raphael [1968] are central to early theory:

- A* is complete even on infinite graphs [Pearl 1984, Theorem 1, p. 77].

- A* is admissible [Pearl 1984, Theorem 2, p. 78].

- If $A^*_2$ is more informed than $A^*_1$, then $A^*_2$ dominates $A^*_1$ [Pearl 1984, Theorem 7, p. 81].

Much of this early work relied on the heuristic being consistent. But as Pearl [1984, p. 111] notes: "The property of monotonicity was introduced by Pohl [1977] to replace that of consistency. Surprisingly, the equivalence of the two have not been previously noted in the literature." These theorems and monotonicity provide a first attempt at a mathematical foundation for heuristic search.

These theorems suggest that A* is robust and that heuristics that are more informed lead to more efficient searches. What is missing is how to relate the effort and accuracy in computing the heuristic to its computational benefit.

## 4 Algorithms, and Numerical Methods, as a model

In AI search is a weak but universal method. To make it efficient we have to give it powerful guidance mechanisms. A well developed theory of search exists in numerical analysis for finding the roots of equations. It represented to me a possible conceptual model for heuristic search.

There are diverse algorithms to solve the root search problem. A sample of methods could include bisection search, Monte Carlo sampling, and Newton-Raphson. Bisection search for root finding is robust. Newton-Raphson (NR) for root finding is efficient. Monte-Carlo root finding is highly robust and very inefficient. How to decide what to use involves efficiency and error concerns. Often efficient methods of root finding require a function be well behaved. So NR converges quadratically, but requires differentiability. Bisection converges linearly but requires continuity. Monte-Carlo in most circumstances works very slowly, but works on discontinuous and non-differentiable functions

In numerical methods error analysis is critical to understanding a method's utility and its efficiency. Also in algorithmic methods one needs to use adversaries to stress test robustness and efficiency. These techniques can be applied to heuristic search algorithms.

Heuristic search theory can be investigated analogously to the theory of root finding. To subsume the various heuristic algorithms, I formulated the following generalization of the search function f, as a linear combination of g and h. Furthermore this combination could be weighted dynamically.

The node selection function is : $f(x) = (1 - w(x))g(x) + w(x)h(x),\ 0 \leq w(x) \leq 1$.

Moore maze/path: $w = 0$, edge costs are 1

Dijkstra: $w = 0$, edge costs are 1

Michie-Doran GT: $w = 1$

HNR A*: $w = 0.5$, h is admissible

This generalization is also adopted in Pearl [1984], Section 3.2.1. Once you have this generalization, one can ask questions that are similar to those studied by numerical analysts. How reliably the heuristic function estimates effort, leads to a notion of error. This error then effects the convergence rate to a solution. This question was first taken up by Pohl [1970a] and later by Gaschnig [1979]. Pearl's results as summarized in [Pearl 1984, Chapters 6–7], provide a detailed treatment of the effects of error on search.

## 5 Variations: Weighted and Bidirectional search

### 5.1 Weighted dynamic search

The four standard weightings for HPA were all static, for example $w = 1/2$ for A*. Dynamic weighting is proposed in [Pohl 1973], where $w(x)$ is dependent on the character of the state. This was inspired by the observation that accuracy of heuristics improved as the search neared its goal. An admissible heuristic is an underestimate. Dynamic weighting can overestimate (see discussion in [Pearl 1984, Chapter 7]) and be not admissible. This technique remains controversial and is underutilized and not extensively researched.

### 5.2 Bidirectional Search

Bidirectional search was originally proposed in optimzation [Nicholson 1966] to improve on unidirectional shortest path algorithms. These implementations presumed a naive termination condition - namely that when an intersection of two paths occurred the resulting path was optimal. Finding this an error and implementing the correct terminating condition led me to consider both practically and theoretically how more efficient bidirectional search was.

Bidirectional search is an attractive approach for several reasons. Searches are normally combinatorially explosive in their depth. Two searches of half the depth ideally save exponential time and space. When run on a parallel architecture they can essentially be done simultaneously. Furthermore, this leads to a natural recursive process of further divide and conquer searches.

The cardinality comparison rule tells us to expand in the sparser hemi-tree and can be important in improving these searches. There is the following intuitive justification as to why such a rule makes more progress than simple alternation. Consider the problem of picking a black ball out of either of two urns. Each urn contains a single black ball and some white balls. The probability of finding a black ball is $1/n$ where n is the number of balls in the urn. In finding a next black ball it is best to pick from the urn with fewest balls. Think of the urn as the collection of open nodes and selection as finding the next node along an optimum path. This leads to the cardinality comparison rule.

Bidirectional search works well for the standard graph shortest path problem. Here, bidirectional search, exclusive of memory constraints, dominates unidirectional search when the metric is nodes expanded. But when search is guided by highly selected heuristics there can be a "wandering in the desert" problem when the two frontiers do not meet in the middle.

To address this problem, I first proposed a parallel computation of all front-to-front node values in 1975. Two of my student's implemented and tested this method [De Champeaux and Sint 1977]. It had some important theoretical advantages, such as retaining admissibility, but it used an expensive front-to-front computation that involved the square of the nodes in the open set. By only looking at nodes expanded as a measure of efficiency it was misleading as to its true computational effort [Davis et al. 1984].

## 6  Judea and Heuristics

In *Heuristics* [Pearl 1984], Judea Pearl insightfully presents the work on heuristic search from 1950-1984. He contextualized and showed it as a mature theory. The book is the defining third generation document that immensely broadens and deepens the mathematical character of the field.

Pearl's book is very important in emphasizing the need for a sophisticated view of computational efficiency. It summarized, systematized and extended the theory of heuristic search. It extensively analyzed A* using both worst-case analysis and expected case analysis. It looked at the case of using nonadmissible heuristics.

Chapter 4 gives pointers to how heuristics can be discovered that will work on difficult problems including NP-Complete problems. An example would be the "out-of-place heuristic" in the sliding blocks puzzles. Here a person would be in one move allowed to swap an out-of-place tile to its final location. The number of out-of-place tiles would be a lower bound on a solution length and easy to compute. A more sophisticated heuristic occurs with respect to the traveling salesman problem (TSP) where the minimum spanning tree is a relaxed constraint version for visiting all nodes in a graph and is $nlog(n)$ in its computation.

In Pearl [1984] Chapter 6, we have a summary of results on complexity versus the precision of the heuristic. Here the work of Pearl and his students Rina Decter and Nam Huyn is presented. Pearl, Dechter, and Huyn [Dechter and Pearl 1985], [Huyn et al. 1980] developed the theory of optimality for A* for the expected case as well as the worst case.

"Theorem 1 [Huyn et al. 1980] For any error distribution, if $A*_2$ is stochastically more informed than $A*_1$, then $A*_2$ is stochastically more efficient than $A*_1$ [Pearl 1984, p. 177]."

This leads to a result by Pearl [1983] that "the exponential relationship established in this section implies that precision-complexity exchange for A* is fairly 'inelastic'." Namely, unless error in the heuristic is better than logarithmic, search branching rates remain exponential.

In Pearl [1984, Chapter 7], there are results on search without admissibility. He provides a formal probabilistic framework to analyze questions of non-admissible heuristics and search efficiency. These results provide one view of dynamic-weighting search.

The impact of the Pearl monograph was transformative. It heralded a third wave of sophistication in the theory of heuristic search. First we had inventive pioneering search algorithms without a theoretical basis, such as GPS or GT. Second we had the Nilsson [1980] work that formulated a basic theory of A* and my work that formulated a very rough theory of efficiency. Third, in 1984, we had Pearl adding depth and breadth to this theory and embedding it in a sophisticated probabilistic reality.

# 7 D-nodes, NP, LPA*

## 7.1 What Pearl inspired

Pearl's synthesis recharged my batteries and led me with several students and colleagues to further examine these problems and algorithms.

NP [Garey and Johnson 1979] complexity is the sine quo non for testing search on known hard problems. The Hamiltonian problem was already known in 1980 as NP, but not sliding tile puzzles. I conjectured that the generalized sliding tile problem was NP in private communications to Ratner and Warmuth [1986]. They in turn produced a remarkable proof that it indeed was NP. This is important in the sense that it puts what some have called "the drosophila of AI" on firm footing as a proper exemplar. In effect it furthers the Nilsson, Pohl, Pearl agenda of having a mathematical foundation for search.

## 7.2 LPA* search

In work with Ratner [Ratner and Pohl 1986], we proposed Local Path A* (LPA*), an algorithm that combines an initial efficient approximation algorithm with local A* improvements. Such an approach retains the computational efficiency of finding the initial solution by carefully constraining improvements to a small computational cost.

The idea behind these algorithms is to combine a fast approximation algorithm with a search method. This idea was first suggested by S. Lin [Lin, 1965], when he used it to find an effective algorithm for the Traveling-Salesman problem (TSP). Our goal was to develop a problem independent approximation method and combine it with search.

An advantage of approximation algorithms is that they execute in polynomial time, where many other algorithms have no such upper bound. The test domain is the 15 puzzle and the approximation algorithm is based on macro-problem-solving

[Korf 1985a]. The empirical results, which come from a test on a standard set of 50 problems [Politowski and Pohl, 1984], show that the algorithms outperform other then published methods within stated time limits.

In order to bound the effort of local search by a constant, each local search will have the start and goal nodes reside on the path, with the distance between them bounded by $d_{max}$, a constant independent of n and the nodes. Then we will apply A * with admissible heuristics to find a shortest path between the two nodes. The above two conditions generally guarantee that each A * use requires less than some constant time. More precisely, if the branching degrees of all the nodes in G are bounded by a constant c which is independent of n then A * will generate at most $c(c-1)^{d_{max}-1}$ nodes.

Theoretically $c(c-1)^{d_{max}-1}$ is a constant, but it can be very large. Nevertheless, most heuristics prune most of the nodes [Pearl 1984]. The fact that not many nodes are generated, is supported by experiments.

The results in [Ratner and Pohl 1986] demonstrate the effectiveness of using LPA * . When applicable, this algorithm achieves a good solution with small execution time. This method require an approximation algorithm as a starting point. Typically, when one has a heuristic function, one has adequate knowledge about the problem to be able to construct an approximation algorithm. Therefore, this method should be preferred in most cases to earlier heuristic search algorithms.

## 7.3 D-node Bidirectional search

Bidirectional heuristic search is potentially more efficient than unidirectional heuristic search. A basic difficulty is that the two search trees do not meet in the middle. This can result in two unidirectional searches and poorer performance. To work around this George Politowski and I implemented a retargeted bidirectional search.

De Champeaux describes a Bidirectional, Heuristic Front-to-Front Algorithm (BHFFA) [De Champeaux, Sint 1977] which is intended to remedy the "meet in the middle" problem. Data is included from a set of sample problems corresponding to those of [Pohl 1971]. The data shows that BHFFA found shorter paths and expanded fewer nodes than Pohl's bidirectional algorithm. However, there are several problems with the data. One is that most of the problems are too easy to constitute a representative sample of the 15-puzzle state space, and this may bias the results. Another is that the overall computational cost of the BHFFA is not adequately measured, although it is of critical importance in evaluating or selecting a search algorithm. A third problem concerns admissibility. Although the algorithm as formally presented is admissible, the heuristics, weightings, termination condition, and pruning involved in the implemented version all violate admissibility. This makes it difficult to determine whether the results which were obtained are a product of the algorithm itself or of the particular implementation. It is also difficult to be sure that the results would hold in the context of admissible search.

The main problem in bidirectional heuristic search is to make the two partial paths meet in the middle. The problem with Pohl's bidirectional algorithm is that each search tree is 'aimed' at the root of the opposite tree. What is needed is some way of aiming at the front of the opposite tree rather than at its root. There are two advantages to this. First, there is a better chance of meeting the opposite front if you are aiming at it. Second, for most heuristics the aim is better when the target is closer. However, aiming at a front rather than a single node is somewhat troublesome since the heuristic function is only designed to estimate the distance between two nodes. One way to overcome this difficulty is to choose from each front a representative node which will be used as a target for nodes in the opposite tree. We call such nodes d-nodes.

Consider a partially developed search tree. The growth of the tree is guided by the heuristic function used in the search, and thus the whole tree is inclined, at least to some degree, towards the goal. This means that one can expect that on the average those nodes furthest from the root will also be closest to the goal. These nodes are the best candidates for the target to be aimed at from the opposite tree. In particular, the very farthest node out from the root should be the one chosen. D-node selection based on this criterion costs only one comparison per node generated.

We incorporated this idea into a bidirectional version of HPA in the following fashion:

1. Let the root node be the initial d-node in each tree.

2. Advance the search n moves in either the forward or backward direction, aiming at the d-node in the opposite tree. At the same time, keep track of the furthest node out, i.e. the one with the highest g value.

3. After n moves, if the g value of the furthest node out is greater than the g value of the last d-node in this tree, then the furthest node out becomes the new d-node. Each time this occurs, all of the nodes in the opposite front should be re-aimed at the new d-node.

4. Repeat steps 2 and 3 in the opposite direction.

The above algorithm does not specify a value for n. Sufficient analysis may enable one to choose a good value based on other search parameters such as branching rate, quality of heuristic, etc. Otherwise, an empirical choice can be made on the basis of some sample problems. In our work good results were obtained with values of n ranging from 25 to 125.

It is instructive to consider what happens when n is too large or too small, because it provides insight into the behavior of the d-node algorithm. A value of n which is too large will lead to performance similar to unidirectional search. This is not surprising since for a sufficiently large n, a path will be found unidirectionally, before any reversal occurs. A value of n which is too small will lead to poor performance in two respects. First, the runtime will be high because the overhead to re-aim the opposite tree is incurred too often. Second, the path quality will be lower (i.e. the The evaluation function used by the d-node search algorithm is the same as that used by HPA, namely $f = (1 - w)g + w * h$, except that h is now the heuristic estimate of the distance from a particular node to the d-node of the opposite tree. This is in contrast to the original bidirectional search algorithm, where h estimates the distance to the root of the opposite tree, and to unidirectional heuristic search, where h estimates the distance to the goal. The d-node algorithm's aim is to perform well for a variety of heuristics and over a range of w values.

The exponential nature of the problem space makes it highly probable that randomly generated puzzles will be relatively hard, i.e. their shortest solution paths will be relatively long with respect to the diameter of the state space. The four functions used to compute h are listed below. These functions were originally developed by Doran and Michie [Doran, Michie 1966], and they are the same functions as those used by Pohl and de Champeaux.

1. h = P
2. h = P +20*R
3. h = S
4. h = S +20*R

The three basic terms P, S, and R have the following definitions.

1. where P is the sum for all tile i of Manhattan distance between the position of tile i in the current state and in the goal.
2. S, a relationship between tile and the blank square defined in [Doran, Michie 1966].
3. R is the number of reversals in the current state with respect to goal. For example if tile 2 is first and tile 1 is next, this is a reversal.

Finally, the w values which we used were 0.5, 0.75, and 1.0. This covers the entire 'interesting' range from w = 0.5, which will result in admissible search with a suitable heuristic, to w = 1.0, which is pure heuristic search.

The detailed results of our test of the d-node algorithm are found in [Politowski 1986]. The most significant result is that the d-node method dominates both previously published bidirectional techniques, regardless of heuristic or weighting. In comparison to de Champeaux's BHFFA, the d-node method is typically 10 to 20 times faster. This is chiefly because the front-to-front calculations required by BHFFA are computationally expensive, even though the number of nodes expanded is roughly comparable for both methods. In comparison to Pohl's bidirectional algorithm, the d-node method typically solves far more problems, and when solving the same problems it expands approximately half as many nodes.

## 8 Next Steps

Improving problem decomposition remains an important underutilized strategy within heuristic search. Divide and conquer remains a central instrument of intelligent deduction in many formats and arenas. Here the bidirectional search theme and the LPA* search are instances of naive but effective first steps. Planning is a further manifestation of divide and conquer. Intuitively planning amounts to a good choice of lemma's when attempting to construct a difficult proof. Korf's [Korf 1985a, Korf 1985b] macro operators can be seen as a first step, or the Pohl-Politowski d-node selection criteria as a an automated attempt at problem decomposition.

Expanding problem selection to hard domains is vital to demonstrating the relevance of heuristic search. Historically these techniques were developed on puzzle domains. Here Korf's [Korf, Zhang 2000] recent work in applying search to genomics problems is welcome. Computational genomics is an unambiguously complex and non-trivial domain that almost certainly requires heuristics search.

Finally, what seems under exploited is the use of a theory of error linked to efficiency. Here my early work and the experimental work of Gaschnig contributed to the deeper studies of Pearl and Davis [1990]. Questions of worst case analysis and average case complexity and its relation to complexity theory while followed up by Pearl, and by Chenoweth and Davis [1992], is only a beginning. A deeper theory that applies both adversary techniques and metric space analysis is needed.

## References

Chenoweth, S., and Davis, H. (1992). New Approaches for Understanding the Asymptotic Complexity of A* Tree Searching. *Annals of Mathematics and AI 5*, 133–162.

Davis, H. (1990). Cost-error Relationships in A* Tree Searching. *Journal of the ACM 37*, 195–199.

Davis, H., Pollack, R., and Sudkamp, T. (1984). Toward a Better Understanding of Bidirectional Search. *Proceedings AAAI*, pp. 68–72.

De Champeaux, D., and Sint, L. (1977). An improved bi-directional heuristic search algorithm. *Journal of the ACM 24*, 177–191.

Dechter, R., and Pearl, J. (1985). Generalized Best-First Search Strategies and the Optimality of A*. *Journal of the ACM 32*, 505–536.

Dijkstra, E. (1959). A Note on Two Problems in Connection with Graphs *Numerische Mathematik 1*, 269–271, 1959.

Doran, J., and Michie, D. (1966). Experiments with the Graph Traverser Program. *Proc.of the Royal Society of London, 294A*, 235–259.

Field, R., Mohyeldin-Said, K., and Pohl, I. (1984). An Investigation of Dynamic Weighting in Heuristic Search, *Proc. 6th ECAI*, pp. 277–278.

Gaschnig, J. (1979). Performance Measurement and Analysis of Certain Search Algorithms. *Ph.D. Dissertation CMU CS 79-124*.

Hart, P. E., Nilsson, N., and Raphael, B. (1968). A Formal Basis for the Heuristic Determination of Minimal Cost Paths Search Reconsidered. *IEEE Trans. Systems Science and Cybernetics SSC-4 2*, 100–07.

Huyn, N., Dechter, R., and Pearl, J. (1980). Probabilistic analysis of the complexity of A*. *Artificial Intelligence:15*, 241–254.

Kaindl, H., and Kainz, G. (1997). Bidirectional Heuristic Search Reconsidered. *Journal of Artificial Intelligence Research 7*, 283–317.

Knuth, D. E. (1994). Leaper graphs. *Mathematical Gazette 78*, 274–297.

Korf, R., E. (1985a). *Learning to Solve problems by Searching Macro-Operators*. Pitman.

Korf, R.,E. (1985b). Iterative Deepening A*. *Proceedings 9th IJCAI, vol. 2*, pp. 1034–1035.

Korf, R.,E., and Zhang, W. (2000). Divide and Conquer Frontier Search Applied to Optimal Sequence Alignment. *AAAI-00*, pp. 910–916.

Korf, R. E. (2004). Best-First Frontier Search with Delayed Duplicate Detection. *AAAI-04*, pp. 650–657.

Lin, S. (1965). Computer solutions of the traveling salesman problem. *Bell Systems Tech. J. 44(10)*, 2245–2269.

Newell, A., and Simon, H. (1972). *Human Problem Solving*. Prentice Hall.

Nilsson, N. (1980). *Principles of Artificial Intelligence*. Palo Alto, CA: Tioga.

Pearl, J. (1983). Knowledge versus search: A quantitative analysis using A*. *Artificial Intelligence:20*, 1–13.

Pearl, J. (1984). *Heuristics: Intelligent Search Strategies for Computer Problem Solving*. Reading, Massachusetts: Addison-Wesley.

Pohl, I. (1967). A method for finding Hamilton paths and knight's tours. *CACM 10*, 446–449.

Pohl, I. (1970a). First Results on the Effect of Error in Heuristic Search. In B. Meltzer and D. Michie (Eds.), *Machine Intelligence 5*, pp. 219–236. Edinburgh University Press.

Pohl, I. (1970b). Heuristic Search Viewed as Path Finding in A Graph. *Artificial Intelligence 1*, 193–204.

Pohl, I. (1971). Bi-directional search. In B. Meltzer and D. Michie (Eds.), *Machine Intelligence 6*, pp. 127–140. Edinburgh University Press.

Pohl, I. (1973). The Avoidance of (Relative) Catastrophe, Heuristic Competence, Genuine Dynamic Weighting and Computational Issues in Heuristic Problem Solving. *IJCAII 3*, pp. 20–23.

Pohl, I. (1977). Practical and Theoretical Considerations in Heuristic Search Algorithms, In E. Elcock and D. Michie (Eds.), *Machine Intelligence 8*, pp. 55–72. New York: Wiley.

Ratner, D., and Pohl, I. (1986). Joint and LPA*: Combination of Approximation and Search. *Proceedings AAAI-86, vol. 1*, pp. 173–177.

Ratner, D., and Warmuth, M. (1986). Finding a Shortest Solution for the $N \times N$ Extension of the 15-Puzzle is Intractable. *Proceedings AAAI-86, vol. 1*, pp. 168–172.

Samuel, A. (1959). Some Studies in Machine Learning Using Checkers. *IBM Research Journal of Research and Development 3*, 211–229.

Slagle, J. R. (1963). A Heuristic Program that Solves Symbolic Integration Problems in Freshman Calculus. *Journal of the ACM 10*, 507–520.

Tutte, W. T. (1946). On Hamiltonian Circuits. *J. London Math. Soc. 21*, 98–101.

# Part II: Probability

# 7
# Inference in Bayesian Networks: A Historical Perspective

ADNAN DARWICHE

## 1 Introduction

Judea Pearl introduced Bayesian networks as a representational device in the early 1980s, allowing one to systematically and locally assemble probabilistic beliefs into a coherent whole. While some of these beliefs could be read off directly from the Bayesian network, many were implied by this representation and required computational work to be made explicit. Computing and explicating such beliefs has been the subject of much research and became known as the problem of *inference* in Bayesian networks. This problem is critical to the practical utility of Bayesian networks as the computed beliefs form the basis of decision making, which typically dictates the need for Bayesian networks in the first place.

Over the last few decades, the interest in inference algorithms for Bayesian networks remained great and has witnessed a number of shifts in emphasis with regards to the adopted computational paradigms and the type of queries addressed. My goal in this paper is to provide a historical perspective on this line of work and the associated shifts, where we shall see the key role that Judea Pearl has played in initiating and inspiring many of the technical developments that have formed and continue to form the basis of work in this area.

## 2 Starting with trees

It all began with trees — and polytrees! These are network structures that permit only one undirected path between any two nodes in the network; see Figure 1. If each node has at most one parent, we have a tree. Otherwise, we have a polytree. Pearl's first inference algorithm — and the very first algorithm for Bayesian networks — was restricted to trees [Pearl 1982] and was immediately followed by a generalization that became known as the *polytree algorithm* [Kim and Pearl 1983; Pearl 1986b]. The goal here was to compute a probability distribution for each node in the network given some evidence, a task which is known as computing node marginals.

The polytree algorithm was based on a message-passing computational paradigm, where nodes in the network send messages to a neighboring node after they have received messages from all other neighbors. Each message can be viewed as summarizing results from one part of the network and passing them on to the rest of the network. Messages that communicated information from parents to their chil-

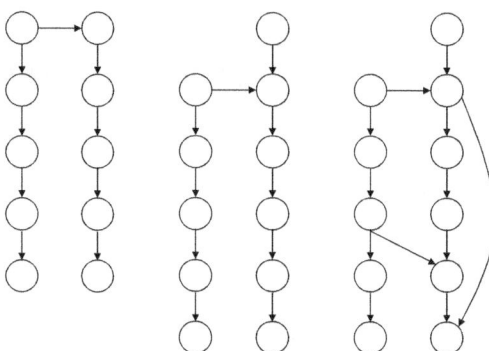

Figure 1. From left to right: a tree, a polytree, and a multiply–connected network.

dren were said to quantify the causal support from parents to these children. On the other hand, messages that communicated information from children to their parents were said to quantify the diagnostic support from children to parents. The notions of causal and diagnostic supports were rooted in the causal interpretation of Bayesian network structures that Pearl insisted on, where parents are viewed as direct causes of their children. According to this interpretation, the distribution associated with a node in the Bayesian network is called the *belief* in that node, and is a function of the causal support it receives from its direct causes, the diagnostic support it receives from its direct effects, and the local information available about that node. This is why the algorithm is also known as the *belief propagation* algorithm, a name which is more common today.

The polytree algorithm has had considerable impact and is of major historical significance for a number of reasons. First, it was the very first *exact* inference algorithm for this class of Bayesian networks. Second, its time and space complexity were quite modest being linear in the size of the network. Third, the algorithm formed the basis for a number of other algorithms, both exact and approximate, that will be discussed later. In addition, the algorithm provided a first example of reading off independence information from a network structure, and then using it to decompose a complex computation into smaller and independent computations. It formally showed the importance of independence, as portrayed by a network structure, in driving computation and in reducing the complexity of inference.

One should also note that, according to Pearl, this algorithm was motivated by the work of [Rumelhart 1976] on reading comprehension, which provided compelling evidence that text comprehension must be a distributed process that combines both top-down and bottom-up inferences. This dual mode of inference, so characteristic of Bayesian analysis, did not match the capabilities of the ruling paradigms for uncertainty management in the 1970s. This led Pearl to develop the polytree algo-

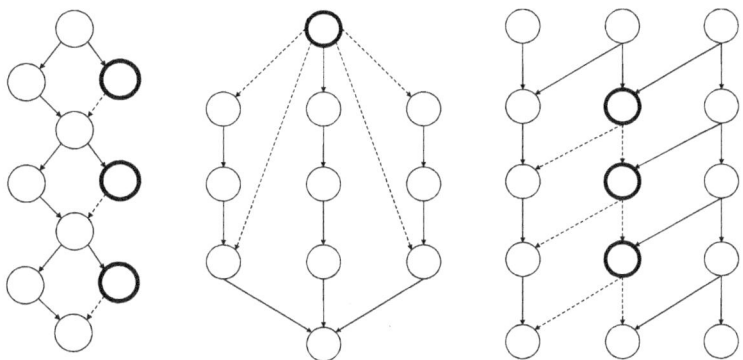

Figure 2. Networks and corresponding loop-cutsets (bold circles).

rithm [Pearl 1986b], which, as mentioned earlier, appeared first in [Pearl 1982] with a restriction to trees, and then in [Kim and Pearl 1983] for polytrees.

## 3 On to more general structures

Soon after the polytree algorithm was introduced, the search began for algorithms that can handle arbitrary network structures. Since polytrees were also referred to as singly–connected networks, arbitrary network structures were said to be multiply–connected; see Figure 1. One of the central ideas for handling these networks is based on the technique of conditioning. That is, one can set variable $X$ to some value $x$ and then solve the problem under that particular condition $X = x$. If this is repeated for all values of $X$, then one can recover the answer to the original problem by assembling the results obtained from the individual cases. The main value of this technique is that by conditioning variables on some values, one can simplify the problem. In Bayesian networks, one can effectively delete edges that are outgoing from a node once the value of that node is known, therefore, creating a simplified structure that can be as informative as the original structure in terms of answering queries.

Pearl used this observation to propose the algorithm of *loop-cutset conditioning* [Pearl 1986a; Pearl 1988], which worked by conditioning on enough network variables to render the network structure singly–connected. The set of variables that needed to be conditioned on is called a *loop–cutset*; see Figure 2. The loop–cutset conditioning algorithm amounted then to a number of invocations to the polytree algorithm, where this number is exponential in the size of the cutset — one invocation for each instantiation of the variables constituting the cutset. A key attraction of this algorithm is its modest space requirements, as it did not need much space beyond that used by the polytree algorithm. The problem with the algorithm, however, was in its time requirements when the size of the loop-cutset was large enough. The algorithm proved impractical in such a case and the search continued for al-

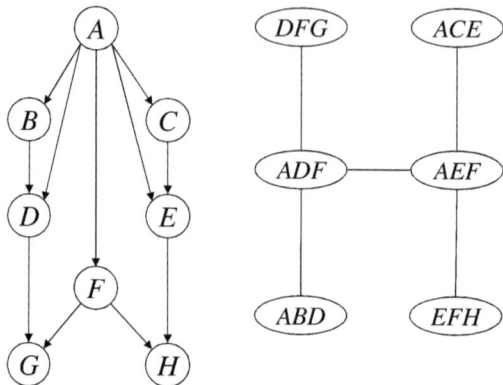

Figure 3. A Bayesian network structure and its corresponding jointree (tree of clusters).

ternative conditioning algorithms that could handle multiply–connected structures more efficiently.

The very first algorithm that found widespread use on multiply–connected networks was the *jointree* algorithm, also known as the *tree clustering* algorithm [Lauritzen and Spiegelhalter 1988]. This algorithm proved quite effective and remains practically influential until today — for example, it is the algorithm of choice in commercial implementations of Bayesian network inference. One way of understanding this algorithm is as a version of the polytree algorithm, invoked on a tree clustering of the multiply–connected network. For an example, consider Figure 3 which depicts a DAG and its corresponding tree of clusters — this is technically known as a *jointree* or a *tree decomposition* [Robertson and Seymour 1986]. One thing to notice here is that each cluster is a set of variables in the original network. The jointree algorithm works by passing messages across the tree of clusters, just as in the polytree algorithm. However, the size of these messages and the amount of work it takes to propagate them is now tied to the size of clusters.

The jointree is not an arbitrary tree of clusters as it must satisfy some conditions to legitimize the message passing algorithm. In particular, every node and its parents in the Bayesian network must belong to some tree cluster. Moreover, if a variable appears in two clusters, it must also appear in every cluster on the path connecting them. Ensuring these conditions may lead to clusters that are large. There is a graph–theoretic notion, known as *treewidth,* which puts a lower bound on the size of largest cluster [Robertson and Seymour 1986]. In particular, if the treewidth of the DAG is $w$, then any jointree of the DAG must have a cluster whose size is at least $w + 1$.[1] In some sense, the treewidth can be viewed as a measure of

---

[1]In graph theory, treewidth is typically defined for undirected graphs. The treewidth of a DAG as used here corresponds to the treewidth of its moralized graph: one which is obtained by

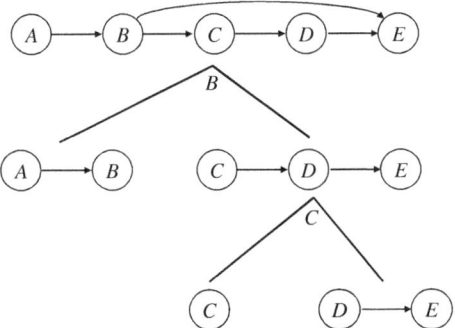

Figure 4. Decomposing a Bayesian network by conditioning on variable $B$ and then on variable $C$.

how similar a DAG structure is to a tree structure as it puts a lower bound on the width of any tree clustering (jointree) of the DAG.

The connection between the complexity of inference algorithms and treewidth is actually the central complexity result that we have today for exact inference [Dechter 1996]. In particular, given a jointree whose width is $w$, node marginals can be computed in time and space that is exponential only in $w$. Note that a network treewidth of $w$ guarantees the existence of such a jointree, but finding it is generally known to be hard. Hence, much work on this topic concerns the construction of jointrees with minimal width using both heuristics and complete search methods (see [Darwiche 2009] for a survey).

## 4 More computational paradigms

Since a typical implementation of the jointree algorithm will indeed use as much time and space as is suggested by the complexity analysis, we will not be able to rely on the jointree algorithm in the case where we do not find a jointree whose width is small enough. To overcome this treewidth barrier, research on inference algorithms continued in a number of directions.

With regards to work on conditioning algorithms, the main breakthrough in this regard was based on observing that one can employ conditioning in other and more effective ways than loop–cutset conditioning. For example, one can condition on enough variables to split the network into disconnected sub–networks, which can then be solved independently. These sub–networks need not be polytrees, as each one of them can be solved recursively using the same method, until sub–networks reduce to a single node each; see Figure 4. With appropriate caching schemes to avoid solving the same sub–network multiple times, this method of *recursive conditioning* can be applied with the same complexity as the jointree algorithm. In

---

connecting every pair of nodes that share a child in the DAG and then dropping the directionality of all edges.

particular, one can guarantee that the space and time requirements of the algorithm are at most exponential in the treewidth of underlying network structure. This result assumes that one has access to a decomposition structure, known as a *dtree*, which is used to control the decomposition process at each level of the recursive process [Darwiche 2001]. Similar to a jointree, finding an optimal dtree (i.e., one that realizes the treewidth guarantee on complexity) is hard. Yet, one can easily construct such a dtree given an optimal jointree, and vice versa [Darwiche 2009].

Even though recursive conditioning and the jointree algorithm are equivalent from this complexity viewpoint, recursive conditioning provided some new contributions to inference. On the theoretical side, it showed that conditioning as an inference paradigm can indeed reach the same complexity as the jointree algorithm — a question that was open for some time. Second, the algorithm provided a flexible paradigm for time-space tradeoffs: by simply controlling the degree of caching, the space requirements of the algorithm can be made to range from being only linear in the network size to being exponential in the network treewidth (given an appropriate dtree). Moreover, the algorithm provided a convenient framework for exploiting local structure as we shall discuss later.

On another front, and in the continued search of an alternative for the jointree algorithm, a sequence of efforts culminated into what is known today as the *variable elimination* algorithm [Zhang and Poole 1994; Dechter 1996]. According to this algorithm, one maintains the probability distribution of the Bayesian network as a set of factors (initially the set of CPTs) and then successively eliminates variables from this set one variable at a time.[2] The elimination of a variable can be implemented by simply combining all factors that mention that variable and then removing the variable from the combined factor. After eliminating a variable, the resulting factors represent a distribution over all remaining (un-eliminated) variables. Hence, by repeating this elimination process, one can obtain the marginal distribution over any subset of variables, including, for example, marginals over single variables.

The main attraction of this computational paradigm is its simplicity — at least as compared to the initial formulations of the jointree algorithm. Variable elimination, however, turned out to be no more efficient than the jointree algorithm in the worst case. In particular, the ideal time and space complexities of the algorithm also depend on the treewidth — in particular, they are exponential in treewidth when computing the marginal over a single variable. To achieve this complexity, however, one needs to use an optimal order for eliminating variables [Bertele and Brioschi 1972]. Again, constructing an optimal elimination order that realizes the treewidth complexity is hard in general. Yet, one can easily construct such an optimal order from an optimal jointree or dtree, and vice versa.

Even though variable elimination proved to have the same treewidth complexity

---

[2] A factor is a function that maps the instantiations of some set of variables into numbers; see Figure 5. In this sense, each probability distribution is a factor and so is the marginal of such a distribution on any set of variables.

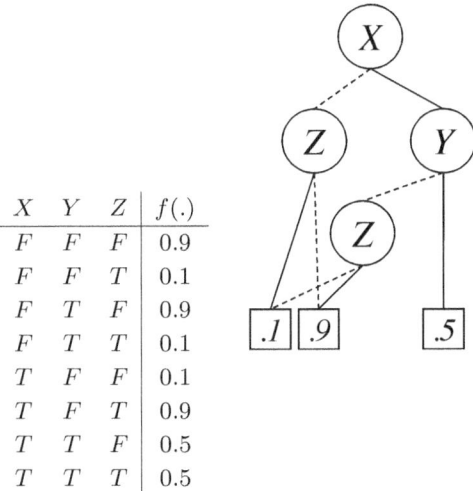

| X | Y | Z | $f(.)$ |
|---|---|---|---|
| F | F | F | 0.9 |
| F | F | T | 0.1 |
| F | T | F | 0.9 |
| F | T | T | 0.1 |
| T | F | F | 0.1 |
| T | F | T | 0.9 |
| T | T | F | 0.5 |
| T | T | T | 0.5 |

Figure 5. A factor over binary variables $X, Y, Z$ with a tabular representation (left) and an ADD representation (right).

as the jointree algorithm, it better explained the semantics of the jointree algorithm, which can now be understood as a sophisticated form of variable elimination. In particular, one can interpret the jointree algorithm as a refinement on variable elimination in which: (1) multiple variables can be eliminated simultaneously instead of one variable at a time; (2) a tree structure is used to control the elimination process and to save the results of intermediate elimination steps. In particular, each message passed by the jointree algorithm can be interpreted as the result of an elimination process, which is saved for re-use when computing marginals over different sets of variables [Darwiche 2009]. As a result of this refinement, the jointree algorithm is able to perform successive invocations of the variable elimination algorithm, for computing multiple marginals, while incurring the cost of only one invocation, due mainly to the re-use of results across multiple invocations.

Given our current understanding of the variable elimination and jointree algorithms, one now speaks of only two main computational paradigms for exact probabilistic inference: conditioning algorithms (including loop-cutset conditioning and recursive conditioning) and elimination algorithms (including variable elimination and the jointree algorithm).

## 5  Beating the treewidth barrier with local structure

Assuming that we ignore the probabilities that quantify a Bayesian network, the treewidth guarantee is the best we have today on the complexity of exact inference. Moreover, the treewidth determines the best-case performance we can expect from the standard algorithms based on conditioning and elimination.

It has long been believed though that exploiting the local structure of a Bayesian network can speed up inference to the point of beating the treewidth barrier, where local structure refers to the specific properties attained by the probabilities quantifying the network. One of the main intuitions here is that local structure can imply independence that is not visible at the structural level and this independence may be utilized computationally [Boutilier et al. 1996]. Another insight is that determinism in the form of 0/1 probabilities can also be computationally useful as it allows one to prune possibilities from consideration [Jensen and Andersen 1990]. There are many realizations of these principles today. For elimination algorithms — which rely heavily on factors and their operations — local structure permits one to have more compact representations of these factors than representations based on tables [Zhang and Poole 1996], leading to a more efficient implementation of the elimination process. One example of this would be the use of Algebraic Decision Diagrams [R.I. Bahar et al. 1993] and associated operations to represent and manipulate factors; see Figure 5. For conditioning algorithms, local structure reduces the number of cases one needs to consider during inference and the number of sub-computations one needs to cache. As an example of the first, suppose that we have an and-gate whose output and one of its inputs belong to a loop cutset. When conditioning the output on 1, both inputs must be 1 as well. Hence, there is no need to consider multiple values for the input in this case during the conditioning process [Allen and Darwiche 2003]. This would no longer be true, however, if we had an or-gate. Moreover, the difference between the two cases is only visible if we exploit the local structure of corresponding Bayesian networks.

Another effective technique for exploiting local structure, which proved to be a turning point in speeding up inference, is based on encoding Bayesian networks using logical constraints and then applying logical inference techniques to the resulting knowledge base [Darwiche 2002]. One can indeed efficiently encode the network structure and some of its local structure, including determinism, using knowledge bases in conjunctive normal form (CNF). One can then either compile the CNF to produce a circuit representation of the Bayesian network (see below), or apply model counting techniques and use the results to recover answers to probabilistic queries [Sang, Beame, and Kautz 2005].

Realizations of the above techniques became practically viable long after the initial observations about local structure, but have allowed one to reason efficiently with some networks whose treewidth can be quite large (e.g., [Chavira, Darwiche, and Jaeger 2006]). Although there is some understanding of the kind of networks that tend to lend themselves to these techniques, we still do not have strong theoretical results that characterize these classes of networks and the savings that one may expect from exploiting their local structure. Moreover, not enough work exists on complexity measures that are sensitive to both network structure and parameters (the treewidth is only sensitive to structure).

One step in this direction has been the use of arithmetic circuits to compactly represent the probability distributions of Bayesian networks [Darwiche 2003]. This

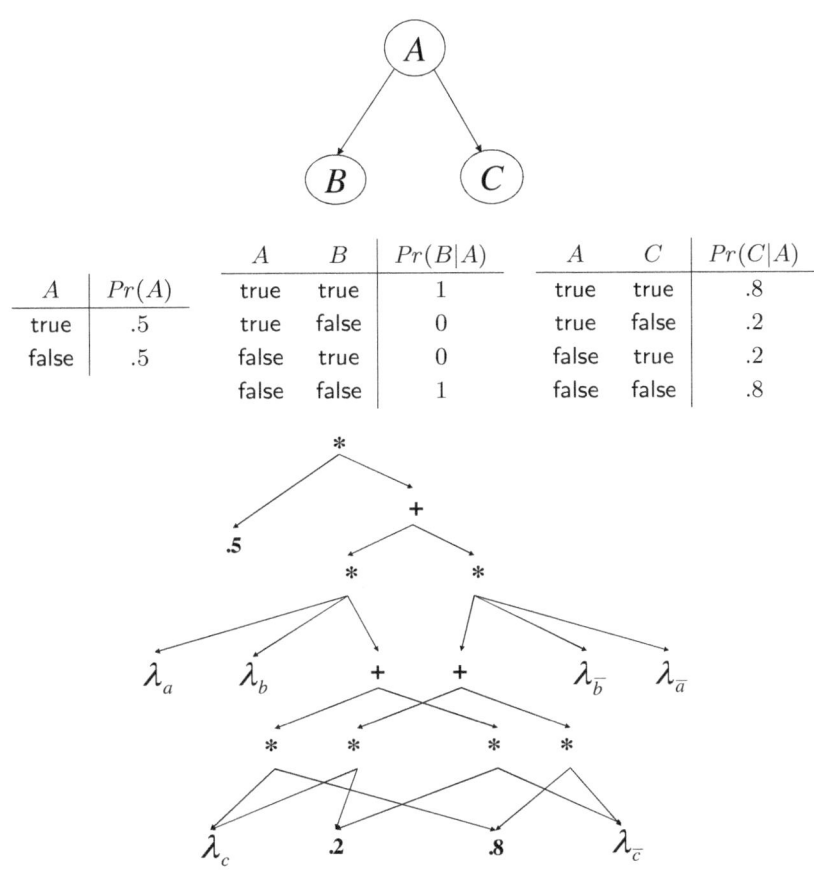

Figure 6. A Bayesian network and a corresponding arithmetic circuit.

representation is sensitive to both network topology and local structure, therefore, allowing for compact circuit representations in some cases where the treewidth of the network can be quite large; see Figure 6. Given a circuit representation, inference can be performed quite efficiently through simple circuit evaluation and differentiation techniques. Hence, the size of a circuit representation can be viewed as an indicator of the complexity of inference with respect to the given network. Again, however, we do not have enough theoretical results to broadly predict the size of these circuit representations or bound the complexity of constructing them.[3]

## 6 More queries for Bayesian networks

Pearl introduced another computational problem for Bayesian networks, known as the MPE for Most Probable Explanations. The goal here is to find the most likely instantiation of the network variables, given that some of these variables are fixed

---

[3]Note, however, that an arithmetic circuit can always be constructed in time which is exponential only in the treewidth, given a jointree of corresponding width.

to some given value. Pearl actually proposed the first algorithm for this purpose, which was a variation on the polytree algorithm [Pearl 1987a].

A more general problem is MAP which stands for Maximum a Posteriori hypothesis. This problem searches for an instantiation of a subset of the network variables that is most probable. Interestingly, MAP and MPE are complete for two different complexity classes, which are also distinct from the class to which node marginals is complete for. In particular, given the standard assumptions of complexity theory, MPE is the easiest and MAP is the most difficult, with node marginals in the middle.[4]

The standard techniques based on variable elimination and conditioning can solve MPE and MAP as well [Dechter 1999]. MPE can be solved with the standard treewidth guarantee. MAP, however, has a worse complexity in terms of what is known as *constrained treewidth,* which depends on both the network topology and MAP variables (that is, variables for which we are trying to find a most likely instantiation of) [Park and Darwiche 2004]. The constrained treewidth can be much larger than treewidth, depending on the set of MAP variables.

MPE and MAP problems have search components which lend themselves to branch-and-bound techniques [Kask and Dechter 2001]. Over the years, many sophisticated MPE and MAP bounds have been introduced, allowing branch-and-bound solvers to prune the search space more effectively. Consequently, this allows one to solve some MPE and MAP problems efficiently, even when the network treewidth or constrained treewidth are relatively high. In fact, only relatively recently did practical MAP algorithms surface, due to some innovative bounds that were employed in branch-and-bound algorithms [Park and Darwiche 2003].

MPE algorithms have traditionally received more attention than MAP algorithms. Recently, techniques based on LP relaxations, in addition to reductions to the MAXSAT problem, have been employed successfully for solving MPE. LP relaxations are based on the observation that MPE has a straightforward formulation in terms of integer programming, which is known to be hard [Wainwright, Jaakkola, and Willsky 2005; Yanover, Meltzer, and Weiss 2006]. By relaxing the integral constraints, the problem becomes a linear program, which is tractable but provides only a bound for MPE. Work in this area has been focused on techniques that compensate partially for the lost integral constraints using larger linear programs, and on developing refined algorithms for handling the resulting "specialized" linear programs.[5] The MAXSAT problem has also been receiving a lot of attention in the logic community [Bonet, Levy, and Manyà 2007; Larrosa, Heras, and de Givry 2008], which developed effective techniques for this purpose. In fact, reductions of certain MPE problems (those with excessive logical constraints) to MAXSAT seem

---

[4]The decision problems for MPE, node marginals, and MAP are $NP$–complete, $PP$–complete, and $NP^{PP}$–complete, respectively.

[5]In the community working on LP relaxations and related methods, "MAP" is used to mean "MPE" as we have discussed it in this article.

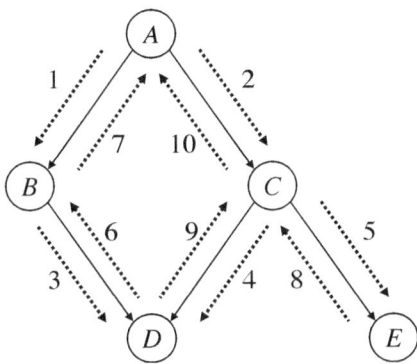

Figure 7. A Bayesian network annotated with an ordering of LBP messages (leading to a sequential message passing schedule).

to be the state of the art for some problems in this category.

## 7 Approximations may be good enough

In addition to work on exact inference algorithms for Bayesian networks, much work has also been dedicated to approximate inference algorithms which are generally more efficient but settle for less than accurate answers. Interestingly enough, the two major paradigms for approximate inference as practiced today were also initiated by Judea Pearl.

In particular, immediately after proposing the polytree algorithm, Pearl also proposed the use of Gibbs sampling as a method for approximate inference in Bayesian networks [Pearl 1987b]. This paper started a tradition in applying MCMC techniques for solving Bayesian networks and is considered as the founding paper in this direction. Further stochastic simulation methods were also proposed after realizing that sampling from Bayesian networks can be done easily by simply traversing the network structure [Henrion 1988].

In his seminal book on Bayesian networks [Pearl 1988], Pearl also proposed applying the belief propagation (polytree) algorithm to networks that have an arbitrary structure (in Exercise 4.7). This proposal required some initialization of network messages and entailed that a node may have to keep sending messages to each of its neighbors until convergence is reached (i.e., the messages are no longer changing); see Figure 7. Interestingly enough, such an algorithm, which is now known as *loopy belief propagation (LBP)*, tends to converge, yielding good approximations to a variety of problems. In fact, this particular algorithm was found to correspond to a state–of–the–art algorithm used in the channel coding community and today is widely viewed as a key method of choice for approximate inference [Frey and MacKay 1997].

This connection and the viability of LBP as an approximation algorithm came to light around the mid 1990s, almost a decade after Pearl first suggested the algorithm. Work on LBP and related methods has been dominating the field of approximate inference for more than a decade now. One of the central questions was: if LBP converges, what is it converging to? This question was answered in a number of ways [Minka 2001; Wainwright, Jaakkola, and Willsky 2003; Choi and Darwiche 2006], but the first characterization was put forth in [Yedidia, Freeman, and Weiss 2000]. According to this characterization, one can understand LBP as approximating the distribution of a Bayesian network by a distribution that has a polytree structure [Yedidia, Freeman, and Weiss 2003]. The iterations of the algorithm can then be interpreted as searching for the node marginals of that approximate distribution, while minimizing the KL–divergence between the original and approximate distributions.

LBP has actually two built-in components. The first corresponds to a particular approximation that it seeks, which is formally characterized as discussed before. The second component is a particular method for seeking the approximation, through a process of message passing. One can try to seek the same approximation using other optimization methods, which has also been the subject of much research. Even the message passing scheme leaves a lot of room for variation, which is captured formally using the notion of a message passing schedule — for example, messages can be passed sequentially, in parallel, or combinations therefore. One therefore talks about the "convergence" properties of such algorithms, where the goal is to seek methods that have better convergence properties.

LBP turns out to be an example of a more general class of approximation algorithms that poses the approximate inference problem as a constrained optimization problem. These methods, which are sometimes known as *variational algorithms*, assume a tractable class of distributions, and seeks to find an instance in this class that best fits the original distribution [Jordan et al. 1999; Jaakkola 2001]. For example, we may want to assume an approximating Bayesian network that is fully-disconnected, and that the distribution it induces should have as small a KL–divergence as possible, when compared to the distribution being approximated. The goal of the constrained optimization problem is then to find the CPT parameters of the approximate network that minimizes the KL–divergence between it and the original network (subject to the appropriate normalization constraints). Work in this area typically varies across two dimensions: proposing forms for the approximating distribution, and devising methods for solving the corresponding optimization problem. Moreover, by varying these two dimensions, we are given access to a spectrum of approximations, where we are able to trade the quality of an approximation with the complexity of computing it.

## 8 Closing Remarks

During the first decade or two after Pearl's introduction of Bayesian networks, inference research was very focused on exact algorithms. The efforts on these algorithms slowed down towards the mid to late 1990s, to pick up again early in the century. The slowdown was mostly due to the treewidth barrier, at a time where large enough networks were being constructed to make standard algorithms impractical at that time. The main developments leading to the revival of exact inference algorithms has been the extended reach of conditioning methods, the deeper understanding of elimination methods, and the more effective exploitation of local structure. Even though these developments have increased the reach of exact algorithms considerably, we still do not understand the extent to which this reach can be pushed further. In particular, the main hope appears to be in further utilization of local structure to speed up inference, but we clearly need better theories for providing guarantees on such speedups and a better characterization of the networks that lend themselves to such techniques.

On the approximate inference side, stochastic simulation methods witnessed a surge after the initial work on this subject, with continued interest throughout, yet not to the level enjoyed recently by methods based on belief propagation and related methods. This class of algorithms remains dominant, with many questions begging for answers. On the theoretical side, we do not seem to know enough on when approximations tend to give good answers, especially that this seems to be tied not only to the given network but also to the posed query. On the practical side, we have yet to translate some of the theoretical results on generalizations of belief propagation — which provides a spectrum that tradeoffs approximation quality with computational resources — into tools that are used routinely by practitioners.

There has been a lot of progress on inference in Bayesian networks since Pearl first made this computational problem relevant. There is clearly a lot more to be done as we seem to always exceed the ability of existing algorithms by building more complex networks. In my opinion, however, what is greatly missed since Pearl's initial work on this subject is his insistence on semantics, where he spared no effort in establishing connections to cognition, and in grounding the most intricate mathematical manipulations in human intuition. The derivation of the polytree algorithm stands as a great example of this research methodology, as it provided high level and cognitive interpretations of almost all intermediate computations performed by the algorithm. It is no wonder then that the polytree algorithm not only started the area of inference in Bayesian networks a few decades ago, but it also remains a basis for some of the latest developments and inspirations in this area of research.

**Acknowledgments:** I wish to thank Arthur Choi for many valuable discussions while writing this article.

# References

Allen, D. and A. Darwiche (2003). New advances in inference by recursive conditioning. In *Proceedings of the Conference on Uncertainty in Artificial Intelligence*, pp. 2–10.

Bertele, U. and F. Brioschi (1972). *Nonserial Dynamic Programming*. Academic Press.

Bonet, M. L., J. Levy, and F. Manyà (2007). Resolution for max-sat. *Artif. Intell. 171*(8-9), 606–618.

Boutilier, C., N. Friedman, M. Goldszmidt, and D. Koller (1996). Context-specific independence in Bayesian networks. In *Uncertainty in Artificial Intelligence: Proceedings of the Twelfth Conference (UAI-96)*, San Francisco, pp. 115–123. Morgan Kaufmann Publishers.

Chavira, M., A. Darwiche, and M. Jaeger (May 2006). Compiling relational Bayesian networks for exact inference. *International Journal of Approximate Reasoning 42*(1–2), 4–20.

Choi, A. and A. Darwiche (2006). An edge deletion semantics for belief propagation and its practical impact on approximation quality. In *Proceedings of the 21st National Conference on Artificial Intelligence (AAAI)*, pp. 1107–1114.

Darwiche, A. (2001). Recursive conditioning. *Artificial Intelligence 126*(1-2), 5–41.

Darwiche, A. (2002). A logical approach to factoring belief networks. In *Proceedings of KR*, pp. 409–420.

Darwiche, A. (2003). A differential approach to inference in Bayesian networks. *Journal of the ACM 50*(3), 280–305.

Darwiche, A. (2009). *Modeling and Reasoning with Bayesian Networks*. Cambridge University Press.

Dechter, R. (1996). Bucket elimination: A unifying framework for probabilistic inference. In *Proceedings of the 12th Conference on Uncertainty in Artificial Intelligence (UAI)*, pp. 211–219.

Dechter, R. (1999). Bucket elimination: A unifying framework for reasoning. *Artificial Intelligence 113*, 41–85.

Frey, B. J. and D. J. C. MacKay (1997). A revolution: Belief propagation in graphs with cycles. In *NIPS*, pp. 479–485.

Henrion, M. (1988). Propagating uncertainty in Bayesian networks by probalistic logic sampling. In *Uncertainty in Artificial Intelligence 2*, New York, N.Y., pp. 149–163. Elsevier Science Publishing Company, Inc.

Jaakkola, T. (2001). Tutorial on variational approximation methods. In D. Saad and M. Opper (Eds.), *Advanced Mean Field Methods*, Chapter 10, pp. 129–160. MIT Press.

Jensen, F. and S. K. Andersen (1990, July). Approximations in Bayesian belief universes for knowledge based systems. In *Proceedings of the Sixth Conference on Uncertainty in Artificial Intelligence (UAI)*, Cambridge, MA, pp. 162–169.

Jordan, M. I., Z. Ghahramani, T. Jaakkola, and L. K. Saul (1999). An introduction to variational methods for graphical models. *Machine Learning 37*(2), 183–233.

Kask, K. and R. Dechter (2001). A general scheme for automatic generation of search heuristics from specification dependencies. *Artificial Intelligence 129*, 91–131.

Kim, J. and J. Pearl (1983). A computational model for combined causal and diagnostic reasoning in inference systems. In *Proceedings IJCAI-83*, Karlsruhe, Germany, pp. 190–193.

Larrosa, J., F. Heras, and S. de Givry (2008). A logical approach to efficient max-sat solving. *Artif. Intell. 172*(2-3), 204–233.

Lauritzen, S. L. and D. J. Spiegelhalter (1988). Local computations with probabilities on graphical structures and their application to expert systems. *Journal of Royal Statistics Society, Series B 50*(2), 157–224.

Minka, T. P. (2001). *A family of algorithms for approximate Bayesian inference.* Ph.D. thesis, MIT.

Park, J. and A. Darwiche (2004). Complexity results and approximation strategies for MAP explanations. *Journal of Artificial Intelligence Research 21*, 101–133.

Park, J. D. and A. Darwiche (2003). Solving MAP exactly using systematic search. In *Proceedings of the 19th Conference on Uncertainty in Artificial Intelligence (UAI-03)*, Morgan Kaufmann Publishers San Francisco, California, pp. 459–468.

Pearl, J. (1982). Reverend Bayes on inference engines: A distributed hierarchical approach. In *Proceedings American Association of Artificial Intelligence National Conference on AI*, Pittsburgh, PA, pp. 133–136.

Pearl, J. (1986a). A constraint-propagation approach to probabilistic reasoning. In L. Kanal and J. Lemmer (Eds.), *Uncertainty in Artificial Intelligence*, pp. 357–369. Amsterdam, North Holland.

Pearl, J. (1986b). Fusion, propagation, and structuring in belief networks. *Artificial Intelligence 29*, 241–288.

Pearl, J. (1987a). Distributed revision of composite beliefs. *Artificial Intelligence 33*(2), 173–215.

Pearl, J. (1987b). Evidential reasoning using stochastic simulation of causal models. *Artificial Intelligence 32*, 245–257.

Pearl, J. (1988). *Probabilistic Reasoning in Intelligent Systems: Networks of Plausible Inference.* Morgan Kaufmann Publishers, Inc., San Mateo, California.

R.I. Bahar, E.A. Frohm, C.M. Gaona, G.D. Hachtel, E. Macii, A. Pardo, and F. Somenzi (1993). Algebraic Decision Diagrams and Their Applications. In *IEEE /ACM International Conference on CAD*, Santa Clara, California, pp. 188–191. IEEE Computer Society Press.

Robertson, N. and P. D. Seymour (1986). Graph minors. II. Algorithmic aspects of tree-width. *J. Algorithms 7*, 309–322.

Rumelhart, D. (1976). Toward an interactive model of reading. Technical Report CHIP-56, University of California, La Jolla, La Jolla, CA.

Sang, T., P. Beame, and H. Kautz (2005). Solving Bayesian networks by weighted model counting. In *Proceedings of the Twentieth National Conference on Artificial Intelligence (AAAI-05)*, Volume 1, pp. 475–482. AAAI Press.

Wainwright, M. J., T. Jaakkola, and A. S. Willsky (2003). Tree-based reparameterization framework for analysis of sum-product and related algorithms. *IEEE Transactions on Information Theory 49*(5), 1120–1146.

Wainwright, M. J., T. Jaakkola, and A. S. Willsky (2005). Map estimation via agreement on trees: message-passing and linear programming. *IEEE Transactions on Information Theory 51*(11), 3697–3717.

Yanover, C., T. Meltzer, and Y. Weiss (2006). Linear programming relaxations and belief propagation — an empirical study. *Journal of Machine Learning Research 7*, 1887–1907.

Yedidia, J. S., W. T. Freeman, and Y. Weiss (2000). Generalized belief propagation. In *NIPS*, pp. 689–695.

Yedidia, J. S., W. T. Freeman, and Y. Weiss (2003). Understanding belief propagation and its generalizations. In G. Lakemeyer and B. Nebel (Eds.), *Exploring Artificial Intelligence in the New Millennium*, Chapter 8, pp. 239–269. Morgan Kaufmann.

Zhang, N. L. and D. Poole (1994). A simple approach to Bayesian network computations. In *Proceedings of the Tenth Conference on Uncertainty in Artificial Intelligence (UAI)*, pp. 171–178.

Zhang, N. L. and D. Poole (1996). Exploiting causal independence in Bayesian network inference. *Journal of Artificial Intelligence Research 5*, 301–328.

# 8
# Graphical Models of the Visual Cortex

THOMAS DEAN

## 1 Pivotal Encounters with Judea

Post graduate school, three chance encounters reshaped my academic career, and all three involved Judea Pearl directly or otherwise. The first encounter was meeting Judea on a visit to the UCLA campus at a time when I was developing what I called temporal Bayesian networks and would later be called *dynamic belief networks* (an unfortunate choice of names for reasons I'll get to shortly). Judea was writing his book *Probabilistic Reasoning in Intelligent Systems: Networks of Plausible Inference* [1988] and his enthusiasm for the subject matter was positively infectious. I determined from that meeting that I was clueless about all things probabilistic and proceeded to read each of Judea's latest papers on Bayesian networks multiple times, gaining an initial understanding of joint and marginal probabilities, conditional independence, etc. In those days, a thorough grounding in probability and statistics was rarely encouraged for graduate students working in artificial intelligence.

The second encounter was with Michael Jordan at a conference where he asked me a question that I was at a loss to answer and made it clear to me that I didn't really understand Bayesian probability theory at all, despite what I'd picked up from Judea's papers. My reaction to that encounter was to read Judea's book cover to cover and discover the work of I.J. Good. Despite being a math major and having met I.J. Good at Virginia Tech where I was an undergraduate and Good was a professor of statistics, I never took a course in probability or statistics. My embarrassment at being flummoxed by Mike's question forced me to initiate a crash course in probability theory based on the textbooks of Morris DeGroot [1970, 1986]. I didn't recognize it at the time, but Judea, Mike and like-minded researchers in central areas of artificial intelligence were in the vanguard of those changing the landscape of our discipline.

The third encounter was with David Mumford when our paths crossed in the midst of a tenure hearing at Brown University and David told me of his work on models of the visual cortex. I read David's paper with Tai Sing Lee [2003] as well as David's earlier related work [1991, 1992] and naïvely set out to implement their ideas as a probabilistic graphical model [Dean 2005]. Indeed, I wanted to extend their work since it did not address the representation of time passing, and I was interested in building a model that dealt with how a robot might make sense of its observations as it explores its environment.

Moreover, the theory makes no mention of how a robot might learn such a model, and, from years of working with robots, I was convinced that building a model by hand would turn out to be a lot of work and very likely prove to be unsuccessful. Here it was Judea's graphical-models perspective that, initially, made it easy for me to think about David's work, and, later, extend it. I also came to appreciate the relevance of Judea's work on causality and, in particular, the role of intervention in thinking about how biological systems engage the world to resolve perceptual ambiguity.

This chapter concerns how probabilistic graphical models might be used to model the visual cortex, and how the challenges faced in developing such models suggest areas where current theory falls short and might be extended. A graphical model is a useful formalism for compactly describing a joint probability distribution characterized by very large number of random variables. We are taking what is known about the anatomy and physiology of the primate visual cortex and attempting to apply that knowledge to construct probabilistic graphical models that we can ultimately use to simulate some functions of primate vision. It may be that the resulting probabilistic model also captures some important characteristics of individual neurons or their ensembles. For practical purposes, this need not be the case, though clearly we believe there are potential advantages to incorporating some lessons from biology into our models. Graphical models also suggest, but do not dictate, how one might use such a model along with various algorithms and computing hardware to perform inference and thereby carry out practical simulations. It is this latter use of graphical models that we refer to when we talk about implementing a model of the visual cortex.

## 2  Primate Visual Cortex

Visual information processing starts in the retina and is routed via the optic tract to the lateral geniculate nuclei (LGN) and then on to the striate cortex also known as visual area one (V1) located in the occipital lobe at the rear of the cortex. There are two primary visual pathways in the primate cortex: The ventral pathway leads from the occipital lobe into the temporal lobe where association areas in the inferotemporal cortex combine visual information with information originating from the auditory cortex. The dorsal pathway leads from the occipital to the parietal lobe which, among other functions, facilitates navigation and manipulation by integrating visual, tactile and proprioceptive signals to provide our spatial sense and perception of shape.

It is only in the earliest portion of these pathways that we have any reasonably accurate understanding of how visual information is processed, and even in the very earliest areas, the striate cortex, our understanding is spotty and subject to debate. It seems that cells in V1 are mapped to cells in the retina so as to preserve spatial relationships, and are tuned to respond to stimuli that appear roughly like oriented bars. Hubel and Wiesel's research on macaque monkeys provides evidence for and

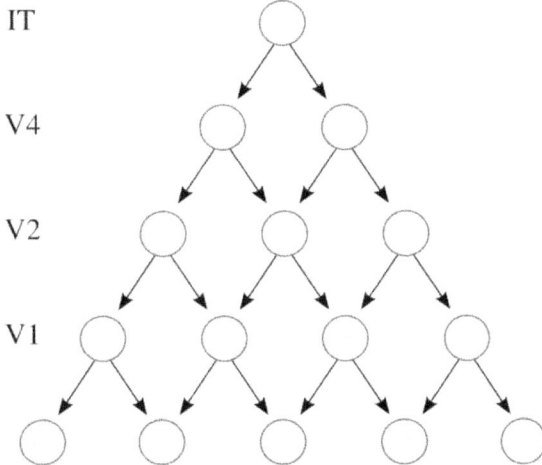

Figure 1. A simple hierarchical model of the ventral visual pathway.

subsequent studies confirm the latter characterization of function in primates [1962, 1968]. That said, there is still a good deal that we don't know about visual processing in V1 [Olshausen and Field 2005], and our understanding gets murkier as we progress along these pathways.

It is said that the ventral pathway is responsible for identifying "what" objects we see, and the dorsal for identifying "where" in our visual field these objects and their various parts are located and "how" we might interact with them by grasping, avoiding, etc. This is sufficiently vague that, given our current understanding of visual processing, it is probably a useful rule of thumb. However, there is ample evidence [Konen and Kastner 2008] to suggest that the "what", "where" and "how" are commingled via a myriad of connections and it is misleading to think of visual processing as a pipeline that leads in pure feed-forward fashion from simple features that subtend small portions of the visual field to more complex features that subtend greater and greater spatial (and temporal) extent. Indeed, there is no conclusive evidence that the brain is organized as a hierarchy of features, despite this being an elegant and comforting hypothesis to entertain.

## 3 Static Graphical Models

Figure 1 depicts a simple graphical model of the ventral visual pathway, where the nodes in the bottom layer are meant to model retinal ganglion cells. Nodes in the second layer model cells in the striate cortex which is also known as Brodmann's area 17 or V1. The next layer corresponds to Brodmann's area 18 or V2 which is responsive to somewhat more complex patterns than V1. The penultimate layer encodes V4 which is tuned to object features of intermediate complexity, and the

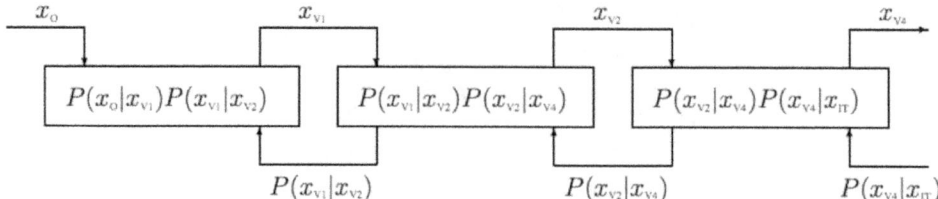

Figure 2. A schematic of the hierarchical Bayesian framework proposed by Lee and Mumford [2003]. The regions of the visual cortex are linked together in a Markov chain. The activity in the $i$th region is influenced by bottom-up feed-forward data $x_{i-1}$ and top-down probabilistic priors $P(x_i|x_{i+1})$ representing feedback from region $i+1$. The Markov property plays an important computational role by allowing units to depend only on their immediate neighbors in the Markov chain.

final layer represents inferotemporal cortex or IT which apparently responds to complex shapes. To get a more realistic picture, imagine that each layer represents a two-dimensional map with correspondence to the surface of the retina. Let's consider several ways in which this simple model falls short of the mark.

The graph in Figure 1 has no edges between nodes in the same layer. We know adjacent cells within each of V1, V2, V4 and IT communicate via many lateral connections. Given such connectivity, it would seem that two nodes representing adjacent cells in a layer are dependent in a statistical sense. However, the model as it stands has the property that any two nodes in a given layer are independent of one another when conditioned on the nodes in the layer immediately above. It turns out that it is very difficult to capture complex spatial relationships among adjacent regions of images using a graphical model having the intra-layer conditional independence implicit in model shown in Figure 1.

How might we capture the statistical properties of adjacent cells in the same cortical layer in a graphical model? First, note that our graphical model, while more complex than a tree, is simpler than an arbitrary directed graph being acyclic. Unfortunately, it is difficult to devise a plausible scheme to connect vertices within layers using directed edges and avoid cycles in the graph, and so we won't even attempt this. One possibility is that we model each layer as a Markov random field with the connectivity of a regular grid, but retain the directional edges between layers. The advantage of this approach is that the graph as a whole is a Markov chain and it is relatively simple to write down the joint distribution. Using the chain rule and the assumption made explicit in the graph shown in Figure 1 that each variable in the sequence $(x_\mathrm{O}, x_\mathrm{V1}, x_\mathrm{V2}, x_\mathrm{V4}, x_\mathrm{IT})$ is independent of the other variables given its immediate neighbors in the sequence, we write the equation relating the one retinal and three cortical regions as

$$P(x_\mathrm{O}, x_\mathrm{V1}, x_\mathrm{V2}, x_\mathrm{V4}, x_\mathrm{IT}) = P(x_\mathrm{O}, x_\mathrm{V1})P(x_\mathrm{V1}, x_\mathrm{V2})P(x_\mathrm{V2}, x_\mathrm{V4})P(x_\mathrm{V4}, x_\mathrm{IT})P(x_\mathrm{IT})$$

where $x_\mathrm{O}$ represents the retinal or *observation* layer. Moreover, we know that, although the edges all point in the same direction, information flows both ways in the hierarchy via Bayes rule (see Figure 2).

Despite the apparent simplicity when we collapse each layer of variables into a single, joint variable, exact inference in such a model is intractable. One might imagine, however, using a variant of the forward-backward algorithm to approximate the joint distribution over all variables. Such an algorithm might work one layer at a time, by isolating each layer in turn, performing an approximation on the isolated Markov network using Gibbs sampling or mean-field approximation, propagating the result either forward or backward and repeating until convergence. Simon Osindero and Geoff Hinton [2008] experimented with just such a model and demonstrated that it works reasonably well at capturing the statistics of patches of natural images.

One major problem with such a graphical model as a model of the visual cortex is that the Markov property of the collapsed-layer simplification fails to capture the inter-layer dependencies implied by the connections observed in the visual cortex. In the cortex as in the rest of the brain, connections correspond to the dendritic branches of one neuron connected at a synaptic cleft to the axonal trunk of a second neuron. We are reasonably comfortable modeling such a *cellular edge* as an edge in a probabilistic graphical model because for every cellular edge running forward along the visual pathways starting from V1 there is likely at least one and probably quite a few cellular edges leading backward along the visual pathways. Not only do these backward-pointing cellular edges far outnumber the forward-pointing ones, they also pay no heed to the Markov property, typically spanning several layers of our erstwhile simple hierarchy. Jin and Geman [2006] address this very problem in their hierarchical, compositional model, but at a considerable computational price. Advances in the development of adaptive Monte Carlo Markov chain (MCMC) algorithms may make inference in such graphical models more practical, but, for the time being, inference on graphical models of a size comparable to the number of neurons in the visual cortex remains out of reach.

## 4 Temporal Relationships

Each neuron in the visual cortex indirectly receives input from some, typically contiguous, region of retinal ganglion cells. This region is called the neuron's *receptive field*. By introducing lags and thereby retaining traces of earlier stimuli, a neuron can be said to have a receptive field that spans both space and time — it has a *spatiotemporal* receptive field. A large fraction of the cells in visual cortex and V1 in particular have spatiotemporal receptive fields. Humans, like most animals, are very attentive to motion and routinely exploit motion to resolve visual ambiguity,

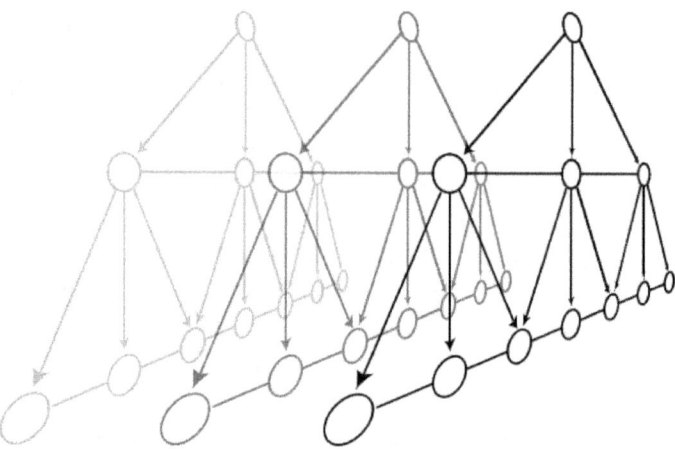

Figure 3. A cartoon of the hierarchical hidden Markov model described in [Dean 2006] showing the same basic structure as in Figure 1, but with additional undirected, intra-layer edges, and replicated for some number of time steps to form a hierarchy of hidden Markov models, so that nodes in a each layer represent different spatial and temporal extent. Not shown, but present in the full model, are edges that span the temporal slices thus modeling neural circuits that have spatiotemporal receptive fields.

and, generally, deal with the four-dimensional, space-time continuum in which we live.

Figure 3 depicts the obvious temporal analog of Figure 1. Dean [2006] presents a model of visual cortex based on the Hierarchical Hidden Markov Model of Fine *et al* [1998]. In the model described in [2006], nodes have edges that span both nodes within the layers of individual time slices and nodes that reside within the same layer of adjacent time slices.

The ability to represent the passage of time is clearly important in enabling us to make predictions and plan our actions, but it also allows us to expedite learning. The cortex evolved to take advantage of *temporal coherence*, the property that, for the most part, the appearance of the objects present in our visual field, animate or otherwise, do not change a great deal from one instant to the next. We can exploit this property to learn about the stable features of our environment. Földiák [Földiák 1991] describes a biologically plausible theory of how neural circuits might exploit temporal coherence to learn useful features by looking for signals that change slowly over time. Wiskott and Sejnowski [2002], Hyvärinen *et al* [2003], George and Hawkins [2005], Dean [2006] and others have proposed various algorithms for improving on this same basic idea.

Using proximity in space and time to group similar visual features was recognized early on by the Gestalt psychologists, and it is one of many characteristics of our visual experience that biology has evolved to exploit to its advantage. However, in this chapter, I want to explore a different facet of how we make sense of and, in some cases, take advantage of spatial and temporal structure to survive and thrive, and how these aspects of our environment offer new challenges for applying graphical models.

## 5 Dynamic Graphical Models

Whether called temporal Bayesian networks [Dean and Wellman 1991] or dynamic Bayesian networks [Russell and Norvig 2003], these graphical models are designed to model properties of our environment that change over time and the events that precipitate those changes. The networks themselves are not dynamic: the numbers of nodes and edges, and the distributions that quantify the dependencies among the random variables that correspond to the nodes are fixed. At first blush, graphical models may seem a poor choice to model the neural substrate of the visual cortex which is anything but static. However, while the graph that comprises a graphical model is fixed, a graphical model can be used to represent processes that are highly dynamic, and contingent on the assignments to observed variables in the model. In the remainder of this section, we describe characteristics of the visual system that challenge our efforts to model the underlying processes required to simulate primate vision well enough to perform such tasks such as object recognition and robot navigation.

The retina and the muscles that control the shape of the lens and the position of the eyes relative to one another and the head comprise a complex system for acquiring and processing visual information. A mosaic of photoreceptors activate several layers of cells, the final layer of which consists of retinal ganglion cells whose axons comprise the optic nerve. This multi-layer extension of the brain performs a range of complex computations ranging from light-dark adaptation to local contrast normalization [Brady and Field 2000]. The information transmitted along the optic tract is already the product of significant computational processing.

Visual information is *retinotopically* mapped from the retinal surface to area V1 so as to preserve the spatial relationships among patches on the retina that comprise the receptive fields of V1 cells. These retinotopic mappings are primarily sorted out *in utero*, but the organization of the visual cortex continues to evolve significantly throughout development — this is particularly apparent when children are learning to read [Dehaene 2009]. Retinotopic maps in areas beyond V1 are more complicated and appear to serve purposes that relate to visual tasks, *e.g.*, the map in V2 anatomically divides the tissue responsible for processing the upper and lower parts of the visual fields. These retinotopic maps, particularly those in area V1, have led some computer-vision researchers to imagine that early visual processing proceeds via transformations on regular grid-like structures with cells analogous to pixels.

The fact is that our eyes, head, and the objects that we perceive are constantly

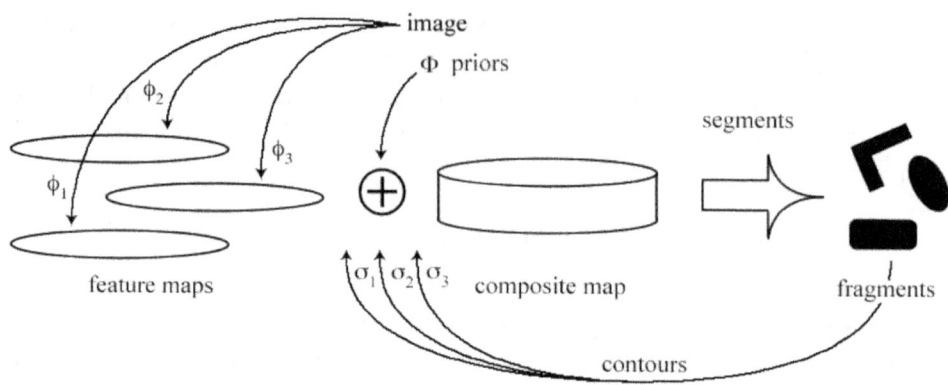

Figure 4. This graphic depicts a variant of the process of segmentation as proposed by Tenenbaum and Barrow [1977] in which the low-level features $\phi_i$ are used to construct feature maps that are combined with priors $\Phi$ derived from context to form composite maps which produce candidate fragments that suggest the boundaries or contours $\sigma_i$ of objects which in turn guide subsequent interpretation and assignment of fragments to objects.

in flux. Even with our head fixed while viewing a still image, our eyes quickly jump or *saccade* up to 90° of visual angle several times a second, and perform much smaller adjustments or *microsaccades* at around 30–50 Hertz. A two-week-old baby can already track objects by *smooth pursuit*, thereby keeping an object centered in the *fovea* which is the central, high-acuity and high-color-sensitivity portion of the retina. Both smooth pursuit and (macro) saccades are driven by attentional mechanisms which combine a bottom-up, small-image-patch, data-driven component with a top-down, whole-image-gist, prior-knowledge component. The important point here is that how we perceive the visual world has little resemblance — beyond the earliest processing stages — to a sequence of orderly transformations performed on a regular grid, and has everything to do with assembling a puzzle out of fragmentary glimpses snatched as our gaze quickly shifts relative to the frames of reference of our head, our body and the ground on which we stand.

Even if we concentrate on the hundred milliseconds or so during which the fovea remains focused on a small patch of a still image between saccades, the representation of this process as a graphical model becomes complicated as we abstract from the "pixel" level. Figure 4 depicts a process whereby the responses of low-level feature detectors are used to construct feature maps. Typically, the feature detectors report information about intensity, color, texture, etc. These maps are combined so that every location in an image is summarized by a vector of features. In bottom-up segmentation, such summaries alone are used to aggregate locations into segments that correspond to object surfaces or at least respect object boundaries. Tenenbaum

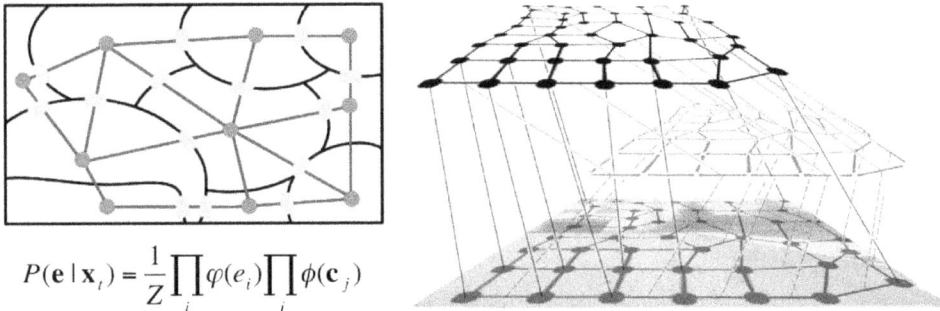

Figure 5. The graphic in the upper left depicts an over-segmentation of an image superimposed with a graph whose nodes (depicted as circles) correspond to segments or *superpixels* and whose edges (marked with rectangular boxes) correspond to boundaries between segments. The graphic on the right (adapted from Saxena *et al* [2007]) shows a graphical model in the form of multi-layer factor graph where nodes in the lowest layer correspond to superpixels, those in the second layer to boundary classes and those in the top to object surfaces.

and Barrow [1977] suggested that this low-level information has to be supplemented by *prior* knowledge, which provides a more global context in which to interpret the low-level information, and combined in an iterative of process of agglomeration.

This process has been realized in a graphical model format using variants of factor graphs [Saxena, Chung, and Ng 2007] and conditional random fields [Hoiem, Efros, and Hebert 2007] and hierarchical Dirichlet-process hidden-Markov trees [Kivinen, Sudderth, and Jordan 2007]. Figure 5 characterizes the basic structure of the algorithm adopted by Saxena *et al* [2007] and by Hoiem *et al* [2007]. First, the raw image is divided into *superpixels* — regions of pixels homogeneous in their intensity, color or texture — which correspond to the segments obtained from an over-segmentation of the image that is assumed to respect image boundaries. Next, the superpixels are used to construct a graphical model consisting of nodes corresponding superpixels, information that pertains to their status as object (occlusion) boundaries, and their relationships *vis a vis* their contribution to the same objects. Inference serves to identify boundaries or the absence thereof, some superpixels can be merged, and the process repeated until no further merging is possible.

In the implementation of this process, the topology of the graphical model is generated on an image-by-image basis, and the iterative refinement process requires adjustments to the size and connectivity of the graph that are particular to the boundaries of object surfaces in the image being observed. Such a process can be implemented within a fixed-graph structure but the model of a dynamically changing graph is simple to implement and apply recursively. One advantage, however, of a graph in which the nodes in a fixed grid of nodes are dynamically assigned to segments as part of inference is that this model is potentially more elegant, and even biologically plausible, in that the recursive process might be represented as a single hierarchical graphical model allowing inference over the entire graph, rather than over sequences of ever more refined graphs.

The above discussion of segmentation is but one example in which nodes in a graphical model might serve as *generic* variables that are bound as required by circumstances. But perhaps this view is short sighted; why not just assume that there are enough nodes that every possible (visual) concept corresponds to a unique combination of existing nodes. In this view, visual interpretation is just mapping visual stimuli to the closest visual "memory". Given the combinatorics, the only way this could be accomplished is to use a hierarchy of features whose base layer consists of small image fragments at many different spatial scales, and all subsequent layers consist of compositions of features at layers lower in the hierarchy [Bienenstock and Geman 1995; Ullman and Soloviev 1999; Ullman, Vidal-Naquet, and Sali 2002]. This view accords well with the idea that most visual stimuli are not determined to be novel and, hence, we construct our reality from bits and pieces of existing memories [Hoffman 1998]. Our visual memories are so extensive that we can almost always create a plausible interpretation by recycling old memories. It may be that in some aspects of cognition we have to employ generic neural structures to perform the analog of binding variables, but for much of visual intelligence this may not be necessary given a large enough memory of reusable fragments. Which raises the question of how we might implement a graphical model that has anywhere near the capacity of the visual cortex.

## 6 Distributed Processing at Cortex Scale

The cortex consists of a layered sheet with a more-or-less uniform cellular structure. Neuroanatomists have identified what are called *columns* corresponding to groups of local cells running perpendicular to the cortical surface. Vernon Mountcastle [2003] writes "The basic unit of cortical operation is the *minicolumn* [...] [containing] on the order of 80–100 neurons [...] The minicolumn measures of the order of 40-50$\mu$ in transverse diameter, separated from adjacent minicolumns by vertical cell-sparse zones which vary in size in different cortical areas." These minicolumns are then grouped into cortical columns which "are formed by the binding together of many minicolumns by common input and short-range horizontal connections."

If we take the cortical column — not the minicolumn — as our basic computational module as in [Anderson and Sutton 1997], then the gross structure of the neocortex consists of a dense mat of inter-columnar connections in the outer-most layer of the cortex and another web of connections at the base of the columns. The inter-columnar connectivity is relatively sparse (something on the order of $10^{15}$ connections spanning approximately $10^{11}$ neurons) and there is evidence [Sporns and Zwi 2004] to suggest that the induced inter-columnar connection graph exhibits the properties of a *small-world graph* [Newman, Watts, and Strogatz 2002]. In particular, evidence suggests the inter-columnar connection graph has low diameter (the length of the longest shortest path separating a pair of vertices in the graph) thereby enabling relatively low-latency communication between any two cortical columns.

It is estimated that there are about a quarter of a billion neurons in the primary visual cortex — think V1 through V4 — counting both hemispheres, but probably only around a million or so cortical columns. If we could roughly model each cortical column with a handful of random variables, then it is at least conceivable that we could implement a graphical model of early vision.

To actually implement a graphical model of visual cortex using current technology, the computations would have to be distributed over many machines. Training such a model might not take as long as raising a child, but it could take many days — if not years — using the current computer technology, and, once trained, we presumably would like to apply the learned model for much longer. Given such extended intervals of training and application, since the mean-time-til-failure for the commodity-hardware-plus-software that comprise most distributed processing clusters is relatively short, we would have to allow for some means of periodically saving local state in the form of the parameters quantifying the model.

The data centers that power the search engines of Google, Yahoo! and Microsoft are the best bet that we currently have for such massive and long-lived computations. Software developed to run applications on such large server farms already have tools that could opportunistically allocate resources to modify the structure of graphical model in an analog of neurogenesis. These systems are also resistant to both software and equipment failures and capable of reallocating resources in the aftermath of catastrophic failure to mimic neural plasticity in the face of cell death.

In their current configuration, industrial data centers may not be well suited to the full range of human visual processing. Portions of the network that handle very early visual processing will undoubtedly require shorter latencies than is typical in such server farms, even among machines on the same rack connected with high-speed Ethernet. Riesenhuber and Poggio [1999] use the term *immediate recognition* to refer to object recognition and scene categorization that occur in the first 100-200ms or so from the onset of the stimuli. In that short span of time — less than the time it takes for a typical saccade, we do an incredibly accurate job of recognizing objects and inferring the gist of a scene. The timing suggests that only a few steps of neural processing are involved in this form of recognition, assuming 10–20ms per synaptic transmission, though given the small diameter of the inter-columnar connection graph, many millions of neurons are likely involved in the processing. It would seem that at least the earliest stages of visual processing will have to be carried out in architectures capable of performing an enormous number of computations involving a large amount of state — corresponding to existing pattern memory — with very low latencies among the processing units. Hybrid architectures that combine conventional processors with co-processors that provide fast matrix-matrix and matrix-vector operations will likely be necessary to handle even a single video stream in real-time.

Geoff Hinton [2005, 2006] has suggested that a single learning rule and a relatively simple layer-by-layer method of training suffices for learning invariant features in text, images, sound and even video. Yoshua Bengio, Yann LeCun and others have also had success with such models [LeCun and Bengio 1995; Bengio, Lamblin, Popovici, and Larochelle 2007; Ranzato, Boureau, and LeCun 2007]. Hyvärinen *et al* [2003], Bruno Olshausen and Charles Cadieu [2007, 2008], Dean *et al* [2009] and others have developed hierarchical generative models to learn sparse codes resembling the responses of neurons in the medial temporal cortex of the dorsal pathway. In each case, the relevant computations can be most easily characterized in terms of linear algebra and implemented using fast vector-matrix operations best carried out on a single machine with lots of memory and many cores (graphics processors are particularly well suited to this sort of computation).

A more vexing problem concerns how we might efficiently implement any of the current models of Hebbian learning in an architecture that spans tens of thousands of machines and incurs latencies measured in terms of milliseconds. Using super computers at the national labs, Eugene Izhikevich and Gerald Edelman [2008] have performed spike-level simulations of millions of so-called *leaky integrate and fire* neurons with fixed, static connections to study the dynamics of learning in such ensembles. Paul Rhodes and his team of researchers at Evolved Machines have taken things a step further in implementing a model that allows for the dynamic creation of edges by simulating dendritic tree growth and the chemical gradients that serve to implement Hebbian learning. In each case, the basic model for a neuron is incredibly simple when compared to the real biology. It is not at all surprising that Henry Markram and his colleagues at EPFL (Ecole Polytechnique Fédérale de Lausanne) require a powerful supercomputer to simulate even a single cortical column at the molecular level. In all three of these examples, the researchers use high-performance computing alternatives to the cluster-of-commodity-computers distributed architectures that characterize most industrial data warehouses. While the best computing architecture for simulating cortical models may not be clear, it is commonly believed that we either how have or soon will have the computing power to simulate significant portions of cortex at some level of abstraction. This assumes, of course, that we can figure out what the cortex is actually computing.

## 7 Beyond Early Visual Processing

The grid of columnar processing units which constitutes the primate cortex and the retinotopic maps that characterize the areas participating in early vision, might suggest more familiar engineered vision systems consisting of frame buffers and graphics processors. But this analogy doesn't even apply to the simplest case in which the human subject is staring at a static image. As pointed out earlier, our eyes make large — up to 90° of visual angle — movements several times a second and tiny adjustments much more often.

A typical saccade of, say, 18° of visual angle takes 60–80ms to complete [Harwood, Mezey, and Harris 1999], a period during which we are essentially blind. During the subsequent 200–500ms interval until the next saccade, the image on the fovea is relatively stable, accounting for small adjustments due to micro saccades. So even a rough model for the simplest sort of human visual processing has to be set against the background of two or three fixations per second, each spanning less than half a second, and separated by short — less than 1/10 of a second — periods of blindness.

During each fixation we have 200–500ms in which to make sense of the events projected on the fovea; simplifying enormously, that's time enough to view around 10–15 frames of a video shown at 30 frames per second. In most of our experience, during such a period there is a lot going on in our visual field; our eyes, head and body are often moving and the many objects in our field of view are also in movement, more often than not, moving independent of one another. Either by focusing on a small patch of an object that is motionless relative to our frame of reference or by performing smooth pursuit, we have a brief period in which to analyze what amounts to a very short movie as seen through a tiny aperture. Most individual neurons have receptive fields that span an even smaller spatial and temporal extent.

If we try to interpret movement with too restrictive a spatial extent, we can mistake the direction of travel of a small patch of texture. If we try to work on too restrictive a temporal extent, then we are inundated with small movements many of which are due to noise or uninteresting as they arise from the analog of smooth camera motion. During that half second or so we need to identify stable artifacts, consisting of the orientation, direction, velocity, etc., of small patches of texture and color, and then combine these artifacts to capture features of the somewhat larger region of the fovea we are fixating on. Such a combination need not entail recognizing shape; it could, for example, consist of identifying a set of candidate patches, that may or may not belong to the same object, and summarizing the processing performed during the fixation interval as a collection of statistics pertaining to such patches, including their relative — but not absolute — positions, velocities, etc.

In parallel with processing foveal stimuli, attentional machinery in several neural circuits and, in particular, the lateral intraparietal cortex — which is retinotopically mapped when the eyes are fixated — estimates the saliency of spatial locations throughout the retina, including its periphery where acuity and color sensitivity are poor. These estimates of "interestingness" are used to decide what location to saccade to next. The oculomotor system keeps track of the dislocations associated with each saccade, and this locational information can be fused together using statistics collected over a series of saccades. How such information is combined and the exact nature of the resulting internal representations is largely a mystery.

The main point of the above discussion is that, while human visual processing may begin early in the dorsal and ventral pathways with something vaguely related to computer image processing using a fixed, spatially-mapped grid of processing and memory units, it very quickly evolves into a process that requires us to combine disjoint intervals of relatively stable imagery into a pastiche from which we can infer properties critical to our survival. Imagine starting with a collection of snapshots taken through a telephoto lens rather than a single high-resolution image taken with a wide-angle lens. This is similar to what several popular web sites do with millions of random, uncalibrated tourist photos.

The neural substrate responsible for performing these combinations must be able to handle a wide range of temporal and spatial scales, numbers and arrangements of inferred parts and surfaces, and a myriad of possible distractions and clutter irrelevant to the task at hand. We know that this processing can be carried out on a more-or-less regular grid of processors — the arrangement of cortical columns is highly suggestive of such a grid. We are even starting to learn the major pathways — bundles of axons sheathed with myelin insulation to speed transmission — connecting these biological processors using diffusion-tensor-imaging techniques. What we don't know is how the cortex allocates its computational resources beyond those areas most directly tied to the peripheral nervous system and that are registered spatially with the locations of the sensors arrayed on the periphery.

From a purely theoretical standpoint, we can simulate any Turing machine with a large enough Boolean circuit, and we can approximate any first-order predicate logic representation that has a finite domain using a propositional representation. Even so, it seems unlikely that even the cortex, with its $10^{11}$ neurons and $10^{15}$ connections, has enough capacity to cover the combinatorially many possible arrangements of primitive features that are likely inferred in early vision. This implies that different portions of the cortex must be allocated dynamically to perform processing on very different arrangements of such features.

Bruno Olshausen [1993] theorized that neural circuits could be used to *route* information so that stimuli corresponding to objects and their parts could be transformed to a standard scale and pose, thereby simplifying pattern recognition. Such transformations could, in principle, be carried out by a graphical model. The neural circuitry that serves as the target of such transformations — think of it as a specialized frame buffer of sorts — could be allocated so that different regions are assigned to different parts — this allocation being an instance of the so-called symbol binding problem in connectionist models [Rumelhart and McClelland 1986] of distributed processing.

## 8 Escaping Retinotopic Tyranny

While much of the computational neuroscience of primate vision seems mired in the first 200 milliseconds or so of early vision when the stimulus is reasonably stable and the image registered on the fovea is mapped retinotopically to areas in V1 through V4, other research on the brain is revealing how we keep track of spatial relationships involving the frames of reference of our head, body, nearby objects, and the larger world in which we operate. The brain maintains detailed maps of the body and its surrounding physical space in the hippocampus and somatosensory, motor, and parietal cortex [Rizzolatti, Sinigaglia, and Anderson 2007; Blakeslee and Blakeslee 2007]. Recall that the dorsal — "where" and "how" — visual pathway leads to the parietal cortex, which plays an important role in visual attention and our perception of shape. These maps are dynamic, constantly adapting to changes in the body as well as reflecting both short- and long-term knowledge of our surroundings and related spatial relationships.

When attempting to gain insight from biology in building engineered vision systems, it is worth keeping in mind the basic *tasks* of evolved biological vision systems. Much of primate vision serves three broad and overlapping categories of tasks: recognition, navigation and manipulation. Recognition for foraging, mating, and a host of related social and survival tasks; navigation for exploration, localization and controlling territory; manipulation for grasping, climbing, throwing, tool making, etc.

The view [Lengyel 1998] that computer vision is really just inverse graphics ignores the fact that most of these tasks don't require you to be able to construct an accurate 3-D representation of your visual experience. For many recognition tasks it suffices to identify objects, faces, and landmarks you've seen before and associate with these items task-related knowledge gained from prior experience. Navigation to avoid obstacles requires the ability to determine some depth information but not necessarily to recover full 3-D structure. Manipulation is probably the most demanding task in terms of the richness of shape information apparently required, but even so it may be that we are over-emphasizing the role of static shape memory and under-emphasizing the role of dynamic visual servoing — see the discussion in [Rizzolatti, Sinigaglia, and Anderson 2007] for an excellent introduction to what is known about how we understand shape in terms of affordances for manipulation.

But when it comes right down to it, we don't know a great deal about how the visual system handles shape [Tarr and Bülthoff 1998] despite some tantalizing glimpses into what might be going on the inferotemporal cortex [Tsunoda, Yamane, Nishizaki, and Tanifuji 2001; Yamane, Tsunoda, Matsumoto, Phillips, and Tanifuji 2006]. Let's suppose for the sake of discussion that we can build a graphical model of the cortex that handles much of the low-level feature extraction managed by the early visual pathways (V1 through V4) using existing algorithms for performing inference on Markov and conditional random fields and related graphical models. How might we construct a graphical model that captures the part of visual memory that pools together all these low-level features to provide us with such a rich visual experience? Lacking any clear direction from computational neuroscience, we'll take a somewhat unorthodox path from here on out.

As mentioned earlier, several popular web sites offer rich visual experiences that are constructed by combining large image corpora. Photo-sharing web sites like Flickr, Google Picasa and Microsoft Live Labs PhotoSynth are able to combine

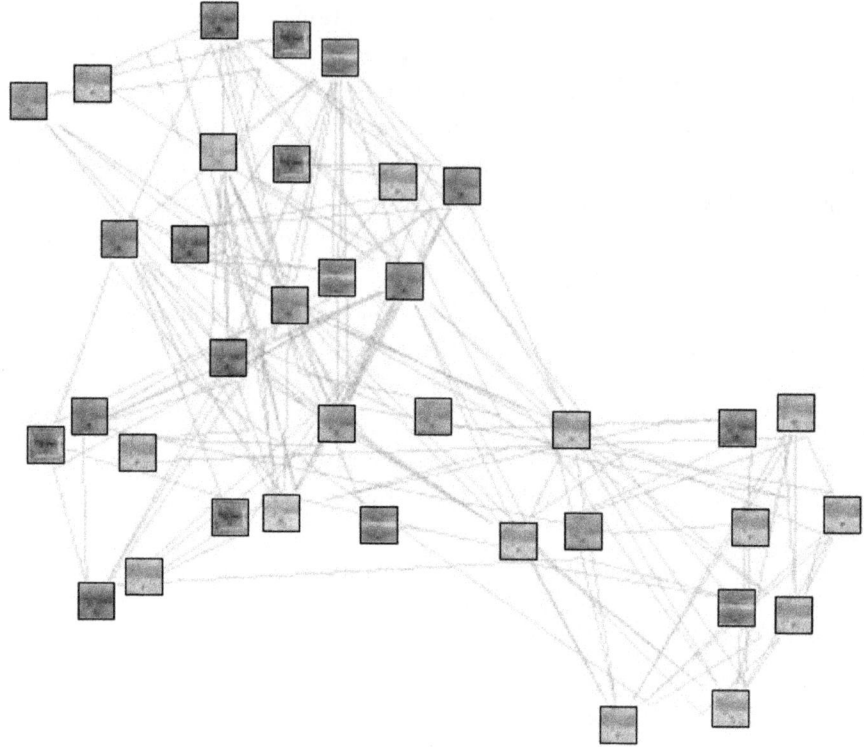

Figure 6. Graphical model with vertices corresponding to image patches and edges representing various relationships among the image patches (after Jing and Baluja [2008]).

multiple snapshots to construct views of popular landmarks from viewpoints not represented by any one snapshot. Google StreetView stitches together video and high-resolution wide-angle images to provide a seamless experience of virtually driving down a street in your home town. Google Earth combines images from satellite, aircraft, ground-based vehicles and, now, sonar-equipped ships and submersibles to allow us explore extensive regions of the planet. These applications are possible due to fast structure-from-motion algorithms that use reliably recoverable and locally distinctive features called *keypoints* extracted from pairs of images to align the images and stitch them together, blending the shared portions and adjusting the color and contrast of the composite to create the illusion of a single image. It is worth noting that these popular web sites facilitate the basic tasks of biological vision that were listed earlier: find me images of a particular famous face, show me a photo of Mount Rainier taken from a site in Tacoma, Washington, tell me what my hotel in New York would look like if I approached it from the direction of Penn Station.

What if the cortex simply memorizes every *novel* fixated foveal patch that spans some fixed-width receptive field and relates them by using low-level features extracted in V1 through V4 as keypoints to estimate geometric and other meaningful relationships among patches? The use of the word "novel" in this context is meant to convey that some method for statistical pooling of similar patches is required to avoid literally storing every possible patch. This is essentially what Jing and Baluja [2008] do by taking a large corpus of images, extracting low-level features from each image, and then quantifying the similarity between pairs of images by analyzing the features that they have in common. The result is a large graph whose vertices are images and whose edges quantify pair-wise similarity (see Figure 6). By using the low-level features as indices, Jing and Baluja only have to search a small subset of the possible pairs of images, and of those only the ones that pass a specified threshold for similarity are connected by edges. Jing and Baluja further enhance the graph by using a form of spectral graph analysis to rank images in much the same way as Google ranks web pages. Torralba *et al* [2007] have demonstrated that even small image patches contain a great deal of useful information, and furthermore that very large collections of images can be quickly and efficiently searched to retrieve semantically similar images given a target image as a query [Torralba, Fergus, and Weiss 2008].

In principle, such a graph could be represented as a probabilistic graphical model and the spectral analysis reformulated in terms of inference on graphical models. The process whereby the graph is grown over time, incorporating new images and new relationships, currently cannot be formulated as inference on a graphical model, but it is interesting to speculate about very large, yet finite graphs that could evolve over time in response to new evidence. Learning the densities used to quantify the edges in graphical models *can* can be formulated in terms of hyper-parameters directly incorporated into the model and carried out by traditional inference algorithms [Buntine 1994; Heckerman 1995]. Learning graphs whose size and topology change over time is somewhat more challenging to cast in terms of traditional methods for learning graphical models. Graph size is probably not the determining technical barrier however. Very large graphical models consisting of documents, queries, genes, and other entities are now quite common, and, while exact inference in such graphs is typically infeasible, approximate inference is often good enough to provide the foundation for industrial-strength tools.

Unfortunately, there is no way to tie up the many loose ends which have been left dangling in this short survey. Progress depends in part on our better understanding the brain and in particular the parts of the brain that are further from the periphery of the body where our senses are directly exposed to external stimuli. Neuroscience has made significant progress in understanding the brain at the cellular and molecular level, even to the point that we are now able to run large-scale simulations with some confidence that our models reflect important properties of the biology. Computational neuroscientists have also made considerable progress developing models — and graphical models in particular — that account for features that appear to play an important role in early visual processing. The barrier to further progress seems to be the same impediment that we run into in so many other areas of computer vision, machine learning and artificial intelligence more generally, namely the problem of representation. How and what does the brain represent about the blooming, buzzing world in which we are embedded? The answer to that question will take some time to figure out, but no doubt probabilistic graphical models will continue to provide a powerful tool in this inquiry, thanks in no small measure to the work of Judea Pearl, his students and his many collaborators.

# References

Anderson, J. and J. Sutton (1997). If we compute faster, do we understand better? *Behavior Ressearch Methods, Instruments and Computers 29*, 67–77.

Bengio, Y., P. Lamblin, D. Popovici, and H. Larochelle (2007). Greedy layer-wise training of deep networks. In *Advances in Neural Information Processing Systems 19*, pp. 153–160. Cambridge, MA: MIT Press.

Bienenstock, E. and S. Geman (1995). Compositionality in neural systems. In M. Arbib (Ed.), *The Handbook of Brain Theory and Neural Networks*, pp. 223–226. Bradford Books/MIT Press.

Blakeslee, S. and M. Blakeslee (2007). *The Body Has a Mind of Its Own*. Random House.

Brady, N. and D. J. Field (2000). Local contrast in natural images: normalisation and coding efficiency. *Perception 29*(9), 1041–1055.

Buntine, W. L. (1994). Operations for learning with graphical models. *Journal of Artificial Intelligence Research 2*, 159–225.

Cadieu, C. and B. Olshausen (2008). Learning transformational invariants from time-varying natural images. In D. Schuurmans and Y. Bengio (Eds.), *Advances in Neural Information Processing Systems 21*. Cambridge, MA: MIT Press.

Dean, T. (2005). A computational model of the cerebral cortex. In *Proceedings of AAAI-05*, Cambridge, Massachusetts, pp. 938–943. MIT Press.

Dean, T. (2006, August). Learning invariant features using inertial priors. *Annals of Mathematics and Artificial Intelligence 47*(3-4), 223–250.

Dean, T., G. Corrado, and R. Washington (2009, December). Recursive sparse, spatiotemporal coding. In *Proceedings of the Fifth IEEE International Workshop on Multimedia Information Processing and Retrieval*.

Dean, T. and M. Wellman (1991). *Planning and Control*. San Francisco, California: Morgan Kaufmann Publishers.

DeGroot, M. (1970). *Optimal Statistical Decisions*. New York: McGraw-Hill.

DeGroot, M. H. (1986). *Probability and Statistics*. Reading, MA: Second edition, Addison-Wesley.

Dehaene, S. (2009). *Reading in the Brain: The Science and Evolution of a Human Invention*. Viking Press.

Fine, S., Y. Singer, and N. Tishby (1998). The hierarchical hidden Markov model: Analysis and applications. *Machine Learning 32*(1), 41–62.

Földiák, P. (1991). Learning invariance from transformation sequences. *Neural Computation 3*, 194–200.

George, D. and J. Hawkins (2005). A hierarchical Bayesian model of invariant pattern recognition in the visual cortex. In *Proceedings of the International Joint Conference on Neural Networks*, Volume 3, pp. 1812–1817. IEEE.

Harwood, M. R., L. E. Mezey, and C. M. Harris (1999). The spectral main sequence of human saccades. *The Journal of Neuroscience 19*, 9098–9106.

Heckerman, D. (1995). A tutorial on learning Bayesian networks. Technical Report MSR-95-06, Microsoft Research.

Hinton, G. and R. Salakhutdinov (2006, July). Reducing the dimensionality of data with neural networks. *Science 313*(5786), 504–507.

Hinton, G. E. (2005). What kind of a graphical model is the brain? In *Proceedings of the 19th International Joint Conference on Artificial Intelligence*.

Hoffman, D. (1998). *Visual Intelligence: How We Create What we See*. New York, NY: W. W. Norton.

Hoiem, D., A. Efros, and M. Hebert (2007). Recovering surface layout from an image. *International Journal of Computer Vision 75*(1), 151–172.

Hubel, D. H. and T. N. Wiesel (1962). Receptive fields, binocular interaction and functional architecture in the cat's visual cortex. *Journal of Physiology 160*, 106–154.

Hubel, D. H. and T. N. Wiesel (1968). Receptive fields and functional architecture of monkey striate cortex. *Journal of Physiology 195*, 215–243.

Hyvärinen, A., J. Hurri, and J. Väyrynen (2003). Bubbles: a unifying framework for low-level statistical properties of natural image sequences. *Journal of the Optical Society of America 20*(7), 1237–1252.

Izhikevich, E. M. and G. M. Edelman (2008). Large-scale model of mammalian thalamocortical systems. *Proceedings of the National Academy of Science 105*(9), 3593–3598.

Jin, Y. and S. Geman (2006). Context and hierarchy in a probabilistic image model. In *Proceedings of the 2006 IEEE Conference on Computer Vision and Pattern Recognition*, Volume 2, pp. 2145–2152. IEEE Computer Society.

Jing, Y. and S. Baluja (2008). Pagerank for product image search. In *Proceedings of the 17th World Wide Web Conference*.

Kivinen, J. J., E. B. Sudderth, and M. I. Jordan (2007). Learning multiscale representations of natural scenes using dirichlet processes. In *Proceedings of the 11th IEEE International Conference on Computer Vision*.

Konen, C. S. and S. Kastner (2008). Two hierarchically organized neural systems for object information in human visual cortex. *Nature Neuroscience 11*(2), 224–231.

LeCun, Y. and Y. Bengio (1995). Convolutional networks for images, speech, and time-series. In M. Arbib (Ed.), *The Handbook of Brain Theory and Neural Networks*. Bradford Books/MIT Press.

Lee, T. S. and D. Mumford (2003). Hierarchical Bayesian inference in the visual cortex. *Journal of the Optical Society of America 2*(7), 1434–1448.

Lengyel, J. (1998). The convergence of graphics and vision. *Computer 31*(7), 46–53.

Mountcastle, V. B. (2003, January). Introduction to the special issue on computation in cortical columns. *Cerebral Cortex 13*(1), 2–4.

Mumford, D. (1991). On the computational architecture of the neocortex I: The role of the thalamo-cortical loop. *Biological Cybernetics 65*, 135–145.

Mumford, D. (1992). On the computational architecture of the neocortex II: The role of cortico-cortical loops. *Biological Cybernetics 66*, 241–251.

Newman, M., D. Watts, and S. Strogatz (2002). Random graph models of social networks. *Proceedings of the National Academy of Science 99*, 2566–2572.

Olshausen, B. and C. Cadieu (2007). Learning invariant and variant components of time-varying natural images. *Journal of Vision 7*(9), 964–964.

Olshausen, B. A., A. Anderson, and D. C. Van Essen (1993). A neurobiological model of visual attention and pattern recognition based on dynamic routing of information. *Journal of Neuroscience 13*(11), 4700–4719.

Olshausen, B. A. and D. J. Field (2005). How close are we to understanding V1? *Neural Computation 17*, 1665–1699.

Osindero, S. and G. Hinton (2008). Modeling image patches with a directed hierarchy of markov random fields. In J. Platt, D. Koller, Y. Singer, and S. Roweis (Eds.), *Advances in Neural Information Processing Systems 20*, pp. 1121–1128. Cambridge, MA: MIT Press.

Pearl, J. (1988). *Probabilistic Reasoning in Intelligent Systems: Networks of Plausible Inference*. San Francisco, California: Morgan Kaufmann.

Ranzato, M., Y. Boureau, and Y. LeCun (2007). Sparse feature learning for deep belief networks. In *Advances in Neural Information Processing Systems 20*. Cambridge, MA: MIT Press.

Riesenhuber, M. and T. Poggio (1999, November). Hierarchical models of object recognition in cortex. *Nature Neuroscience 2*(11), 1019–1025.

Rizzolatti, G., C. Sinigaglia, and F. Anderson (2007). *Mirrors in the Brain How Our Minds Share Actions, Emotions, and Experience*. Oxford, UK: Oxford University Press.

Rumelhart, D. E. and J. L. McClelland (Eds.) (1986). *Parallel Distributed Processing: Explorations in the Microstructure of Cognition, Volume I: Foundations*. Cambridge, Massachusetts: MIT Press.

Russell, S. and P. Norvig (2003). *Artificial Intelligence: A Modern Approach*. Upper Saddle River, NJ: Second edition, Prentice Hall.

Saxena, A., S. Chung, and A. Ng (2007). 3-D depth reconstruction from a single still image. *International Journal of Computer Vision 76*(1), 53–69.

Sporns, O. and J. D. Zwi (2004). The small world of the cerebral cortex. *Neuroinformatics 2*(2), 145–162.

Tarr, M. and H. Bülthoff (1998). Image-based object recognition in man, monkey and machine. *Cognition 67*, 1–20.

Tenenbaum, J. and H. Barrow (1977). Experiments in interpretation-guided segmentation. *Artificial Intelligence 8*, 241–277.

Torralba, A., R. Fergus, and W. Freeman (2007). Object and scene recognition in tiny images. *Journal of Vision 7*(9), 193–193.

Torralba, A., R. Fergus, and Y. Weiss (2008). Small codes and large image databases for recognition. In *Proceedings of IEEE Computer Vision and Pattern Recognition*, pp. 1–8. IEEE Computer Society.

Tsunoda, K., Y. Yamane, M. Nishizaki, and M. Tanifuji (2001). Complex objects are represented in macaque inferotemporal cortex by the combination of feature columns. *Nature Neuroscience 4*, 832–838.

Ullman, S. and S. Soloviev (1999). Computation of pattern invariance in brain-like structures. *Neural Networks 12*, 1021–1036.

Ullman, S., M. Vidal-Naquet, and E. Sali (2002). Visual features of intermediate complexity and their use in classification. *Nature Neuroscience 5*(7), 682–687.

Wiskott, L. and T. Sejnowski (2002). Slow feature analysis: Unsupervised learning of invariances. *Neural Computation 14*(4), 715–770.

Yamane, Y., K. Tsunoda, M. Matsumoto, N. A. Phillips, and M. Tanifuji (2006). Representation of the spatial relationship among object parts by neurons in macaque inferotemporal cortex. *Journal Neurophysiology 96*, 3147–3156.

# 9
# On the Power of Belief Propagation: A Constraint Propagation Perspective

R. Dechter, B. Bidyuk, R. Mateescu and E. Rollon

## 1 Introduction

In his seminal paper, Pearl [1986] introduced the notion of Bayesian networks and the first processing algorithm, *Belief Propagation (BP)*, that computes posterior marginals, called beliefs, for each variable when the network is singly connected. The paper provided the foundation for the whole area of Bayesian networks. It was the first in a series of influential papers by Pearl, his students and his collaborators that culminated a few years later in his book on probabilistic reasoning [Pearl 1988]. In his early paper Pearl showed that for singly connected networks (e.g., polytrees) the distributed message-passing algorithm converges to the correct marginals in a number of iterations equal to the diameter of the network. In his book Pearl goes further to suggest the use of BP for loopy networks as an approximation algorithm (see page 195 and exercise 4.7 in [Pearl 1988]). During the decade that followed researchers focused on extending BP to general loopy networks using two principles. The first is tree-clustering, namely, the transformation of a general network into a tree of large-domain variables called clusters on which BP can be applied. This led to the join-tree or junction-tree clustering and to the bucket-elimination schemes [Pearl 1988; Dechter 2003] whose time and space complexity is exponential in the tree-width of the network. The second principle is that of cutset-conditioning that decomposes the original network into a collection of independent singly-connected networks all of which must be processed by BP. The cutset-conditioning approach is time exponential in the network's loop-cutset size and require linear space [Pearl 1988; Dechter 2003].

The idea of applying belief propagation directly to multiply connected networks caught up only a decade after the book was published, when it was observed by researchers in coding theory that high performing probabilistic decoding algorithms such as turbo codes and low density parity-check codes, which significantly outperformed the best decoders at the time, are equivalent to an iterative application of Pearl's belief propagation algorithm [McEliece, MacKay, and Cheng 1998]. This success intrigued researchers and started massive explorations of the potential of these local computation algorithms for general applications. There is now a significant body of research seeking the understanding and improvement of the inference power of iterative belief propagation (IBP).

The early work on IBP showed its convergence for a single loop, provided empirical evidence of its successes and failures on various classes of networks [Rish, Kask, and Dechter 1998; Murphy, Weiss, and Jordan 2000] and explored the relationship between energy minimization and belief-propagation shedding light on convergence and stable points [Yedidia, Freeman, and Weiss 2000]. Current state of the art in convergence analysis are the works by [Ihler, Fisher, and Willsky 2005; Mooij and Kappen 2007] that characterize convergence in networks having no determinism. The work by [Roosta, Wainwright, and Sastry 2008] also includes an analysis of the possible effects of strong evidence on convergence which can act to suppress the effects of cycles. As far as accuracy, the work of [Ihler 2007] considers how weak potentials can make the graph sufficiently tree-like to provide error bounds, a work which is extended and improved in [Mooij and Kappen 2009]. For additional information see [Koller 2010].

While a significant progress has been made in understanding the relationship between belief propagation and energy minimization, and while many extensions and variations were proposed, some with remarkable performance (e.g., survey propagation for solving satisfiability for random SAT problems), the following questions remain even now:

- Why does belief propagation work so well on coding networks?

- Can we characterize additional classes of problems for which IBP is effective?

- Can we assess the quality of the algorithm's performance once and if it converges.

In this paper we try to shed light on the power (and limits) of belief propagation algorithms and on the above questions by explicating its relationship with constraint propagation algorithms such as arc-consistency. Our results are relevant primarily to networks that have determinism and extreme probabilities. Specifically, we show that: (1) Belief propagation converges for zero beliefs; (2) All IBP-inferred zero beliefs are correct; (3) IBP's power to infer zero beliefs is as weak and as strong as that of arc-consistency; (4) Evidence and inferred singleton beliefs act like cutsets during IBP's performance. From points (2) and (4) it follows that if the inferred evidence breaks all the cycles, then IBP converges to the exact beliefs for all variables.

Subsequently, we investigate empirically the behavior of IBP for inferred near-zero beliefs. Specifically, we explore the hypothesis that: (5) If IBP infers that the belief of a variable is close to zero then this inference is relatively accurate. We will see that while our empirical results support the hypothesis on benchmarks having no determinism, the results are quite mixed for networks with determinism.

Finally, (6) We investigate if variables that have extreme probabilities in all its domain values (i.e., extreme support) also nearly cut off information flow. If that hypothesis is true, whenever the set of variables with extreme support constitute a loop-cutset, IBP is likely to converge and, if the inferred beliefs for those variables are sound, it will converge to accurate beliefs throughout the network.

On coding networks that posses significant determinism, we do see this desired behavior. So, we could view this hypothesis as the first to provide a plausible explanation to the success of belief propagation on coding networks. In coding networks the channel noise is modeled through a normal distribution centered at the transmitted character and controlled by a small standard deviation. The problem is modeled as a layered belief network whose sink nodes are all evidence that transmit extreme support to their parents, which constitute all the rest of the variables. The remaining dependencies are functional and arc-consistency on this type of networks is strong and often complete. Alas, as we show, on some other deterministic networks IBP's performance inferring near zero values is utterly inaccurate, and therefore the strength of this explanation is questionable.

The paper is based for the most part on [Dechter and Mateescu 2003] and also on [Bidyuk and Dechter 2001]. The empirical portion of the paper includes significant new analysis of recent empirical evaluations carried on in UAI 2006 and UAI 2008[1].

## 2 Arc-consistency

DEFINITION 1 (constraint network). A constraint network $\mathcal{C}$ is a triple $\mathcal{C} = \langle X, D, C \rangle$, where $X = \{X_1, ..., X_n\}$ is a set of variables associated with a set of discrete-valued domains $D = \{D_1, ..., D_n\}$ and a set of constraints $C = \{C_1, ..., C_r\}$. Each constraint $C_i$ is a pair $\langle S_i, R_i \rangle$ where $R_i$ is a relation $R_i \subseteq D_{S_i}$ defined on a subset of variables $S_i \subseteq X$ and $D_{S_i}$ is the Cartesian product of the domains of variables $S_i$. The relation $R_i$ denotes all tuples of $D_{S_i}$ allowed by the constraint. The projection operator $\pi$ creates a new relation, $\pi_{S_j}(R_i) = \{x \mid x \in D_{S_j}$ and $\exists y, y \in D_{S_i \setminus S_j}$ and $x \cup y \in R_i\}$, where $S_j \subseteq S_i$. Constraints can be combined with the join operator $\bowtie$, resulting in a new relation, $R_i \bowtie R_j = \{x \mid x \in D_{S_i \cup S_j}$ and $\pi_{S_i}(x) \in R_i$ and $\pi_{S_j}(x) \in R_j\}$.

DEFINITION 2 (constraint satisfaction problem). The constraint satisfaction problem (CSP) defined over a constraint network $\mathcal{C} = \langle X, D, C \rangle$, is the task of finding a solution, that is, an assignment of values to all the variables $x = (x_1, ..., x_n), x_i \in D_i$, such that $\forall C_i \in C, \pi_{S_i}(x) \in R_i$. The set of all solutions of the constraint network $\mathcal{C}$ is $sol(\mathcal{C}) = \bowtie R_i$.

### 2.1 Describing Arc-Consistency Algorithms

*Arc-consistency* algorithms belong to the well-known class of constraint propagation algorithms [Mackworth 1977; Dechter 2003]. All constraint propagation algorithms are polynomial time algorithms that are at the center of constraint processing techniques.

DEFINITION 3 (arc-consistency). [Mackworth 1977] Given a binary constraint net-

---

[1] http://graphmod.ics.uci.edu/uai08/Evaluation/Report

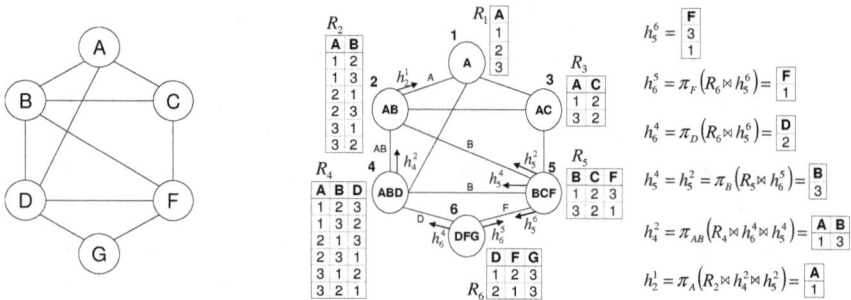

Figure 1. Part of the execution of RDAC algorithm

work $\mathcal{C} = \langle X, D, C \rangle$, $\mathcal{C}$ is arc-consistent iff for every binary constraint $R_i \in C$ s.t. $S_i = \{X_j, X_k\}$, every value $x_j \in D_j$ has a value $x_k \in D_k$ s.t. $(x_j, x_k) \in R_i$.

When a binary constraint network is not arc-consistent, arc-consistency algorithms remove values from the domains of the variables until an arc-consistent network is generated. A variety of such algorithms were developed over the past three decades [Dechter 2003]. We will consider here a simple and not the most efficient version, which we call *relational distributed arc-consistency* algorithm. Rather than defining it on binary constraint networks we will define it directly over the dual graph, extending the arc-consistency condition to non-binary networks.

DEFINITION 4 (dual graph). Given a set of functions/constraints $F = \{f_1, ..., f_r\}$ over scopes $S_1, ..., S_r$, the dual graph of $F$ is a graph $\mathcal{D}_F = (V, E, L)$ that associates a node with each function, namely $V = F$, and an arc connects any two nodes whose scope share a variable, $E = \{(f_i, f_j) | S_i \cap S_j \neq \emptyset\}$. $L$ is a set of labels for the arcs, where each arc is labeled by the shared variables of its nodes, $L = \{l_{ij} = S_i \cap S_j | (i, j) \in E\}$.

Algorithm *Relational distributed arc-consistency* (RDAC) is a message passing algorithm defined over the dual graph $\mathcal{D}_C$ of a constraint network $\mathcal{C} = \langle X, D, C \rangle$. It enforces what is known as *relational arc-consistency* [Dechter 2003]. Each node (a constraint) in $\mathcal{D}_{C_i}$, for a constraint $C_i \in \mathcal{C}$ maintains a current set of viable tuples $R_i$. Let $ne(i)$ be the set of neighbors of $C_i$ in $\mathcal{D}_C$. Every node $C_i$ sends a message to any node $C_j \in ne(i)$, which consists of the tuples over their label variables $l_{ij}$ that are allowed by the current relation $R_i$. Formally, let $R_i$ and $R_j$ be two constraints sharing scopes, whose arc in $\mathcal{D}_C$ is labeled by $l_{ij}$. The message that $R_i$ sends to $R_j$ denoted $h_i^j$ is defined by:

(1) $\quad h_i^j \leftarrow \pi_{l_{ij}}(R_i \bowtie (\bowtie_{k \in ne(i)} h_k^i))$

and each node updates its current relation according to:

(2) $\quad R_i \leftarrow R_i \bowtie (\bowtie_{k \in ne(i)} h_k^i)$

EXAMPLE 5. Figure 1 describes part of the execution of RDAC for a graph coloring problem, having the constraint graph shown on the left. All variables have the same domain, {1,2,3}, except for variable $C$ whose domain is 2, and variable $G$ whose domain is 3. The arcs correspond to *not equal* constraints, and the relations are $R_A$, $R_{AB}$, $R_{AC}$, $R_{ABD}$, $R_{BCF}$, $R_{DFG}$, where the subscript corresponds to their scopes. The dual graph of this problem is given on the right side of the figure, and each table shows the initial constraints (there are unary, binary and ternary constraints). To initialize the algorithm, the first messages sent out by each node are universal relations over the labels. For this example, RDAC actually solves the problem and finds the unique solution A=1, B=3, C=2, D=2, F=1, G=3.

Relational distributed arc-consistency algorithm converges after $O(r \cdot t)$ iterations to the largest relational arc-consistent network that is equivalent to the original network, where $r$ is the number of constraints and $t$ bounds the number of tuples in each constraint. Its complexity can be shown to be $O(r^2 t^2 \log t)$ [Dechter 2003].

## 3 Iterative Belief Propagation

DEFINITION 6 (belief network). A belief network is a quadruple $\mathcal{B} = \langle X, D, G, P \rangle$ where $X = \{X_1, \ldots, X_n\}$ is a set of random variables, $D = \{D_1, \ldots, D_n\}$ is the set of the corresponding domains, $G = (X, E)$ is a directed acyclic graph over $X$ and $P = \{p_1, \ldots, p_n\}$ is a set of conditional probability tables (CPTs) $p_i = P(X_i | pa(X_i))$, where $pa(X_i)$ are the parents of $X_i$ in $G$. The belief network represents a probability distribution over $X$ having the product form $P(x_1, \ldots, x_n) = \prod_{i=1}^{n} P(x_i | x_{pa(X_i)})$. An evidence set $e$ is an instantiated subset of variables. The family of $X_i$, denoted by $fa(X_i)$, includes $X_i$ and its parent variables. Namely, $fa(X_i) = \{X_i\} \cup pa(X_i)$.

DEFINITION 7 (belief updating problem). The belief updating problem defined over a belief network $\mathcal{B} = \langle X, D, G, P \rangle$ is the task of computing the posterior probability $P(Y|e)$ of *query* nodes $Y \subseteq X$ given evidence $e$. We will sometime denote by $P_B$ the exact probability according the Baysian network $B$. When $Y$ consists of a single variable $X_i$, $P_B(X_i|e)$ is also denoted as $Bel(X_i)$ and called belief, or posterior marginal, or just marginal.

### 3.1 Describing Iterative Belief Propagation

*Iterative belief propagation* (IBP) is an iterative application of Pearl's algorithm that was defined for poly-trees [Pearl 1988]. Since it is a distributed algorithm, it is well defined for any network. We will define IBP as operating over the belief network's dual join-graph.

DEFINITION 8 (dual join-graph). Given a belief network $\mathcal{B} = \langle X, D, G, P \rangle$, a dual join-graph is an arc subgraph of the dual graph $\mathcal{D}_B$ whose arc labels are subsets of the labels of $\mathcal{D}_B$ satisfying the *running intersection property*, namely, that any two nodes that share a variable in the dual join-graph be connected by a path of arcs whose labels *contain* the shared variable. An *arc-minimal* dual join-graph is one for which none of the labels can be further reduced while maintaining the running

---

**Algorithm IBP**
**Input:** An arc-labeled dual join-graph $DJ = (V, E, L)$ for a belief network $\mathcal{B} = \langle X, D, G, P \rangle$. Evidence $e$.
**Output:** An augmented graph whose nodes include the original CPTs and the messages received from neighbors. Approximations of $P(X_i|e)$ and $P(fa(X_i)|e)$, $\forall X_i \in X$.
Denote by: $h_u^v$ the message from $u$ to $v$; $ne(u)$ the neighbors of $u$ in $V$; $ne_v(u) = ne(u) - \{v\}$; $l_{uv}$ the label of $(u,v) \in E$; $elim(u,v) = fa(X_i) - fa(X_j)$, where $u$ and $v$ are the vertexs of family $fa(X_i)$ and $fa(X_j)$ in $DJ$, respectively.

- **One iteration of IBP**
  For every node $u$ in $DJ$ in a topological order and back, do:
  1. **Process observed variables**
     Assign evidence variables to each $p_i$ and remove them from the labeled arcs.
  2. **Compute and send to $v$ the function:**
     $$h_u^v = \sum_{elim(u,v)} (p_u \cdot \prod_{\{h_i^u, i \in ne_v(u)\}} h_i^u)$$
  **EndFor**
- **Compute approximations of $P(X_i|e)$ and $P(fa(X_i)|e)$:**
  For every $X_i \in X$ (let $u$ be the vertex of family $fa(X_i)$ in $DJ$), do:
  $P(fa(X_i)|e) = \alpha (\prod_{h_i^u, u \in ne(i)} h_i^u) \cdot p_u$
  $P(X_i|e) = \alpha \sum_{fa(X_i) - \{X_i\}} P(fa(X_i)|e)$
  **EndFor**

---

Figure 2. Algorithm Iterative Belief Propagation

intersection property.

In IBP each node in the dual join-graph sends a message over an adjacent arc whose scope is identical to its label. Pearl's original algorithm sends messages whose scopes are singleton variables only. It is easy to show that any dual graph (which itself is a dual join-graph) has an arc-minimal singleton dual join-graph which can be constructed directly by labeling the arc between the CPT of a variable and the CPT of its parent, by its parent variable. Algorithm IBP defined for any dual join-graph is given in Figure 2. One iteration of IBP is time and space linear in the size of the belief network, and when IBP is applied to the singleton labeled dual graph it coincides with Pearl's belief propagation. The inferred approximation of belief $P(X|e)$ output by IBP, will be denoted by $P_{IBP}(X|e)$.

## 4  Belief Propagation's Inferred Zeros

We will now make connections between distributed relational arc-consistency and iterative belief propagation. We first associate any belief network with a constraint network that captures its zero probability tuples and define algorithm IBP-RDAC, an IBP-like algorithm that achieves relational arc-consistency on the associated constraint network. Then, we show that IBP-RDAC and IBP are equivalent in terms of removing inconsistent domain values and computing zero marginal probabilities, respectively. Since arc-consistency algorithms are well understood, this correspondence between IBP-RDAC and IBP yields the main claims and provides insight into the behavior of IBP for inferred zero beliefs. In particular, this relationship justifies the iterative application of belief propagation algorithms, while also illuminates their "distance" from being complete.

More precisely, in this section we will show that: (a) If a variable-value pair is assessed in some iteration by IBP as having a zero-belief, it remains zero in subsequent iterations; (b) Any IBP-inferred zero-belief is correct with respect to the corresponding belief network's marginal; and (c) IBP converges in finite time for all its inferred zeros.

### 4.1 Flattening the Belief Network

Given a belief network $\mathcal{B} = \langle X, D, G, P \rangle$, we define the flattening of a belief network $\mathcal{B}$, called $flat(\mathcal{B})$, as the constraint network where all the zero entries in a probability table are removed from the corresponding relation. Formally,

DEFINITION 9 (flattening). *Given a belief network $\mathcal{B} = \langle X, D, G, P \rangle$, its flattening is a constraint network $flat(\mathcal{B}) = \langle X, D, flat(P) \rangle$. Each CPT $p_i \in P$ over $fa(X_i)$ is associated with a constraint $\langle S_i, R_i \rangle$ s.t. $S_i = fa(X_i)$ and $R_i = \{(x_i, x_{pa(X_i)}) \in D_{S_i} | P(x_i|x_{pa(X_i)}) > 0\}$. The set $flat(P)$ is the set of the constraints $\langle S_i, R_i \rangle$, $\forall p_i \in P$.*

EXAMPLE 10. Figure 3 shows (a) a belief network and (b) its corresponding flattening.

THEOREM 11. *Given a belief network $\mathcal{B} = \langle X, D, G, P \rangle$, where $X = \{X_1, \ldots, X_n\}$, for any tuple $x = (x_1, \ldots, x_n)$: $P_\mathcal{B}(x) > 0 \Leftrightarrow x \in sol(flat(\mathcal{B}))$, where $sol(flat(\mathcal{B}))$ is the set of solutions of $flat(B)$.*

**Proof.** $P_\mathcal{B}(x) > 0 \Leftrightarrow \Pi_{i=1}^n P(x_i|x_{pa(X_i)}) > 0 \Leftrightarrow \forall i \in \{1, \ldots, n\}, \ P(x_i|x_{pa(X_i)}) > 0 \Leftrightarrow \forall i \in \{1, \ldots, n\}, \ (x_i, x_{pa(X_i)}) \in R_{F_i} \Leftrightarrow x \in sol(flat(\mathcal{B}))$. □

Clearly this can extend to Bayesian networks with evidence:

COROLLARY 12. *Given a belief network $\mathcal{B} = \langle X, D, G, P \rangle$, and evidence $e$ $P_B(x|e) > 0 \Leftrightarrow x \in sol(flat(B) \wedge e)$.*

We next define algorithm IBP-RDAC and show that it achieves relational arc-consistency on the flat network.

DEFINITION 13 (Algorithm IBP-RDAC). *Given $\mathcal{B} = \langle X, D, G, P \rangle$ and evidence $e$, let $\mathcal{D}_\mathcal{B}$ be a dual join-graph and $\mathcal{D}_{flat(\mathcal{B})}$ be a corresponding dual join-graph of the constraint network $flat(\mathcal{B})$. Algorithm IBP-RDAC applied to $\mathcal{D}_{flat(\mathcal{B})}$ is defined using IBP's specification in Figure 2 with the following modifications:*

1. *Pre-processing evidence: when processing evidence, we remove from each $R_i \in flat(P)$ those tuples that do not agree with the assignments in evidence $e$.*

2. *Instead of $\prod$, we use the join operator $\bowtie$.*

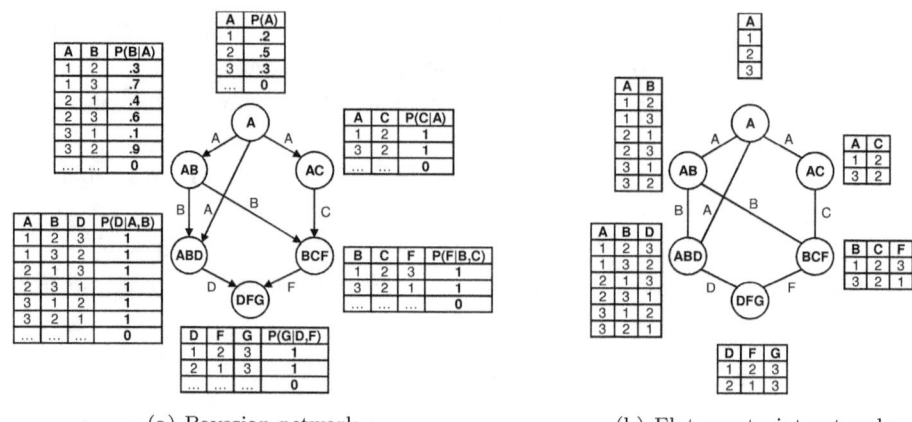

(a) Bayesian network  (b) Flat constraint network

Figure 3. Flattening of a Bayesian network

3. Instead of $\sum$, we use the projection operator $\pi$.

4. At the termination, we update the domains of variables by:

$$D_i \leftarrow D_i \cap \pi_{X_i}((\bowtie_{v \in ne(u)} h_{(v,u)}) \bowtie R_i)$$

By construction, it should be easy to see that,

PROPOSITION 14. *Given a belief network* $\mathcal{B} = \langle X, D, G, P \rangle$, *algorithm IBP-RDAC is identical to algorithm RDAC when applied to* $\mathcal{D}_{flat(\mathcal{B})}$. *Therefore, IBP-RDAC enforces relational arc-consistency over* $flat(\mathcal{B})$.

Due to the convergence of RDAC, we get that:

PROPOSITION 15. *Given a belief network* $\mathcal{B}$, *algorithm IBP-RDAC over* $flat(\mathcal{B})$ *converges in* $O(n \cdot t)$ *iterations, where n is the number of nodes in* $\mathcal{B}$ *and t is the maximum number of tuples over the labeling variables between two nodes that have positive probability.*

## 4.2 The Main Claim

In the following we will establish an equivalence between IBP and IBP-RDAC in terms of zero probabilities.

PROPOSITION 16. *When IBP and IBP-RDAC are applied in the same order of computation to* $\mathcal{B}$ *and* $flat(\mathcal{B})$ *respectively, the messages computed by IBP are identical to those computed by IBP-RDAC in terms of zero / non-zero probabilities. That is, for any pair of corresponding messages,* $h_{(u,v)}(t) \neq 0$ *in IBP iff* $t \in h_{(u,v)}$ *in IBP-RDAC.*

**Proof.** The proof is by induction. The base case is trivially true since messages $h$ in IBP are initialized to a uniform distribution and messages $h$ in IBP-RDAC are initialized to complete relations.

The induction step. Suppose that $h_{(u,v)}^{IBP}$ is the message sent from $u$ to $v$ by IBP. We will show that if $h_{(u,v)}^{IBP}(x) \neq 0$, then $x \in h_{(u,v)}^{IBP-RDAC}$ where $h_{(u,v)}^{IBP-RDAC}$ is the message sent by IBP-RDAC from $u$ to $v$. Assume that the claim holds for all messages received by $u$ from its neighbors. Let $f \in u$ in IBP and $R_f$ be the corresponding relation in IBP-RDAC, and $t$ be an assignment of values to variables in $elim(u,v)$. We have $h_{(u,v)}^{IBP}(x) \neq 0 \Leftrightarrow \sum_{elim(u,v)} \prod_f f(x) \neq 0 \Leftrightarrow \exists t, \prod_f f(x,t) \neq 0 \Leftrightarrow \exists t, \forall f, f(x,t) \neq 0 \Leftrightarrow \exists t, \forall f, \pi_{scope(R_f)}(x,t) \in R_f \Leftrightarrow \exists t, \pi_{elim(u,v)}(\bowtie_{R_f} \pi_{scope(R_f)}(x,t)) \in h_{(u,v)}^{IBP-RDAC} \Leftrightarrow x \in h_{(u,v)}^{IBP-RDAC}$. □

Moving from tuples to domain values, we will show that whenever IBP computes a marginal probability $P_{IBP}(x_i|e) = 0$, IBP-RDAC removes $x_i$ from the domain of variable $X_i$, and vice-versa.

PROPOSITION 17. *Given a belief network $\mathcal{B}$ and evidence $e$, IBP applied to $\mathcal{B}$ derives $P_{IBP}(x_i|e) = 0$ iff IBP-RDAC over $flat(\mathcal{B})$ decides that $x_i \notin D_i$.*

**Proof.** According to Proposition 16, the messages computed by IBP and IBP-RDAC are identical in terms of zero probabilities. Let $f \in cluster(u)$ in IBP and $R_f$ be the corresponding relation in IBP-RDAC, and $t$ be an assignment of values to variables in $\chi(u) \setminus X_i$. We will show that when IBP computes $P(X_i = x_i) = 0$ (upon convergence), then IBP-RDAC computes $x_i \notin D_i$. We have $P(X_i = x_i) = \sum_{X \setminus X_i} \prod_f f(x_i) = 0 \Leftrightarrow \forall t, \prod_f f(x_i,t) = 0 \Leftrightarrow \forall t, \exists f, f(x_i,t) = 0 \Leftrightarrow \forall t, \exists R_f, \pi_{scope(R_f)}(x_i,t) \notin R_f \Leftrightarrow \forall t, (x_i,t) \notin (\bowtie_{R_f} R_f(x_i,t)) \Leftrightarrow x_i \notin D_i \cap \pi_{X_i}(\bowtie_{R_f} R_f(x_i,t)) \Leftrightarrow x_i \notin D_i$. Since arc-consistency is sound, so is the decision of zero probabilities. □

We can now conclude that:

THEOREM 18. *Given evidence $e$, whenever IBP applied to $\mathcal{B}$ infers $P_{IBP}(x_i|e) = 0$, the marginal $Bel(x_i) = P_{\mathcal{B}}(x_i|e) = 0$.*

**Proof.** By Proposition 17, if IBP over $\mathcal{B}$ computes $P_{IBP}(x_i|e) = 0$, then IBP-RDAC over $flat(\mathcal{B})$ removes the value $x_i$ from the domain $D_i$. Therefore, $x_i \in D_i$ is a no-good of the constraint network $flat(\mathcal{B})$ and from Theorem 11 it follows that $Bel(x_i) = 0$. □

Next, we show that the time it takes IBP to find its inferred zeros is bounded.

PROPOSITION 19. *Given a belief network $\mathcal{B}$ and evidence $e$, IBP finds all its $x_i$ for which $P_{IBP}(x_i|e) = 0$ in finite time, that is, there exists a number $k$ such that no $P_{IBP}(x_i|e) = 0$ will be generated after $k$ iterations.*

**Proof.** This follows from the fact that the number of iterations it takes for IBP to compute $P_{IBP}(X_i = x_i|e) = 0$ over $\mathcal{B}$ is exactly the same number of iterations IBP-RDAC needs to remove $x_i$ from the domain $D_i$ over $flat(\mathcal{B})$ (Propositions 16 and 17) and the fact that IBP-RDAC's number of iterations is bounded (Proposition 15). □

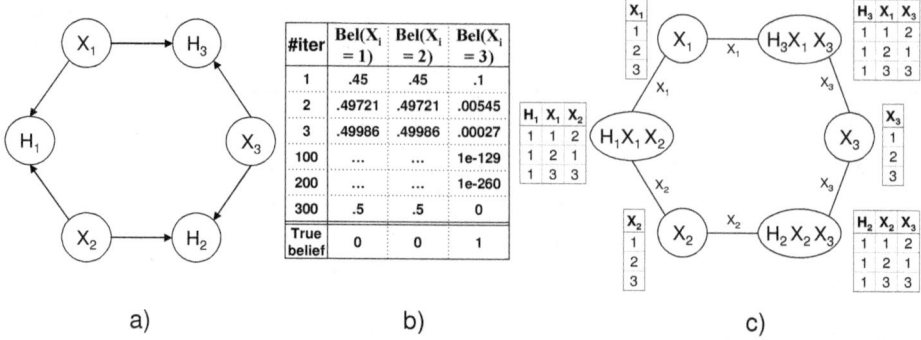

Figure 4. a) A belief network; b) Example of a finite precision problem; and (c) An arc-minimal dual join-graph.

### 4.3 A Finite Precision Problem

Algorithms should always be implemented with care on finite precision machines. In the following example we show that IBP's messages converge in the limit (i.e. in an infinite number of iterations), but they do not stabilize in any finite number of iterations.

EXAMPLE 20. Consider the belief network in Figure 4a defined over 6 variables $X_1, X_2, X_3, H_1, H_2, H_3$. The domain of the $X$ variables is $\{1,2,3\}$ and the domain of the $H$ variables is $\{0,1\}$. The priors on $X$ variables are:

$$P(X_i) = \begin{cases} 0.45, & if \ X_i = 1; \\ 0.45, & if \ X_i = 2; \\ 0.1, & if \ X_i = 3; \end{cases}$$

There are three CPTs over the scopes: $\{H_1, X_1, X_2\}$, $\{H_2, X_2, X_3\}$, and $\{H_3, X_1, X_3\}$. The values of the CPTs for every triplet of variables $\{H_k, X_i, X_j\}$ are:

$$P(h_k = 1|x_i, x_j) = \begin{cases} 1, & if \ (3 \neq x_i \neq x_j \neq 3); \\ 1, & if \ (x_i = x_j = 3); \\ 0, & otherwise \ ; \end{cases}$$
$$P(h_k = 0|x_i, x_j) = 1 - P(h_k = 1|x_i, x_j).$$

Consider the evidence set $e = \{H_1 = H_2 = H_3 = 1\}$. This Bayesian network expresses the probability distribution that is concentrated in a single tuple:

$$P(x_1, x_2, x_3|e) = \begin{cases} 1, & if \ x_1 = x_2 = x_3 = 3; \\ 0, & otherwise. \end{cases}$$

The belief for any of the $X$ variables as a function of the number of iteration is given in Figure 4b. After about 300 iterations, the finite precision of our computer is not able to represent the value for $Bel(X_i = 3)$, and this appears to be zero, yielding the final updated belief $(.5, .5, 0)$, when in fact the true updated belief should be $(0, 0, 1)$. Notice that $(.5, .5, 0)$ cannot be regarded as a legitimate fixed point for IBP. Namely, if we would initialize IBP with the values $(.5, .5, 0)$, then the algorithm would maintain them, appearing to have a fixed point. However, initializing IBP with zero values cannot be expected to be correct. Indeed, when we initialize with zeros we forcibly introduce determinism in the model, and IBP will always maintain it afterwards.

However, this example does not contradict our theory because, mathematically, $Bel(X_i = 3)$ never becomes a true zero, and IBP never reaches a quiescent state. The example shows however that a close to zero inferred belief by IBP can be arbitrarily inaccurate. In this case the inaccuracy seems to be due to the initial prior belief which are so different from the posterior ones.

### 4.4 Zeros Inferred by Generalized Belief Propagation

Belief propagation algorithms were extended yielding the class of *generalized belief propagation* (GBP) algorithms [Yedidia, Freeman, and Weiss 2000]. These algorithms fully process subparts of the networks, transforming it closer to a tree structure on which IBP can be more effective [Dechter, Mateescu, and Kask 2002; Mateescu, Kask, Gogate, and Dechter 2010]. The above results for IBP can now be extended to GBP and in particular to the variant of *iterative join-graph propagation*, IJGP [Dechter, Mateescu, and Kask 2002]. The algorithm applies message passing over a partition of the CPTs into clusters, called a join-graph, rather than over the dual graph. The set of clusters in such a partition defines a unique dual graph (i.e., each cluster is a node). This dual graph can be associated with various dual join-graphs, each defined by the labeling on the arcs between neighboring cluster nodes.

Algorithm IJGP has an accuracy parameter $i$, called $i$-bound, which restricts the maximum number of variables that can appear in a cluster and it is more accurate as $i$ grows. The extension of all the previous observations regarding zeros to IJGP is straightforward and is summarized next, where the inferred approximation of the belief $P_{calB}(X_i|e)$ computed by IJGP is denoted by $P_{IJGP}(X_i|e)$.

THEOREM 21. *Given a belief network $\mathcal{B}$ to which IJGP is applied then:*

1. *IJGP generates all its $P_{IJGP}(x_i|e) = 0$ in finite time, that is, there exists a number $k$, such that no $P_{IJGP}(x_i) = 0$ will be generated after $k$ iterations.*

2. *Whenever IJGP determines $P_{IJGP}(x_i|e) = 0$, it stays 0 during all subsequent iterations.*

3. *Whenever IJGP determines $P_{IJGP}(x_i|e) = 0$, then $Bel(x_i) = 0$.*

## 5 The Impact of IBP's Inferred Zeros

This section discusses the ramifications of having sound inferred zero beliefs.

## 5.1 The Inference Power of IBP

We now show that the inference power of IBP for zeros is sometimes very limited and other times strong, exactly wherever arc-consistency is weak or strong.

**Cases of weak inference power.** Consider the belief network described in Example 20. The flat constraint network of that belief network is defined over the scopes $S_1=\{H_1, X_1, X_2\}$, $S_2=\{H_2, X_2, X_3\}$, $S_3=\{H_3, X_1, X_3\}$. The constraints are defined by: $R_{S_i} = \{(1,1,2), (1,2,1), (1,3,3), (0,1,1), (0,1,3), (0,2,2), (0,2,3), (0,3,1), (0,3,2)\}$. The prior probabilities for $X_i$'s imply unary constraints equal to the full domain $\{1,2,3\}$. An arc-minimal dual join-graph that is identical to the constraint network is given in Figure 4b. In this case, IBP-RDAC sends as messages the full domains of the variables and thus no tuple is removed from any constraint. Since IBP infers the same zeros as arc-consistency, IBP will also *not* infer any zeros. Since the true probability of most tuples is zero, we can conclude that the inference power of IBP on this example is weak or non-existent.

The weakness of arc-consistency in this example is not surprising. Arc-consistency is known to be far from complete. Since every constraint network can be expressed as a belief network (by adding a variable for each constraint as we did in the above example) and since arc-consistency can be arbitrarily weak on some constraint networks, so could be IBP.

**Cases of strong inference power.** The relationship between IBP and arc-consistency ensures that IBP is zero-complete, whenever arc-consistency is. In general, if for a flat constraint network of a belief network $\mathcal{B}$, arc-consistency removes all the inconsistent domain values, then IBP will also discover all the true zeros of $\mathcal{B}$. Examples of constraint networks that are complete for arc-consistency are max-closed constraints. These constraints have the property that if 2 tuples are in the relation so is their intersection. Linear constraints are often max-closed and so are Horn clauses (see [Dechter 2003]). Clearly, IBP is zero complete for acyclic networks which include binary trees, polytrees and networks whose dual graph is a hypertree [Dechter 2003]. This is not too illuminating though as we know that IBP is fully complete (not only for zeros) for such networks.

An interesting case is when the belief network has no evidence. In this case, the flat network always corresponds to the *causal constraint network* defined in [Dechter and Pearl 1991]. The inconsistent tuples or domain values are already explicitly described in each relation and no new zeros can be inferred. What is more interesting is that in the absence of evidence IBP is also complete for non-zero beliefs for many variables as we show later.

## 5.2 IBP and Loop-Cutset

It is well-known that if evidence nodes form a loop-cutset, then we can transform any multiply-connected belief network into an equivalent singly-connected network which can be solved by belief propagation, leading to the loop-cutset conditioning method [Pearl 1988]. Now that we established that inferred zeros, and in particular inferred evidence (i.e., when only a single value in the domain of a variable has a non-zero probability) are sound, we show that evidence play the cutset role automatically during IBP's performance. Indeed, we can show that during IBP's operation, an observed node $X_i$ in a Bayesian network blocks the path between its parents and its children as defined in the d-separation criteria. All the proofs of claims appearing in Section 5.2 and Section 5.3 can be found in [Bidyuk and Dechter 2001].

PROPOSITION 22. *Let $X_i$ be an observed node in a belief network $\mathcal{B}$. Then for any child $Y_j$ of node $X_i$, the belief of $Y_j$ computed by IBP is not dependent on the messages that $X_i$ receives from its parents $pa(X_i)$ or the messages that node $X_i$ receives from its other children $Y_k, k \neq j$.*

From this we can conclude that:

THEOREM 23. *If evidence nodes, original or inferred, constitute a loop-cutset, then IBP converges to the correct beliefs in linear time.*

## 5.3 IBP on Irrelevant Nodes

An orthogonal property is that unobserved nodes that have only unobserved descendents are irrelevant to the beliefs of the remaining nodes and therefore, processing can be restricted to the relevant subgraphs. In IBP, this property is expressed by the fact that irrelevant nodes send messages to their parents that equally support each value in the domain of a parent and thus do not affect the computation of marginal posteriors of its parents.

PROPOSITION 24. *Let $X_i$ be an unobserved node without observed descendents in $\mathcal{B}$ and let $\mathcal{B}'$ be a subnetwork obtained by removing $X_i$ and its descendents from $\mathcal{B}$. Then, $\forall Y \in \mathcal{B}'$ the belief of $Y$ computed by IBP over $\mathcal{B}$ equals the belief of $Y$ computed by IBP over $\mathcal{B}'$.*

Thus, in a loopy network without evidence, IBP always converges after 1 iteration since only propagation of top-down messages affects the computation of beliefs and those messages do not change. Also in that case, IBP converges to the correct marginals for any node $X_i$ such that there exists only one directed path from any ancestor of $X_i$ to $X_i$. This is because the relevant subnetwork that contains only the node and its ancestors is singly-connected and by Proposition 24 they are the same as the beliefs computed by applying IBP to the complete network. In summary,

THEOREM 25. *Let $\mathcal{B}'$ be a subnetwork obtained from $\mathcal{B}$ by recursively eliminating all its unobserved leaf nodes. If observed nodes constitute a loop-cutset of $\mathcal{B}'$, then IBP applied to $\mathcal{B}$ converges to the correct beliefs for all nodes in $\mathcal{B}'$.*

THEOREM 26. *If a belief network does not contain any observed nodes or only has observed root nodes, then IBP always converges.*

In summary, in Sections 5.2 and 5.3 we observed that IBP exploits the two properties of observed and unobserved nodes, *automatically*, without any outside intervention for network transformation. As a result, the correctness and convergence of IBP on a node $X_i$ in a multiply-connected belief network will be determined by the structure restricted to $X_i$'s relevant subgraph. If the relevant subnetwork of $X_i$ is singly-connected relative to the evidence (observed or inferred), IBP will converge to the correct beliefs for node $X_i$.

## 6 Experimental Evaluation

The goal of the experiments is two-fold. First, since zero values inferred by IBP/IJGP are proved correct, we want to explore the behavior of IBP/IJGP for near zero inferred beliefs. Second, we want to explore the hypothesis that the loop-cutset impact on IBP's performance, as discussed in Section 5.2, also extends to variables with extreme support. The next two subsections are devoted to these two issues, respectively.

### 6.1 On the Accuracy of IBP in Near Zero Marginals

We test the performance of IBP and IJGP both on cases of strong and weak inference power. In particular, we look at networks where probabilities are extreme and investigate empirically the accuracy of IBP/IJGP across the range of belief values from 0 to 1. Since zero values inferred by IBP/IJGP are proved correct, we focus especially on the behavior of IBP/IJGP for near zero inferred beliefs.

Using names inspired by the well known measures in information retrieval, we report *Recall Absolute Error* and *Precision Absolute Error* over small intervals spanning $[0, 1]$. *Recall* is the absolute error averaged over all the exact beliefs that fall into the interval, and can therefore be viewed as capturing the level of completeness. For *precision*, the average is taken over all the belief values computed by IBP/IJGP that fall into the interval, and can be viewed as capturing soundness.

The $X$ coordinate in Figure 5 and Figure 10 denotes the interval $[X, X + 0.05)$. For the rest of the figures, the $X$ coordinate denotes the interval $(X - 0.05, X]$, where the 0 interval is $[0, 0]$. The left Y axis corresponds to the histograms (the bars), while the right Y axis corresponds to the absolute error (the lines). For problems with binary variables, we only show the interval $[0, 0.5]$ because the graphs are symmetric around 0.5. The number of variables, number of evidence variables and induced width $w^*$ are reported in each graph.

Since the behavior within each benchmark is similar, we report a subset of the results (for an extended report see [Rollon and Dechter 2009].

**Coding networks.** Coding networks are the famous case where IBP has impressive performance. The instances are from the class of linear block codes, with 50 nodes per layer and 3 parent nodes for each variable. We experiment with instances having three different values of channel noise: 0.2, 0.4 and 0.6. For each channel value, we generate 1000 samples.

Figure 5 shows the results. When the noise level is 0.2, all the beliefs computed

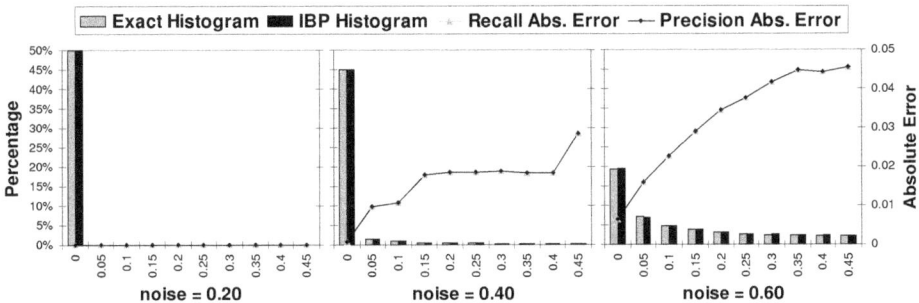

Figure 5. Coding, N=200, evidence=100, w*=15, 1000 instances.

by IBP are extreme. The Recall and Precision are very small, of the order of $10^{-11}$. So, in this case, all the beliefs are very small (i.e., $\epsilon$ small) and IBP is able to infer them correctly, resulting in almost perfect accuracy (IBP is indeed perfect in this case for the bit error rate). As noise increases, the Recall and Precision get closer to a bell shape, indicating higher error for values close to 0.5 and smaller error for extreme values. The histograms show that fewer belief values are extreme as noise increases.

**Linkage Analysis networks**. Genetic linkage analysis is a statistical method for mapping genes onto a chromosome. The problem can be modeled as a belief network. We experimented with four *pedigree* instances from the UAI08 competition. The domain size ranges between 1 to 4. For these instances exact results are available. Figure 6 shows the results. We observe that the number of exact 0 beliefs is small and IJGP correctly infers all of them. The behavior of IJGP for $\epsilon$ small beliefs varies accross instances. For *pedigree1*, the Exact and IJGP histograms are about the same (for all intervals). Moreover, Recall and Precision errors are relatively small. For the rest of the instances, the accuracy of IJGP for extreme inferred marginals decreases. Notice that IJGP infers more $\epsilon$ small beliefs than the number of exact extremes in the corresponding intervals, leading to relatively high Precision error while small Recall error. The behaviour for beliefs in the 0.5 interval is reversed, leading to high Recall error while small Precision error. As expected, the accuracy of IJGP improves as the value of the control parameter $i$-bound increases.

**Grid networks**. Grid networks are characterized by two parameters $(N, D)$, where $N \times N$ is the size of the network and $D$ is the percentage of determinism (i.e., the percentage of values in all CPTs assigned to either 0 or 1). We experiment with *grids2* instances from the UAI08 competition. They are characterized by parameters $(\{16, \ldots, 42\}, \{50, 75, 90\})$. For each parameter configuration, there are samples of size 10 generated by randomly assigning value 1 to one leaf node.

Figure 7 and Figure 8 report the results. IJGP correctly infers all 0 beliefs.

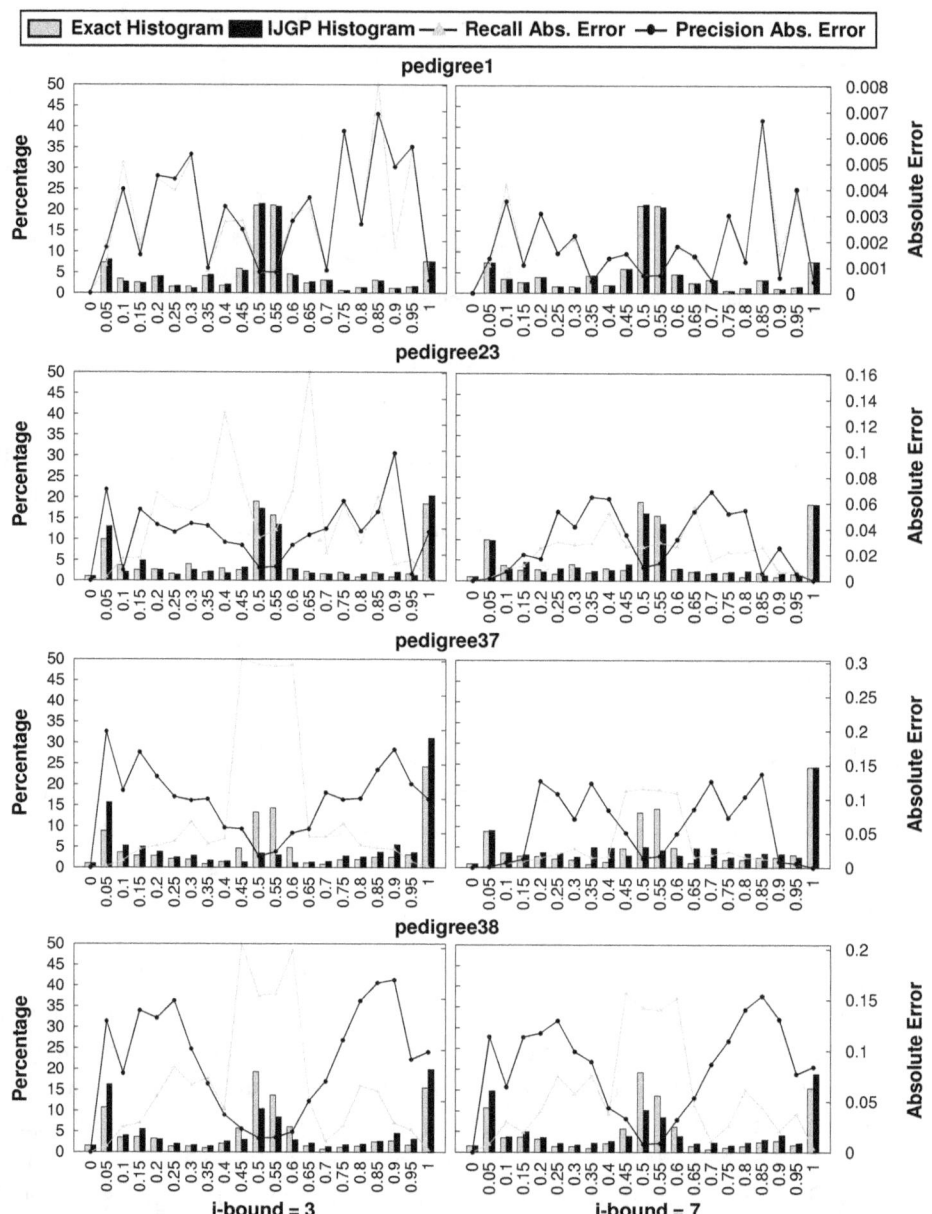

Figure 6. Results on pedigree instances. Each row is the result for one instance. Each column is the result of running IJGP with $i$-bound equal to 3 and 7, respectively. The number of variables $N$, number of evidence variables $NE$, and induced width w* of each instance is as follows. Pedigree1: $N = 334$, $NE = 36$ and w*=21; pedigree23: $N = 402$, $NE = 93$ and w*=30; pedigree37: $N = 1032$, $NE = 306$ and w*=30; pedigree38: $N = 724$, $NE = 143$ and w*=18.

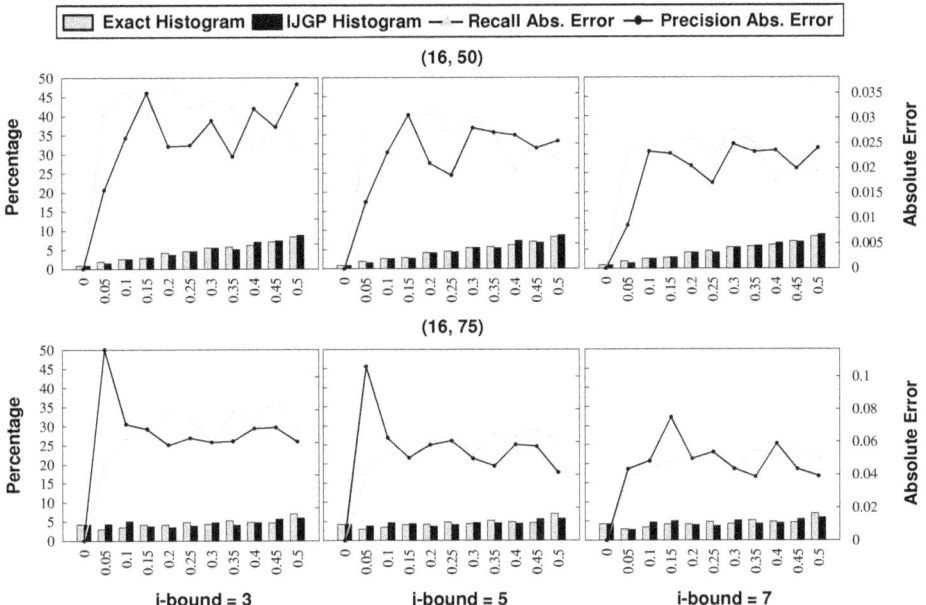

Figure 7. Results on grids2 instances. First row shows the results for parameter configuration (16, 50). Second row shows the results for (16, 75). Each column is the result of running IJGP with $i$-bound equal to 3, 5, and 7, respectively. Each plot indicates the mean value for up to 10 instances. Both parameter configurations have 256 variables, one evidence variable, and induced width w*=22.

However, its performance for $\epsilon$ small beliefs is quite poor. Only for networks with parameters (16, 50) the Precision error is relatively small (less than 0.05). If we fix the size of the network and the $i$-bound, both Precision and Recall errors increase as the determinism level $D$ increases. The histograms clearly show the gap between the number of true $\epsilon$ small beliefs and the ones inferred by IJGP. As before, the accuracy of IJGP improves as the value of the control parameter $i$-bound increases.

**Two-layer noisy-OR networks.** Variables are organized in two layers where the ones in the second layer have 10 parents. Each probability table represents a noisy OR-function. Each parent variable $y_j$ has a value $P_j \in [0..P_{noise}]$. The CPT for each variable in the second layer is then defined as, $P(x = 0|y_1, \ldots, y_P) = \prod_{y_j=1} P_j$ and $P(x = 1|y_1, \ldots, y_P) = 1 - P(x = 0|y_1, \ldots, y_P)$. We experiment on $bn2o$ instances from the UAI08 competition.

Figure 9 reports the results for 3 instances. In this case, IJGP is very accurate for all instances. In particular, the accuracy in $\epsilon$ small beliefs is very high.

**CPCS networks.** These are medical diagnosis networks derived from the Computer-Based Patient Care Simulation system (CPCS) expert system. We tested on two

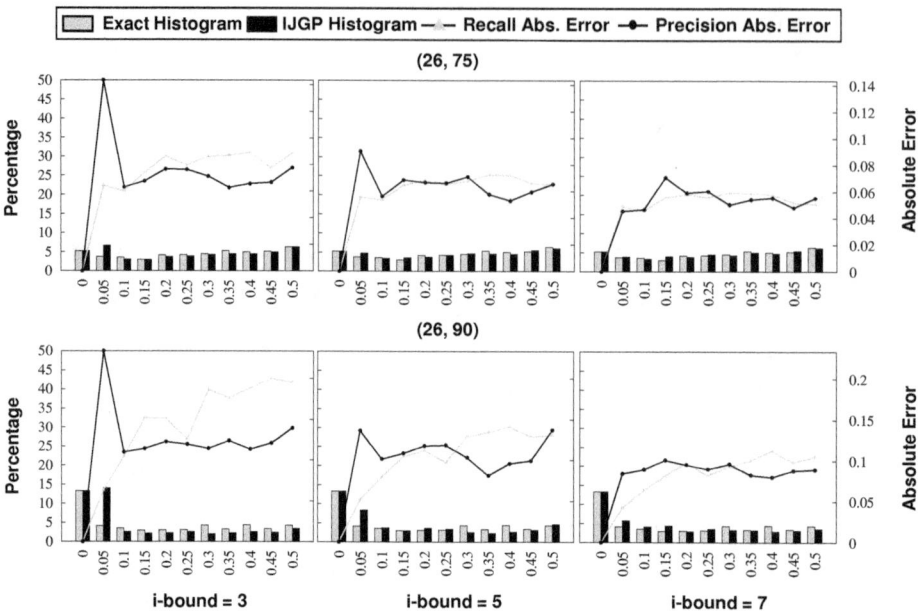

Figure 8. Results on grids2 instances. First row shows the results for parameter configuration (26, 75). Second row shows the results for (26, 90). Each column is the result of running IJGP with $i$-bound equal to 3, 5 and 7, respectively. Each plot indicates the mean value for up to 10 instances. Both parameter configurations have 676 variables, one evidence variable, and induced width w*=40.

networks, *cpcs54* and *cpcs360*, with 54 and 360 variables, respectively. For the first network, we generate samples of size 100 by randomly assigning 10 variables as evidence. For the second network, we also generate samples of the same size by randomly assigning 20 and 30 variables as evidence.

Figure 10 shows the results. The histograms show opposing trends in the distribution of beliefs. Although irregular, the absolute error tends to increase towards 0.5 for cpcs54. In general, the error is quite small throughout all intervals and, in particular, for inferred extreme marginals.

## 6.2 On the Impact of Epsilon Loop-Cutset

In [Bidyuk and Dechter 2001] we explored also the hypothesis that the loop-cutset impact on IBP's performance, as discussed in Section 5.2, extends to variables with extreme support. Extreme support is expressed in the form of either extreme prior value $P(x_i) < \epsilon$ or strong correlation with an observed variable. We hypothesize that a variable $X_i$ with extreme support nearly-cuts the information flow from its parents to its children similar to an observed variable. Subsequently, we conjecture that when a subset of variables with extreme support, called $\epsilon$-*cutset*, form a loop-

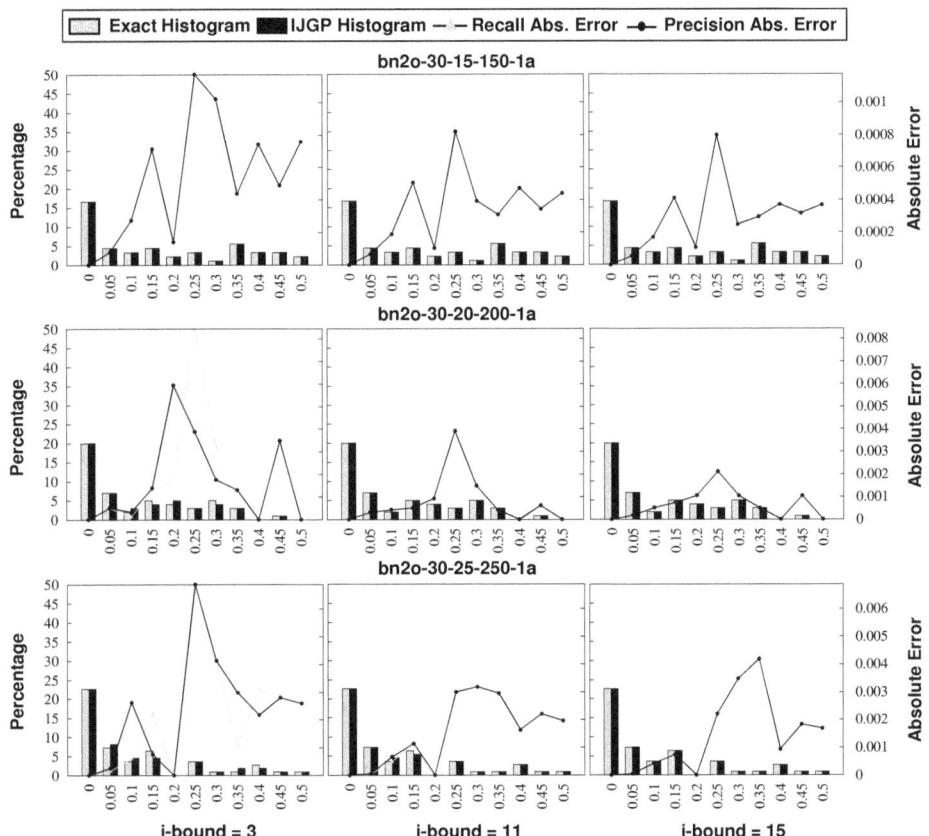

Figure 9. Results on bn2o instances. Each row is the result for one instance. Each column in each row is the result of running IJGP with $i$-bound equal to 3, 5 and 7, respectively. The number of variables $N$, number of evidence variables $NE$, and induced width w* of each instance is as follows. bn2o-30-15-150-1a: $N = 45$, $NE = 15$, and w*=24; bn2o-30-20-200-1a: $N = 50$, $NE = 20$, and w*=27; bn2o-30-25-250-1a: $N = 55$, $NE = 25$, and w*=26.

cutset of the graph, IBP converges and computes beliefs that approach exact ones.

We will briefly recap the empirical evidence supporting the hypothesis in 2-layer noisy-OR networks. The number of root nodes $m$ and total number of nodes $n$ was fixed in each test set (indexed $m - n$). Generating the networks, each leaf node $Y_j$ was added to the list of children of a root node $U_i$ with probability 0.5. All nodes were bi-valued. All leaf nodes were observed. We used average absolute error in the posterior marginals (averaged over all unobserved variables) to measure IBP's accuracy and the percent of variables for which IBP converged as a measure of convergence. In each group of experiments, the results were averaged over 100 instances.

In one set of experiments, we measured the performance of IBP while changing

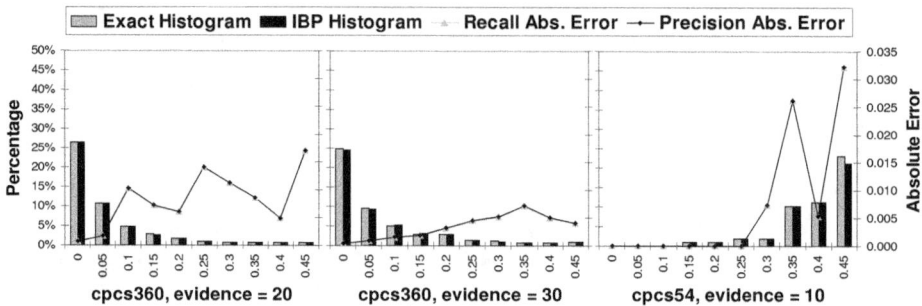

Figure 10. CPCS54, 100 instances, w*=15; CPCS360, 5 instances, w*=20

the number of observed loop-cutset variables (we fixed all priors to (.5, .5) and picked observed value for loop-cutset variables at random). The results are shown in Figure 11, top. As expected, the number of converged nodes increased and the absolute average error decreased monotonically as number of observed loop-cutset nodes increased.

Then, we repeated the experiment except now, instead of instantiating a loop-cutset variable, we set its priors to extreme ($\epsilon$, 1-$\epsilon$) with $\epsilon$=$1E-10$, i.e., instead of increasing the number of observed loop-cuset variables, we increased the number of $\epsilon$-cutset variables. If our hypothesis is correct, increasing the size of $\epsilon$-cutset should produce an effect similar to increasing the number of observed loop-cutset variables, namely, improved convergence and better accuracy in IBP computed beliefs. The results, in Figure 11, bottom, demonstrate that initially, as the number of $\epsilon$-cutset variables grows, the performance of IBP improves just as we conjectured. However, the percentage of nodes with converged beliefs never reaches 100% just like the average absolute error converges to some $\delta > 0$. In the case of 10-40 network, the number of converged beliefs (average absolute error) reaches maximum of $\approx$ 95% (minimum of $\approx$ .001) at 3 $\epsilon$-cutset nodes and then drops to $\approx$ 80% (increases to $\approx$ .003) as the size of $\epsilon$-cutset increases.

To further investigate the effect of the *strength* of $\epsilon$-support on the performance of IBP, we experimented on the same 2-layer networks varying the prior values of the loop-cutset nodes from ($\epsilon$, 1-$\epsilon$) to (1-$\epsilon$, $\epsilon$) for $\epsilon \in [1E-10, .5]$. As shown in Figure 12, initially, as $\epsilon$ decreased, the convergence and accuracy of IBP worsened. This effect was previously reported by Murphy, Weiss, and Jordan [Murphy, Weiss, and Jordan 2000]. However, as the priors of loop-cutset nodes continue to approach 0 and 1, the average error value approaches 0 and the number of converged nodes reaches 100%. Note that convergence is not symmetric with respect to $\epsilon$. The average absolute error and percentage of converged nodes approach 0 and 1 respectively for $\epsilon$=1-(1E-10) but not for $\epsilon$=1E-10 (which we also observed in Figure 11, bottom).

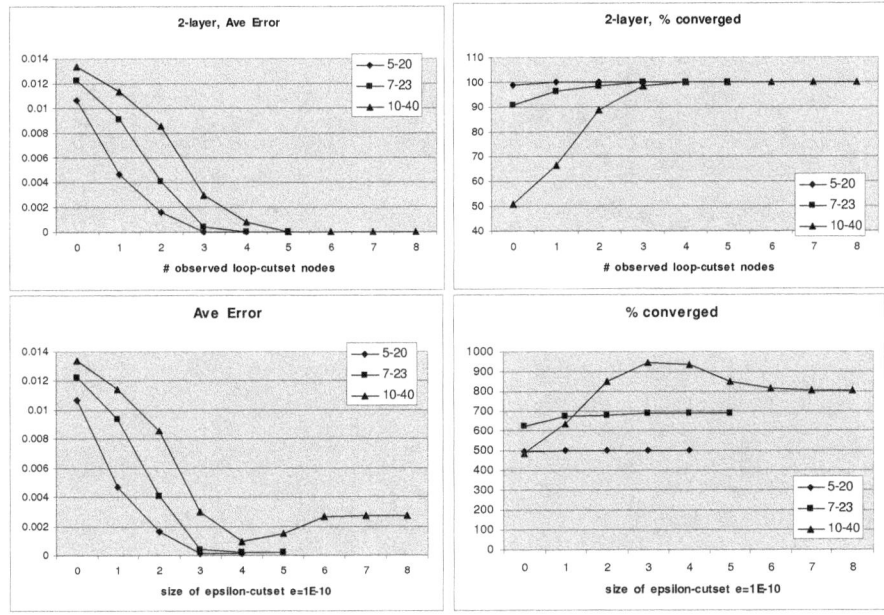

Figure 11. Results for 2-layer Noisy-OR networks. The average error and the number of converged nodes vs the number of truly observed loop-cutset nodes (top) and the size of of $\epsilon$-cutset (bottom).

## 7 Conclusion

The paper provides insight into the power of the Iterative Belief Propagation (IBP) algorithm by making its relationship with constraint propagation explicit. We show that the power of belief propagation for zero beliefs is identical to the power of arc-consistency in removing inconsistent domain values. Therefore, the strength and weakness of this scheme can be gleaned from understanding the inference power of arc-consistency. In particular we show that the inference of zero beliefs (marginals) by IBP and IJGP is always sound. These algorithms are guaranteed to converge for inferred zeros and are as efficient as the corresponding constraint propagation algorithms.

Then the paper empirically investigates whether the sound inference of zeros by IBP is extended to near zeros. We show that while the inference of near zeros is often quite accurate, it can sometimes be extremely inaccurate for networks having significant determinism. Specifically, for networks without determinism IBP's near zero inference was sound in the sense that the average absolute error was contained within the length of the 0.05 interval (see *two layer noisy-OR* and *CPCS* benchmarks). However, the behavior was different on benchmark networks having determinism. For example, experiments on *coding* networks show that IBP is almost perfect, while for *pedigree* and *grid* networks the results are quite inaccurate

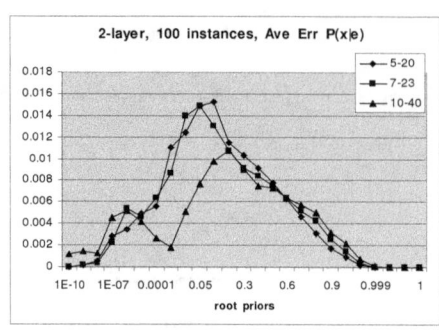 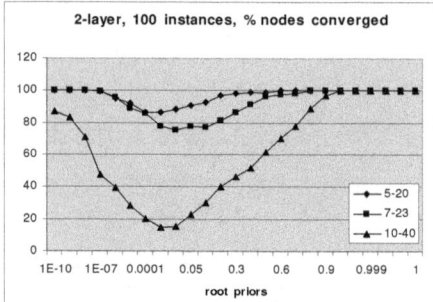

Figure 12. Results for 2-layer Noisy-OR networks. The average error and the percent of converged nodes vs $\epsilon$-support.

near zeros.

Finally, we show that evidence, observed or inferred, automatically acts as a cycle-cutting mechanism and improves the performance of IBP. We also provide preliminary empirical evaluation showing that the effect of loop-cutset on the accuracy of IBP extends to variables that have extreme probabilities.

**Acknowledgement:** This work was supported in part by NSF grant IIS-0713118 and by NIH grant 5R01HG004175-03.

# References

Bidyuk, B. and R. Dechter (2001). The epsilon-cutset effect in Bayesian networks, r97, r97a in http://www.ics.uci.edu/ dechter/publications. Technical report, University of California, Irvine.

Dechter, R. (2003). *Constraint Processing.* Morgan Kaufmann Publishers.

Dechter, R. and R. Mateescu (2003). A simple insight into iterative belief propagation's success. In *Proceedings of the Nineteenth Conference on Uncertainty in Artificial Intelligence (UAI'03)*, pp. 175–183.

Dechter, R., R. Mateescu, and K. Kask (2002). Iterative join-graph propagation. In *Proceedings of the Eighteenth Conference on Uncertainty in Artificial Intelligence (UAI'02)*, pp. 128–136.

Dechter, R. and J. Pearl (1991). Directed constraint networks: A relational framework for causal reasoning. In *Proceedings of the Twelfth International Joint Conferences on Artificial Intelligence (IJCAI'91)*, pp. 1164–1170.

Ihler, A. T. (2007). Accuracy bounds for belief propagation. In *Proceedings of the Twenty Third Conference on Uncertainty in Artificial Intelligence (UAI'07)*.

Ihler, A. T., J. W. Fisher, III, and A. S. Willsky (2005). Loopy belief propagation: Convergence and effects of message errors. *J. Machine Learning Research 6*, 905–936.

Koller, D. (2010). Belief propagation in loopy graphs. In *Heuristics, Probabilities and Causality: A tribute to Judea Pearl, Editors, R. Dechter, H. Gefner and J. Halpern*.

Mackworth, A. K. (1977). Consistency in networks of relations. *Artificial Intelligence 8*(1), 99–118.

Mateescu, R., K. Kask, V. Gogate, and R. Dechter (2010). Iterative join-graph propagation. *Journal of Artificial Intelligence Research (JAIR) (accepted, 2009)*.

McEliece, R. J., D. J. C. MacKay, and J. F. Cheng (1998). Turbo decoding as an instance of Pearl's belief propagation algorithm. *IEEE J. Selected Areas in Communication 16*(2), 140–152.

Mooij, J. M. and H. J. Kappen (2007). Sufficient conditions for convergence of the sum-product algorithm. *IEEE Trans. Information Theory 53*(12), 4422–4437.

Mooij, J. M. and H. J. Kappen (2009). Bounds on marginal probability distributions. In *Advances in Neural Information Processing Systems 21 (NIPS'08)*, pp. 1105–1112.

Murphy, K., Y. Weiss, and M. Jordan (2000). Loopy-belief propagation for approximate inference: An empirical study. In *Proceedings of the Sixteenth Conference on Uncertainty in Artificial Intelligence (UAI'00)*, pp. 467–475.

Pearl, J. (1986). Fusion, propagation, and structuring in belief networks. *Artifical. Intelligence 29*(3), 241–288.

Pearl, J. (1988). *Probabilistic Reasoning in Intelligent Systems*. Morgan Kaufmann Publishers.

Rish, I., K. Kask, and R. Dechter (1998). Empirical evaluation of approximation algorithms for probabilistic decoding. In *Proceedings of the Fourteenth Conference on Uncertainty in Artificial Intelligence (UAI'98)*, pp. 455–463.

Rollon, E. and R. Dechter (December, 2009). Some new empirical analysis in iterative join-graph propagation, r170 in http://www.ics.uci.edu/ dechter/publications. Technical report, University of California, Irvine.

Roosta, T. G., M. J. Wainwright, and S. S. Sastry (2008). Convergence analysis of reweighted sum-product algorithms. *IEEE Trans. Signal Processing 56*(9), 4293–4305.

Yedidia, J. S., W. T. Freeman, and Y. Weiss (2000). Generalized belief propagation. In *Advances in Neural Information Processing Systems 13 (NIPS'00)*, pp. 689–695.

# 10
# Bayesian Nonparametric Learning: Expressive Priors for Intelligent Systems

MICHAEL I. JORDAN

## 1 Introduction

One of the milestones in the development of artificial intelligence (AI) is the embrace of uncertainty and inductive reasoning as primary concerns of the field. This embrace has been a surprisingly slow process, perhaps because the naive interpretation of "uncertain" seems to convey an image that is the opposite of "intelligent." That the field has matured beyond this naive opposition is one of the singular achievements of Judea Pearl. While the pre-Pearl AI researcher tended to focus on mimicking the deductive capabilities of human intelligence, a post-Pearl researcher has been sensitized to the inevitable uncertainty that intelligent systems face in any realistic environment, and the need to explicitly represent that uncertainty so as to be able to mitigate its effects. Not only does this embrace of uncertainty accord more fully with the human condition, but it also recognizes that the first artificially intelligent systems—necessarily limited in their cognitive capabilities—will be if anything *more* uncertain regarding their environments than us humans. It is only by embracing uncertainty that a bridge can be built from systems of limited intelligence to those having robust human-level intelligence.

A computational perspective on uncertainty has two aspects: the explicit representation of uncertainty and the algorithmic manipulation of this representation so as to transform and (often) to reduce uncertainty. In his seminal 1988 book, *Probabilistic Reasoning in Intelligent Systems*, Pearl showed that these aspects are intimately related. In particular, obtaining a compact representation of uncertainty has important computational consequences, leading to efficient algorithms for marginalization and conditioning. Moreover, marginalization and conditioning are the core inductive operations that tend to reduce uncertainty. Thus, by developing an effective theory of the representation of uncertainty, Pearl was able to also develop an effective computational approach to probabilistic reasoning.

Uncertainty about an environment can also be reduced by simply observing that environment; i.e., by learning from data. Indeed, another response to the early focus on deduction in AI has been to emphasize learning as a pathway to the development of intelligent systems. In the 1980's, concurrently with Pearl's work on probabilistic expert systems, this perspective was taken up in earnest, building on an earlier tradition in pattern recognition (which itself built on even earlier traditions in statistics). The underlying inductive principle was essentially the law of large numbers, a principle of probability theory which states that the statistical aggregation of independent, identically distributed samples yields a decrease of uncertainty that goes (roughly speaking) at a rate inversely proportional to the square root of the number of samples. The question has been how to perform this "aggregation," and the learning field has been avidly empirical, exploring a variety of computational architectures, including extremely simple representations (e.g., nearest neighbor), ideas borrowed from deductive traditions (e.g., decision trees), ideas closely related to classical statistical models (e.g., boosting and the support vector machine), and architectures motivated at least in part by complex biological and physical systems (e.g., neural networks). Several of these architectures have factorized or graphical representations, and numerous connections to graphical models have been made.

A narrow reader of Pearl's book might wish to argue that learning is not distinct from the perspective on reasoning presented in that book; in particular, observing the environment is simply a form of conditioning. This perspective on learning is indeed reasonable if we assume that a learner maintains an explicit probabilistic model of the environment; in that case, making an observation merely involves instantiating some variable in the model. However, many learning researchers do not wish to make the assumption that the learner maintains an explicit probabilistic model of the environment, and many algorithms developed in the learning field involve some sort of algorithmic procedure that is not necessarily interpretable as computing a conditional probability. These procedures are instead justified in terms of their unconditional performance when used again and again on various data sets.

Here we are of course touching on the distinction between the Bayesian and the frequentist approaches to statistical inference. While this is not the place to develop that distinction in detail, it is worth noting that statistics—the field concerned with the theory and practice of inference—involves the interplay of the conditional (Bayesian) and the unconditional (frequentist) perspectives and this interplay also underlies many developments in AI research. Indeed, the trend since Pearl's work in the 1980's has been to blend reasoning and learning: put simply, one does not need to learn (from data) what one can infer (from the current model). Moreover, one does not need to infer what one can learn (intractable inferential procedures can be circumvented by collecting data). Thus learning (whether conditional or not) and reasoning interact. The most difficult problems in AI are currently being approached with methods that blend reasoning with learning. While the extremes of classical expert systems and classical tabula rasa learning are still present and still have their value in specialized situations, they are not the centerpieces of the field. Moreover, the caricatures of probabilistic reasoning and statistical inference that fed earlier ill-informed debates in AI have largely vanished. For this we owe much to Judea Pearl.

There remain, however, a number of limitations—both perceived and real—of probabilistic and statistical approaches to AI. In this essay, I wish to focus on some of these limitations and provide some suggestions as to the way forward.

It is both a perception and reality that to use probabilistic methods in AI one is generally forced to write down long lists of assumptions. This is often a helpful exercise, in that it focuses a designer to bring hidden assumptions to the foreground. Moreover, these assumptions are often qualitative in nature, with the quantitative details coming from elicitation methods (i.e., from domain experts) and learning methods. Nonetheless, the assumptions are not always well motivated. In particular, independence assumptions are often imposed for reasons of computational convenience, not because they are viewed as being true, and the effect on inference is not necessarily clear. More subtly, and thus of particular concern, is the fact that the tail behavior of probability distributions is often not easy to obtain (from elicitation or from data), and choices of convenience are often made.

A related issue is that probabilistic methods are often not viewed as sufficiently expressive. One common response to this issue has involved trying to bring ideas from first-order logic to bear on probabilistic modeling. This line of work has, however, mainly involved using logical representations as a high-level interface for model specification and then compiling these representations down to flat probabilistic representations for inference. It is not yet clear how to bring together the powerful inferential methods of logic and probability into an effective computational architecture.

In the current paper, we will pursue a different approach to expressive probabilistic representation and to a less assumption-laden approach to inference. The idea is to move beyond the simple fixed-dimensional random variables that have been generally used in graphical models (multinomials, Gaussians and other exponential family distributions) and to consider a wider range of probabilistic representations. We are motivated by the ubiquity of flexible data structures in computer science—the field is based heavily on objects such as trees, lists and collections of sets that are able to expand and contract as needed. Moreover, these data structures are often associated with combinatorial and algebraic identities that lead to efficient algorithms. We would like to mimic this flexibility within the world of probabilistic representations.

In fact, the existing field of *stochastic processes* provides essentially this kind of flexibility. Recall that a stochastic process is an indexed collection of random variables, where the index set can be infinite (countably infinite or uncountably infinite) [Karlin and Taylor 1975]. Within the general theory of stochastic processes it is quite natural to define probability distributions on objects such trees, lists and collections of sets. It is also possible to define probability distributions on spaces of probability distributions, yielding an appealing recursivity. Moreover, many stochastic processes have interesting ties to combinatorics (and to other areas of mathematics concerned with compact structure, such as algebra). Probability theorists have spent many decades developing these ties and a rich literature on "combinatorial stochastic processes" has emerged [Pitman 2002]. It is natural to take this literature as a point of departure for the development of expressive data structures for computationally efficient reasoning and learning.

One general way to use stochastic processes in inference is to take a Bayesian perspective and replace the parametric distributions used as priors in classical Bayesian analysis with stochastic processes. Thus, for example, we could consider a model in which the prior distribution is a stochastic process that ranges over trees of arbitrary depth and branching factor. Combining this prior with a likelihood, we obtain a posterior distribution that is also a stochastic process that ranges over trees of arbitrary depth and branching factor. Bayesian learning amounts to updating one flexible representation (the *prior stochastic process*) into another flexible representation (the *posterior stochastic process*).

This idea is not new, indeed it is the core idea in an area of research known as *Bayesian nonparametrics*, and there is a small but growing community of researchers who work in the area. The word "nonparametrics" needs a bit of explanation. The word does not mean "no parameters"; indeed, many stochastic processes can be usefully viewed in terms of parameters (often, infinite collections of parameters). Rather, it means "not parametric," in the sense that Bayesian nonparametric inference is not restricted to objects whose dimensionality stays fixed as more data is observed. The spirit of Bayesian nonparametrics is that of flexible data structures—representations can grow as needed. Moreover, stochastic processes yield a much broader class of probability distributions than the class of exponential family distributions that is the focus of the graphical model literature. In this sense, Bayesian nonparametric learning is less assumption-laden than classical Bayesian parametric learning.

In this paper we offer an invitation to Bayesian nonparametrics. Our presentation is meant to evoke Pearl's presentation of Bayesian networks in that our focus is on foundational representational issues. As in the case of graphical models, if the representational issues are handled well, then there are favorable algorithmic consequences. Indeed, the parallel is quite strong—in the case of graphical models, these algorithmic consequences are combinatorial in nature (they involve the combinatorics of sums and products), and in the case of Bayesian nonparametrics favorable algorithmic consequences also arise from the combinatorial properties of certain stochastic process priors.

## 2 De Finetti's theorem and the foundations of Bayesian inference

A natural point of departure for our discussion is a classical theorem due to Bruno De Finetti that is one of the pillars of Bayesian inference. This core result not only suggests the need for prior distributions in statistical models but it also leads directly to the consideration of stochastic processes as Bayesian priors.

Consider an infinite sequence of random variables, $(X_1, X_2, \ldots)$. To simplify our discussion somewhat, let us assume that these random variables are discrete. We say that such a sequence is *infinitely exchangeable* if the joint probability distribution of any finite subset of those random variables is invariant to permutation. That is, for any $N$, we have $p(x_1, x_2, \ldots, x_N) = p(x_{\pi(1)}, x_{\pi(2)}, \ldots, x_{\pi(N)})$, where $\pi$ is a permutation and $p$ is a probability mass function. De Finetti's theorem states that $(X_1, X_2, \ldots)$ are infinitely exchangeable if and only the joint probability distribution of any finite subset can be written as a marginal probability in the following way:

$$p(x_1, x_2, \ldots, x_N) = \int \prod_{i=1}^{N} p(x_i \mid G) P(dG). \qquad (1)$$

In one direction this theorem is straightforward: If the joint distribution can be written as in integral in this way, then we clearly have invariance to permutation (because the product is invariant to permutation). It is the other direction that is non-trivial. It states that for exchangeable random variables, there necessarily exists an underlying random element $G$, and a probability distribution $P$, such that the random variables $X_i$ are conditionally independent given $G$, and such that their joint distribution is obtained by integrating over the distribution $P$. If we view $G$ as a "parameter," then this theorem can be interpreted as stating that exchangeability implies the existence of an underlying parameter and a prior distribution on that parameter. As such, De Finetti's theorem is often viewed as providing foundational support for the Bayesian paradigm.

We placed "parameter" in quotes in the preceding paragraph because there is no restriction that $G$ should be a finite-dimensional object. Indeed, the full import of De Finetti's theorem is clear when we realize that in many instances $G$ is in fact an infinite-dimensional object, and $P$ defines a stochastic process.

Let us give a simple example. The *Pólya urn model* is a simple probability model for sequentially labeling the balls in an urn. Consider an empty urn and a countably infinite collection of colors. Pick a color at random according to some fixed distribution $G_0$ and place a ball having that color in the urn. For all subsequent balls, either choose a ball from the urn (uniformly at random) and return that ball to the urn with another ball of the same color, or choose a new color from $G_0$ and place a ball of that color in the urn. Mathematically, we have:

$$p(X_i = k \mid x_1, \ldots x_{i-1}) \propto \begin{cases} n_k & \text{if } x_j = k \text{ for some } j \in \{1, \ldots, i-1\} \\ \alpha_0 & \text{otherwise,} \end{cases} \qquad (2)$$

where $\alpha_0 > 0$ is a parameter of the process.

It turns out that the Pólya urn model is exchangeable. That is, even though we defined the model by picking a particular ordering of the balls, the resulting distribution is independent of the order. This is proved by writing the joint distribution $p(x_1, x_2, \ldots, x_N)$ as a product of conditionals of the form in Eq. (2) and noting (after some manipulation) that the resulting expression is independent of order.

While the Pólya urn model defines a distribution on labels, it can also be used to induce a distribution on partitions. This is achieved by simply partitioning the balls into groups that have the same color. This distribution on partitions is known as the *Chinese restaurant process* [Aldous 1985]. As we discuss in more detail in Section 4, the Chinese restaurant process and the Pólya urn model can be used as the basis of a Bayesian nonparametric model of clustering where the random partition provides a prior on clusterings and the color associated with a given cell can be viewed as a parameter vector for a distribution associated with a given cluster.

The exchangeability of the Pólya urn model implies—by De Finetti's theorem—the existence of an underlying random element $G$ that renders the ball colors conditionally independent. This random element is not a classical fixed-dimension random variable; rather, it is a stochastic process known as the *Dirichlet process*. In the following section we provide a brief introduction to the Dirichlet process.

## 3 The Dirichlet process

In thinking about how to place random distributions on infinite objects, it is natural to begin with the special case of the positive integers. A distribution $\pi = (\pi_1, \pi_2, \ldots)$ on the integers can be viewed as a sequence of nonnegative numbers that sum to one. How can we obtain *random* sequences that sum to one?

One solution to this problem is provided by a procedure known as "stick-breaking." Define an infinite sequence of independent random variables as follows:

$$\beta_k \sim \text{Beta}(1, \alpha_0) \qquad k = 1, 2, \ldots, \qquad (3)$$

where $\alpha_0 > 0$ is a parameter. Now define an infinite random sequence as follows:

$$\pi_1 = \beta_1, \qquad \pi_k = \beta_k \prod_{l=1}^{k-1}(1 - \beta_l) \qquad k = 2, 3, \ldots. \qquad (4)$$

It is not difficult to show that $\sum_{k=1}^{\infty} \pi_k = 1$ (with probability one).

We can exploit this construction to generate a large class of random distributions on sets other than the integers. Consider an arbitrary measurable space $\Omega$ and let $G_0$ be a probability distribution on $\Omega$. Draw an infinite sequence of points $\{\phi_k\}$ independently from $G_0$. Now define:

$$G = \sum_{k=1}^{\infty} \pi_k \delta_{\phi_k}, \qquad (5)$$

where $\delta_{\phi_k}$ is a unit mass at the point $\phi_k$. Clearly $G$ is a measure. Indeed, for any measurable subset $B$ of $\Omega$, $G(A)$ just adds up the values $\pi_k$ for those $k$ such that $\phi_k \in B$, and this process satisfies the countable additivity needed in the definition of a measure. Moreover, $G$ is a probability measure, because $G(\Omega) = 1$.

Note that $G$ is random in two ways—the weights $\pi_k$ are obtained by a random process, and the locations $\phi_k$ are also obtained by a random process. While it seems clear that such an object is not a classical finite-dimensional random variable, in what sense is $G$ is a stochastic process; i.e., an indexed collection of random variables? The answer is that $G$ is a stochastic process where the indexing variables are the measurable subsets of $\Omega$. Indeed, for any fixed $A \subseteq \Omega$, $G(A)$ is a random variable. Moreover (and this is not an obvious fact), ranging over sets of subsets, $\{A_1, A_2, \ldots, A_K\}$, the joint distributions on the collections of random variables $\{G(A_i)\}$ are consistent with each other. This shows, via an argument in the spirit of the Kolmogorov theorem that $G$ is a stochastic process. A more concrete understanding of this fact can be obtained by specializing to sets $\{A_1, A_2, \ldots, A_K\}$ that form a partition of $\Omega$. In this case, the random vector $(G(A_1), G(A_2), \ldots, G(A_K))$ can be shown to have a classical finite-dimensional Dirichlet distribution:

$$(G(A_1), \ldots, G(A_K)) \sim \text{Dir}(\alpha_0 G_0(A_1), \ldots, \alpha_0 G_0(A_K)), \tag{6}$$

from which the needed consistency properties follow immediately from classical properties of the Dirichlet distribution. For this reason, the stochastic process defined by Eq. (5) is known as a *Dirichlet process*. Eq. (6) can be summarized as saying that a Dirichlet process has Dirichlet marginals.

Having defined a stochastic process $G$, we can now turn De Finetti's theorem around and ask what distribution is induced on a sequence $(X_1, X_2, \ldots, X_N)$ if we draw these variables independently from $G$ and then integrate out $G$. The answer: the Pólya urn. We say that the Dirichlet process is the De Finetti mixing distribution underlying the Pólya urn.

In the remainder of this chapter, we denote the stochastic process defined by Eq. (5) as follows:

$$G \sim \mathcal{DP}(\alpha_0, G_0). \tag{7}$$

The Dirichlet process has two parameters, a *concentration parameter* $\alpha_0$, which is proportional to the probability of obtaining a new color in the Pólya urn, and the *base measure* $G_0$, which is the source of the "atoms" $\phi_k$.

The set of ideas introduced in this section emerged slowly over several decades. The basic definition of the Dirichlet process as a stochastic process is due to Ferguson [1973], based on earlier work by Freedman [1963]. The fact that the Dirichlet process is the De Finetti mixing distribution underlying the Pólya urn model is due to Blackwell and MacQueen [1973]. The stick-breaking construction of the Dirichlet process was presented by Sethuraman [1994]. The application of these ideas to Bayesian modeling and inference required some additional work as described in the following section.

The Dirichlet process and the stick-breaking process are essential tools in Bayesian nonparametrics. It is as important for a Bayesian nonparametrician to master them as it is for a graphical modeler to master Pearl's book. See Hjort et al. [2010] for a book-length treatment of the Dirichlet process and related ideas.

## 4 Dirichlet process mixtures

With an interesting class of stochastic process priors in hand, let us now describe an application of these priors to a Bayesian nonparametric modeling problem. In particular, as alluded to in the previous section, the Dirichlet process defines a prior on partitions of objects, and this prior can be used to develop a Bayesian nonparametric approach to clustering. A notable aspect of this approach is that one does not have to fix the number of clusters a priori.

Let $(X_1, X_2, \ldots, X_N)$ be a sequence of random vectors, whose realizations we want to model in terms of an underlying set of clusters. We treat these variables as exchangeable (i.e., as embedded in an infinitely-exchangeable sequence) and, as suggested by De Finetti's theorem, treat these variables as conditionally independent given an underlying random element $G$. In particular, letting $G$ be a draw from a Dirichlet process, we define a *Dirichlet process mixture model* (DP-MM) [Antoniak 1974; Lo 1984] as follows:

$$\begin{aligned} G &\sim \mathcal{DP}(\alpha_0, G_0) \\ \theta_i \,|\, G &\sim G, \quad i = 1, \ldots, N \\ x_i \,|\, \theta_i &\sim p(x_i \,|\, \theta_i), \quad i = 1, \ldots, N, \end{aligned}$$

where $p(x_i \,|\, \theta_i)$ is a cluster-specific distribution (e.g., a Gaussian distribution, where $\theta_i$ is a mean vector and covariance matrix). This probabilistic specification is indeed directly related to De Finetti's theorem—the use of the intermediate variable $\theta_i$ is simply an expanded way to write the factor $p(x_i \,|\, G)$ in Eq. (1). In particular, $G$ is a sum across atoms, and thus $\theta_i$ is simply one of the atoms in $G$, chosen with probability equal to the weight associated with that atom.

We provide a graphical model representation of the DP-MM in Figure 1. As this figure suggests, it is entirely possible to use the graphical model formalism to display Bayesian nonparametric models. Nodes in such a graph are associated with general random elements, and the distributions on these random elements can be general stochastic processes. By going to stochastic process priors we have not strayed beyond probability theory, and all of the conditional independence semantics of graphical models continue to apply.

## 5 Inference for Dirichlet process mixtures

Inference with stochastic processes is an entire topic of its own, and we limit ourselves here to a brief description of one particular Markov chain Monte Carlo (MCMC) inference procedure for the DP-MM. This particular procedure is due to Escobar [1994], and its virtue is simplicity of exposition, but it should not be viewed as the state of the art. See Neal [2000] for a discussion of a variety of other MCMC inference procedures for DP-MMs.

We begin by noting that the specification in Eq. (8) induces a Pólya urn marginal distribution on $\theta = (\theta_1, \theta_2, \ldots, \theta_N)$. The joint distribution of $\theta$ and $X = (X_1, X_2, \ldots, X_N)$

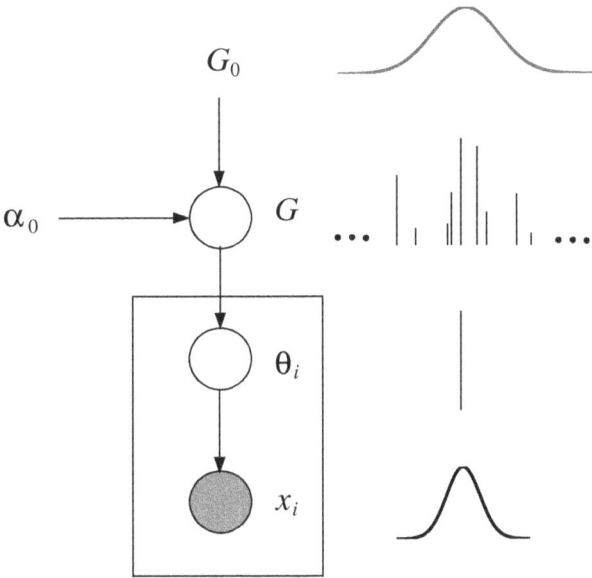

Figure 1. A graphical model representation of the Dirichlet process mixture model. Recall that the plate representation means that the parameters $\theta_i$ are drawn independently conditional on $G$. On the right side of the figure we have depicted specific instantiations of the random elements $G$ and $\theta_i$ and the distribution of the observation $x_i$.

is thus the following product:

$$p(\theta, x) = p(\theta_1, \theta_2, \ldots, \theta_N) \prod_{i=1}^{N} p(x_i \mid \theta_i), \qquad (8)$$

where the first factor is the Pólya urn model. This can be viewed as a product of a prior (the first factor) and a likelihood (the remaining factors).

The variable $x$ is held fixed in inference (it is the observed data) and the goal is to sample $\theta$. We develop a Gibbs sampler for this purpose. The main problem is to sample a particular component $\theta_i$ while holding all of the other components fixed. It is here that the property of exchangeability is essential. Because the joint probability of $(\theta_1, \ldots, \theta_N)$ is invariant to permutation, we can permute the vector to move $\theta_i$ to the end of the list. But the prior probability of the last component given all of the preceding variables is given by the urn model specification in Eq. (2). We multiply each of the distributions in this expression by the likelihood $p(x_i \mid \theta)$ and integrate with respect to $\theta$. (We are assuming that $G_0$ and the likelihood are conjugate that this integral can be done in closed form.) The result is the conditional distribution of $\theta_i$ given the other components and given $x_i$. This conditional is sampled to yield the updated value of $\theta_i$. This is done for all of the indices $i \in \{1, \ldots, N\}$ and the process iterates.

This link between exchangeability and an efficient inference algorithm is an important one. In other more complex Bayesian nonparametric models, while we may no longer assume exchangeability, we generally aim to maintain some weaker notion (e.g., partial exchangeability) so as to have some hope of tractable inference.

## 6 Hierarchical Dirichlet processes

The spirit of the graphical model formalism—in particular the Bayesian network formalism based on directed graphs—is that of hierarchical Bayesian modeling. In a hierarchical Bayesian model, the joint distribution of all of the variables in the model is obtained as a product over conditional distributions, where each conditional may depend on other variables in the model. While the graphical model literature has focused almost exclusively on parametric hierarchies—where each of the conditionals is a finite-dimensional distribution—it is also possible to build hierarchies in which the components are stochastic processes. In this section we consider how to do this for the Dirichlet process.

One of the simplest and most useful ways in which hierarchies arise in Bayesian models is in the form of a conditional independence motif in which a set of variables, $(\theta_1, \theta_2, \ldots, \theta_m)$, are coupled via an underlying variable $\theta_0$. For example, $\theta_i$ might be a Gaussian variable whose mean is equal to $\theta_0$, which is also Gaussian; moreover, the $\theta_i$ are conditionally independent given $\theta_0$. The inferential effect of this construction is to "shrink" the posterior distributions of $\theta_i$ towards each other. This is often a desirable effect, particularly when $m$ is large relative to the number of observed data points.

The same tying of distributions can be done with Dirichlet processes. Recall that a Dirichlet process, $G_i \sim \mathcal{DP}(\alpha_0, G_0)$, is a random measure $G_i$ that has a "parameter" $G_0$ that is itself a measure. If we treat $G_0$ as itself a draw from a Dirichlet process, and let the measures $\{G_1, G_2, \ldots, G_m\}$ be conditionally independent given $G_0$, we obtain the following hierarchy:

$$
\begin{aligned}
G_0 \,|\, \gamma, H &\sim \mathcal{DP}(\gamma, H) \\
G_i \,|\, \alpha, G_0 &\sim \mathcal{DP}(\alpha_0, G_0) \qquad i = 1, \ldots, m,
\end{aligned}
$$

where $\gamma$ and $H$ are concentration and base measure parameters at the top of the hierarchy. This construction—which is known as a *hierarchical Dirichlet process* (HDP)—yields an interesting kind of "shrinkage." Recall that $G_0$ is a discrete random measure, with its support on a countably infinite set of atoms. Drawing $G_i \sim \mathcal{DP}(\alpha_0, G_0)$ means that $G_i$ will also have its support on the same set of atoms, and this will be true for each of $\{G_1, G_2, \ldots, G_m\}$. Thus these measures will share atoms. They will differ in the weights assigned to these atoms. The weights are obtained via conditionally independent stick-breaking processes.

One application of this sharing of atoms is to share mixture components across multiple clustering problems. Consider in particular a problem in which we have

$m$ groups of data, $\{(x_{11}, x_{12}, \ldots, x_{1N_1}), \ldots, (x_{m1}, x_{m2}, \ldots x_{mN_m})\}$, where we wish to cluster the points $\{x_{ij}\}$ in the $i$th group. Suppose, moreover, that we view the groups as related, and we think that clusters discovered in one group might also be useful in other groups. To achieve this, we define the following *hierarchical Dirichlet process mixture model* (HDP-MM):

$$\begin{aligned}
G_0 \,|\, \gamma, H &\sim \mathcal{DP}(\gamma, H) \\
G_i \,|\, \alpha, G_0 &\sim \mathcal{DP}(\alpha_0, G_0) \quad i = 1, \ldots, m, \\
\theta_{ij} \,|\, G_i &\sim G_i \quad j = 1, \ldots, N_i, \\
x_{ij} \,|\, \theta_{ij} &\sim F(x_{ij}, \theta_{ij}) \quad j = 1, \ldots, N_i.
\end{aligned}$$

This model is shown in graphical form in Figure 2. To see how the model achieves our goal of sharing clusters across groups, recall that the Dirichlet process clusters points within a single group by assigning the same parameter vector to those points. That is, if $\theta_{ij} = \theta_{ij'}$, the points $x_{ij}$ and $x_{ij'}$ are viewed as belonging to the same cluster. This equality of parameter vectors is possible because both $\theta_{ij}$ and $\theta_{ij'}$ are drawn from $G_i$, and $G_i$ is a discrete measure. Now if $G_i$ and $G_{i'}$ share atoms, as they do in the HDP-MM, then points in different groups can be assigned to the same cluster. Thus we can share clusters across groups.

The HDP was introduced by Teh, Jordan, Beal and Blei [2006] and it has since appeared as a building block in a variety of applications. One application is to the class of models known as *grade of membership models* [Erosheva 2003], an instance of which is the *latent Dirichlet allocation* (LDA) model [Blei, Ng, and Jordan 2003]. In these models, each entity is associated not with a single cluster but with a set of clusters (in LDA terminology, each "document" is associated with a set of "topics"). To obtain a Bayesian nonparametric version of these models, the DP does not suffice; rather, the HDP is required. In particular, the topics for the $i$th document are drawn from a random measure $G_i$, and the random measures $G_i$ are drawn from a DP with a random base measure $G_0$; this allows the same topics to appear in multiple documents.

Another application is to the hidden Markov model (HMM) where the number of states is unknown a priori. At the core of the HMM is the transition matrix, each row of which contains the conditional probabilities of transitioning to the "next state" given the "current state." Viewing states as clusters, we obtain a set of clustering problems, one for each row of the transition matrix. Using a DP for each row, we obtain a model in which the number of next states is open-ended. Using an HDP to couple these DPs, the same pool of next states is available from each of the current states. The resulting model is known as the *HDP-HMM* [Teh, Jordan, Beal, and Blei 2006]. Marginalizing out the HDP component of this model yields an urn model that is known as the *infinite HMM* [Beal, Ghahramani, and Rasmussen 2002].

Similarly, it is also possible to use the HDP to define an architecture known as

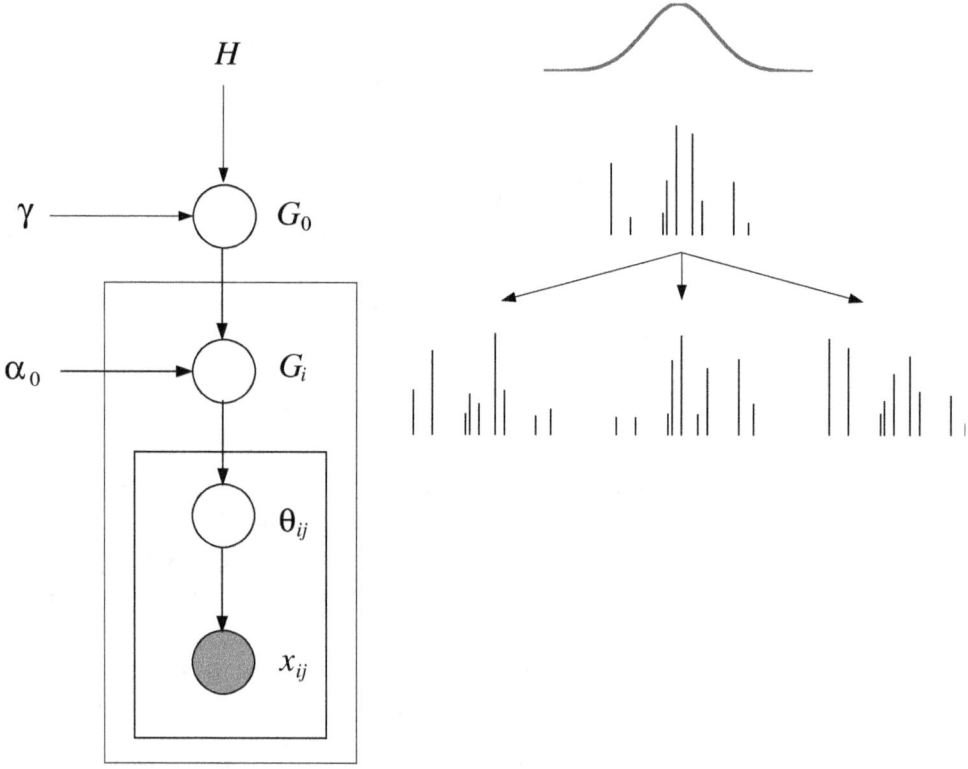

Figure 2. A graphical model representation of the hierarchical Dirichlet process mixture model. The nested plate representation means that $G_0$ is first drawn and held fixed, then the random measures $\{G_i\}$ are drawn independently (conditional on $G_0$), and finally the parameters $\{\theta_{ij}\}$ are drawn independently (conditional on $G_i$). On the right side of the figure we have depicted draws from $G_0$ and the $\{G_i\}$. Note that the atoms in these measures are at the same locations; only the weights associated with the atoms differ.

the *HDP hidden Markov tree* (HDP-HMT), a Markovian tree in which the number of states at each node in the tree is unknown a priori and the state space is shared across the nodes. The HDP-HMT has been shown to be useful in image denoising and scene recognition problems [Kivinen, Sudderth, and Jordan 2007].

Let us also mention that the HDP can be also used to develop a Bayesian nonparametric approach to probabilistic context free grammars. In particular, the HDP-PCFG of Liang, Jordan and Klein [2010] involves an HDP-based lexicalized grammar in which the number of nonterminal symbols is open-ended and inferred from data (see also Finkel, Grenager and Manning [2007] and Johnson, Griffiths and Goldwater [2007]). When a new nonterminal symbol is created at some location in a parse tree, the tying achieved by the HDP makes this symbol available at other locations in the parse tree.

There are other ways to connect multiple Dirichlet processes. One broadly useful idea is to use a Dirichlet process to define a distribution on Dirichlet processes. In particular, let $\{G_1^*, G_2^*, \ldots\}$ be independent draws from a Dirichlet process, $\mathcal{DP}(\gamma, H)$, and then let $G$ be equal to $G_k^*$ with probability $\pi_k$, where the weights $\{\pi_k\}$ are drawn from the stick-breaking process in Eq. (4). This construction (which can be extended to multiple levels) is known as a *nested Dirichlet process* [Rodríguez, Dunson, and Gelfand 2008]. Marginalizing over the Dirichlet processes the resulting urn model is known as the *nested Chinese restaurant process* [Blei, Griffiths, and Jordan 2010], which is a model that can be viewed as a tree of Chinese restaurants. A customer enters the tree at a root Chinese restaurant and sits at a table. This points to another Chinese restaurant, where the customer goes to dine on the following evening. The construction then recurses. Thus a given customer follows a path through the tree of restaurants, and successive customers tend to follow the same paths, eventually branching off.

These nested constructions differ from the HDP in that they do not share atoms among the multiple instances of lower-level DPs. That is, the draws $\{G_1^*, G_2^*, \ldots\}$ involve disjoint sets of atoms. The higher-level DP involves a choice among these disjoint sets.

A general discussion of some of these constructions involving multiple DPs and their relationships to directed graphical model representations can be found in Welling, Porteous and Bart [2008]. Finally, let us mention the work of MacEachern [1999], whose *dependent Dirichlet processes* provide a general formalism for expressing probabilistic dependencies among both the stick-breaking weights and the atom locations in the stick-breaking representation of the Dirichlet process.

# 7 Completely random measures

The Dirichlet process is not the only tool in the Bayesian nonparametric toolbox. In this section we briefly consider another class of stochastic processes that significantly expands the range of models that can be considered.

From the graphical model literature we learn that probabilistic independence of random variables has desirable representational and computational consequences. In the Bayesian nonparametric setting, random variables arise by evaluating a random measure $G$ on subsets of a measurable space $\Omega$; in particular, for fixed subsets $A_1$ and $A_2$, $G(A_1)$ and $G(A_2)$ are random variables. If $A_1$ and $A_2$ are disjoint it seems reasonable to ask that $G(A_1)$ and $G(A_2)$ be independent. Such an independence relation would suggest a divide-and-conquer approach to inference.

The class of stochastic processes known as *completely random measures* are characterized by this kind of independence—for a completely random measure the random masses assigned to disjoint subsets of the sample space $\Omega$ are independent [Kingman 1967]. Note that the Dirichlet process is *not* a completely random measure—the fact that the total mass is one couples the random variables $\{G(A_i)\}$.

The Dirichlet process provides a latent representation for a clustering problem, where each entity is assigned to one and only cluster. This couples the cluster assignments and suggests (correctly) that the underlying stochastic process is not completely random. If, on the other hand, we consider a latent trait model—one in which entities are described via a set of non-mutually-exclusive binary traits—it is natural to consider completely random processes as latent representations. In particular, the *beta process* is a completely random measure in which a draw consists of a countably infinite collection of atoms, each associated with a probability, where these probabilities are independent [Hjort 1990; Thibaux and Jordan 2007]. In effect, a draw from a beta process yields an infinite collection of independent coins. Tossing these coins once yields a binary featural representation for a single entity. Tossing the coins multiple times yields an exchangeable featural representation for a set of entities.

The beta process arises via the following general construction. Consider the product space $\Omega \otimes (0,1)$. Place a product measure on this space, where the measure associated with $\Omega$ is the *base measure* $B_0$, and the measure associated with $(0,1)$ is obtained from the improper beta density, $cp^{-1}(1-p)^{c-1}$, where $c > 0$ is a parameter. Treating this product measure as a rate measure for a nonhomogeneous Poisson process, draw a set of points $\{(\omega_i, p_i)\}$ in the product space $\Omega \otimes (0,1)$. From these points, form a random measure on $\Omega$ as follows:

$$B = \sum_{i=1}^{\infty} p_i \delta_{\omega_i}. \qquad (9)$$

The fact that we obtain an infinite collection of atoms is due to the fact that we have used a beta density that integrates to infinity. This construction is depicted graphically in Figure 3.

If we replace the beta density in this construction with other densities (generally defined on the positive real line rather than the unit interval (0,1)), we obtain other completely random measures. In particular, we obtain the *gamma process* by using an improper gamma density in place of the beta density. The gamma process provides a natural latent representation for models in which entities are

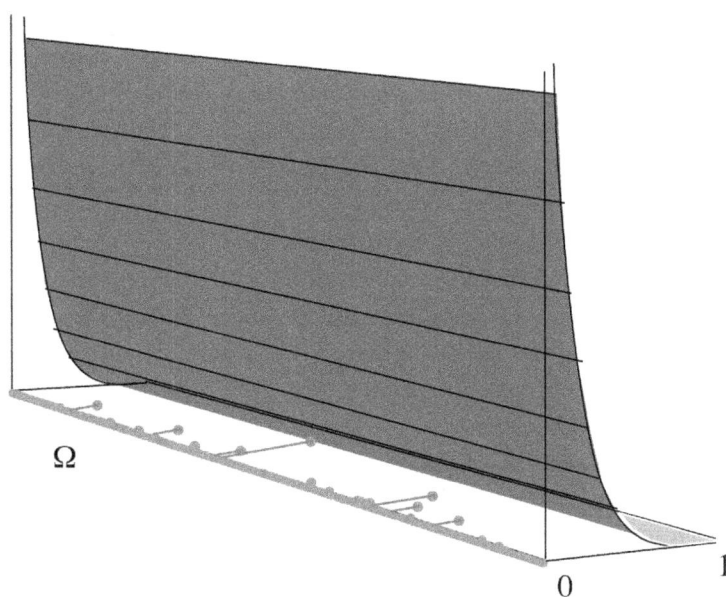

Figure 3. The construction of the beta process from a Poisson process. In this example, $\Omega$ is a bounded interval. The rate measure for the Poisson process is the shaded surface—it is the product of a uniform distribution on $\Omega$ with an improper beta distribution on $(0,1)$. Sampling the Poisson process yields the red points in the plane, and these points are connected by line segments to the $\Omega$-axis interval to form the random measure $B = \sum_{i=1}^{\infty} p_i \delta_{\omega_i}$.

represented by a countably infinite set of counts or rates. It is also worth noting that the Dirichlet process can be obtained by normalizing the gamma process.

Recall from our discussion in Section 2 that the Chinese restaurant process can be obtained by integrating out the Dirichlet process in a conditional independence hierarchy. In the other direction, the Dirichlet process is the random measure that is guaranteed (by exchangeability and De Finetti's theorem) to underlie the Chinese restaurant process. Given the importance of the latter model in Bayesian nonparametric modeling and computation, it is of interest to ask if there is a corresponding probability law on binary matrices obtained by integrating out the beta process. As shown by Thibaux and Jordan [2007], the answer is yes, where the probability law is the *Indian buffet process* (IBP) of Griffiths and Ghahramani [2006].

To describe the IBP, consider an Indian buffet with a countably infinite number of dishes. Let $N$ customers arrive in sequence in the buffet line. Let $Z$ denote a binary-valued matrix in which the rows are customers and the columns are the dishes, and where $Z_{n,k} = 1$ if customer $n$ samples dish $k$. Customer $n$ samples dish $k$ with probability $m_k/n$, where $m_k$ is the number of customers who have previously sampled dish $k$; that is, $Z_{n,k} \sim \text{Ber}(m_k/n)$. (Note that this rule can be interpreted in terms of classical Bayesian analysis as sampling the predictive distribution obtained from a sequence of Bernoulli draws based on an improper beta prior.) Having sampled from the dishes previously sampled by other customers, customer $n$ then goes on to sample an additional number of new dishes determined by a draw from a $\text{Poiss}(\alpha/n)$ distribution.

The connection to the beta process delineated by Thibaux and Jordan [2007] is as follows (see Teh and Jordan [2010] for an expanded discussion). Dishes in the IBP correspond to atoms in the beta process, and the independent beta/Bernoulli updating of the dish probabilities in the IBP reflects the independent nature of the atoms in the beta process. Moreover, the fact that a Poisson distribution is adopted for the number of dishes in the IBP reflects the fact that the beta process is defined in terms of an underlying Poisson process. The exchangeability of the IBP (which requires considering equivalence classes of matrices if argued directly on the IBP representation) follows immediately from the beta process construction (by the conditional independence of the rows of $Z$ given the underlying draw from the beta process).

It is also possible to define *hierarchical beta processes* for models involving multiple beta processes that are tied in some manner [Thibaux and Jordan 2007]. This is done by simply letting the base measure for the beta process itself be drawn from the beta process:

$$B_0 \sim \text{BP}(c_0, B_{00})$$
$$B \sim \text{BP}(c, B_0),$$

where $\text{BP}(c, B_0)$ denotes the beta process with concentration parameter $c$ and base measure $B_0$. This construction can be used in a manner akin to the hierarchical Dirichlet process; for example, we can use it to model groups of entities that are described by sparse binary vectors, where we wish to share the sparsity pattern among groups.

## 8 Conclusions

Judea Pearl's work on probabilistic graphical models yielded a formalism that was significantly more expressive than existing probabilistic representations in AI, but yet retained enough mathematical structure that it was possible to design efficient computational procedures for a wide class of useful models. In this short article, we have argued that Bayesian nonparametrics provides a framework in which this agenda can be taken further. By replacing the traditional parametric prior distributions of Bayesian analysis with stochastic processes, we obtain a rich vocabulary, encompassing probability distributions on objects such as trees of infinite depth, partitions, subsets of features, measures and functions. We also obtain natural notions of recursion. In addition to this structural expressiveness, the Bayesian nonparametric framework also permits a wide range of distributional shapes. Finally, although we have devoted little attention to computation in this article, the stochastic processes that have been used in Bayesian nonparametrics have properties (e.g., exchangeability, independence of measure on disjoint sets) that permit the design of efficient inference algorithms. Certainly the framework is rich enough to design some intractable models, but the same holds true for graphical models. The point is that the Bayesian nonparametric framework opens the door to a richer class of useful models for AI. The growing list of successful applications of Bayesian nonparametrics testifies to the practical value of the framework [Hjort, Holmes, Mueller, and Walker 2010].

A skeptical reader might question the value of Bayesian nonparametric modeling given that for any given finite data set the posterior distribution of a Bayesian nonparametric model will concentrate on a finite set of degrees of freedom, and it would be possible in principle to build a parametric model that mimics the nonparametric model on those degrees of freedom. While this skepticism should not be dismissed out of hand—and we certainly do not wish to suggest that parametric modeling should be abandoned—this skeptical argument has something of the flavor of a computer scientist arguing that data structures such as linked lists and heaps are not needed because they can always be mimicked by fixed-dimension arrays. The nonparametric approach can lead to conceptual insights that are only available at the level of an underlying stochastic process. Moreover, by embedding a model for a fixed number of data points in a sequence of models for a growing number of data points, one can often learn something about the statistical properties of the model—this is the spirit of nonparametric statistics in general. Finally, infinite limits often lead to simpler mathematical objects.

In short, we view Bayesian nonparametrics as providing an expressive, useful language for probabilistic modeling, one which follows on directly from the tradition of graphical models. We hope and expect to see Bayesian nonparametrics have as broad of an effect on AI as that of graphical models.

# References

Aldous, D. (1985). Exchangeability and related topics. In *Ecole d'Eté de Probabilités de Saint-Flour XIII–1983*, pp. 1–198. Springer, Berlin.

Antoniak, C. E. (1974). Mixtures of Dirichlet processes with applications to Bayesian nonparametric problems. *Annals of Statistics 2*, 1152–1174.

Beal, M. J., Z. Ghahramani, and C. E. Rasmussen (2002). The infinite hidden Markov model. In *Advances in Neural Information Processing Systems*, Volume 14, Cambridge, MA. MIT Press.

Blackwell, D. and J. B. MacQueen (1973). Ferguson distributions via Pólya urn schemes. *Annals of Statistics 1*, 353–355.

Blei, D. M., T. L. Griffiths, and M. I. Jordan (2010). The nested Chinese restaurant process and Bayesian inference of topic hierarchies. *Journal of the ACM 57*.

Blei, D. M., A. Y. Ng, and M. I. Jordan (2003). Latent Dirichlet allocation. *Journal of Machine Learning Research 3*, 993–1022.

Erosheva, E. A. (2003). Bayesian estimation of the grade of membership model. In *Bayesian Statistics*, Volume 7, Oxford, UK, pp. 501–510. Oxford University Press.

Escobar, M. D. (1994). Estimating normal means with a Dirichlet process prior. *Journal of the American Statistical Association 89*, 268–277.

Ferguson, T. S. (1973). A Bayesian analysis of some nonparametric problems. *Annals of Statistics 1*, 209–230.

Finkel, J. R., T. Grenager, and C. D. Manning (2007). The infinite tree. In *Proceedings of the Annual Meeting of the Association for Computational Linguistics*, Prague, Czech Republic.

Freedman, D. (1963). On the asymptotic behavior of Bayes estimates in the discrete case. *Annals of Mathematical Statistics 34*, 1386–1403.

Griffiths, T. L. and Z. Ghahramani (2006). Infinite latent feature models and the Indian buffet process. In *Advances in Neural Information Processing Systems*, Volume 18, Cambridge, MA. MIT Press.

Hjort, N., C. Holmes, P. Mueller, and S. Walker (2010). *Bayesian Nonparametrics: Principles and Practice*. Cambridge, UK: Cambridge University Press.

Hjort, N. L. (1990). Nonparametric Bayes estimators based on beta processes in models for life history data. *Annals of Statistics 18*, 1259–1294.

Johnson, M., T. L. Griffiths, and S. Goldwater (2007). Adaptor grammars: A framework for specifying compositional nonparametric Bayesian models. In *Advances in Neural Information Processing Systems*, Volume 19, Cambridge, MA. MIT Press.

Karlin, S. and H. M. Taylor (1975). *A First Course in Stochastic Processes*. New York, NY: Springer.

Kingman, J. F. C. (1967). Completely random measures. *Pacific Journal of Mathematics 21*, 59–78.

Kivinen, J., E. Sudderth, and M. I. Jordan (2007). Learning multiscale representations of natural scenes using Dirichlet processes. In *IEEE International Conference on Computer Vision (ICCV)*, Rio de Janeiro, Brazil.

Liang, P., M. I. Jordan, and D. Klein (2010). Probabilistic grammars and hierarchical Dirichlet processes. In *The Handbook of Applied Bayesian Analysis*, Oxford, UK. Oxford University Press.

Lo, A. (1984). On a class of Bayesian nonparametric estimates: I. Density estimates. *Annals of Statistics 12*, 351–357.

MacEachern, S. (1999). Dependent nonparametric processes. In *Proceedings of the Section on Bayesian Statistical Science*. American Statistical Association.

Neal, R. M. (2000). Markov chain sampling methods for Dirichlet process mixture models. *Journal of Computational and Graphical Statistics 9*, 249–265.

Pitman, J. (2002). Combinatorial stochastic processes. Technical Report 621, Department of Statistics, University of California at Berkeley.

Rodríguez, A., D. B. Dunson, and A. E. Gelfand (2008). The nested Dirichlet process. *Journal of the American Statistical Association 103*, 1131–1154.

Sethuraman, J. (1994). A constructive definition of Dirichlet priors. *Statistica Sinica 4*, 639–650.

Teh, Y. W. and M. I. Jordan (2010). Hierarchical Bayesian nonparametric models with applications. In *Bayesian Nonparametrics: Principles and Practice*. Cambridge, UK: Cambridge University Press.

Teh, Y. W., M. I. Jordan, M. J. Beal, and D. M. Blei (2006). Hierarchical Dirichlet processes. *Journal of the American Statistical Association 101*, 1566–1581.

Thibaux, R. and M. I. Jordan (2007). Hierarchical beta processes and the Indian buffet process. In *Proceedings of the International Workshop on Artificial Intelligence and Statistics*, Volume 11, San Juan, Puerto Rico.

Welling, M., I. Porteous, and E. Bart (2008). Infinite state Bayesian networks for structured domains. In *Advances in Neural Information Processing Systems*, Volume 20, Cambridge, MA. MIT Press.

# 11
# Judea Pearl and Graphical Models for Economics

MICHAEL KEARNS

Judea Pearl's tremendous influence on the fields of artificial intelligence and machine learning began with the fundamental insight that much of traditional statistical modeling lacked expressive means for articulating known or learned structure and relationships between probabilistic entities. Judea and his early colleagues focused their efforts on a type of structure that proved to be particularly important — namely network structure, or the graph-theoretic structure that arises from pairwise influences between random variables. Judea's legacy includes not only the introduction of Bayesian networks — perhaps the most important class of probabilistic graphical models — but a rich series of results establishing firm semantics for inference, independence and causality, and efficient algorithms and heuristics for fundamental probabilistic computations. His body of work is one of those rare instances in which the contributions range from the most conceptual and philosophical to the eminently practical.

Inspired by the program established by Judea for statistical models, about a decade ago a number of us became intrigued by the possibility of replicating it in the domains of strategic, economic and game-theoretic modeling. At its highest level, the proposed metaphor was both simple and natural. Rather than a large number of random variables related by a joint distribution, imagine we have a large number of players in a (normal-form) game. Instead of the edges of a network representing direct probabilistic influences between random variables, they represent direct influences on payoffs by the actions of neighboring players. As opposed to being concerned primarily with conditional inferences on the joint distribution, we are interested in the computation of Nash and other types of equilibrium for the game. As with probabilistic graphical models, although the network succinctly articulates only local influences, in the game-theoretic setting, at equilibrium there are certainly global influences and coordination via the propagation of local effects. And finally, if we were lucky, we might hope to capture for game theory some of the algorithmic benefits that models like Bayesian networks brought to statistical modeling.

The early work following this metaphor was broadly successful in its goals. The first models proposed, which included graphical games [Kearns, Littman, and Singh 2001; Kearns 2007] and Multi-Agent Influence Diagrams [Koller and Milch 2003; Vickrey and Koller 2002], provided succinct languages for expressing strategic structure in the form of networks over the players. The NashProp algorithm for computing (approximate) Nash equilibria in graphical games was the strategic analogue of the belief propagation algorithm developed by Judea and others, and like that algorithm it came in both provably efficient form for restricted network topologies, or in more heuristic but more general form for "loopy" or highly cyclical structures [Ortiz and Kearns 2003]. There are also works carefully relating probabilistic and game-theoretic graphical models in interesting ways, as in a result showing that the distributions forming the correlated equlibria of a graphical game can be succinctly represented by a (probabilistic) Markov network using (almost) the same underlying graph structure [Kakade, Kearns, Langford, and Ortiz 2003]. Graphical games have also played an important role in some recent complexity-theoretic work, most notably the breakthrough proof establishing that the problem of computing Nash equilibria in general games for even 2 players is PPAD-complete and thus potentially intractable [Daskalakis, Goldberg, and Papadimitriou 2006].

In short, we now have a rather rich set of network-based models for game theory, and a firm understanding of their semantic and algorithmic properties. The execution of this agenda relied on Judea's work in many places for inspiration and guidance, from the very conception of the models studied to the usage of cutset conditioning and distributed dynamic programming techniques in the development of NashProp and its variants.

Encouraged by this success, more recent works have sought to expand its scope to include more specifically economic models, developing networked variants of the classical exchange economies studied by Arrow and Debreu, Fisher, and others [Kakade, Kearns, and Ortiz 2004]. Now network structure represents permissible trading partners or relationships, and again the primary solution concept of interest is an equilibrium — but now an equilibrium in prices or exchange rates that permits self-interested traders to clear the market in all goods. While, as to be expected, there are different technical details, we can again establish the algorithmic benefits of such models in the form of a price propagation algorithm for computing an approximate equilibrium. Perhaps more interesting are examinations of how network topology and equilibrium properties interact. It is worth noting that for probabilistic graphical models such as Bayesian networks, the question of what the "typical" structure looks like is somewhat nonsensical — the reply might be that there is no "typical" structure, and topology will depend highly on the domain (whether it be machine vision, medical diagnosis, and so on). In contrast, the emerging literature on social and economic networks is indeed beginning to establish at least broad topological features that arise frequently in empirical networks. This invites, for example, results establishing that if the network structure exhibits a heavy-tailed distribution of connectivity (degrees), agent wealths at equilibrium will also be distributed in highly unequal fashion [Kakade, Kearns, Ortiz, Pemantle, and Suri 2005] . Thus social network structure may be (just one) explanation for observed disparities in wealth.

The lines of research sketched above continue to grow and deepen, and have become one of the many topics of mutual interest between computer scientists, economists and sociologists. Those of us who were exposed to and inspired by Judea's work in probabilistic graphical models were indeed most fortunate to have had the opportunity to help initiate a fundamental and interdisciplinary subject only shortly before social, economic and technological network structure became a topic of such general interest.

Thank you Judea!

# References

Daskalakis, C., P. Goldberg, and C. Papadimitriou (2006). The complexity of computing a Nash equilibrium. In *Proceedings of the Thirty-Eighth ACM Symposium on the Theory of Computing*, pp. 71–78. ACM Press.

Kakade, S., M. Kearns, J. Langford, and L. Ortiz (2003). Correlated equilibria in graphical games. In *Proceedings of the 4th ACM Conference on Electronic Commerce*, pp. 42–47. ACM Press.

Kakade, S., M. Kearns, and L. Ortiz (2004). Graphical economics. In *Proceedings of the 17th Annual Conference on Learning Theory*, pp. 17–32. Springer Berlin.

Kakade, S., M. Kearns, L. Ortiz, R. Pemantle, and S. Suri (2005). Economic properties of social networks. In L. Saul, Y. Weiss, and L. Bottou (Eds.), *Advances in Neural Information Processing Systems 17*, pp. 633–640. MIT Press.

Kearns, M. (2007). Graphical games.

Kearns, M., M. Littman, and S. Singh (2001). Graphical models for game theory. In *Proceedings of the 17th Annual Conference on Uncertainty in Artificial Intelligence*, pp. 253–260. Morgan Kaufmann.

Koller, D. and B. Milch (2003). Multi-agent influence diagrams for representing and solving games. *Games and Economic Behavior 45*(1), 181–221.

Ortiz, L. and M. Kearns (2003). Nash propagation for loopy graphical games. In S. Becker, S. Thrun, and K. Obermayer (Eds.), *Advances in Neural Information Processing Systems 15*, pp. 793–800. MIT Press.

Vickrey, D. and D. Koller (2002). Multi-agent algorithms for solving graphical games. In *Proceedings of the 18th National Conference on Artificial Intelligence*, pp. 345–351. AAAI Press.

# 12
# Belief Propagation in Loopy Graphs

DAPHNE KOLLER

## 1 Introduction and Historical Perspective

Of Judea Pearl's many seminal contributions, perhaps the one that has had the greatest impact (so far) is the development of key ideas in the representation, semantics, and inference of *probabilistic graphical models*. This formalism provides an elegant and practical framework for representing a probability distribution over a high-dimensional space defined as the set of possible assignments to a set of *random variables* $X_1, \ldots, X_n$. The number of such assignments grows exponentially in $n$, but due to the key insights of Pearl and others, we now understand how *conditional independence* properties of the joint probability distribution $P(X_1, \ldots, X_n)$ allow it to be represented compactly and naturally using a graph annotated with local probabilistic interactions (see section 2). The family of probabilistic graphical models includes *Bayesian networks*, which are based on directed graphs, and *Markov networks* (also called *Markov random fields*), which use undirected graphs.

The number of applications of this framework is far too large to enumerate. One of the earliest applications is in the area of medical diagnosis. Here, we might have hundreds of random variables, representing predisposing factors, possible diseases, symptoms, and test results. The framework of Bayesian networks allows such a distribution to be encoded using a limited set of local (directed) interactions, such as those between a disease and its predisposing factors, or those between a symptom and the diseases that cause it (e.g., [Heckerman, Horvitz, and Nathwani 1992; Shwe, Middleton, Heckerman, Henrion, Horvitz, Lehmann, and Cooper 1991]). In a very different application, we might want to encode a probability distribution over possible segmentations of an image — labelings of the pixels in the image into different semantic categories (such as sky, grass, building, person, etc.). Here, we have a random variable for each pixel in the image (hundreds of thousands even for the smallest images), representing its possible labels. And yet, the distribution over the space of possible segmentations is often well-represented in a Markov network, using only terms that encode each pixel's individual preferences over possible labels and (undirected) local interactions between the labels of adjacent pixels (see Szeliski, Zabih, Scharstein, Veksler, Kolmogorov, Agarwala, Tappen, and Rother [2008] for a survey).

A key question, however, is how to use this compact representation to answer questions about the distribution. The most common types of questions are *condi-*

*tional probability queries*, where we wish to infer the probability distribution over some (small) subset of variables given evidence concerning some of the others; for example, in the medical diagnosis setting, we might want to infer the distribution over each possible disease given observations about the patient's predisposing factors, symptoms, and some test results. A second common type of query is the *maximum a posteriori* (or *MAP*) query, where we wish to find the most likely joint assignment to all of our random variables; for example, we often wish to find the most likely joint segmentation to all of the pixels in an image.

In general, it is not difficult to show that both of these inference problems are NP-hard [Cooper 1990; Shimony 1994], yet (as always) this is not end of the story. In their seminal paper, Kim and Pearl [1983] presented an algorithm that passes messages between the nodes in the Bayesian network graph to propagate beliefs between them. The algorithm was developed in the context of singly connected directed graphs, also known as *polytrees*, where there is at most one path (ignoring edge directionality) between each pair of nodes. In this case, the message passing process produces correct posterior beliefs for each node in the graph.

Pearl also considered what happens when the algorithm is executed (without change) over a loopy (multiply connected) graph. In his seminal book, Pearl [1988] says:

> When loops are present, the network is no longer singly connected and local propagation schemes will invariably run into trouble ... If we ignore the existence of loops and permit the nodes to continue communicating with each other as if the network were singly connected, messages may circulate indefinitely around the loops and the process may not converge to a stable equilibrium ... Such oscillations do not normally occur in probabilistic networks ... which tend to bring all messages to some stable equilibrium as time goes on. However, this asymptotic equilibrium is not coherent, in the sense that it does not represent the posterior probabilities of all nodes of the networks.

As a consequence of these problems, the idea of *loopy belief propagation* was largely abandoned for many years.

The revival of this approach is surprisingly due to a seemingly unrelated advance in coding theory. The area of coding addresses the problem of sending messages over a noisy channel, and recovering it from the garbled result. We send a $k$-bit message, redundantly coded using $n$ bits. These $n$ bits are sent over the noisy channel, so the received bits are possibly corrupted. The decoding task is to recover the original message from the bits received. The *bit error rate* is the probability that a bit is ultimately decoded incorrectly. This error rate depends on the code and decoding algorithm used and on the amount of noise in the channel. The *rate* of a code is $k/n$ — the ratio between the number of bits in the message and the number of bits used to transmit it. In 1948, Claude Shannon provided a theoretical analysis of the coding problem [Shannon 1948]. For a given rate, Shannon provided an upper bound on the maximum noise level that can be tolerated while still achieving a certain bit error rate, no matter which code is used. Shannon also showed that there exist channel codes that achieve this limit, but his proof was nonconstructive — he did not present practical encoders and decoders that achieve this limit.

Since Shannon's landmark result, multiple codes were suggested. However, despite a gradual improvement in the quality of the code (bit-error rate for a given noise level), none of the codes even came close to the Shannon limit. The big breakthrough came in the early 1990s, when Berrou, Glavieux, and Thitimajshima [1993] came up with a new scheme that they called a *turbocode*, which, empirically, came much closer to achieving the Shannon limit than any other code proposed up to that point. However, their decoding algorithm had no theoretical justification, and, while it seemed to work well in real examples, could be made to diverge or converge to the wrong answer. The second big breakthrough was the subsequent realization, due to McEliece, MacKay, and Cheng [1998] and Frey and MacKay [1997] that the turbocoding procedure was simply performing loopy belief propagation message passing on a Bayesian network representing the probability model for the code and the channel noise!

This revelation had a tremendous impact on both the coding theory community and the graphical models community. For the former, loopy belief propagation provides a general-purpose algorithm for decoding a large family of codes. By separating the algorithmic question of decoding from the question of the code design, it allowed the development of many new coding schemes with improved properties. These codes have come much, much closer to the Shannon limit than any previous codes, and they have revolutionized both the theory and the practice of coding. For the graphical models community, it was the astounding success of loopy belief propagation for this application that led to the resurgence of interest in these approaches. Subsequent work showed that this algorithm works very well in practice on a broad range of other problems (see, for example, Weiss [1996] and Murphy, Weiss, and Jordan [1999]), leading to a large amount of work on this topic. In this short paper, we review only some of the key ideas underlying this important class of methods; see section 6 for some discussion and further references.

## 2 Background

### 2.1 Probabilistic Graphical Models

Probabilistic graphical models are a general family of representations for probability distributions over high-dimensional spaces. Specifically, our goal is to encode a joint probability distribution over the possible assignments to a set of random variables $\boldsymbol{X} = \{X_1, \ldots, X_n\}$. We focus on the discrete setting, where each random variable $X_i$ takes values in some set $Val(X_i)$. In this case, the number of possible assignments grows exponentially with the number of variables $n$, making an explicit enumeration of the joint distribution infeasible.

Probabilistic graphical models use a factored representation to avoid the exponential representation of the joint distribution. In the most general setting, the distribution is defined via a set of *factors* $\Phi$. A factor $\phi_k$ is defined over a *scope* $Scope[\phi_k] = \boldsymbol{X}_k \subseteq \boldsymbol{X}$; the factor is a function $\phi_k : Val(\boldsymbol{X}_k) \mapsto I\!R^+$. The joint distribution $P_\Phi$ is defined by multiplying together all the factors in $\Phi$, and renormalizing to form a distribution:

$$\tilde{P}_\Phi(\boldsymbol{x}) = \prod_k \phi_k(\boldsymbol{x}_k)$$

$$Z = \sum_{\boldsymbol{x} \in Val(\boldsymbol{X})} \tilde{P}_\Phi(\boldsymbol{x})$$

$$P_\Phi(\boldsymbol{x}) = \frac{1}{Z}\tilde{P}_\Phi(\boldsymbol{x}).$$

For example, if we have a distribution over $\{X_1, \ldots, X_3\}$, defined by two pairwise factors $\phi_1(X_1, X_2)$ and $\phi_2(X_2, X_3)$, then $\tilde{P}_\Phi(x_1, x_2, x_3) = \phi_1(x_1, x_2) \cdot \phi_2(x_2, x_3)$. The normalizing constant $Z$ is historically called the *partition function*.

This factorization is generally tied to a graph whose nodes represent the variables $X_1, \ldots, X_n$ and whose edge structure corresponds to the factorization of the distribution. In particular, the *Markov network* representation uses an undirected graph $\mathcal{H}$ over the nodes $X_1, \ldots, X_n$. A factorized distribution $P_\Phi$ is said to *factorize* over $\mathcal{H}$ if, for every factor $\phi_k \in P_\Phi$, we have that $Scope[\phi_k]$ is a completely connected subgraph in $\mathcal{H}$ (so that every $X_i, X_j \in Scope[\phi_k]$ are connected by an undirected edge in $\mathcal{H}$). A *Bayesian network* uses a directed acyclic graph $\mathcal{G}$ to represent the distribution. In this case, each variable $X_i$ has a set of *Parents* $\mathbf{Pa}^{\mathcal{G}}_{X_i}$. The distribution is now parameterized using a set of factors $\Phi$ which take the form $P(X_i \mid \mathbf{Pa}^{\mathcal{G}}_{X_i})$. In other words, in this factorization, we have precisely one factor for each variable $X_i$ containing $\{X_i\} \cup \mathbf{Pa}^{\mathcal{G}}_{X_i}$, and this factor is locally normalized so that $\sum_{x_i \in Val(X_i)} P(x_i \mid \boldsymbol{u}_i) = 1$ for each assignment $\boldsymbol{u}_i \in Val(\mathbf{Pa}^{\mathcal{G}}_{X_i})$. For this set of factors, the partition function is guaranteed to be 1, and so we can now say that a distribution $P$ *factorizes* over $\mathcal{G}$ if it can be written as:

$$P(X_1, \ldots, X_n) = \prod_i P(X_i \mid \mathbf{Pa}^{\mathcal{G}}_{X_i}),$$

a formula typically known as the *chain rule for Bayesian networks*.

We note that the graph structure associated with a distribution $P_\Phi$ reveal independence properties that hold in $P_\Phi$. That is, an examination of the network structure over which $P_\Phi$ factorizes provides us with a set of independencies that are guaranteed to hold for $P_\Phi$, regardless of the specific parameterization. The connection between the graph structure and the independencies in the distribution was a large focus of the early work on graphical models, and many of the key contributions were developed by Pearl and his students. However, this topic is outside the scope of this paper.

## 2.2 Inference Tasks

Our probabilistic model $P(X_1, \ldots, X_n)$ often defines a general-purpose distribution that can be applied in multiple cases. For example, in a medical diagnosis setting, we typically have a distribution over diseases, symptoms, and test results that might hold for an entire patient population. Given a particular patient, we might observe values values for some subset of the variables (say some symptoms and test results), so that we know $\boldsymbol{E} = \boldsymbol{e}$. Thus, we now have a conditional distribution $P(\boldsymbol{W} \mid \boldsymbol{E} = \boldsymbol{e})$, where $\boldsymbol{W} = \boldsymbol{X} - \boldsymbol{E}$. This conditional distribution has the form $P(\boldsymbol{W}, \boldsymbol{e})/P(\boldsymbol{e})$, where $P(\boldsymbol{e}) = \sum_{\boldsymbol{W}} P(\boldsymbol{W}, \boldsymbol{e})$ is a normalizing constant.

Importantly, if our distribution is derived as $P_\Phi$ for some set of factors $\Phi$, we can easily obtain a factored form for the numerator by simply *reducing* each factor in $\Phi$ to contain only those entries that are consistent with $\boldsymbol{E} = \boldsymbol{e}$. The resulting reduced factors can be multiplied to produce $P(\boldsymbol{W}, \boldsymbol{e})$. If the original distribution $P_\Phi$ factorizes over a Markov network $\mathcal{H}$, the conditional distribution now factorizes over the Markov network where we simply remove the nodes in $\boldsymbol{E}$ from the graph. If the original distribution $P_\Phi$ factorizes over a Bayesian network, the resulting reduced factors no longer satisfy the local normalization requirements defined by the directed graph. Since these local normalization requirements (even if they hold) do not play a role in most inference algorithms, it is generally easier to ignore them and simply consider a distribution defined by a set of (possibly reduced) factors $\Phi$. This will be our focus for the rest of the discussion.

In this setting, we generally consider two main inference tasks. The first is computing the marginal distribution over one or more query variables; for example, we might want to compute

$$P_\Phi(\boldsymbol{Y}) = \sum_{\boldsymbol{W}} P_\Phi(\boldsymbol{Y}, \boldsymbol{W}) = \frac{1}{Z} \sum_{\boldsymbol{W}} \prod_k \phi_k(\boldsymbol{Y}_k, \boldsymbol{W}_k),$$

where $\boldsymbol{Y}_k, \boldsymbol{W}_k$ represents the assignment in $\boldsymbol{Y}, \boldsymbol{W}$ to $\boldsymbol{X}_k = Scope[\phi_k]$. The form of this expression gives rise to the name *sum-product* for this type of inference task. This task is used in the many settings (such as medical diagnosis, for example) where we wish to compute the posterior distribution over some small subset of variables given our current observations.

A second task is computing a single joint assignment $\boldsymbol{x}$ to all variables $\boldsymbol{X}$ that achieves the highest joint probability:

$$\boldsymbol{x}^{map} = \mathrm{argmax}_{\boldsymbol{x}} P_\Phi(\boldsymbol{x}) = \mathrm{argmax}_{\boldsymbol{x}} \frac{1}{Z} \tilde{P}_\Phi(\boldsymbol{x}) = \mathrm{argmax}_{\boldsymbol{x}} \tilde{P}_\Phi(\boldsymbol{x}), \qquad (1)$$

where the partition function cancels since it has no effect on the choice of maximizing assignment. This assignment $\boldsymbol{x}^{map}$ is known as the *maximum a posteriori*, or *MAP*, assignment. The form of this expression gives rise to the name *max-product* for this inference task. MAP queries are used in tasks where we wish to find a single consistent joint hypothesis about the unobserved variables in our domain, for example, a single consistent segmentation of an image or the most likely utterance in a speech recognition system.

## 3 Exact Inference: Clique Trees

One approach to addressing the problem of exact inference in a graphical model is by using a graphical structure called a *clique tree*. Let $\mathcal{T}$ be an undirected graph, each of whose nodes $i$ is associated with a subset $\boldsymbol{C}_i \subseteq \mathcal{X}$. We say that $\mathcal{T}$ is *family-preserving* with respect to $\Phi$ if each factor $\phi \in \Phi$ must be associated with a cluster $\boldsymbol{C}$, denoted $\alpha(\phi)$, such that $Scope[\phi] \subseteq \boldsymbol{C}_i$. Each edge between a pair of clusters $\boldsymbol{C}_i$ and $\boldsymbol{C}_j$ is associated with a *sepset* $\boldsymbol{S}_{i,j} = \boldsymbol{C}_i \cap \boldsymbol{C}_j$. We say that $\mathcal{T}$ satisfies the *running intersection property* if, whenever there is a variable $X$ such that $X \in \boldsymbol{C}_i$ and $X \in \boldsymbol{C}_j$, then for every edge $e$ in the unique path between $\boldsymbol{C}_i$ and $\boldsymbol{C}_j$, we have that $X \in \boldsymbol{S}_e$. If $\mathcal{T}$ satisfies the family-preservation and running-intersection properties, we say that it is a *clique tree* for the graphical model defined by $\Phi$.

We can now specify a general inference algorithm that can be implemented via *message passing* in a clique tree. Let $\mathcal{T}$ be a clique tree with the cliques $\boldsymbol{C}_1, \ldots, \boldsymbol{C}_k$. Roughly speaking, we begin by multiplying the factors assigned to each clique, resulting in our initial potentials. We then use the clique-tree data structure to pass messages between neighboring cliques.

More precisely, recall that each factor $\phi \in \Phi$ is assigned to some clique $\alpha(\phi)$. We define the *initial potential* of $\boldsymbol{C}_j$ to be:

$$\pi_j^0[\boldsymbol{C}_j] = \prod_{\phi \,:\, \alpha(\phi)=j} \phi.$$

Because each factor is assigned to exactly one clique, we have that

$$\prod_\phi \phi = \prod_j \pi_j^0.$$

We now use the clique tree structure to pass messages. The message from $\boldsymbol{C}_i$ to another clique $\boldsymbol{C}_j$ is computed using the following *sum-product message passing* operation:

$$\delta_{i \to j} = \sum_{\boldsymbol{C}_i - \boldsymbol{S}_{i,j}} \pi_i^0 \times \prod_{k \in (\mathcal{N}_i - \{j\})} \delta_{k \to i}. \tag{2}$$

In words, the clique $\boldsymbol{C}_i$ multiplies all incoming messages from its other neighbors with its initial clique potential, resulting in a factor $\psi$ whose scope is the clique. It then sums out all variables except those in the sepset between $\boldsymbol{C}_i$ and $\boldsymbol{C}_j$, and sends the resulting factor as a message to $\boldsymbol{C}_j$.

This computation can be scheduled in a variety of ways. Most generally, we say that $\boldsymbol{C}_i$ is ready to transmit a message to $\boldsymbol{C}_j$ when $\boldsymbol{C}_i$ has messages from all of its neighbors except from $\boldsymbol{C}_j$. In such a setting, $\boldsymbol{C}_i$ can compute the message $\delta_{i \to j}(\boldsymbol{S}_{i,j})$ by multiplying its initial potential with all of its incoming messages except the one from $\boldsymbol{C}_j$, and then eliminate the variables in $\boldsymbol{C}_i - \boldsymbol{S}_{i,j}$. Although the algorithm is defined asynchronously, the message-passing process performed by the algorithm is equivalent to a much more systematic process that consists of an *upward pass* where all messages are sent toward a clique known as the *root*, and then a *downward pass* where messages are sent to all the leaves.

At the end of this process, all cliques have all of their incoming messages, at which point each clique can compute a factor called the *beliefs*:

$$\pi_r[\boldsymbol{C}_i] = \pi_i^0 \times \prod_{k \in \mathcal{N}_{\boldsymbol{C}_i}} \delta_{k \to i}.$$

This algorithm, when applied to a clique tree that satisfies the family preservation and running intersection property, computes messages and beliefs repesenting well-defined expressions. In particular, we can show that the message passed from $\boldsymbol{C}_i$ to $\boldsymbol{C}_j$ is the product of all the factors in $\mathcal{F}_{\prec(i \to j)}$, marginalized over the variables in the sepset (that is, summing out all the others):

$$\delta_{i \to j}(\boldsymbol{S}_{i,j}) = \sum_{\mathcal{V}_{\prec(i \to j)}} \prod_{\phi \in \mathcal{F}_{\prec(i \to j)}} \phi.$$

It then follows that, when the algorithm terminates, we have, for each clique $i$

$$\pi_i[\boldsymbol{C}_i] = \sum_{\mathcal{X} - \boldsymbol{C}_i} \tilde{P}_\Phi(\mathcal{X}), \tag{3}$$

that is, the value of the unnormalized measure $\tilde{P}_\Phi$, marginalized over the variables in $\boldsymbol{C}_i$.

We note that this expression holds for all cliques; thus, in one upward-downward pass of the algorithm, we obtain all of the marginals of all of the cliques in the network, from which we can also obtain the marginals over all variables: to compute the marginal probability over a particular variable $X$, we can select a clique whose scope contains $X$, and marginalize all variables other than $X$. This capability is very valuable in many applications; for example, in a medical-diagnosis setting, we generally want the probability of several possible diseases.

An important consequence of (3) is that we obtain the same marginal distribution over $X$ regardless of the from which we extracted it. More generally, for any two adjacent cliques $\boldsymbol{C}_i$, we must have that

$$\sum_{\boldsymbol{C}_i - \boldsymbol{S}_{i,j}} \pi_i[\boldsymbol{C}_i] = \sum_{\boldsymbol{C}_j - \boldsymbol{S}_{i,j}} \pi_j[\boldsymbol{C}_j].$$

In this case, we say that $\boldsymbol{C}_i$ and $\boldsymbol{C}_j$ are *calibrated*.

These message passing rules, albeit in a simplified form, were first developed in Pearl's analysis [Kim and Pearl 1983; Pearl 1988] on inference in singly connected (polytree) Bayesian networks. In this case, each clique represents a family — a set comprising an individual variable and its parents — and the connections between the cliques follow the structure of the original Bayesian network. With that mapping,

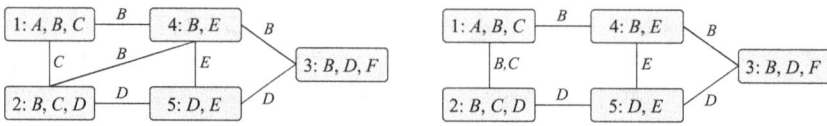

Figure 1. **Two examples of generalized cluster graph for an MRF** with potentials over $\{A, B, C\}$, $\{B, C, D\}$, $\{B, D, F\}$, $\{B, D\}$ and $\{D, E\}$.

the clique tree message passing algorithm we described is precisely Pearl's belief propagation algorithm. The more general case of this particular algorithm was developed by Shafer and Shenoy [1990], who described it in a much broader form that applies to many factored models other than probabilistic graphical models. An alternative but ultimately equivalent message passing scheme (which uses a sum-product-divide sequence for each message passing step) was was developed in parallel, in a series of papers by Lauritzen and Spiegelhalter [1988] and Jensen, Olesen, and Andersen [1990].

## 4 Belief Propagation in Loopy Graphs

While very compelling, the clique tree algorithm often hits against significant computational barriers. There are many graphical models for which any legal clique tree — one that satisfies family preservation and running intersection — has cliques that are very large. For example, any clique tree for a pairwise Markov network encoding an $n \times n$ grid (a class of network commonly used in computer vision applications) has cliques involving at least $n$ variables. In such cases, inference in a clique tree requires computation that is exponential in the size of the graphical model. In a sense, this is inevitable in the worst case, given that the exact inference problem is NP-hard. However, since this exponential blowup arises in many applications of significant practical impact, another solution is necessary.

### 4.1 Cluster Graphs

One generalization of the basic algorithm relaxes the requirements on the message passing structure. In particular, we generalize the clique tree structure to that of a *cluster graph*. This structure is also comprised of a set of clusters $\boldsymbol{C}_i \subseteq \mathcal{X}$ connected by edges. There are three important differences: (1) a cluster graph need not be a tree; (2) the sepsets are required only to satisfy $\boldsymbol{S}_{i,j} \subseteq \boldsymbol{C}_i \cap \boldsymbol{C}_j$; and (3) we have a modified version of the running intersection property, where we require that whenever $X \in \boldsymbol{C}_i$ and $X \in \boldsymbol{C}_j$, there is exactly one path between $\boldsymbol{C}_i$ and $\boldsymbol{C}_j$ for which $X \in \boldsymbol{S}_e$ for all edges $e$ in the path. The generalized running intersection property implies that all edges associated with $X$ form a tree that spans all the clusters that contain $X$. Thus, intuitively, there is only a single path by which information *that is directly about* $X$ can flow in the graph. Both parts of this assumption are significant. The fact that some path must exist forces information about $X$ to flow between all clusters that contain it, so that, in a calibrated cluster graph, all clusters must agree about the marginal distribution of $X$. The fact that there is at most one path prevents loops in the cluster graph where all of the clusters contain $X$. In graphs that contain such loops, a message passing algorithm can propagate information about $X$ endlessly around the loop, making the beliefs more extreme due to "cyclic arguments."

Importantly, however, since the graph is not necessarily a tree, the same pair of clusters might also be connected by other paths. For example, in the cluster graph of figure 1a, we see that the edges labeled with $B$ form a subtree that spans all the clusters that contain $B$. However, there are loops in the graph. For example, there are two paths from $\boldsymbol{C}_3 = \{B, D, F\}$ to $\boldsymbol{C}_2 = \{B, C, D\}$. The first, through $\boldsymbol{C}_4$, propagates information about $B$, and the second, through $\boldsymbol{C}_5$, propagates information about $D$. Thus, we can still get circular reasoning, albeit less directly than we would in a graph that did not satisfy the running intersection property. Note that while in the case of trees the definition of running intersection implied that $\boldsymbol{S}_{i,j} = \boldsymbol{C}_i \cap \boldsymbol{C}_j$, in a graph this equality is no longer enforced by the running intersection property. For example, cliques $\boldsymbol{C}_1$ and $\boldsymbol{C}_2$ in figure 1a have $B$ in common, but $\boldsymbol{S}_{1,2} = \{C\}$.

We note that there are many possible choices for the cluster graph, and the decision on which to use can make a significant difference to the algorithm. In particular, different graphs can lead to very different computational cost, different convergence behavior and even different answers.

EXAMPLE 1. Consider, for example, the cluster graphs $\mathcal{U}_1$ and $\mathcal{U}_2$ of figure 1a and figure 1b. Both are fairly similar, yet in $\mathcal{U}_2$ the edge between $\boldsymbol{C}_1$ and $\boldsymbol{C}_2$ involves the marginal distribution over $B$ and $C$. On the other hand, in $\mathcal{U}_1$, we propagate the marginal only over $C$. Intuitively, we expect inference in $\mathcal{U}_2$ to better capture the dependencies between $B$ and $C$. For example, assume that the potential of $\boldsymbol{C}_1$ introduces strong correlations between $B$ and $C$ (say $B = C$). In $\mathcal{U}_2$, this correlation is conveyed to $\boldsymbol{C}_2$ directly. In $\mathcal{U}_1$, the marginal on $C$ is conveyed on the edge (1–2), while the marginal on $B$ is conveyed through $\boldsymbol{C}_4$. In this case, the strong dependency between the two variables is lost. In particular, if the marginal on $C$ is diffuse (close to uniform), then the message $\boldsymbol{C}_1$ sends to $\boldsymbol{C}_4$ will also have a uniform distribution on $B$, and from $\boldsymbol{C}_2$'s perspective the messages on $B$ and $C$ will appear as two independent variables.

One class of networks for which a simple cluster graph construction exists is the class of *pairwise Markov networks*. In these networks, we have a univariate potential $\phi_i[X_i]$ over each variable $X_i$, and in addition a pairwise potential $\phi_{(i,j)}[X_i, X_j]$ over some pairs of variables. These pairwise potentials correspond to edges in the Markov network. Many problems are naturally formulated as pairwise Markov networks, such as the grid networks common in computer vision applications. Indeed, if we are willing to transform our variables, any distribution can be reformulated as a pairwise Markov network.

One straightforward transformation of a pairwise Markov network into a cluster graph is as follows: For each potential, we introduce a corresponding cluster, and put edges between the clusters that have overlapping scope. In other words, there is an edge between the cluster $C_{(i,j)}$ that corresponds to the edge $X_i$—$X_j$ and the clusters $C_i$ and $C_j$ that correspond to the univariate factors over $X_i$ and $X_j$. Because there is a direct correspondence between the clusters in the cluster graphs and variables or edges in the original Markov network, it is often convenient to think of the propagation steps as operations on the original network. Moreover, since each pairwise cluster has only two neighbors, we can consider two propagation steps along the path $C_i$—$C_{(i,j)}$—$C_j$ as propagating information between $X_i$ and $X_j$.

A highly related transformation applies to Bayesian networks. Here, as in the case of polytrees, we define a cluster $C_i$ for each family $\{X_i\}\cup \mathbf{Pa}_{X_i}$. For every edge $X_i \to X_j$, we connect $C_i$ to $C_j$ via a sepset whose scope is $X_i$. With this cluster graph construction, the message passing algorithm described below is performing precisely the loopy belief propagation for Bayesian networks first proposed by Pearl.

A related but more general construction that applies to arbitrary sets of factors is the *Bethe cluster graph*. This construction uses a bipartite graph: The first layer consists of "large" clusters, with one cluster for each factor $\phi$ in $\Phi$, whose scope is $Scope[\phi]$. These clusters ensure that we satisfy the family-preservation property. The second layer consists of "small" univariate clusters, one for each random variable. Finally, we place an edge between each univariate cluster $X$ on the second layer and each cluster in the first layer that includes $X$; the scope of this edge is $X$ itself. We can easily verify that this cluster graph is a proper one. First, by construction, it satisfies the family preservation property. Second, the edges that mention a variable $X$ form a star-shaped subgraph with edges from the univariate cluster for $X$ to all the large clusters that contain $X$. The construction of this cluster graph is simple and can easily be automated.

The broader notion of message passing on a more general cluster graph was first proposed by Yedidia, Freeman, and Weiss [2000] and Dechter, Kask, and Mateescu [2002]. Indeed, Yedidia, Freeman, and Weiss [2000, 2005] defined an even more general notion of message passing on a *region graph*, which is outside the scope of this paper.

## 4.2  Message Passing in Cluster Graphs

How do we perform inference in a cluster graph rather than a clique tree? From the local perspective of a single cluster $C_i$, there is not much difference between a cluster graph and a clique tree. The cluster is related to each neighbor through an edge that conveys information on variables in the sepset. Thus, we can transmit information by simply having one cluster pass a message to the other. Of course, as the graph may have no leaves, we might initially not have any cliques that are ready to transmit. We address this issue by initializing all messages $\delta_{i\to j} = \mathbf{1}$. Clusters

**Procedure** CGraph-SP-Calibrate (
   $\Phi$,   // Set of factors
   $\mathcal{U}$   // Generalized cluster graph $\Phi$
)

```
1    for each cluster C_i
2        π_i ← ∏_{φ : α(φ)=i} φ
3        for each edge (i–j) ∈ E_U
4            δ_{i→j} ← 1; δ_{j→i} ← 1
5
6    while graph is not calibrated
7        Select (i–j) ∈ E_U
8        δ_{i→j}(S_{i,j}) ← ∑_{C_i−S_{i,j}} ( π_i^0 × ∏_{k∈(N_i−{j})} δ_{k→i} )
9
10   for each clique i
11       π_i ← π_i^0 × ∏_{k∈N_i} δ_{k→i}
12   return {π_i}
```

Figure 2. Calibration using sum-product belief propagation in a cluster graph

then pass messages to their neighbors, summarizing the current beliefs derived from their own initial potentials and from the messages received by their neighbors. The algorithm is shown in figure 2. Convergence is achieved when the cluster graph is *calibrated*; that is, if for each edge $(i-j)$, connecting the clusters $C_i$ and $C_j$, we have that

$$\sum_{C_i - S_{i,j}} \pi_i = \sum_{C_j - S_{i,j}} \pi_j.$$

Note that this definition is weaker than cluster tree calibration, since the clusters do not necessarily agree on the joint marginal of all the variables they have in common, but only on those variables in the sepset. However, if a calibrated cluster graph satisfies the running intersection property, then the marginal of a variable $X$ is identical in all the clusters that contain it. This algorithm clearly generalizes the clique-tree message-passing algorithm described earlier.

EXAMPLE 2. With this framework in hand, we can now revisit the message decoding task. Assume that we wish to send a $k$-bit message $u_1, \ldots, u_k$. We *code* the message using a number of bits $x_1, \ldots, x_n$, which are then sent over the noisy channel, resulting in a set of (possibly corrupted) outputs $y_1, \ldots, y_n$. The *message decoding* task is to recover an estimate $\hat{u}_1, \ldots, \hat{u}_k$ from $y_1, \ldots, y_n$. We first observe that message decoding can easily be reformulated as a probabilistic inference task: We have a prior over the message bits $U = \langle U_1, \ldots, U_k \rangle$, a (usually deterministic) function that defines how a message is converted into a sequence of transmitted bits $X_1, \ldots, X_n$, and another (stochastic) model that defines how the channel ran-

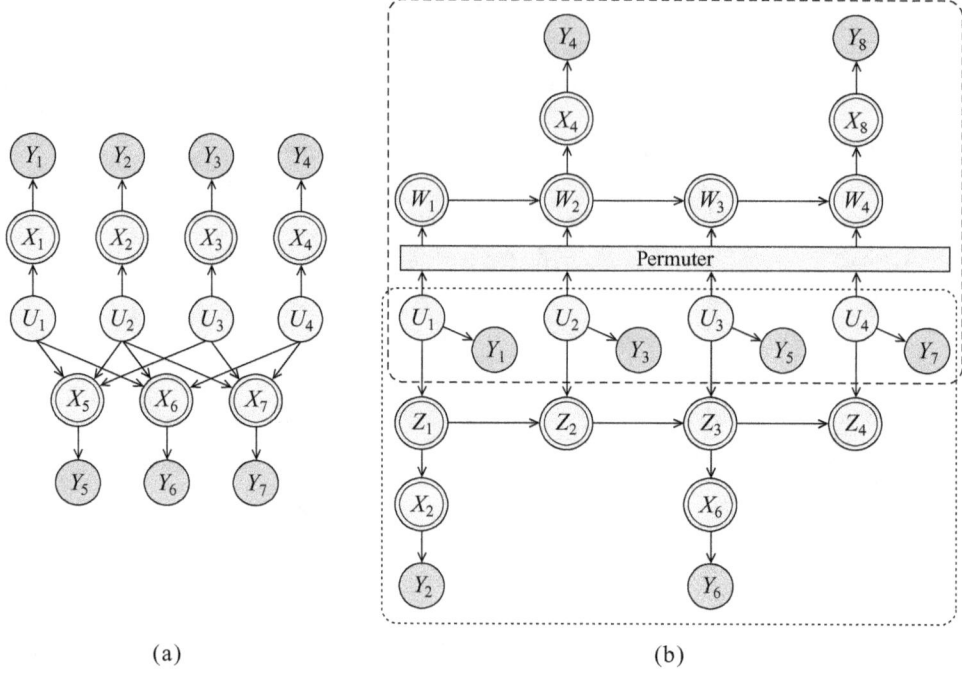

Figure 3. **Two examples of codes** (a) A $k = 4, n = 7$ parity check code, where every four message bits are sent along with three bits that encode parity checks. (b) A $k = 4, n = 8$ turbocode. Here, the $\boldsymbol{X}^a$ bits $X_1, X_3, X_5, X_7$ are simply the original bits $U_1, U_2, U_3, U_4$ and are omitted for clarity of the diagram; the $\boldsymbol{X}^b$ bits use a *shift register* — a state bit that changes with each bit of the message, where the $i$th state bit depends on the $(i-1)$st state bit and on the $i$th message bit. The code uses two shift registers, one applied to the original message bits and one to a set of permuted message bits (using some predetermined permutations). The sent bits contain both the original message bits and some number of the state bits.

domly corrupts the $X_i$'s to produce $Y_i$'s. The decoding task can then be viewed as finding the most likely joint assignment to $\boldsymbol{U}$ given the observed message bits $\boldsymbol{y} = \langle y_1, \ldots, y_n \rangle$, or (alternatively) as finding the posterior $P(U_i \mid \boldsymbol{y})$ for each bit $U_i$. The first task is a MAP inference task, and the second task one of computing posterior probabilities. Unfortunately, the probability distribution is of high dimension, and the network structure of the associated graphical model is quite densely connected and with many loops.

The turbocode approach, as first proposed, comprised both a particular coding scheme, and the use of a message passing algorithm to decode it. The coding scheme transmits two sets of bits: one set comprises the original message bits $\boldsymbol{X}^a = \langle X_1^a, \ldots, X_k^a \rangle = \boldsymbol{u}$, and the second some set $\boldsymbol{X}^b = \langle X_1^b, \ldots, X_k^b \rangle$ of transformed bits (like the parity check bits, but more complicated). The received bits then can also be partitioned into the noisy $\boldsymbol{y}^a, \boldsymbol{y}^b$. Importantly, the code is designed so that the message can be decoded (albeit with errors) using either $\boldsymbol{y}^a$ or $\boldsymbol{y}^b$. The turbocoding algorithm then works as follows: It uses the model of $\boldsymbol{X}^a$ (trivial in this case) and of the channel noise to compute a posterior probability over $\boldsymbol{U}$ given $\boldsymbol{y}^a$. It then uses that posterior $\pi_a(U_1), \ldots, \pi_a(U_k)$ as a prior over $\boldsymbol{U}$ and computes a new posterior over $\boldsymbol{U}$, using the model for $\boldsymbol{X}^b$ and the channel, and $\boldsymbol{y}^b$ as the evidence, to compute a new posterior $\pi_b(U_1), \ldots, \pi_b(U_k)$. The "new information," which is $\pi_b(U_i)/\pi_a(U_i)$, is then transmitted back to the first decoder, and the process repeats until a stopping criterion is reached. In effect, the turbocoding idea was to use two weak coding schemes, but to "turbocharge" them using a feedback loop. Each decoder is used to decode one subset of received bits, generating a more informed distribution over the message bits to be subsequently updated by the other. The specific method proposed used particular coding scheme for the $\boldsymbol{X}^b$ bits, illustrated in figure 3b.

This process looked a lot like black magic, and in the beginning, many people did not even believe that the algorithm worked. However, when the empirical success of these properties was demonstrated conclusively, an attempt was made to understand its theoretical properties. McEliece, MacKay, and Cheng [1998] and Frey and MacKay [1997] subsequently showed that the specific message passing procedure proposed by Berrou et al. is precisely an application of belief propagation (with a particular message passing schedule) to the Bayesian network representing the turbocode (as in figure 3b).

## 4.3 Convergence of Loopy Belief Propagation

Pearl's main reason for rejecting the loopy belief propagation algorithm was the fact that it may fail to converge. Indeed, this is one of the thorniest issues associated with the use of belief propagation in practical applications — much more so than the fact that the resulting beliefs may not be exact. Nonconvergence is particularly problematic when we build systems that use inference as a subroutine within other tasks, for example, as the inner loop of a learning algorithm. Much work has been done on analyzing the convergence properties of generalized belief propagation algorithms, producing some valuable theoretical insights into its properties (such as the recent work of Ihler, Fisher, and Willsky [2005] and Mooij and Kappen [2007]). In practice, several approaches have been used for addressing the nonconvergence issue, some of which we now describe.

A first observation is that nonconvergence is often a local problem. In many practical cases, most of the beliefs in the network do converge, and only a small portion of the network remains problematic. In such cases, it is often quite reasonable simply to stop the algorithm at some point (for example, when some predetermined amount of time has elapsed) and use the beliefs at that point, or a running average of the beliefs over some time window. This heuristic is particularly reasonable when we are not interested in individual beliefs, but rather in some aggregate over the entire network, for example, in a learning setting.

A second observation is that nonconvergence is often due to oscillations in the beliefs. As proposed by Murphy, Weiss, and Jordan [1999] and Heskes [2002], we can dampen the oscillations by reducing the difference between two subsequent updates. In particular, we can replace the update rule in (2) by a *smoothed* version that averages the update $\delta_{i \to j}$ with the previous message between the two cliques:

$$\delta_{i \to j} \leftarrow \lambda \left( \sum_{C_i - S_{i,j}} \prod_{k \neq j} \delta_{k \to i} \right) + (1 - \lambda) \delta_{i \to j}^{\text{old}}, \qquad (4)$$

where $\lambda$ is the damping weight and $\delta_{i \to j}^{\text{old}}$ is the previous value of the message. When $\lambda = 1$, this update is equivalent to standard belief propagation. For $0 < \lambda < 1$, the update is partial and although it shifts $\pi_j$ toward agreement with $\pi_i$, it leaves some momentum for the old value of the belief, a dampening effect that in turn reduces the fluctuations in the beliefs. It turns out that this smoothed update rule is "equivalent" to the original update rule, in that a set of beliefs is a convergence point of the smoothed update if and only if it is a convergence point of standard updates. Moreover, one can show that, if run from a point close enough to a *stable convergence point* of the algorithm, with a sufficiently small $\lambda$, this smoothed update rule is guaranteed to converge. Of course, this guarantee is not very useful in practice, but there are indeed many cases where the smoothed update rule is convergent, whereas the original update rule oscillates indefinitely (see figure 4).

A broader-spectrum heuristic, which plays an important role not only in ensuring convergence but also in speeding it up considerably, is intelligent *message scheduling*. The simplest and perhaps most natural approach is to implement BP message passing as a *synchronous* algorithm, where all messages are updated at once. *Asynchronous* message passing updates messages one at a time, using the most recent version of the incoming messages to generate the outgoing message. It turns out that, in most cases, the synchronous schedule is far from optimal, both in terms of reaching convergence, and in the number of messages required for convergence. As one simple example, consider a cluster graph with $m$ edges, and diameter $d$, synchronous message passing requires $m(d-1)$ messages to pass information from one side of the graph to the other. By contrast, *asynchronous* message passing, appropriately scheduled, can pass information between two clusters at opposite ends of the graph using $d-1$ messages. Moreover, the fact that, in synchronous message passing, each cluster uses messages from its neighbors that are based on their previous beliefs appears to increase the chances of oscillatory behavior and nonconvergence in general.

In practice, an *asynchronous* message passing schedule works significantly better than the synchronous approach (see figure 4). Moreover, even greater improvements can be obtained by scheduling messages in a guided way. One approach, called *tree reparameterization (TRP)* [Wainwright, Jaakkola, and Willsky 2003], selects a set of trees, each of which spans a large number of the clusters, and whose union covers all of the edges in the network. The TRP algorithm then iteratively selects a tree and does an upward-downward calibration of the tree, keeping all other messages fixed. Of course, calibrating this tree has the effect of "uncalibrating" other trees, and so this process repeats. This approach has the advantage of passing information more globally within the graph. It therefore converges more often, and more quickly, than other asynchronous schedules, particularly if the trees are selected using a careful design that accounts for the properties of the problem.

An even more flexible approach attempts to detect dynamically in which parts of the network messages would be most useful. Specifically, as we observed, often some parts of the network converge fairly quickly, whereas others require more messages. We can schedule messages in a way that accounts for their potential usefulness; for example, we can pass a message between clusters where the beliefs disagree most strongly on the sepset. This approach, called *residual belief propagation* [Elidan, McGraw, and Koller 2006] is convenient, since it is fully general and does not require a deep understanding of the properties of the network. It also works well across a range of different real-world networks.

To illustrate these issues, we show the behavior of loopy belief propagation on an $11 \times 11$ grid with binary-valued variables; the network is parameterized as an *Ising model* — one where the pairwise potentials are defined as: $\phi_{i,j}(x_i, x_j) = \exp^{w_{i,j} x_i x_j}$. The network potentials were randomly sampled as follows: Each univariate potential was sampled uniformly in the interval $[0, 1]$; for each pair of variables $X_i, Z_j$, $w_{i,j}$ is sampled uniformly in the range $[-C, C]$. This sampling process creates an energy function where some potentials are attractive ($w_{i,j} > 0$), causing adjacent variables to prefer taking the same value, and some are repulsive ($w_{i,j} < 0$). This regime can result in very difficult inference problems. The magnitude of $C$ (11 in this example) controls the magnitude of the forces and higher values correspond, on average, to more challenging inference problems.

Figure 4 illustrates the convergence behavior on this problem. (a) shows the percentage of messages converged as a function of time for three variants of the belief propagation algorithm: synchronous BP with smoothing (dashed line), where only a small fraction of the messages ever converge; asynchronous BP with smoothing that converges (solid line); asynchronous BP with no smoothing (dash-dot line) that does not fully converge. The benefit of using asynchronous propagation over synchronous updating is obvious. At early round, smoothing tends to slow convergence, because some messages converge quickly when updates are not slowed down by smoothing. However, as the algorithm progresses, smoothing allows all messages to achieve convergence, whereas the unsmoothed algorithm never converges. We note that smoothing is equally beneficial for synchronous updates; indeed, the graph for unsmoothed synchronous updates is not shown because virtually none of the messages achieve convergence.

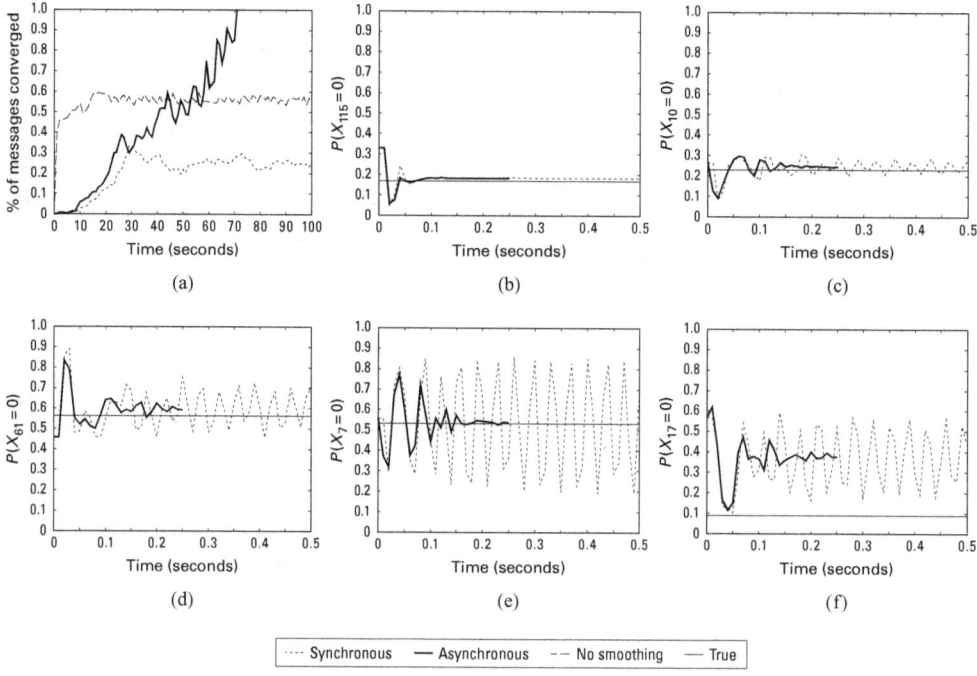

Figure 4. **Example of behavior of BP in practice on an** $11 \times 11$ **Ising grid.** Comparison of three different BP variants: synchronous BP with smoothing (dashed line), asynchronous BP with smoothing (solid line), and asynchronous BP with no smoothing (dash-dot line — only shown in (a)). (a) Percentage of messages converged as a function of time. (b) A marginal where both variants converge rapidly. (c–e) Marginals where the synchronous BP marginals oscillate around the asynchronous BP marginals. (f) A marginal where both variants are inaccurate.

The remaining panels illustrate the progression of the marginal beliefs over the course of the algorithm. (b) shows a marginal where both the synchronous and asynchronous updates converge quite rapidly and are close to the true marginal (thin solid black). Such behavior is atypical, and it comprises only around 10 percent of the marginals in this example. In the vast majority of the cases (almost 80 percent in this example), the synchronous beliefs oscillate around the asynchronous ones ((c)–(e)). In many cases, such as the ones shown in (e), the entropy of the synchronous beliefs is quite significant. For about 10 percent of the marginals (for example (f)), both the asynchronous and synchronous marginals are inaccurate. In these cases, using more informed message schedules can significantly improve the algorithms performance.

These qualitative differences between the BP variants are quite consistent across many random and real-life models. Typically, the more complex the inference problem, the larger the gaps in performance. For very complex real-life networks involving tens of thousands of variables and multiple cycles, even asynchronous BP is not very useful and more elaborate propagation methods or convergent alternatives must be adopted.

## 5 Max-Product Message Passing for MAP Inference

We now consider the application of belief propagation algorithms to the task of computing the MAP assignment, as in (1).

### 5.1 Computing Max-Marginals

The MAP task goes hand in hand with finding the value of the unnormalized probability of the most likely assignment: $\max_{\boldsymbol{x}} \tilde{P}_\Phi(\boldsymbol{x})$. We note that, given an assignment $\boldsymbol{x}$, we can easily compute its unnormalized probability simply by multiplying all of the factors in $\Phi$, evaluated at $\boldsymbol{x}$. However, we cannot retrieve the *actual* probability of $\boldsymbol{x}$ without computing the partition function, a problem that requires that we also solve the sum-product task. Because $\tilde{P}_\Phi$ is a product of factors, tasks that involve maximizing $\tilde{P}_\Phi$ are often called *max-product* inference tasks.

A large subset of algorithms for the MAP problem operate by first computing a set of factors that are *max-marginals*. For a general function $f$, we define the *max-marginal* of $f$ relative to a set of variables $\boldsymbol{Y}$ as

$$MaxMarg_f(\boldsymbol{y}) = \max_{\boldsymbol{x}\langle \boldsymbol{Y}\rangle=\boldsymbol{y}} f(\boldsymbol{x}), \qquad (5)$$

for any assignment $\boldsymbol{y} \in Val(\boldsymbol{Y})$. For example, the max-marginal $MaxMarg_{\tilde{P}_\Phi}(\boldsymbol{Y})$ is a factor that determines a value for each assignment $\boldsymbol{y}$ to $\boldsymbol{Y}$; this value is the unnormalized probability of the most likely joint assignment consistent with $\boldsymbol{y}$.

The same belief propagation algorithms that we showed for sum-product can easily be adapted to the case of max-product. In particular, the *max-product belief propagation* algorithm in clique trees executes precisely the same initialization and overall message scheduling as in the sum-product clique tree algoirthm; the only difference is that we replace (2) with the following:

$$\delta_{i \to j} = \max_{\boldsymbol{C}_i - \boldsymbol{S}_{i,j}} \pi_i^0 \times \prod_{k \in (\mathcal{N}_i - \{j\})} \delta_{k \to i}. \qquad (6)$$

As for sum-product message passing, the algorithm will converge after a single upward and downward pass. After those steps, the resulting clique tree $\mathcal{T}$ will contain the appropriate max-marginal in every clique. In particular, for each clique $\boldsymbol{C}_i$ and each assignment $\boldsymbol{c}_i$ to $\boldsymbol{C}_i$, we will have that

$$\pi_i[\boldsymbol{c}_i] = MaxMarg_{\tilde{P}_\Phi}(\boldsymbol{c}_i). \qquad (7)$$

That is, the clique belief contains, for each assignment $\boldsymbol{c}_i$ to the clique variables, the (unnormalized) measure $\tilde{P}_\Phi(\boldsymbol{x})$ of the most likely assignment $\boldsymbol{x}$ consistent with $\boldsymbol{c}_i$. Note that, because the max-product message passing process does not compute the partition function, we *cannot* derive from these max-marginals the actual probability of any assignment; however, because the partition function is a constant, we can still compare the values associated with different assignments, and therefore compute the assignment $x$ that maximizes $\tilde{P}_\Phi(x)$.

Because max-product message passing over a clique tree produces max-marginals in every clique, and because max-marginals must agree, it follows that any two adjacent cliques must agree on their sepset:

$$\max_{C_i - S_{i,j}} \pi_i = \max_{C_j - S_{i,j}} \pi_j = \mu_{i,j}(S_{i,j}). \tag{8}$$

In this case, the clusters are said to be *max-calibrated*. We say that a clique tree is *max-calibrated* if all pairs of adjacent cliques are max-calibrated.

The same transformation from sum-product to max-product can be applied to the case of loopy belief propagation. Here, the algorithm is the same as in figure 2, except that we replace the sum-product message computation with the max-product computation of (6). As for sum-product, there are no guarantees that this algorithm will converge. Indeed, in practice, it tends to converge somewhat less often than the sum-product algorithm, perhaps because the averaging effect of the summation operation tends to smooth out messages, and reduce oscillations. The same ideas that we discussed in section 4.3 can be used to improve convergence in this algorithm as well.

At convergence, the result will be a set of calibrated clusters: As for sum-product, if the clusters are not calibrated, convergence has not been achieved, and the algorithm will continue iterating. However, the resulting beliefs will not generally be the exact max-marginals; these beliefs are often called *pseudo-max-marginals*.

## 5.2 Locally Optimal Assignments

How do we go from a set of (approximate) max-marginals to a consistent joint assignment that has high probability? One obvious solution is to use the max-marginal for each variable $X_i$ to compute its own optimal assignment, and thereby compose a full joint assignment to all variables. However, this simplistic approach may not always work, even if we have exact max-marginals.

EXAMPLE 3. Consider a simple XOR-like distribution $P(X_1, X_2)$ that gives probability 0.1 to the assignments where $X_1 = X_2$ and 0.4 to the assignments where $X_1 \neq X_2$. In this case, for each assignment to $X_1$, there is a corresponding assignment to $X_2$ whose probability is 0.4. Thus, the max-marginal of $X_1$ is the symmetric factor $(0.4, 0.4)$, and similarly for $X_2$. Indeed, we can choose either of the two values for $X_1$ and complete it to a MAP assignment, and similarly for $X_2$. However, if we choose the values for $X_1$ and $X_2$ in an *inconsistent* way, we may get an assignment whose probability is much lower. Thus, our joint assignment cannot be chosen by separately optimizing the individual max-marginals.

Such examples cannot arise if the max-marginals are *unambiguous*: For each variable $X_i$, there is a unique $x_i^*$ that maximizes:

$$x_i^* = \max_{x_i \in Val(X_i)} MaxMarg_f(x_i). \qquad (9)$$

This condition prevents symmetric cases like the one in the preceding example. Indeed, it is not difficult to show that the following two conditions are equivalent:

- The set of node beliefs $\{MaxMarg_{\tilde{P}_\Phi}(X_i) : X_i \in \mathcal{X}\}$ is unambiguous, with

$$x_i^* = \mathrm{argmax}_{x_i} MaxMarg_{\tilde{P}_\Phi}(X_i)$$

  the unique optimizing value for $X_i$;

- $\tilde{P}_\Phi$ has a unique MAP assignment $(x_1^*, \ldots, x_n^*)$.

For generic probability measures, the assumption of unambiguity is not overly stringent, since we can always break ties by introducing a slight random perturbation into all of the factors, making all of the elements in the joint distribution have slightly different probabilities. However, if the distribution has special structure — deterministic relationships or shared parameters — that we want to preserve, this type of ambiguity may be unavoidable.

The situation where there are ties in the node beliefs is more complex. In this case, we say that an assignment $\boldsymbol{x}^*$ has the *local optimality property* if, for each cluster $\boldsymbol{C}_i$ in the tree, we have that

$$\boldsymbol{x}^*\langle \boldsymbol{C}_i \rangle \in \mathrm{argmax}_{\boldsymbol{c}_i} \pi_i[\boldsymbol{c}_i], \qquad (10)$$

that is, the assignment to $\boldsymbol{C}_i$ in $\boldsymbol{x}^*$ optimizes the $\boldsymbol{C}_i$ belief. The task of finding a locally optimal assignment $\boldsymbol{x}^*$ given a max-calibrated set of beliefs is called the *decoding* task.

Importantly, for approximate max-marginals derived from loopy belief propagation, a locally optimal joint assignment may not exist:

EXAMPLE 4. Consider a cluster graph with the three clusters $\{A, B\}, \{B, C\}, \{A, C\}$ and the beliefs

|       | $a^1$ | $a^0$ |
|-------|-------|-------|
| $b^1$ | 1     | 2     |
| $b^0$ | 2     | 1     |

|       | $b^1$ | $b^0$ |
|-------|-------|-------|
| $c^1$ | 1     | 2     |
| $c^0$ | 2     | 1     |

|       | $a^1$ | $a^0$ |
|-------|-------|-------|
| $c^1$ | 1     | 2     |
| $c^0$ | 2     | 1     |

These beliefs are max-calibrated, in that all messages are $(2, 2)$. However, there is no joint assignment that maximizes all of the cluster beliefs simultaneously. For example, if we select $a^0, b^1$, we maximize the value in the $A, B$ belief. We can now select $c^0$ to maximize the value in the $B, C$ belief. However, we now have a nonmaximizing assignment $a^0, c^0$ in the $A, C$ belief. No matter which assignment of values we select in this example, we do not obtain a single joint assignment that maximizes all three beliefs. Loops such as this are often called *frustrated*.

How do we find a locally optimal joint assignment, if one exists? Recall from the definition that an assignment is locally optimal if and only if it selects one of the optimizing assignments in every single cluster. Thus, we can essentially label the assignments in each cluster as either "legal" if they optimize the belief or "illegal" if they do not. We now must search for an assignment to $\mathcal{X}$ that results in a legal value for each cluster. This problem is precisely an instance of a *constraint satisfaction problem (CSP)*. A constraint satisfaction problem can be defined in terms of a Markov network (or factor graph) where all of the entries in the beliefs are either 0 or 1. The CSP problem is now one of finding an assignment whose (unnormalized) measure is 1, if one exists, and otherwise reporting failure. In other words, the CSP problem is simply that of finding the MAP assignment in this model with $\{0, 1\}$-valued beliefs. The field of CSP algorithms is a large one, and a detailed survey is outside the scope of the paper; see Dechter [2003] for a recent survey. Interestingly, it is an area to which Pearl also made important early contributions [Dechter and Pearl 1987]. Recent work has reinvigorated this trajectory, studying the surprisingly deep connections between CSP methods and belief propagation, and exploiting it (for example, within the context of the *survey propagation* algorithm [Maneva, Mossel, and Wainwright 2007]).

Thus, given a max-product calibrated cluster graph, we can convert it to a discrete-valued CSP by simply taking the belief in each cluster, changing each assignment that locally optimizes the belief to 1 and all other assignments to 0. We then run some CSP solution method. If the outcome is an assignment that achieves 1 in every belief, this assignment is guaranteed to be a locally optimal assignment. Otherwise, there is no locally optimal assignment. Importantly, as we discuss below, for the case of calibrated clique trees, we are guaranteed that this approach finds a globally optimal assignment.

In the case where there is no locally optimal assignment, we must resort to the use of alternative solution methods. One heuristic in this latter situation is to use information obtained from the max-product propagation to construct a partial assignment. For example, assume that a variable $X_i$ is unambiguous in the calibrated cluster graph, so that the only value that locally optimizes its node marginal is $x_i$. In this case, we may decide to restrict attention only to assignments where $X_i = x_i$. In many real-world problems, a large fraction of the variables in the network are unambiguous in the calibrated max-product cluster graph. Thus, this heuristic can greatly simplify the model, potentially even allowing exact methods (such as clique tree inference) to be used for the resulting restricted model. We note, however, that the resulting assignment would not necessarily satisfy the local optimality condition, and all of the guarantees we will present hold only under that assumption.

### 5.3 Optimality Guarantees

The local optimality property comes with some fairly strong guarantees. In particular, for exact max-marginals, one can show the following result:

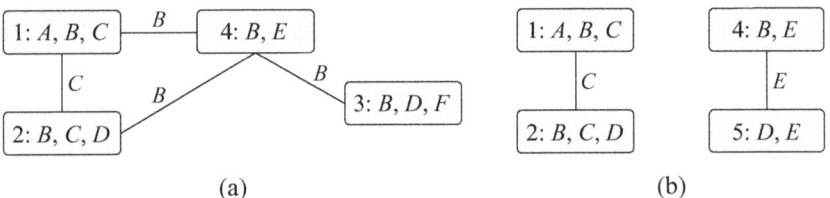

Figure 5. **Two induced subgraphs derived from figure 1a.** (a) Graph over $\{B, C\}$; (b) Graph over $\{C, E\}$.

THEOREM 5. *Let $\pi_i[C_i]$ be a set of max-marginals for the distribution $\tilde{P}_\Phi$, and let $\mu_{i,j}$ be the associated sepset beliefs. Then an assignment $x^*$ satisfies the local optimality property relative to the beliefs $\{\pi_i[C_i]\}_{i \in \mathcal{V}_\mathcal{T}}$ if and only if it is the global MAP assignment relative to $\tilde{P}_\Phi$.*

What type of guarantee can we provide for a decoded assignment from the pseudo-max-marginals produced by the max-product belief propagation algorithm? It is certainly not the case that this assignment is the MAP assignment; nor is it even the case that we can guarantee that the probability of this assignment is "close" in any sense to that of the true MAP assignment. However, if we can construct a locally optimal assignment $x^*$ relative to the beliefs produced by max-product BP, we can prove that $x^*$ is a *strong local maximum*, in the following sense: For certain subsets of variables $Y \subset \mathcal{X}$, there is no assignment $x'$ that is higher-scoring than $x^*$ and differs from it only in the assignment to $Y$. These subsets $Y$ are those that induce any disjoint union of subgraphs each of which contains at most a single loop (including trees, which contain no loops).

More precisely, for a subset of variables $Y$, we define the *induced subgraph* $\mathcal{U}_Y$ to be the subgraph of clusters and sepsets in $\mathcal{U}$ that contain some variable in $Y$. In the straightforward cluster graph for a pairwise Markov network (as described earlier), the induced subgraph for a set $Y$ is simply the set of nodes corresponding to $Y$ and any edges that contain them. Figure 5 shows two examples of an induced subgraph for a more general cluster graph.

We can now state the following important theorem:

THEOREM 6. *Let $\mathcal{U}$ be a max-product calibrated cluster graph for $\tilde{P}_\Phi$, and let $x^*$ be a locally optimal assignment for $\mathcal{U}$. Let $Z$ be any set of variables for which $\mathcal{U}_Z$ is a collection of disjoint subgraphs each of which contains at most a single loop. Then for any assignment $x'$ which is the same as $x^*$ except for the assignment to the variables in $Z$, we have that $\tilde{P}_\Phi(x') \leq \tilde{P}_\Phi(x^*)$.*

This result generalizes one by Weiss and Freeman [2001], who showed a corresponding version in the unambiguous case, for a pairwise Markov network. Its proof in the more general case rests heavily on the analysis of Wainwright, Jaakkola, and Willsky [2005], who proved that a different variant of max-product message passing, if it converges to an unambiguous solution, is guaranteed to produce the true MAP assignment.

This theorem implies as a corollary the (well-known) result that, for a max-product calibrated clique tree, the decoding process is guaranteed to produce a globally optimal (MAP) assignment. However, its more important implications are in the context of a loopy cluster graph.

EXAMPLE 7. Consider a $4 \times 4$ grid network, and assume that we use the pairwise cluster graph construction described earlier. In this case, theorem 6 implies that the MAP solution found by max-product belief propagation has higher probability than any assignment obtained by changing the assignment to any of the following subsets of variables $Y$: a set of variables in any single row, such as $Y = \{A_{1,1}, A_{1,2}, A_{1,3}, A_{1,4}\}$; a set of variables in any single column; a "comb" structure such as the variables in row 1, column 2 and column 4; a single loop, such as $Y = \{A_{1,1}, A_{1,2}, A_{2,2}, A_{2,1}\}$; or a collection of disconnected subsets of the preceding form.

This result is a powerful one, inasmuch as it shows that the solution obtained from max-product belief propagation is robust against large perturbations. Thus, although one can construct examples where max-product belief propagation obtains the wrong solutions, these solutions are strong local maxima, and therefore they often have high probability. Conversely, it is important to realize the limitations of this result. For one, it only applies if the max-product belief propagation algorithm converges to a fixed point, which is not always the case; indeed, as we mentioned earlier, convergence here is generally harder to achieve than in the sum-product variant. Second, even if convergence is achieved, one has to be able to decode the resulting pseudo-max-marginals in order to obtain a locally-optimal joint assignment. It is only if these two conditions hold that this result can be brought to bear.

## 6 Conclusions

This paper has reviewed a small fraction of the recent results regarding the belief propagation algorithm. This line of work has been hugely influential in the area of probabilistic modeling, both in practice and in theory. On the practical side, belief propagation algorithms are among the most commonly used for inference in graphical models for which exact inference is intractable. They have been used successfully for a broad range of applications, including message decoding, natural language processing, computer vision, computational biology, web analysis, and many more. There have also been tremendous developments on the algorithmic side, with many important extensions to the basic approach.

On the theoretical side, the work of many people has served to provide a much deeper understanding of the theoretical foundations of this algorithm, which has tremendously influenced our entire perspective on probabilistic inference. One seminal line of work along these lines was initiated by the landmark paper of Yedidia, Freeman, and Weiss [2000, 2005], showing that beliefs obtained as fixed points of the belief propagation algorithm are also solutions to an optimization problem; this problem is an approximation to another optimization problem whose solutions are the exact marginals that would be obtained from clique tree inference. Thus, both exact (clique tree) and approximate (cluster graph) inference can be viewed in terms of optimization of an objective. This observation was the basis for the development of a whole range of novel methods that explored different variations on the formulation of the optimization problem, or different algorithms for performing the optimization. One such line of work uses *convex* versions of the optimization problem underlying belief propagation, a trajectory initiated by Wainwright, Jaakkola, and Willsky [2002]. Algorithms based on this approach (e.g., [Heskes 2006; Hazan and Shashua 2008]) can also guarantee convergence as well as provide bounds on the partition function.

For the MAP problem, a similar optimization-based view has also recently come to dominate the field. Here, the original MAP problem is reformulated as an *integer programming* problem, where the (discrete-valued) variables in the optimization represent the space of possible assignments $x$. This discrete optimization is then relaxed to produce a continuous-valued optimization problem that is a *linear program* (LP). This LP-relaxation approach was first proposed by Schlesinger [1976], and then subsequently rediscovered independently by several researchers. Most notably, Wainwright, Jaakkola, and Willsky [2005] established the first connection between the dual problem to this LP and message passing algorithms, and proposed a new message-passing algorithm (TRW) based on this connection. Many recent works build on these ideas and develop a suite of increasingly better algorithms for solving the MAP inference problem. Some of these algorithms utilize message-passing techniques; others merely adopt the idea of using the LP dual but utilize other optimization methods for solving it. Importantly, for several of these algorithms, one can guarantee that a solution, if one is found, is guaranteed to be the optimal MAP assignment.

In summary, the simple message-passing algorithm first proposed by Pearl has recently returned to revolutionize the world of inference in graphical models. It has dramatically affected both the practice in the field and has led to a new, optimization-based perspective on the foundations of the inference task. This new understanding has, in turn, given rise to the development of much better algorithms, which continue to improve our ability to apply probabilistic graphical models to challenging, real-world applications.

**Acknowledgments** This material in this review paper is extracted from the book of Koller and Friedman [2009], published by MIT Press. Some of this material is based on contributions by Nir Friedman and Gal Elidan. I also thank Amir Globerson, David Sontag, and Yair Weiss for useful discussions regarding MAP inference.

# References

Berrou, C., A. Glavieux, and P. Thitimajshima (1993). Near Shannon limit error-correcting coding: Turbo codes. In *Proc. International Conference on Communications*, pp. 1064–1070.

Cooper, G. (1990). Probabilistic inference using belief networks is NP-hard. *Artificial Intelligence 42*, 393–405.

Dechter, R. (2003). *Constraint Processing*. Morgan Kaufmann.

Dechter, R., K. Kask, and R. Mateescu (2002). Iterative join-graph propagation. In *Proc. 18th Conference on Uncertainty in Artificial Intelligence (UAI)*, pp. 128–136.

Dechter, R. and J. Pearl (1987). Network-based heuristics for constraint-satisfaction problems. *Artificial Intelligence 34*(1), 1–38.

Elidan, G., I. McGraw, and D. Koller (2006). Residual belief propagation: Informed scheduling for asynchronous message passing. In *Proc. 22nd Conference on Uncertainty in Artificial Intelligence (UAI)*.

Frey, B. and D. MacKay (1997). A revolution: Belief propagation in graphs with cycles. In *Proc. 11th Conference on Neural Information Processing Systems (NIPS)*.

Hazan, T. and A. Shashua (2008). Convergent message-passing algorithms for inference over general graphs with convex free energies. In *Proc. 24th Conference on Uncertainty in Artificial Intelligence (UAI)*.

Heckerman, D., E. Horvitz, and B. Nathwani (1992). Toward normative expert systems: Part I. The Pathfinder project. *Methods of Information in Medicine 31*, 90–105.

Heskes, T. (2002). Stable fixed points of loopy belief propagation are minima of the Bethe free energy. In *Proc. 16th Conference on Neural Information Processing Systems (NIPS)*, pp. 359–366.

Heskes, T. (2006). Convexity arguments for efficient minimization of the Bethe and Kikuchi free energies. *Journal of Machine Learning Research 26*, 153–190.

Ihler, A. T., J. W. Fisher, and A. S. Willsky (2005). Loopy belief propagation: Convergence and effects of message errors. *Journal of Machine Learning Research 6*, 905–936.

Jensen, F. V., K. G. Olesen, and S. K. Andersen (1990, August). An algebra of Bayesian belief universes for knowledge-based systems. *Networks 20*(5), 637–659.

Kim, J. and J. Pearl (1983). A computational model for combined causal and diagnostic reasoning in inference systems. In *Proc. 7th International Joint Conference on Artificial Intelligence (IJCAI)*, pp. 190–193.

Koller, D. and N. Friedman (2009). *Probabilistic Graphical Models: Principles and Techniques.* MIT Press.

Lauritzen, S. L. and D. J. Spiegelhalter (1988). Local computations with probabilities on graphical structures and their application to expert systems. *Journal of the Royal Statistical Society, Series B B 50*(2), 157–224.

Maneva, E., E. Mossel, and M. Wainwright (2007, July). A new look at survey propagation and its generalizations. *Journal of the ACM 54*(4), 2–41.

McEliece, R., D. MacKay, and J.-F. Cheng (1998, February). Turbo decoding as an instance of Pearl's "belief propagation" algorithm. *IEEE Journal on Selected Areas in Communications 16*(2).

Mooij, J. M. and H. J. Kappen (2007). Sufficient conditions for convergence of the sum-product algorithm. *IEEE Trans. Information Theory 53*, 4422–4437.

Murphy, K. P., Y. Weiss, and M. Jordan (1999). Loopy belief propagation for approximate inference: an empirical study. In *Proc. 15th Conference on Uncertainty in Artificial Intelligence (UAI)*, pp. 467–475.

Pearl, J. (1988). *Probabilistic Reasoning in Intelligent Systems.* San Mateo, California: Morgan Kaufmann.

Schlesinger, M. (1976). Sintaksicheskiy analiz dvumernykh zritelnikh singnalov v usloviyakh pomekh (syntactic analysis of two-dimensional visual signals in noisy conditions). *Kibernetika 4*, 113–130.

Shafer, G. and P. Shenoy (1990). Probability propagation. *Annals of Mathematics and Artificial Intelligence 2*, 327–352.

Shannon, C. (1948). A mathematical theory of communication. *Bell System Technical Journal 27*, 379–423; 623–656.

Shimony, S. (1994). Finding MAPs for belief networks in NP-hard. *Artificial Intelligence 68*(2), 399–410.

Shwe, M., B. Middleton, D. Heckerman, M. Henrion, E. Horvitz, H. Lehmann, and G. Cooper (1991). Probabilistic diagnosis using a reformulation of the INTERNIST-1/QMR knowledge base. I. The probabilistic model and inference algorithms. *Methods of Information in Medicine 30*, 241–55.

Szeliski, R., R. Zabih, D. Scharstein, O. Veksler, V. Kolmogorov, A. Agarwala, M. Tappen, and C. Rother (2008, June). A comparative study of energy minimization methods for Markov random fields with smoothness-based priors. *IEEE Trans. on Pattern Analysis and Machine Intelligence 30*(6), 1068–1080. See http://vision.middlebury.edu/MRF for more detailed results.

Wainwright, M., T. Jaakkola, and A. Willsky (2003). Tree-based reparameterization framework for analysis of sum-product and related algorithms. *IEEE Transactions on Information Theory 49*(5).

Wainwright, M., T. Jaakkola, and A. Willsky (2005). MAP estimation via agreement on trees: Message-passing and linear programming. *IEEE Transactions on Information Theory*.

Wainwright, M., T. Jaakkola, and A. S. Willsky (2002). A new class of upper bounds on the log partition function. In *Proc. 18th Conference on Uncertainty in Artificial Intelligence (UAI)*.

Weiss, Y. (1996). Interpreting images by propagating bayesian beliefs. In *Proc. 10th Conference on Neural Information Processing Systems (NIPS)*, pp. 908–914.

Weiss, Y. and W. Freeman (2001). On the optimality of solutions of the max-product belief propagation algorithm in arbitrary graphs. *IEEE Transactions on Information Theory 47(2)*, 723–735.

Yedidia, J., W. Freeman, and Y. Weiss (2005). Constructing free-energy approximations and generalized belief propagation algorithms. *IEEE Trans. Information Theory 51*, 2282–2312.

Yedidia, J. S., W. T. Freeman, and Y. Weiss (2000). Generalized belief propagation. In *Proc. 14th Conference on Neural Information Processing Systems (NIPS)*, pp. 689–695.

# 13
# Extending Bayesian Networks to the Open-Universe Case

BRIAN MILCH AND STUART RUSSELL

## 1 Introduction

One of Judea Pearl's now-classic examples of a Bayesian network involves a home alarm system that may be set off by a burglary or an earthquake, and two neighbors who may call the homeowner if they hear the alarm. Like most scenarios modeled with BNs, this example involves a known set of objects (one house, one alarm, and two neighbors) with known relations between them (the alarm is triggered by events that affect this house; the neighbors can hear this alarm). These objects and relations determine the relevant random variables and their dependencies, which are then represented by nodes and edges in the BN.

In many real-world scenarios, however, the relevant objects and relations are initially unknown. For instance, suppose we have a set of ASCII strings containing irregularly formatted and possibly erroneous academic citations extracted from online documents, and we wish to make a list of the distinct publications that are referred to, with correct author names and titles. In this case, the publications, authors, venues, and so on are not known in advance, nor is the mapping between publications and citations. The same challenge of making inferences about unknown objects is called *coreference resolution* in natural language processing, *data association* in multitarget tracking, and *record linkage* in database systems. The issue is actually much more widespread than this short list suggests; it arises in any data interpretation problem in which objects or events come without unique identifiers.

In this chapter, we show how the Bayesian network (BN) formalism that Judea Pearl pioneered has been extended to handle such scenarios. The key contribution on which we build is the use of acyclic directed graphs of local conditional distributions to generate well-defined, global probability distributions. We begin with a review of *relational probability models* (RPMs), which specify how to construct a BN for a given set of objects and relations. We then describe *open-universe probability models*, or OUPMs, which represent uncertainty about what objects exist. OUPMs may not boil down to finite, acyclic BNs; we present results from Milch [2006] showing how to extend the factorization and conditional independence semantics of BNs to models that are only *context-specifically* finite and acyclic. Finally, we discuss how Markov chain Monte Carlo (MCMC) methods can be used to perform approximate inference on OUPMs and briefly describe some applications.

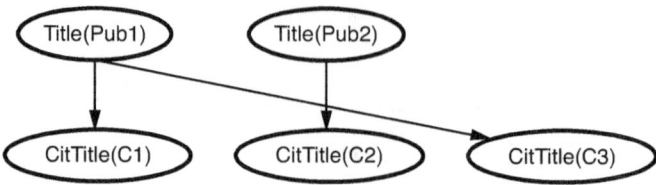

Figure 1. A BN for a bibliography scenario where we know that citations Cit1 and Cit3 refer to Pub1, while citation Cit2 refers to Pub2.

$$\text{Title}(p) \sim \text{TitlePrior}()$$
$$\text{CitTitle}(c) \sim \text{TitleEditCPD}(\text{Title}(\text{PubCited}(c)))$$

Figure 2. Dependency statements for a bibliography scenario, where $p$ ranges over publications and $c$ ranges over citations. This model assumes that the PubCited function and the sets of publications and citations are known.

## 2 Relational probability models

Suppose we are interested in inferring the true titles of publications given some observed citations, and we know the set of publications and the mapping from citations to publications. Assuming we have a prior distribution for true title strings (perhaps a word $n$-gram model) and a conditional probability distribution (CPD) for citation titles given true titles, we can construct a BN for this scenario, as shown in Figure 1.

### 2.1 The RPM formalism

A relational probability model represents such a BN compactly using *dependency statements* (see Figure 2), which specify the CPDs and parent sets for whole classes of variables at once. In this chapter, we will not specify any particular syntax for dependency statements, although we use a syntax based loosely on BLOG [Milch et al. 2005]. The important point is that dependencies are specified via relations among objects. For example, the dependency statement for CitTitle in Figure 2 specifies that each CitTitle($c$) variable depends (according to the conditional distribution TitleEditCPD that describes how titles may be erroneously transcribed) on Title(PubCited($c$))—that is, on the true title of the publication that $c$ cites. The PubCited relation is nonrandom, and thus forms part of the known *relational skeleton* of the RPM. In this case, the skeleton also includes the sets of citations and publications.

Formally, it is convenient to think of an RPM $M$ as defining a probability distribution over a set of model structures of a typed first-order logical language. These structures are called the *possible worlds* of $M$ and denoted $\Omega_M$. The function symbols of the logical language (including constant and predicate symbols) are divided into a set of *nonrandom* function symbols whose interpretations are specified by the relational skeleton, and a set of *random* function symbols whose interpretations vary between possible worlds. An RPM includes one dependency statement for each random function symbol.

Each RPM $M$ defines a set of *basic random variables* $\mathbf{V}_M$, one for each application of a random function to a tuple of arguments. We will write $X(\omega)$ for the value of a random variable $X$ in world $\omega$. If $X$ represents the value of the random function $f$ on some arguments, then the dependency statement for $f$ defines a parent set and CPD for $X$. The parent set for $X$, denoted $\text{Pa}(X)$, is the set of basic variables that are needed to evaluate the expressions in the dependency statement in any possible world. For instance, if we know that PubCited(Cit1) = Pub1, then the dependency statement in Figure 2 yields the single parent Title(Pub1) for the variable CitTitle(Cit1). The CPD for a basic variable $X$ is a function $\varphi_X(x, \mathbf{pa})$ that defines a conditional probability distribution over values $x$ of $X$ given each instantiation $\mathbf{pa}$ of $\text{Pa}(X)$. We obtain this CPD by evaluating the expressions in the dependency statement (such as Title(PubCited(Cit1))) and passing them to an *elementary distribution function* such as TitleEditCPD.

Thus, an RPM defines a BN over its basic random variables. If this BN is acyclic, it defines a joint distribution for the basic RVs. Since there is a one-to-one correspondence between full instantiations of $\mathbf{V}_M$ and worlds in $\Omega_M$, this BN also gives us a probability measure over $\Omega_M$. We define this to be the probability measure represented by the RPM.

## 2.2 Relational uncertainty

This formalism also allows us to model cases of *relational uncertainty*, such as a scenario where the mapping from citations to publications is unknown. We can handle this by making PubCited a random function and giving it a dependency statement such as:

$$\text{PubCited}(c) \sim \text{Uniform}(\{\text{Pub } p\}) \ .$$

This statement says that each citation refers to a publication chosen uniformly at random from the set of all publications $p$. The dependency statement for CitTitle in Figure 2 now represents a *context-specific* dependency: for a given citation $C_i$, the Title($p$) variable that CitTitle($C_i$) depends on varies from world to world.

In the BN defined by this model, shown in Figure 3, the parents of each CitTitle($c$) variable include all variables that might be needed to evaluate the dependency statement for CitTitle($c$) in any possible world. This includes PubCited($c$) and all the Title($p$) variables. The CPD in the BN is a multiplexer that conditions on the appropriate Title($p$) variable for each value of PubCited($c$). If the BN constructed this way is still finite and acyclic, the usual BN semantics hold.

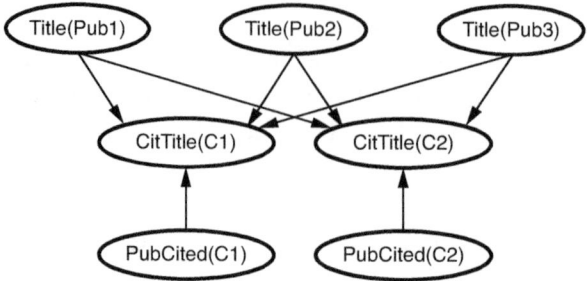

Figure 3. A BN arising from an RPM with relational uncertainty.

## 2.3 Names, objects, and identity uncertainty

We said earlier that the function symbols of an RPM include the constants and predicate symbols. For predicates, this simply means that a predicate can be thought of as a Boolean function that returns *true* or *false* for each tuple of arguments. The constants, on the other hand, are 0-ary functions that refer to objects. In most RPM languages, all constants are nonrandom and assumed to refer to distinct objects—the *unique names* assumption for constants. With this assumption, there is no need to distinguish between constant symbols and the objects they refer to, which is why we are able to name the basic random variables Title(Pub1), Title(Pub2) and so on, even though, strictly speaking, the arguments should be objects in the domain rather than constant symbols.

If the RPM language allows constants to be random functions, then the equivalent BN will include a node for each such constant. For example, suppose that Milch asks Russell to "fix the typo in the Pfeffer citation." Russell's mental software may already have formed nonrandom constant symbols C1, C2, and so on for all the citations at the end of the chapter, and these are in one-to-one correspondence with all the objects in this particular universe. It may then form a new constant symbol ThePfefferCitation, which co-refers with one of these. Because there is more than one citation to a work by Pfeffer, there is *identity uncertainty* concerning which citation object the new symbol refers to. Identity uncertainty is a degenerate form of relational uncertainty, but often has a quite distinct flavor.

## 3 Open-universe probability models

For all RPMs, even those with relational and identity uncertainty, the *objects* are known and are the same across all possible worlds. If the set of objects is unknown, however—e.g., if we don't know the set of publications that exist and might be cited—then RPMs as we have described them do not suffice. Whereas an RPM can be seen as defining a generative process that chooses a value for each random function on each tuple of arguments, an *open-universe probability model* (OUPM) includes generative steps that add objects to the world. These steps set the values of *number variables*.

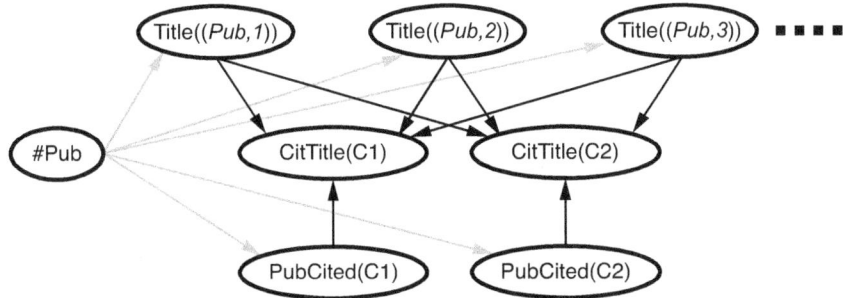

Figure 4. A BN that defines a probability distribution over worlds with unbounded numbers of publications.

## 3.1 Number variables

For the bibliography example, we introduce just one number variable, defining the total number of publications. There is no reason to place any *a priori* upper bound on the number of publications; we might be interested in asking how many publications there are for which we have found no citations (this question becomes more well-defined and pressing if we ask, say, how many aircraft are in an area but have not generated a blip on our radar screens). Thus, this number variable may have a distribution that assigns positive probability to all natural numbers.

We can specify the conditional probability distribution for a number variable using a dependency statement. In our bibliography example, we might use a very simple statement:

$$\#\mathsf{Pub} \sim \mathsf{NumPubsPrior}() \ .$$

Number variables can also depend on other variables; we will consider an example of this below.

In the RPM where we had a fixed set of publications, the relational skeleton specified a constant symbol such as Pub1 for each publication. In an OUPM where the set of publications is unknown, it does not make sense for the language to include such constant symbols. The possible worlds contain publication objects—which will assume are pairs $\langle Pub, 1\rangle$, $\langle Pub, 2\rangle$, etc.—but now they are not necessarily in one-to-one correspondence with any constant symbols.

The set of basic variables now includes the number variable #Pub itself, and variables for the application of each random function to all arguments that exist in any possible world. Figure 4 shows the BN over these variables. Note that we have an infinite sequence of Title variables: if we had a finite number, our BN would not define probabilities for worlds with more than that number of publications. We stipulate that if a basic variable has an object $o$ as an argument, then in worlds where $o$ does not exist, the variable takes on the special value null. Thus, #Pub is a parent of each Title($p$) variable, determining whether that variable takes the value null or not. The set of publications available for selection in the dependency

> \#Researcher $\sim$ NumResearchersPrior()
>
> Position($r$) $\sim$ [0.7 : GradStudent, 0.2 : PostDoc, 0.1 : Prof]
>
> \#Pub(FirstAuthor $= r$) $\sim$ NumPubsCPD(Position($r$))

Figure 5. Dependency statements that augment our bibliography model to represent a set of researchers, the position of each researcher, and the set of first-authored publications by each researcher.

statement for PubCited($c$) also depends on the number variable.

Objects of a given type may be generated by more than one event in the generative process. For instance, if we include objects of type Researcher in our model and add a function FirstAuthor($p$) that maps publications to researchers, we may wish to say that each researcher independently generates a crop of papers on which he or she is the first author. The number of papers generated may depend on the researcher's position (graduate student, professor, etc.). We now get a family of number variables \#Pub(FirstAuthor $= r$), where $r$ ranges over researchers. The number of researchers may itself be governed by a number variable. Figure 5 shows the dependency statements for these aspects of the scenario.

In this model, FirstAuthor($p$) is an *origin function*: in the generative model underlying the OUPM, it is set when $p$ is created, not in a separate generative step. The values of origin functions on an object tell us which number variable governs that object's existence; for example, if FirstAuthor($p$) is $\langle Researcher, 5\rangle$, then \#Pub(FirstAuthor $= \langle Researcher, 5\rangle$) governs the existence of $p$. Origin functions can also be used in dependency statements, just like any other function: for instance, we might change the dependency statement for PubCited($c$) so that more significant publications are more likely to be cited, with the significance of a publication $p$ being influenced by Position(FirstAuthor($p$)).

In the scenario we have considered so far, each possible world contains finitely many Researcher and Pub objects. OUPMs can also accommodate infinite numbers of objects. For instance, we could define a model for academia where each researcher $r$ generates a random number of new researchers $r'$ such that Advisor($r'$) = $r$. Some possible worlds in this model may contain infinitely many researchers.

## 3.2 Possible worlds and basic random variables

In defining the semantics of RPMs, we said that a model $M$ defines a BN over its basic random variables $\mathbf{V}_M$, and then we exploited the one-to-one correspondence between full instantiations of those variables and possible worlds. In an OUPM, however, there may be instantiations of the basic random variables that do not correspond to any possible world. An example in our bibliography scenario is an instantiation where #Pub = 100, but Title($p$) takes on a non-null value for 200 publications.

To facilitate using the basic variables to define a probability measure over the possible worlds, we would like to have a one-to-one mapping between $\Omega_M$ and a set of *achievable* instantiations of $\mathbf{V}_M$. This is straightforward in cases like our first OUPM, where there is only one number variable for each type of object. Then our semantics specifies that the *non-guaranteed objects* of each type—that is, the objects that exist in some possible worlds and not others, like the publications in our example—are pairs $\langle Pub, 1\rangle, \langle Pub, 2\rangle, \ldots$. In each world, the set of non-guaranteed objects of each type that exist is required to be a prefix of this numbered sequence. Thus, if we know that #Pub = 4 in a world $\omega$, we know that the publications in $\omega$ are $\langle Pub, 1\rangle$ through $\langle Pub, 4\rangle$, not some other set of non-guaranteed objects.

Things are more complicated when we have multiple number variables for a type, as in our example with researchers generating publications. Given values for all the number variables of the form #Pub(FirstAuthor = $r$), we do not want there to be any uncertainty about *which* non-guaranteed objects have each FirstAuthor value. We can achieve this by letting the non-guaranteed objects be nested tuples that encode their generation history. For the publications with $\langle Researcher, 5\rangle$ as their first author, we use tuples

$$\langle Pub, \langle FirstAuthor, \langle Researcher, 5\rangle\rangle, 1\rangle$$
$$\langle Pub, \langle FirstAuthor, \langle Researcher, 5\rangle\rangle, 2\rangle$$

and so on. As before, in each possible world, the set of tuples in each sequence must form a prefix of the sequence. This construction yields the following lemma.

LEMMA 1. *In any OUPM $M$, each complete instantiation of $\mathbf{V}_M$ is consistent with at most one possible world in $\Omega_M$.*

Section 4.3 of Milch [2006] gives a more rigorous formulation and proof of this result. Given this lemma, the probability measure defined by an OUPM $M$ on $\Omega_M$ is well-defined if the OUPM specifies a joint probability distribution for $\mathbf{V}_M$ that is concentrated on the set of achievable instantiations. Since the OUPM's CPDs implicitly force a variable to take the value null when any of its arguments do not exist, any distribution consistent with the CPDs will indeed put probability one on achievable instantiations.

Informally, the probability distribution for the basic random variables can be defined by a generative process that builds up an instantiation step-by-step, sampling a value for each variable according to its dependency statement. In the next section, we show how this intuitive semantics can be formalized using an extended version of Bayesian networks.

## 4 Extending BN semantics

There are two equivalent ways of defining the probability distribution represented by a BN $\mathbf{B}$. The first is based on conditional independence statements; specifically,

```
#Pub ~ NumPubsPrior()

Title(p) ~ TitlePrior()

Date(c) ~ DatePrior()

SourceCopied(c) ~ [0.9 : null,
                   0.1 : Uniform({Citation c_2 :
                        (PubCited(c_2) = PubCited(c))
                        ∧ (Date(c_2) < Date(c))})]

CitTitle(c) ~ if SourceCopied(c) = null
              then TitleEditCPD(Title(PubCited(c)))
              else TitleEditCPD(CitTitle(SourceCopied(c)))
```

Figure 6. Dependency statements for a model where each citation was written on some date, and a citation may copy the title from an earlier citation of the same publication rather than copying the publication title directly.

the directed local Markov property: each variable is conditionally independent of its non-descendants given its parents. The second is based on a product expression for the joint distribution; if $\sigma$ is any instantiation of the full set of variables $\mathbf{V_B}$ in the BN, then

$$P(\sigma) = \prod_{X \in \mathbf{V_B}} \varphi_X \left( \sigma[X], \sigma\left[\text{Pa}(X)\right] \right) .$$

The remarkable property of BNs is that if the graph is finite and acyclic, then there is guaranteed to be exactly one joint distribution that satisfies these conditions.

## 4.1 Infinite sets of variables

Note that in the BN in Figure 4, the CitTitle($c$) variables have infinitely many parents. The fact that the BN has infinitely many nodes means that we can no longer use the standard product-expression semantics for the BN, because the product of the CPDs for all variables is an infinite product, and will typically be zero for all values of the variables. We would like to specify probabilities for certain partial, finite instantiations of the variables that are sufficient to define the joint distribution. As noted by Kersting and DeRaedt [2001], if it is possible to number the nodes of the BN in topological order, then it suffices to specify the product expression for each finite prefix of this numbering. However, if a variable has infinitely many parents, then the BN has no topological numbering—if we try numbering the nodes in topological order, we will spend forever on $X$'s parents and never reach $X$.

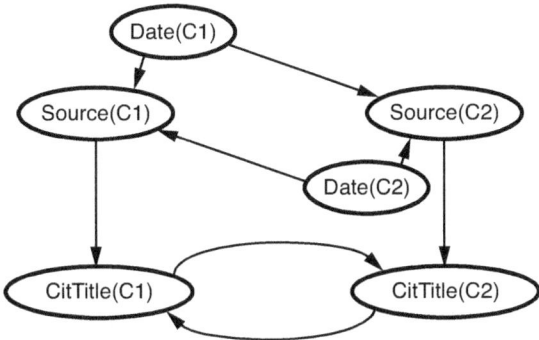

Figure 7. Part of the BN defined by the OUPM in Figure 6, for two citations.

### 4.2 Cyclic sets of potential dependencies

In OUPMs and even RPMs with relational uncertainty, it is fairly easy to write dependency statements that define a cyclic BN. For instance, suppose that some citations are composed by copying another citation, and we do not know who copied whom. We can specify a model where each citation was written at some unknown date, and with probability 0.1, a citation copies an earlier citation to the same publication if one exists. Figure 6 shows the dependency statements for this model. (Note that Date here is the date the citation was written, i.e., the date of the citing paper, not the date of the paper being cited.)

The BN defined by this OUPM is cyclic, as shown in Figure 7. In general, a cyclic BN may fail to define a distribution; there may be no joint distribution with the specified CPDs. However, in this case, it is intuitively clear that that cannot happen. Since a citation can only copy another citation with a strictly earlier date, the dependencies that are active in any positive-probability world must be acyclic. There are actually elements of the possible world set $\Omega_M$ where the dependencies are cyclic: these are worlds where, for some citation $c$, SourceCopied($c$) does not have an earlier date than $c$. But the CPD for SourceCopied forces these worlds to have probability zero.

The difficult aspect of semantics for this class of cyclic BNs is the directed local Markov property. It is no longer sufficient to assert that $X$ is independent of its non-descendants in the full BN given its parents, because its set of non-descendants in the full BN may be too small. In this model, all the CitTitle nodes are descendants of each other, so the standard directed local Markov property would yield no assertions of conditional independence between them.

## 4.3 Partition-based semantics for OUPMs

We can solve these difficulties by exploiting the context-specific nature of dependencies in an OUPM, as revealed by dependency statements.[1] For each basic random variable $X$, an OUPM defines a partition $\Lambda_X$ of $\Omega_M$. Two worlds are in the same block of this partition if evaluating the dependency statement for $X$ in these two worlds yields the same conditional distribution for $X$. For instance, in our OUPM for the bibliography domain, the partition blocks for CitTitle(Cit1) are sets of worlds that agree on the value of Title(PubCited(Cit1)). For each block $\lambda \in \Lambda_X$, the OUPM defines a probability distribution $\varphi_X(x, \lambda)$ over values of $X$.

One defining property of the probability measure $P_M$ specified by an OUPM $M$ is that for each basic random variable $X \in \mathbf{V}_M$ and each partition block $\lambda \in \Lambda_X$,

$$P_M(X = x \mid \lambda) = \varphi_X(x, \lambda) \tag{1}$$

To fully define $P_M$, however, we need to make an assertion analogous to a BN's factorization property or directed local Markov property. We will say that a partial instantiation $\sigma$ *supports* a random variable $X$ if there is some block $\lambda \in \Lambda_X$ such that $\sigma \subseteq \lambda$. An instantiation that supports $X$ in an OUPM is an analogous to an instantiation that assigns values to all the parents of $X$ in a BN. We define an instantiation $\sigma$ to be *self-supporting* if for each variable $X \in \text{vars}(\sigma)$, the restriction of $\sigma$ to $\text{vars}(\sigma) \setminus \{X\}$ (denoted $\sigma_{-X}$) supports $X$. We can now state a factorization property for OUPMs.

PROPERTY 2 (Factorization property for an OUPM $M$). For each finite, self-supporting instantiation $\sigma$ on $\mathbf{V}_M$,

$$P_M(\sigma) = \prod_{X \in \text{vars}(\sigma)} \varphi_X(\sigma[X], \lambda_X(\sigma_{-X}))$$

where $\lambda_X(\sigma_{-X})$ is the partition block in $\Lambda_X$ that has $\sigma_{-X}$ as a subset.

We can also define an analogue of the directed local Markov property for OUPMs. Recall that in the BN case, the directed local Markov property asserts that $X$ is conditionally independent of every subset of its non-descendants given $\text{Pa}(X)$. In fact, it turns out to be sufficient to make this assertion for only a special class of non-descendant subsets, namely those that are *ancestral* (closed under the parent relation). Any ancestral set of variables that does not contain $X$ contains only non-descendants of $X$. So in the BN case, we can reformulate the directed local Markov property to assert that given $\text{Pa}(X)$, $X$ is conditionally independent of any ancestral set of variables that does not contain $X$.

In OUPMs, the equivalent of a variable set that is closed under the parent relation is a self-supporting instantiation. We can formulate the directed local Markov property for an OUPM $M$ as follows:

---

[1] We will assume all random variables are discrete in this treatment, but the ideas can be extended to the continuous case.

PROPERTY 3 (Directed local Markov property for an OUPM $M$). *For each basic random variable $X \in \mathbf{V}_M$, each block $\lambda \in \Lambda_X$, and each self-supporting instantiation $\sigma$ on $\mathbf{V}_M$ such that $X \notin \text{vars}(\sigma)$, $X$ is conditionally independent of $\sigma$ given $\lambda$.*

Under what conditions is there a unique probability measure $P_M$ on $\Omega_M$ that satisfies Properties 2 and 3? In the BN case, it suffices for the graph to admit a topological numbering. We can define a similar notion that is specific to individual worlds: a *supportive numbering* for a world $\omega \in \Omega_M$ is a numbering $X_0, X_1, \ldots$ of $\mathbf{V}_M$ such that for each natural number $n$, the instantiation $(X_0(\omega), \ldots, X_{n-1}(\omega))$ supports $X_n$.

THEOREM 4. *Let $M$ be an OUPM such that for every world $\omega \in \Omega_M$, either:*

- *$\omega$ has a supportive numbering, or*

- *for some basic random variable $X \in \mathbf{V}_M$, $\varphi_X(X(\omega), \lambda_X(\omega)) = 0$.*

*Then there is exactly one probability measure on $\Omega_M$ satisfying the factorization property (Property 2), and it is also the unique probability measure that satisfies both Equation 1 and the directed local Markov property (Property 3).*

This theorem follows from Lemma 1 and results proved in Section 3.4 of Milch [2006]. Note that the theorem does not require supportive numberings for worlds that are *directly disallowed*—that is, those that are forced to have probability zero by the CPD for some variable.

In our basic bibliography scenario with unknown publications, we can construct a supportive numbering for each possible world $\omega$ by taking first the number variable #Pub, then the PubCited($c$) variables, then the Title($p$) variables for the publications that serve as values of PubCited($c$) variables in $\omega$, then the CitTitle($c$) variables, and finally the infinitely many Title($p$) variables for publications that are uncited or do not exist in $\omega$. For the scenario where citation titles can be copied from earlier citations, we have to add the Date($c$) variables and then the SourceCopied($c$) variables before the CitTitle($c$) variables. We order the CitTitle($c$) variables in a way that is consistent with Date($c$). This procedure yields a supportive numbering in all worlds except those where $\exists c\, \text{Date}(\text{SourceCopied}(c)) \geq \text{Date}(c)$, but such worlds are directly disallowed by the CPD for SourceCopied($c$).

## 4.4 Representing OUPMs as contingent Bayesian networks

The semantics we have given for OUPMs so far does not make reference to any graph. But we can also view an OUPM as defining a *contingent Bayesian network* (CBN) [Milch et al. 2005], which is a BN where each edge is labeled with an event. The event indicates when the edge is active, in a sense we will soon make precise. Figures 8 and 9 show CBNs corresponding to the infinite BN in Figure 4 and the cyclic BN in Figure 7, respectively.

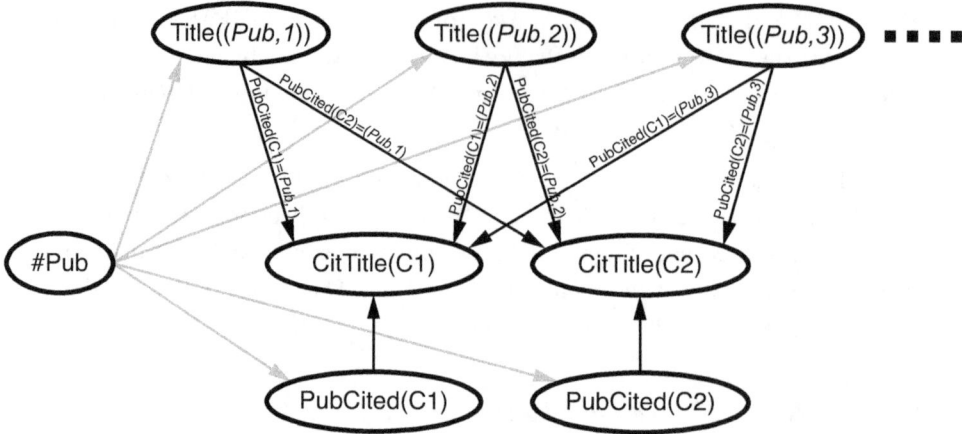

Figure 8. A contingent BN for the bibliography scenario with unknown objects.

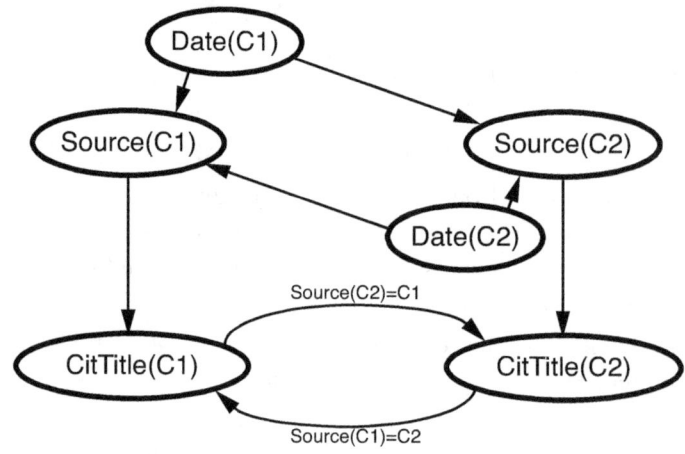

Figure 9. Part of a contingent BN for the OUPM in Figure 6.

A CBN can be viewed as a partition-based model where the partition $\Lambda_X$ for each random variable $X$ is defined by a decision tree. The internal nodes in this decision tree are labeled with random variables; the edges are labeled with variable values; and the leaves specify conditional probability distributions for $X$. The blocks in $\Lambda_X$ correspond to the leaves in this tree (we assume the tree has no infinite paths, so the leaves cover all possible worlds). The restriction to decision trees allows us to define a notion of a parent being active in a particular world: if we walk along $X$'s tree from the root, following edges consistent with a given world $\omega$, then the random variables on the nodes we visit are the active parents of $X$ in $\omega$. The label on an edge $W \to X$ in a CBN is the event consisting of those worlds where $W$ is an active parent of $X$. (In diagrams, we omit the trivial label $A = \Omega_M$, which indicates that the dependency is always active.)

The abstract notions of a self-supporting instantiation and a supportive numbering have simple graphical analogues in a CBN. We will use $\mathbf{B}^\sigma$ to denote the BN obtained from a CBN $\mathbf{B}$ by keeping only those edges whose conditions are entailed by $\sigma$. An instantiation $\sigma$ supports a variable $X$ if and only if all the parents of $X$ in $\mathbf{B}^\sigma$ are in vars($\sigma$), and it is self-supporting if and only if vars($\sigma$) is an ancestral set in $\mathbf{B}^\sigma$. A supportive numbering for a world $\omega$ is a topological numbering of the BN $\mathbf{B}^\omega$ obtained by keeping only those edges whose conditions are satisfied by $\omega$. Thus, the well-definedness condition in Theorem 4 can be stated for CBNs as follows: for each world $\omega \in \Omega_M$ that is not directly disallowed, $\mathbf{B}^\omega$ must have a topological numbering.

Not all partitions can be represented exactly as the leaves of a decision tree, so there are sets of context-specific independence properties that can be captured by OUPMs and not CBNs. However, when we perform inference on an OUPM, we typically use a function that evaluates the dependency statement for each variable, looking up the values of other random variables in a given world (or partial instantiation) as needed. For example, a function evaluating the dependency statement for CitTitle(Cit1) will always access PubCited(Cit1), and then it will access a particular Title variable depending on the value of the PubCited variable. This evaluation process implicitly defines a decision tree; the order of splits in the tree depends on the evaluation order used. When we discuss inference for OUPMs, we will assume that we are operating on the CBN implicitly defined by some evaluation function.

## 5 Inference

Given an OUPM, we would like to be able to compute the probability of a query event $Q$ given an evidence event $E$. For example, $Q$ could be the event that PubCited(Cit1) = PubCited(Cit2) and $E$ could be the event that CitTitle(Cit1) = "Learning Probabilistic Relational Models" and CitTitle(Cit2) = "Learning Probabilitsic Relation Models". The ideas we present can be extended to other tasks such as computing the posterior distribution of a random variable, or finding the *maximum a posteriori* (MAP) assignment of values to a set of random variables.

### 5.1 MCMC over partial worlds

Sampling-based or Monte Carlo inference algorithms are well-suited for OUPMs because each sample specifies what objects exist and what relations hold among them. We focus on Markov chain Monte Carlo (MCMC), where we simulate a Markov chain over possible worlds consistent with the evidence $E$, such that the stationary distribution of the chain is the posterior distribution over worlds given $E$. Such a chain can be constructed using the Metropolis–Hastings method, where we use an arbitrary *proposal distribution* $q(\omega'|\omega_t)$, but accept or reject each proposal based on the relative probabilities of $\omega'$ and $\omega_t$.

Specifically, at each step $t$ in our Markov chain, we sample $\omega'$ from $q(\omega'|\omega_t)$ and then compute the *acceptance probability*:

$$\alpha = \min\left(1, \frac{P_M(\omega')q(\omega_t|\omega')}{P_M(\omega_t)q(\omega'|\omega_t)}\right).$$

With probability $\alpha$, we accept the proposal and let $\omega_{t+1} = \omega'$; otherwise we reject the proposal and let $\omega_{t+1} = \omega_t$.

The difficulty in OUPMs is that each world may be very large. For instance, if we have a world where $\#Pub = 1000$, but only 100 publications are referred to by our observed citations, then the world must also specify the titles of the 900 unobserved publications. Sampling values for these 900 Title variables and computing their probabilities will slow down our algorithm unnecessarily. In scenarios where some possible worlds have infinitely many objects, specifying a possible world completely may be impossible.

Thus, we would like to run MCMC over *partial* descriptions that specify values only for certain random variables. The set of instantiated variables may vary from world to world. Since a partial instantiation $\sigma$ defines an event (the set of worlds that are consistent with it), a Markov chain over partial instantiations can be viewed as a chain over events. Thus, we use the acceptance probability:

$$\alpha = \min\left(1, \frac{P_M(\sigma')q(\sigma_t|\sigma')}{P_M(\sigma_t)q(\sigma'|\sigma_t)}\right)$$

where $P_M(\sigma)$ is the probability of the event $\sigma$. As long as the set $\Sigma$ of partial instantiations that can be returned by $q$ forms a partition of $E$, and each partial instantiation is specific enough to determine whether $Q$ is true, we can estimate $P(Q|E)$ using a Markov chain on $\Sigma$ with stationary distribution proportional to $P_M(\sigma)$ [Milch and Russell 2006].

In general, computing the probability $P_M(\sigma)$ involves summing over all the variables not instantiated in $\sigma$—which is precisely what we want to avoid by using a Monte Carlo inference algorithm. Fortunately, if each instantiation in $\Sigma$ is self-supporting, we can compute its probability using the product expression from Property 2. Thus, our partial worlds are self-supporting instantiations that include the query and evidence variables. We also make sure to use *minimal* instantiations satisfying this condition—that is, instantiations that would cease to be self-supporting if we removed any non-query, non-evidence variable. It can be shown that in a CBN, such minimal self-supporting instantiations are mutually exclusive . So if our set of partial worlds $\Sigma$ covers all of $E$, we are guaranteed to have a partition of $E$, as required. An example of a partial world in our bibliography scenario is:

$\#\mathsf{Pub} = 50, \mathsf{CitTitle}(\mathsf{Cit1}) =$ "Calculus", $\mathsf{CitTitle}(\mathsf{Cit2}) =$ "Intro to Calculus", $\mathsf{PubCited}(\mathsf{Cit1}) = \langle Pub, 17\rangle, \mathsf{PubCited}(\mathsf{Cit2}) = \langle Pub, 31\rangle,$
$\mathsf{Title}(\langle Pub, 17\rangle) =$ "Calculus", $\mathsf{Title}(\langle Pub, 31\rangle) =$ "Intro to Calculus"

## 5.2 Abstract partial worlds

In the partial instantiation above, we specify the tuple representation of each publication, as in PubCited(Cit1) = $\langle Pub, 17 \rangle$. If partial worlds are represented this way, then the code that implements the proposal distribution has to choose numbers for any new objects it adds, keep track of the probability of its choices, and compute the probability of the reverse proposal. Some kinds of moves are impossible unless the proposer renumbers the objects: for instance, the total number of publications cannot be decreased from 1000 to 900 when publication 941 is in use.

To simplify the proposal distribution, we can use partial worlds that abstract away the identities of objects using existential quantifiers:

$\exists$ distinct $x, y$
#Pub = 50, CitTitle(Cit1) = "Calculus", CitTitle(Cit2) = "Intro to Calculus",
PubCited(Cit1) = $x$, PubCited(Cit2) = $y$,
Title($x$) = "Calculus", Title($y$) = "Intro to Calculus"

The probability of the event corresponding to an abstract partial world depends on the number of ways the logical variables can be mapped to distinct objects. For simplicity, we will assume that there is only one number variable for each type. If an abstract partial world $\sigma$ uses logical variables for a type $\tau$, we require it to instantiate the number variable for that type. We also require that for each logical variable $x$, there is a distinct ground term $t_x$ such that $\sigma$ implies $t_x = x$; this ensures that each mapping from logical variables to tuple representations yields a distinct possible world. Let $T$ be the set of types of logical variables in $\sigma$, and for each type $\tau \in T$, let $n_\tau$ be the value of $\#\tau$ in $\sigma$ and $\ell_\tau$ be the number of logical variables of type $\tau$ in $\sigma$. Then we have:

$$P(\sigma) = P_c(\sigma) \prod_{\tau \in T} \frac{n_\tau!}{(n_\tau - \ell_\tau)!}$$

where $P_c(\sigma)$ is the probability of any one of the "concrete" instantiations obtained by substituting distinct tuple representations for the logical variables in $\sigma$.

## 5.3 Locality of computation

Given a current instantiation $\sigma_t$ and a proposed instantiation $\sigma'$, computing the acceptance probability involves computing the ratio:

$$\frac{P_M(\sigma')q(\sigma_t|\sigma')}{P_M(\sigma_t)q(\sigma'|\sigma_t)} = \frac{q(\sigma_t|\sigma') \prod_{X \in \text{vars}(\sigma')} \varphi_X(\sigma'[X], \sigma'[\text{Pa}_{\sigma'}(X)])}{q(\sigma'|\sigma_t) \prod_{X \in \text{vars}(\sigma_t)} \varphi_X(\sigma_t[X], \sigma_t[\text{Pa}_{\sigma_t}(X)])}$$

where $\text{Pa}_\sigma(X)$ is the set of parents of $X$ whose edge conditions are entailed by $\sigma$. This expression is daunting, because even though the instantiations $\sigma_t$ and $\sigma'$ are only partial descriptions of possible worlds, they may still assign values to large sets of random variables — and the number of instantiated variables grows at least linearly with the number of observations we have. Since we may want to run millions of MCMC steps, having each step take time proportional to the number of observations would make inference prohibitively expensive.

Fortunately, with most proposal distributions used in practice, each step changes the values of only a small set of random variables. Furthermore, if the edges that are active in any given possible world are fairly sparse, then $\sigma\left[\text{Pa}_{\sigma'}(X)\right]$ will also be the same as $\sigma_t\left[\text{Pa}_{\sigma_t}(X)\right]$ for many variables $X$. Thus, many factors will cancel out in the ratio above.

We need to compute the "new" and "old" probability factors for a variable $X$ only if either $\sigma'[X] \neq \sigma_t[X]$, or there is some active parent $W \in \text{Pa}_{\sigma_t}(X)$ such that $\sigma'[W] \neq \sigma_t[W]$. (We take these inequalities to include the case where $\sigma'$ assigns a value to the variable and $\sigma_t$ does not, or vice versa.) Note that it is not possible for $\text{Pa}_{\sigma'}(X)$ to be different from $\text{Pa}_{\sigma_t}(X)$ unless one of the "old" active parents in $\text{Pa}_{\sigma_t}(X)$ has changed: given that $\sigma_t$ is a self-supporting instantiation, the values of $X$'s instantiated parents in $\sigma_t$ determine the truth values of the conditions on all the edges into $X$, so the set of active edges into $X$ cannot change unless one of these parent variables changes.

This fact is exploited in the BLOG system [Milch and Russell 2006] to efficiently detect which probability factors need to be computed for a given proposal. The system maintains a graph of the edges that are active in the current instantiation $\sigma_t$. The proposer provides a list of the variables that are changed in $\sigma'$, and the system follows the active edges in the graph to identify the children of these variables, whose probability factors also need to be computed. Thus, the graphical locality that is central to many other BN inference algorithms also plays a role in MCMC over relational structures.

## 6 Related work

The connection between probability and first-order languages was first studied by Carnap [1950]. Gaifman [1964] and Scott and Krauss [1966] defined a formal semantics whereby probabilities could be associated with first-order sentences and for which models were probability measures on possible worlds. Within AI, this idea was developed for propositional logic by Nilsson [1986] and for first-order logic by Halpern [1990]. The basic idea is that each sentence *constrains* the distribution over possible worlds; one sentence entails another if it expresses a stronger constraint. For example, the sentence $\forall x\, P(\mathsf{Hungry}(x)) > 0.2$ rules out distributions in which any object is hungry with probability less than 0.2; thus, it entails the sentence $\forall x\, P(\mathsf{Hungry}(x)) > 0.1$. Bacchus [1990] investigated knowledge representation issues in such languages. It turns out that writing a *consistent* set of sentences in these languages is quite difficult and constructing a unique probability model nearly impossible unless one adopts the representational approach of Bayesian networks by writing suitable sentences about conditional probabilities.

The impetus for the next phase of work came from researchers working with BNs directly. Rather than laboriously constructing large BNs by hand, they built them by composing and instantiating "templates" with logical variables that described local causal models associated with objects [Breese 1992; Wellman et al. 1992]. The most important such language was BUGS (Bayesian inference Using Gibbs Sampling) [Gilks et al. 1994], which combined Bayesian networks with the indexed-random-variable notation common in statistics. These languages inherited the key property of Bayesian networks: every well-formed knowledge base defines a unique, consistent probability model. Languages with well-defined semantics based on unique names and domain closure drew on the representational capabilities of logic programming [Poole 1993; Sato and Kameya 1997; Kersting and De Raedt 2001] and semantic networks [Koller and Pfeffer 1998; Pfeffer 2000]. Initially, inference in these models was performed on the equivalent Bayesian network. *Lifted* inference techniques borrow from first-order logic the idea of performing an inference once to cover an entire equivalence class of objects [Poole 2003; de Salvo Braz et al. 2007; Kisynski and Poole 2009]. MCMC over relational structures was introduced by Pasula and Russell [2001]. Getoor and Taskar [2007] collect many important papers on first-order probability models and their use in machine learning.

Probabilistic reasoning about identity uncertainty has two distinct origins. In statistics, the problem of *record linkage* arises when data records do not contain standard unique identifiers—for example, in financial, medical, census, and other data [Dunn 1946; Fellegi and Sunter 1969]. In control theory, the problem of *data association* arises in multitarget tracking when each detected signal does not identify the object that generated it [Sittler 1964]. For most of its history, work in symbolic AI assumed erroneously that sensors could supply sentences with unique identifiers for objects. The issue was studied in the context of language understanding by Charniak and Goldman [1993] and in the context of surveillance by Huang and Russell [1998] and Pasula *et al.* [1999]. Pasula *et al.* [2003] developed a complex generative model for authors, papers, and citation strings, involving both relational and identity uncertainty, and demonstrated high accuracy for citation information extraction. The first formally defined language for open-universe probability models was BLOG [Milch et al. 2005], from which the material in the current chapter was developed. Laskey [2008] describes another open-universe modeling language called multi-entity Bayesian networks.

Another important thread goes under the name of *probabilistic programming languages*, which include IBAL [Pfeffer 2007] and CHURCH [Goodman et al. 2008]. These languages represent first-order probability models using a programming language extended with a randomization primitive; any given "run" of a program can be seen as constructing a possible world, and the probability of that world is the probability of all runs that construct it.

The OUPMs we have described here bear some resemblance to probabilistic programs, since each dependency statement can be viewed as a program fragment for sampling a value for a child variable. However, expressions in dependency statements have different semantics from those in a probabilistic functional language such as IBAL: if an expression such as Title(Pub5) is evaluated in several dependency statements in a given possible world, it returns the same value every time, whereas the value of an expression in IBAL is sampled independently each time it appears. The CHURCH language incorporates aspects of both approaches: it includes a *stochastic memoization* construct that lets the programmer designate certain expressions as having values that are sampled once and then reused. McAllester *et al.* [2008] define a probabilistic programming language that makes sources of random bits explicit and has a possible-worlds semantics similar to OUPMs.

This chapter has described generative, directed models. The combination of relational and first-order notations with (undirected) Markov networks is also interesting [Taskar et al. 2002; Richardson and Domingos 2006]. Undirected formalisms are convenient because there is no need to avoid cycles. On the other hand, an essential assumption underlying relational probability models is that one set of CPD parameters is appropriate for a wide range of relational structures. For instance, in our RPMs, the prior for a publication's title does not depend on how many citations refer to it. But in an undirected model, adding more citations to a publication (and thus more potentials linking Title($p$) to CitTitle($c$) variables) will usually change the marginal on Title($p$), even when none of the CitTitle($c$) values are observed. This suggests that all the potentials must be learned jointly on a training set with roughly the same distribution of relational structures as the test set; in the directed case, we are free to learn different CPDs from different data sources.

## 7 Discussion

This chapter has stressed the importance of unifying probability theory with first-order logic—particularly for cases with unknown objects—and has presented one possible approach based on open-universe probability models, or OUPMs. OUPMs draw on the key idea introduced into AI by Judea Pearl: generative probability models based on local conditional distributions. Whereas BNs generate worlds by assigning values to variables one at a time, relational models can assign values to a whole class of variables through a single dependency assertion, while OUPMs add object creation as one of the generative steps.

OUPMs appear to enable the straightforward representation of a wide range of situations. In addition to the citation model mentioned in this chapter (see Milch [2006] for full details), models have been written for multitarget tracking, plan recognition, sibyl attacks (a security threat in which a reputation system is compromised by individuals who create many fake identities), and detection of nuclear explosions using networks of seismic sensors [Russell and Vaidya 2009]. In each case, the model is essentially a transliteration of the obvious English description of the generative process.

Inference, however, is another matter. The generic Metropolis–Hastings inference engine written for BLOG in 2006 is far too slow to support any of the applications described in the preceding paragraph. For the citation problem, Milch [2006] describes an application-specific proposal distribution for the generic M–H sampler that achieves speeds comparable to a completely hand-coded, application-specific inference engine. This approach is feasible in general but requires a significant coding effort by the user. Current efforts in the BLOG project are aimed instead at improving the generic engine: implementing a generalized Gibbs sampler for structurally varying models; enabling the user to specify blocks of variables that are to be sampled jointly to avoid problems with slow mixing; borrowing compiler techniques from the logic programming field to reduce the constant factors; and building in parametric learning. With these changes, we expect BLOG to be usable over a wide range of applications with only minimal user intervention.

# References

Bacchus, F. (1990). *Representing and Reasoning with Probabilistic Knowledge*. MIT Press.

Breese, J. S. (1992). Construction of belief and decision networks. *Computational Intelligence 8*(4), 624–647.

Carnap, R. (1950). *Logical Foundations of Probability*. Univ. of Chicago Press.

Charniak, E. and R. P. Goldman (1993). A Bayesian model of plan recognition. *Artificial Intelligence 64*(1), 53–79.

de Salvo Braz, R., E. Amir, and D. Roth (2007). Lifted first-order probabilistic inference. In L. Getoor and B. Taskar (Eds.), *Introduction to Statistical Relational Learning*. MIT Press.

Dunn, H. L. (1946). Record linkage. *Am. J. Public Health 36*(12), 1412–1416.

Fellegi, I. and A. Sunter (1969). A theory for record linkage. *J. Amer. Stat. Assoc. 64*, 1183–1210.

Gaifman, H. (1964). Concerning measures in first order calculi. *Israel J. Math. 2*, 1–18.

Getoor, L. and B. Taskar (Eds.) (2007). *Introduction to Statistical Relational Learning*. MIT Press.

Gilks, W. R., A. Thomas, and D. J. Spiegelhalter (1994). A language and program for complex Bayesian modelling. *The Statistician 43*(1), 169–177.

Goodman, N. D., V. K. Mansinghka, D. Roy, K. Bonawitz, and J. B. Tenenbaum (2008). Church: A language for generative models. In *Proc. 24th Conf. on Uncertainty in AI*.

Halpern, J. Y. (1990). An analysis of first-order logics of probability. *Artificial Intelligence 46*, 311–350.

Huang, T. and S. J. Russell (1998). Object identification: A Bayesian analysis with application to traffic surveillance. *Artificial Intelligence 103*, 1–17.

Kersting, K. and L. De Raedt (2001). Adaptive Bayesian logic programs. In *Proc. 11th International Conf. on Inductive Logic Programming*, pp. 104–117.

Kisynski, J. and D. Poole (2009). Lifted aggregation in directed first-order probabilistic models. In *Proc. 21st International Joint Conf. on Artificial Intelligence*, pp. 1922–1929.

Koller, D. and A. Pfeffer (1998). Probabilistic frame-based systems. In *Proc. 15th AAAI National Conf. on Artificial Intelligence*, pp. 580–587.

Laskey, K. B. (2008). MEBN: A language for first-order Bayesian knowledge bases. *Artificial Intelligence 172*, 140–178.

McAllester, D., B. Milch, and N. D. Goodman (2008). Random-world semantics and syntactic independence for expressive languages. Technical Report MIT-CSAIL-TR-2008-025, Massachusetts Institute of Technology.

Milch, B., B. Marthi, S. Russell, D. Sontag, D. L. Ong, and A. Kolobov (2005). BLOG: Probabilistic models with unknown objects. In *Proc. 19th International Joint Conf. on Artificial Intelligence*, pp. 1352–1359.

Milch, B., B. Marthi, D. Sontag, S. Russell, D. L. Ong, and A. Kolobov (2005). Approximate inference for infinite contingent Bayesian networks. In *Proc. 10th International Workshop on Artificial Intelligence and Statistics*.

Milch, B. and S. Russell (2006). General-purpose MCMC inference over relational structures. In *Proc. 22nd Conf. on Uncertainty in Artificial Intelligence*, pp. 349–358.

Milch, B. C. (2006). *Probabilistic Models with Unknown Objects*. Ph.D. thesis, Univ. of California, Berkeley.

Nilsson, N. J. (1986). Probabilistic logic. *Artificial Intelligence 28*(1), 71–87.

Pasula, H., B. Marthi, B. Milch, S. Russell, and I. Shpitser (2003). Identity uncertainty and citation matching. In *Advances in Neural Information Processing Systems 15*. MIT Press.

Pasula, H. and S. Russell (2001). Approximate inference for first-order probabilistic languages. In *Proc. 17th International Joint Conf. on Artificial Intelligence*, pp. 741–748.

Pasula, H., S. J. Russell, M. Ostland, and Y. Ritov (1999). Tracking many objects with many sensors. In *Proc. 16th International Joint Conf. on Artificial Intelligence*, pp. 1160–1171.

Pfeffer, A. (2000). *Probabilistic Reasoning for Complex Systems*. Ph.D. thesis, Stanford Univ.

Pfeffer, A. (2007). The design and implementation of IBAL: A general-purpose probabilistic language. In L. Getoor and B. Taskar (Eds.), *Introduction to Statistical Relational Learning*. MIT Press.

Poole, D. (1993). Probabilistic Horn abduction and Bayesian networks. *Artificial Intelligence 64*(1), 81–129.

Poole, D. (2003). First-order probabilistic inference. In *Proc. 18th International Joint Conf. on Artificial Intelligence*, pp. 985–991.

Richardson, M. and P. Domingos (2006). Markov logic networks. *Machine Learning 62*, 107–136.

Russell, S. and S. Vaidya (2009). Machine learning and data mining for Comprehensive Test Ban Treaty monitoring. Technical Report LLNL-TR-416780, Lawrence Livermore National Laboratory.

Sato, T. and Y. Kameya (1997). PRISM: A symbolic–statistical modeling language. In *Proc. 15th International Joint Conf. on Artificial Intelligence*, pp. 1330–1335.

Scott, D. and P. Krauss (1966). Assigning probabilities to logical formulas. In J. Hintikka and P. Suppes (Eds.), *Aspects of Inductive Logic*. North-Holland.

Sittler, R. W. (1964). An optimal data association problem in surveillance theory. *IEEE Trans. Military Electronics MIL-8*, 125–139.

Taskar, B., P. Abbeel, and D. Koller (2002). Discriminative probabilistic models for relational data. In *Proc. 18th Conf. on Uncertainty in Artificial Intelligence*, pp. 485–492.

Wellman, M. P., J. S. Breese, and R. P. Goldman (1992). From knowledge bases to decision models. *Knowledge Engineering Review 7*, 35–53.

# 14
# A Heuristic Procedure for Finding Hidden Variables

AZARIA PAZ

## 1 Introduction

This paper investigates Probabilistic Distribution (PD) induced independency relations which are representable by Directed Acyclic Graphs (DAGs), and are marginalized over a subset of their variables. PD-induced relations have been shown in the literature to be representable as relations that can be defined on various graphical models. All those graphical models have two basic properties: They are *compact*, i.e., the space required for storing such a model is polynomial in the number of variables, and they are *decidable*, i.e., a polynomial algorithm exists for testing whether a given independency is represented in the model. In particular, two such models will be encountered in this paper; the DAG model and the Annotated Graph (AG) model. The reader is supposed to be familiar with the DAG-model which was studied extensively in the literature. An ample introduction to the DAG model is included in Pearl [7, 1988], Pearl [8, 2000], and Lauritzen [2, 1996].[1]

The AG-model in a general form was introduced by Paz, Geva, and Studeny in [5, 2000] and a restricted form of this model, which is all we need for this paper, was introduced by Paz [3, 2003a] and investigated further in Paz [4, 2003b]. For the sake of completeness, we shall reproduce here some of the basic definitions and properties of those models which are relevant for this paper.

Given a DAG-representable PD-induced relation it is often the case that we need to marginalize the relation over a subset of variables. Unfortunately it is seldom the case that such a marginalized relation can be represented by a DAG, which is an easy to manage and a well understood model.

In the paper [4, 2003] the author proved a set of necessary corelations for a given AG to be equivalent to a DAG (see Lemma 1 in the next section). In the same paper a decision procedure is given for checking whether a given AG which satisfies the necessary conditions is equivalent to a DAG. Moreover, if the answer is "yes", the procedure constructs an equivalent DAG to the given AG. In a subsequent paper [6, 2006], the author generalizes the AG model and gives a procedure which enables the representation of any marginalized DAG representable relation by a generalized model.

---
[1] The main part of this work was done while the author visited the Cognitive Systems Laboratory at UCLA and was supported in part by grants from Air Force, NSF and ONR (MURI).

## 2 Preliminaries

### 2.1 Definitions and notations

*UGs* will denote undirected graphs $G = (V, E)$ where $V$ is a set of *vertices* and $E$ is a set of undirected *edges* connecting between two vertices. Two vertices connected by an edge are *adjacent* or *neighbors*. A path in $G$ of length $k$ is a sequence of vertices $v_1 \ldots v_{k+1}$ such that $(v_i, v_{i+1})$ is an edge in $E$ for $i < 1, \ldots, k$. A DAG is an acyclic directed graph $D = (V, E)$ where $V$ is a set of vertices and $E$ is a set of directed *arcs* connecting between two vertices in $V$. A *trail* of length $k$ in $D$ is a sequence $v_1 \ldots v_{k+1}$ of vertices in $V$ such that $(v_i, v_{i+1})$ is an arc in $E$ for $i = 1 \ldots k$. If all the arcs on the trail are directed in the same direction then the trail is called a directed path. If a directed path exists in $D$ from $v_i$ to $v_j$ then $v_j$ is a *descendant* of $v_i$ and $v_i$ is a *predecessor* or *ancestor* of $v_j$. If the path is of length one then $v_i$ is a *parent* of $v_j$ who is a *child* of $v_i$.

The *skeleton* of a DAG is the UG derived from the DAG when the orientations of the arcs are removed. A pattern of the form $v_i \to v_j \leftarrow v_k$ is a *collider pattern* where $v_j$ is the *collider*. If there is no arc between $v_i$ and $v_k$ then $v_j$ is an *uncoupled collider*. The moralizing procedure is the procedure generating a UG from a DAG, by first joining both parents of uncoupled colliders in the DAG by an arc, and then removing the orientation of all arcs. The edges resulting from the coupling of the uncoupled collider are called *moral edges*. As mentioned in the introduction UG's and DAG's represent PD-induced relations whose elements are triplets $t = (X; Y|Z)$ over the set of vertices of the graphs. For a given triplet $t$ we denote by $v(t)$ the set of vertices $v(t) = X \cup Y \cup Z$. Two graph models are *equivalent* if they represent the *same* relation.

### 2.2 DAG-model

Let $D = (V, E)$ be a DAG whose vertices are $V$ and whose arcs are $E$. $D$ represents the relation $R(D) = \{t = (X; Y|Z)|t \in D\}$ where $X, Y, Z$ are disjoint subsets of $V$, the vertices in $V$ represent variables in a PD, $t$ is interpreted as "$X$ is independent of $Y$ given $Z$" and $t \in D$ means: $t$ is represented in $D$. To check whether a given triplet $t$ is represented in $D$ we use the Algorithm L1 below due to Lauritzen et al. [1, 1990].

**Algorithm L1:**
*Input*: $D = (V, E)$ and $t = (X; Y|Z)$.

1. Let $V'$ be the set of ancestors of $v(t) = X \cup Y \cup Z$ and let $D'(t)$ be the subgraph of $D$ over $V'$.

2. Moralize $D'(t)$ (i.e., join all uncoupled parents of uncoupled colliders in $D'(t)$). Denote the resulting graph by $D''(t)$.

3. Remove all orientations in $D''(t)$ and denote the resulting UG by $G(D''(t))$.

4. $t \in G(D''(t))$ iff $t \in D$.

REMARK 1. $t \in G$ where $G$ is a UG if and only if $Z$ is a cutset in $G$ (not necessarily minimal) between $X$ and $Y$.

The definition above and the L1 Algorithm show that the DAG model is both compact and decidable.

## 2.3  Annotated Graph – model

Let $D = (V, E)$ be a DAG. We derive from $D$ an AG $A = (G, K)$ where $G$ is a UG And $K$ is a set of elements $K = \{e = (d, r(d))\}$ as follows: $G$ is derived from $D$ by moralizing $D$ and removing all orientations from it.

For every moral edge $d$ in $G$ we put an element $e = (d, r(d))$ in $K$ such that $d(a, b)$, the *domain* of $e$, is the pair of endpoints of the moral edge and $r(d)$, the *range* of $e$, is the set of vertices including all the uncoupled colliders in $D$ whose parents are $a$ and $b$, and all the successors of those colliders. Notice that $d$ denotes both a moral edge and the pair of its endpoints. The relation $R(A)$ defined by the AG A is the relation below:

$$R(A) = \{t = (X; Y|Z) | t \in A\}$$

In order to check whether $t \in A$ we use the algorithm L2 due to Paz [3, 2003a] below.

**Algorithm** L2
*Input*: An AG $A = (G, K)$.

1. For every element $e = (d, r(d))$ in $K$ such that $r(d) \cap v(t) = \emptyset$ ($v(t) = X \cup Y \cup Z$). Disconnect the edge $(a, b)$ in $G$ corresponding to $d$ and remove from $G$ all the vertices in $r(d)$ and incident edges. Denote the resulting UG by $G(t)$.

2. $t \in A$ if and only if $t \in G(t)$.

REMARK 2. It is clear from the definitions and from the L2 Algorithm that the AG model is both compact and decidable. In addition, it was shown in [3, 2003a] that the AG model has the following uniqueness property: $R(A_1) = R(A_2)$ implies that $A_1 = A_2$ when $A_1$ and $A_2$ are AG's. This property does not hold for DAG models where it is possible for two different (and equivalent) DAGs to define the same relation. In fact the AG $(D)$ derived from a DAG $D$ represents the equivalence class of all DAGs which are equivalent to the given DAG $D$.

REMARK 3. The AGs derived from DAG's are a particular case of AGs as defined in Paz et al. [5, 2000] and there are additional ways to derive AGs that represent PD-induced relations which are not DAG-representable. Consider e.g., the example below. It was shown by Pearl [7, 1988 Ch. 3] that every DAG representable relation is a PD-induced relation. Therefore the relation defined by the DAG in Fig. 1 represents a PD-induced relation.

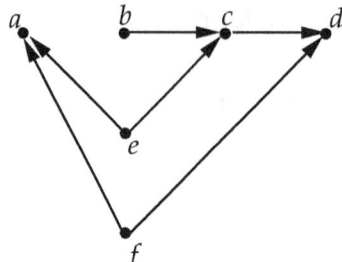

Figure 1. DAG representing relation

If we marginalize this relation over the vertices $e$ and $f$ we get another relation, PD-induced, that can be represented by the AG $A$ in Fig. 2, as will be shown in the sequel, under the semantics of the L2 Algorithm, with $R(A) =$

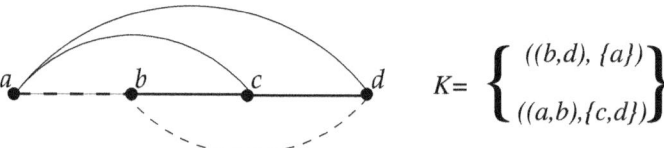

Figure 2. AG $A$ representing a marginalized relation

$\{(a; b|\emptyset), (b; d|c) +$ symmetric images$\}$. But $R(A)$ above cannot be represented by a DAG. This follows from the following lemma that was proven in [4, 2003b].

LEMMA 1. *Let $(G(D), K(D))$ be the annotated graph representation of a DAG $D$. $K(D)$ has the following properties:*

1. *For every element $((a,b), r) \in K(D)$, there is a vertex $v \in r$ which is a child of both $a$ and $b$ and every vertex $w \in r$ is connected to some vertex $v$ in $r$ whose parents are both $a$ and $b$.*

2. *For any two elements $(d_1, r_1), (d_2, r_2)$ in $K(D)$, if $d_1 = d_2$ then $r_1 = r_2$.*

3. *For every $((a,b), r) \in K(D)$, $(a,b)$ is an edge in $G(D)$.*

4. *The set of elements $K(D)$ is a poset (=partially ordered set) with regards to the relation "$\succ$" defined as follows: For any two elements $(d_p, r_p)$ and $(d_q, r_q)$. If $d_p \cap r_q \neq \emptyset$ then $(d_p, r_p) \succ (d_q, r_q)$, in words "$(d_p, r_p)$ is strictly greater than $(d_q, r_q)$". Moreover $(d_p, r_p) \succ (d_q, r_q)$ implies that $r_p \subset r_q$.*

5. *For any two elements $(d_1, r_1)$ and $(d_2, r_2)$ If $r_1 \cap r_2 \neq \emptyset$ and $r_1, r_2$ are not a subset of one another, then there is an element $(d_3, r_3)$ in $K(D)$ such that $r_3 \subseteq r_1 \cap r_2$.*

As is easy to see the annotation, $K$ in Fig. 2 does not satisfy the condition 4 of the lemma since the first element in $K$ is bigger than the second but it's range is not a subset of the range of the second element. Therefore $A$ is not DAG-representable.

REMARK 4. An algorithm is provided in [3, 2003a] that tests whether a given AG, possibly derived from a marginalized DAG relation, which satisfies the (necessary but not sufficient) conditions in Lemma 1 above, is DAG-representable. The main result of this work is to provide a polynomial algorithm which generates a "generalized annotated graph" representation (concept to be defined in the sequel) which is both compact and decidable. In some cases the generalized annotated graph reduces to a regular annotated graph which satisfies the condition of Lemma 1. If this is the case than, using the testing algorithm in [4, 2003b] we can check whether the given AG is DAG-representable. It is certainly not DAG-representable if the generalized annotated graph is not a regular AG or is a regular AG but does not satisfy the conditions of Lemma 1.

REMARK 5. When a given AG $A$ is derived from a DAG then the annotation set $K = \{(d, r(d))\}$ can be interpreted as follows: The edge $(a, b)$, in $G$, corresponding to $d$, (a moral edge) represents a conditional dependency. That is: there is some set of vertices, disjoint of $r(d)$, $S_{ab}$ such that $(a_i b | S_{ab})$ is represented in $A$ but $a$ and $b$ become dependent if any proper subset of $r(d)$ is observed i.e., $\neg(a; b | S)$ if $\emptyset \neq S \subseteq r(d)$.

In this paper we are concerned with the following problem: Given a relation which is represented by a generalized UG model and is not representable by a DAG (see [4, 2003]). Is it possible to find hidden variables such that the given relation results from the marginalization of the DAG representable relation over the expanded set of variables, including the hidden variables in addition to the given AG variables. We do not have a full solution to this problem which is so far an open problem. We present only a heuristic procedure, illustrated by several examples, for partially solving the problem.

## 3 PD-induced relations not representable as a marginalized DAG-representable relations-an example

While DAG's are widely used as a model that can represent PD-induced relations one may ask whether it might be possible to represent every PD-induced relation either by a DAG or, assuming the existence of latent variables, by a marginalized DAG. The answer to this question is negative as should be expected. A counterexample is given below.

Consider the following PD-induced relation, over 3 variable $x, y$, and $z$, consisting of two triplets only:

$$R = \{(x; y | \emptyset), (x; y | z) + \text{symmetric triplets}\}$$

Then $R$ cannot be represented by a marginalized DAG. To prove this claim assume that there is a DAG $D$ with $n$ variables, including $x, y$ and $z$ such that when $D$ is marginalized over $\{x, y, z\}$, the marginalized DAG represents $R$. This assumption leads to a contradiction: Since $(x; z|\emptyset)$ and $(y; z|\emptyset)$ are not in $R$ there must be trails $\pi_{xz}$ and $\pi_{yz}$ in $D$ with no colliders included in them. Let $\pi_{xy}$ be the concatenation of the two trails $\pi_{xz}$ and $\pi_{zy}$ (which is the trail $\pi_{yz}$ reversed). Then $\pi_{xy}$ connects between $x$ and $y$ and has no colliders on it except perhaps the vertex $z$. If $z$ is a collider then $(x; y|z)$ is not represented in $D$. If $z$ is not a collider then $\pi_{xy}$ has no colliders on it and therefore $(x; y|\emptyset)$ is not represented in $D$. Therefore $R$ cannot be represented by marginalizing $D$ over $\{x, y, z\}$, a contradiction. That $R$ is a PD-induced relation was shown by Milan Studeny [9, private communication, 2000] as follows:

Consider the PD over the binary variables $x, y$ and the ternary variable $z$. The probability of the three variables for the different values of $x, y, z$ is given below

$$\begin{aligned} p(0,0,0) &= p(0,0,1) = p(1,0,1) = p(1,0,2) = \tfrac{1}{8} \\ p(0,1,0) &= p(1,1,1) = \tfrac{1}{4} \\ p(x,y,z) &= 0 \text{ for all other configurations} \end{aligned}$$

The reader can convince himself that the relation induced by the above PD is the relation $R = \{(x; y|\emptyset), (x; y|z)\}$. Notice however that the relation $R$ above is represented by the annotated graph below

$$G : x\text{—}z\text{—}y \qquad K = \{((x,y), \{z\})\}$$

see Paz [4, 2003b].

## 4 Finding Hidden Variables

In this section we consider the following problem. Given an AG which represents a relation that is not DAG representable, e.g. the relation represented by the AG shown in Fig. 2. Is it possible to find hidden variables which when added to the variables of the given relation will enable the representation of the given relation as a marginalized DAG representable relation, over the extra hidden variables. At this stage of the research we do not have an algorithmic procedure for solving this problem,a nd we do not have a characterization lemma, similar to Lemma 1 for AGs representing marginalized DAG representable relations. We can however present a heuristic procedure for tackling with this problem. We hope that it will be possible to extend the procedure into a full algorithm in the future. The procedure will be illustrated here by examples. The examples we will use however are such that we know in advance that the problem can be solved for them. This is due to the fact that we can not characterize so far the AG's for which the problem is solvable, as mentioned above. On the other hand we should keep in mind that not every PD-induced relation can be represented as a marginalized DAG-representable relation, as shown in the previous section.

## 4.1 Example 1

Consider the DAG $D$ shown in Fig. 3. An equivalent AG, representing the same relation [4, 2003b] is shown in Fig. 4 where the broken lines (edges) represent conditional dependencies and correspond to uncoupled parents of colliders in Fig. 3.

Assume now that we want to marginalize the relation represented by the AG shown in Fig. 4 over the variables $p$ and $q$. Using the procedure given in [6, 2006] we get a relation represented by the AG shown in Fig. 5 below.

The Derivation of the AG in Fig. 5 can be explained as follows:

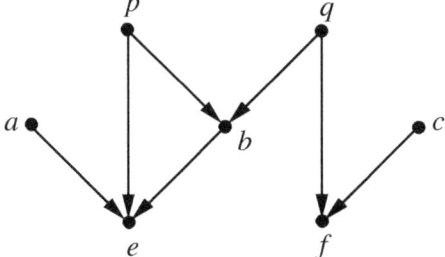

Figure 3. DAG $D$

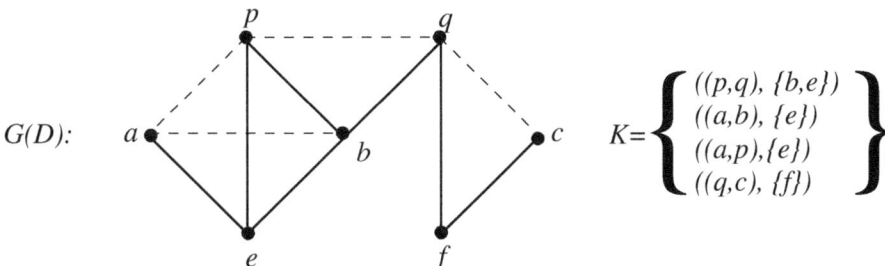

Figure 4. The annotated graph $A(D)$

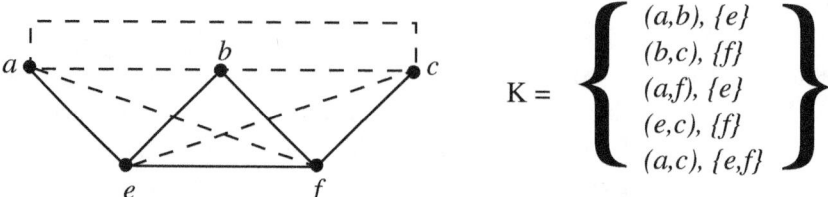

Figure 5. $AG$ for the marginalized relation

- The solid edge $b$-$f$ is induced by the path, in the AG shown in Fig. 4, $b$-$q$-$f$, through the marginalized variable $q$.

- Similarly, the solid edge $e$-$f$ in Fig. 5 is induced by the path, in Fig. 4, $e - p - q - f$ through the marginalized variables $p$ and $q$.

- The element $((a, b), \{e\})$ in Fig. 5 was transferred from Fig. 4 since it involves only non marginalized variables.

- The element $((b, c), \{f\})$ in Fig. 5 is induced by the path $b - q - c$ in Fig. 4 which is activated if $f$ is observed.

- $((a, f)\{e\})$ is induced by the path $a - p - q - f$ which is activated in Fig. 4 if $e$ is observed.

- $((e, c), \{f\})$ is induced by the path in Fig. 4 $e - p - q - c$ which is activated if $f$ is observed.

- Finally $((a, c), \{e \wedge f\}$ is induced by the path $a - p - q - c$ in Fig. 4 which is activated if *both* $e$ and $f$ are observed.

REMARK 6. The dependency conditional shown in the fifth element of $K$ in Fig. 5 is different from the conditionals in any AG representing DAG-representable relations in the following sense. The conditionals for DAG-representable relations consist of a set of variables such that, if any variable in the set is observed then the conditional is activated. In Fig. 5, in order to activate the conditional of the fifth element in $K$ *both* variables $e$ and $f$ must be observed.

Assume now that we are given the AG shown in Fig. 5 with no prior knowledge that it represents a marginalized DAG-representable relation. We can see immediately that the relation represented in it is not DAG-representable. This follows from the remark above and also from the fact that $K$ does not satisfy the necessary conditions of Lemma 1. E.g. the conditional $\{f\}$ in the second element is included in the pair $\{a, f\}$ of the third element but $\{e\}$ is not a subset of the conditional $\{f\}$ as required by Lemma 1, part 4.

A Heuristic Procedure for Finding Hidden Variables

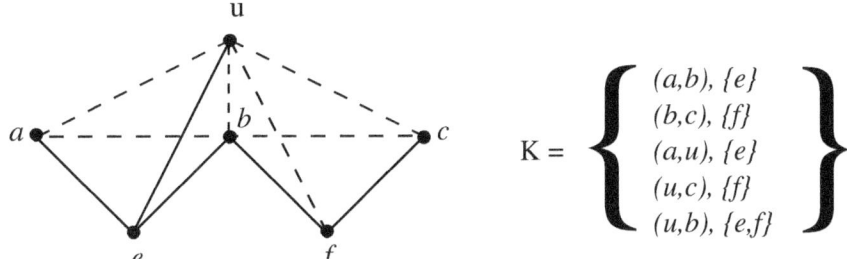

Figure 6. Extended AG

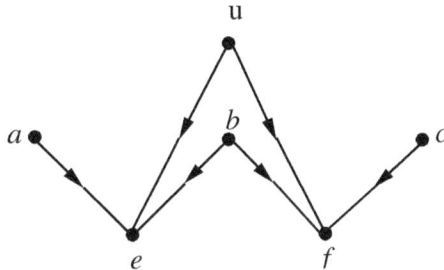

Figure 7. DAG which is equivalent to the AG in Fig. 6

Given the fact that the relation is not DAG-representable, we ask ourselves whether it is possible to add hidden variables to the variables of the relation such that the extended relation is DAG-representable and is such that when it is marginalized over the added (hidden) variables, it reduces to the given no DAG-representable relation. Consider first the fifth element (see Fig. 5) $(a,c)\{e \wedge f\}$ of $K$ for the given relation. The conditional $\{e \wedge f\}$ of this element does not fit DAG-representable relation.

We notice that this element can be eliminated if we add a hidden variable $u$ that is connected by solid lines (unconditional dependant) to the variables $e$ and $f$ and is conditionally dependant on $a$ with conditional $e$ and is conditionally dependant on $c$ with conditional $f$. The resulting graph is shown in Fig. 6.

The reader will easily convince himself that the annotation $K$ shown in Fig. 6 fits the extended graph, where the first 2 elements are inherited from Fig. 5 and the other 3 elements are induced by the (hidden) new variable $u$. The solid edge $(e, f)$ in Fig. 5 is removed in Fig. 6 since it is implied by the path $e - u - f$ when the extended relation is marginalized over $u$. The reader will also easily convince himself that if the relation shown in Fig. 6 is marginalized over $u$ we get back the relation shown in Fig. 5. Moreover the annotation $K$ in Fig. 6 complies with the necessary conditions of Lemma 1. Indeed the relation represented by the AG in Fig. 6 is DAG representable: Just direct all solid edges incident with $e$ into $e$, direct all

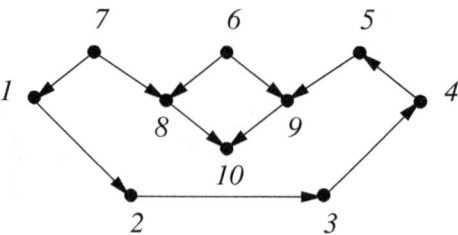

Figure 8. DAG for Example 2

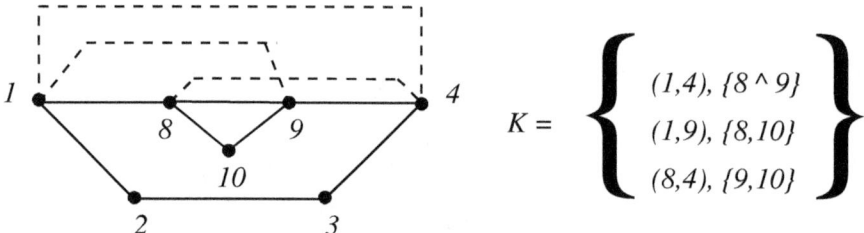

Figure 9. UG equivalent with the DAG in Fig. 8

solid edges incident with $f$ into $f$ and remove all broken edges. The result is shown in Fig. 7.

Notice that the DAG in Fig. 7 is quite different from the DAG shown in Fig. 3, but both reduce to the same relation when marginalized over their extra variables.

## 4.2 Example 2

Consider the DAG shown in Fig. 8. Using methods similar to the methods we used in the previous example we can get an equivalent AG which when marginalized over the vertices 5, 6 and 7 results in the UG shown in Fig. 9.

Here again we can check and verify that $K$ does not satisfy the conditions of Lemma 1, in particular the first element in $K$ is a compound statement that does not fit DAG representable relations. So the relation represented by the AG in Fig. 9 is not DAG representable. Trying to eliminate the first element in $K$, we may assume the existence of a hidden variable $u$ which is connected by solid edges to both 8 and 9, but is conditionally dependent on 1 with conditional 8 and is conditionally dependent on 4 with conditional 9. The resulting extended AG is shown in Fig. 10. The $K$ in Fig. 10 satisfies the conditions of Lemma 1 and one can see that the AG in DAG is equivalent. Moreover marginalizing it over $u$ reduces to the UG shown in Fig. 9. The equivalent DAG is shown in Fig. 11.

Notice that the DAG in Fig. 11 is quite different from the DAG in Fig. 8, but both result in the same AG when marginalizing over their corresponding extra variables.

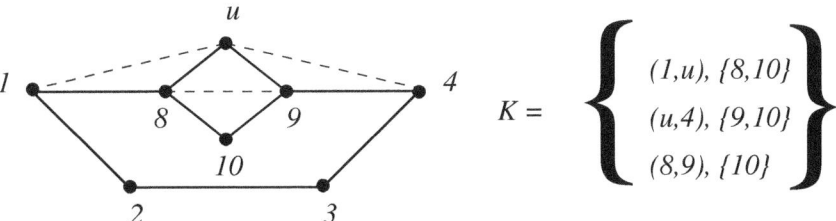

Figure 10. Expanded DAG equivalent AG

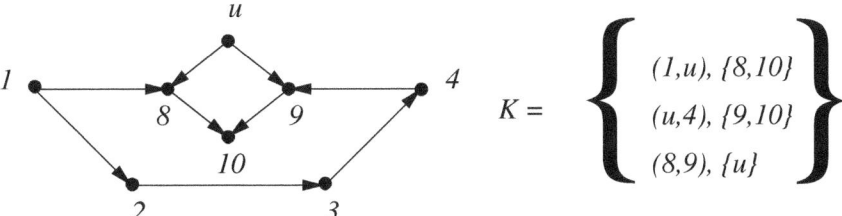

Figure 11. DAG equivalent to the AG in Fig. 10

### 4.3 Example 3

In this last example we consider the AG shown in Fig. 2 which is reproduced here, for convenience as Fig. 12.

While the $K$ sets in the previous examples included elements with compound conditionals, both conditionals in this example are simple but the conditions of Lemma 1 are not satisfied: $d$ in the first element is included in the second conditional and $a$ of the second element is included in the first conditional, but the conditionals are no a subset of each other. Consider first the second element. We can introduce an additional variable $u$ so that $u$ will be conditionally dependent on $b$, and the element $(u,b),\{c,d\}$ will replace the element $(a,b),\{c,d\}$. The resulting AG is shown in Fig. 13 below.

It is easy to see that the graph in Fig. 13 reduces to graph in Fig. 12 when marginalized over $u$. We still need to take care of the first element since it is not satisfying the conditions of Lemma 1. We can now add additional new variable $v$ and replace the first element $(b,d),\{a\}$ by the element $(u,v),\{a\}$. The resulting larger AG is shown in Fig. 14. Notice that the graph in Fig. 14 will reduce to the graph in Fig. 12.

To verify this we observe that marginalization over $u$ and $v$ induces the element $(b,d),\{a\}$ since when $a$ and $d$ are observed $b$ gets connected to $u$ (by $d$ and the second element) and $u$ is connected to $v$ by the first element so that the path $b-u-v-d$ is activated through the extra variables ($b,d$ and $a$ exist in Fig. 12).

One can also verify that the AG in Fig. 14 is DAG equivalent after some simples

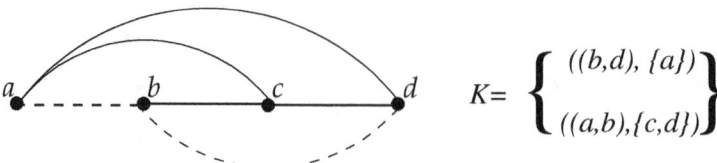

Figure 12. AG from Fig. 2

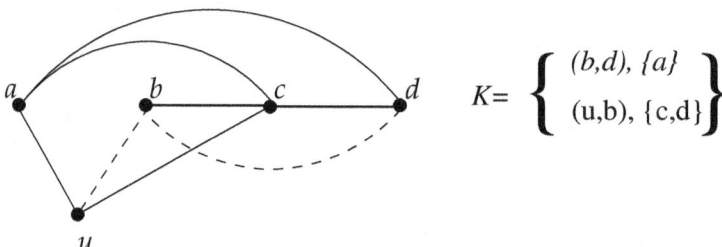

Figure 13. First extension of the AG in Fig. 12

modifications: we can remove the edges $(a,c)$ and $(a,d)$ since they are implied when marginalizing over $u$ and $v$ and we need to add the element $(v,c), \{d\}$ and a corresponding broken edge between $v$ and $c$, which will be discarded when marginalizing. The equivalent DAG is shown in Fig. 15 and is identical with the DAG shown in Fig. 1.

## Acknowledgment

I would like to thank the editors for allowing me to contribute this paper to the book honoring Judea Pearl, a friend and collaborator for more than 20 years now. Much of my scientific work in the past 20 years was influenced by discussions I had with him. His sharp intuition and his ability to pinpoint important problems and ask the right questions helped me in choosing the problems to investigate and finding their solutions, resulting in several scientific papers, including this one.

## References

[1] S .L. Lauritzen, A. P. Dawid, B. N. Larsen, and H. G. Leimer. Independence properties of directed Markov fields. *Networks*, 20:491–505, 1990.

[2] S.L. Lauritzen. *Graphical Models*. Claredon, Oxford, U.K., 1996.

[3] A. Paz. An alternative version of Lauritzen et al's algorithm for checking representation of independencies. *Journal of Soft Computing*, pages 491–505, 2003a.

[4] A. Paz. The annotated graph model for representing DAG-representable relations – algorithmic approach. Technical Report Technical Report R-312,

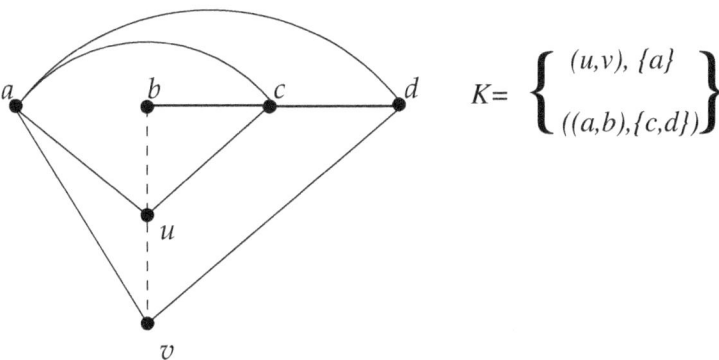

Figure 14. Extended AG

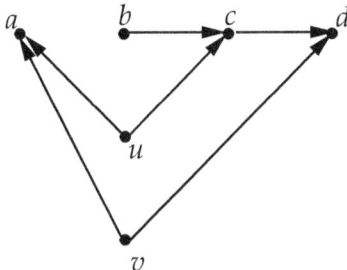

Figure 15. DAG which is equivalent to the AG in Fig. 14

Computer Science Department, UCLA, 2003b.

[5] A. Paz, R.Y. Geva, and M. Studeny. Representation of irrelevance relations by annotated graphs. *Fundamenta Informaticae*, 48:149–199, 2000.

[6] A. Paz. A New Graphical Model for the Representation of Marginalized DAG-Representable Relations. prodeedings of the 7th Workshop on Uncertainty Processing, pages 111–137, 2006, Mikulov, The Check Republic.

[7] J. Pearl. *Probabilistic Reasoning in Intelligent Systems*. Morgan Kaufmann, San Mateo, CA, 1988.

[8] J. Pearl. *Causality: Models, Reasoning, and Inference*. Cambridge University Press, New York, 2000.

[9] M. Studeny, 2000. Private communication.

[10] T. Verma and J. Pearl. Equivalence and synthesis of causal models. In L.N. Kanal P. Bonissone, M. Henrion and J.F. Lemmer, editors, *Uncertainty in Artificial Intelligence 6*, pages 225–268, B.V., 1991. Elsevier Science Publishers.

# 15
# Probabilistic Programming Languages: Independent Choices and Deterministic Systems

DAVID POOLE

Pearl [2000, p. 26] attributes to Laplace [1814] the idea of a probabilistic model as a deterministic system with stochastic inputs. Pearl defines causal models in terms of deterministic systems with stochastic inputs. In this paper, I show how deterministic systems with (independent) probabilistic inputs can also be seen as the basis of modern probabilistic programming languages. Probabilistic programs can be seen as consisting of independent choices (over which there are probability distributions) and deterministic programs that give the consequences of these choices. The work on developing such languages has gone in parallel with the development of causal models, and many of the foundations are remarkably similar. Most of the work in probabilistic programming languages has been in the context of specific languages. This paper abstracts the work on probabilistic programming languages from specific languages and explains some design choices in the design of these languages.

Probabilistic programming languages have a rich history starting from the use of simulation languages such as Simula [Dahl and Nygaard 1966]. Simula was designed for discrete event simulations, and the built-in random number generator allowed for stochastic simulations. Modern probabilistic programming languages bring three extra features:

**conditioning:** the ability to make observations about some variables in the simulation and to compute the posterior probability of arbitrary propositions given these observations. The semantics can be seen in terms of rejection sampling: accept only the simulations that produce the observed values, but there are other (equivalent) semantics that have been developed.

**inference:** more efficient inference for determining posterior probabilities than rejection sampling.

**learning:** the ability to learn probabilities from data.

In this paper, I explain how we can get from Bayesian networks [Pearl 1988] to independent choices plus a deterministic system (by augmenting the set of variables). I explain the results from [Poole 1991; Poole 1993b], abstracted to be language independent, and show how they can form the foundations for a diverse set of probabilistic programming languages.

Consider how to represent a probabilistic model in terms of a deterministic system with independent inputs. In essence, given the probabilistic model, we construct a random variable for each free parameter of the original model. A deterministic system can be used to obtain the original variables from the new variables. There are two possible worlds structures, the original concise set of possible worlds in terms of the original variables, and the augmented set of possible worlds in terms of the new random variables. The dimensionality of the augmented space is the number of free parameters which is greater than the dimensionality of the original space (unless all variables were independent). However, the variables in the augmented worlds can be assumed to be independent, which makes them practical to use in a programming environment. The original worlds can be obtained using abduction.

Independent choices with a deterministic programming language can be seen as the basis for most of the probabilistic programming languages, where the deterministic system can be a logic program [Poole 1993b; Sato and Kameya 1997; De Raedt, Kimmig, and Toivonen 2007], a functional program [Koller, McAllester, and Pfeffer 1997; Pfeffer 2001; Goodman, Mansinghka, Roy, Bonawitz, and Tenenbaum 2008], or even a low-level language like C [Thrun 2000].

There had been parallel developments in the development of causality [Pearl 2000], with causal models being deterministic systems with stochastic inputs. The augmented variables in the probabilistic programming languages are the variables needed for counterfactual reasoning.

## 1 Probabilistic Models and Deterministic Systems

In order to understand probabilistic programming languages, it is instructive to see how a probabilistic model in terms of a Bayesian network [Pearl 1988] can be represented as a deterministic system with probabilistic inputs.

Consider the following simple belief network, with Boolean random variables:

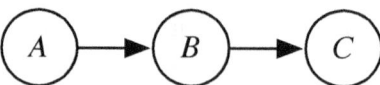

There are 5 free parameters to be assigned for this model; for concreteness assume the following values (where $A = true$ is written as $a$, and similarly for the other variables):

$$P(a) = 0.1$$
$$P(b|a) = 0.8$$
$$p(b|\neg a) = 0.3$$
$$P(c|b) = 0.4$$
$$p(c|\neg b) = 0.75$$

To represent such a belief network in a probabilistic programming language, there are probabilistic inputs corresponding to the free parameters, and the programming language specifies what follows from them. For example, in Simula [Dahl and Nygaard 1966], this could be represented as:

```
begin
   Boolean a,b,c;
   a := draw(0.1);
   if a then
      b := draw(0.8);
   else
      b := draw(0.3);
   if b then
      c := draw(0.4);
   else
      c := draw(0.75);
end
```

where $draw(p)$ is a Simula system predicate that returns true with probability $p$; each time it is called, there is an independent drawing.

Suppose $c$ was observed, and we want the posterior probability of $b$. The conditional probability $P(b|c)$ is the proportion of those runs with $c$ true that have $b$ true. This could be computed using the Simula compiler by doing rejection sampling: running the program many times, and rejecting those runs that do not assign $c$ to true. Out of the non-rejected runs, it returns the proportion that have $b$ true. Of course, conditioning does not need to implemented that way; much of the development of probabilistic programming languages over the last twenty years is in devising more efficient ways to implement conditioning.

Another equivalent model to the Simula program can be given in terms of logic. There can be 5 random variables, corresponding to the independent draws, let's call them $A$, $Bifa$, $Bifna$, $Cifb$ $Cifnb$. These are independent with $P(a) = 0.1$, $P(bifa) = 0.8$, $P(bifna) = 0.3$, $P(cifb) = 0.4$, and $P(cifnb) = 0.75$. The other variables can be defined in terms of these:

$$b \iff (a \land bifa) \lor (\neg a \land bifna) \tag{1}$$

$$c \iff (b \land cifb) \lor (\neg b \land cifnbc) \tag{2}$$

These two formulations are essentially the same, they differ in how the deterministic system is specified, whether it is in Simula or in logic.

Any discrete belief network can be represented as a deterministic system with independent inputs. This was proven by Poole [1991, 1993b] and Druzdzel and Simon [1993]. These papers used different languages for the deterministic systems, but gave essentially the same construction.

## 2 Possible Worlds Semantics

A probabilistic programming language needs a specification of a deterministic system (given in some programming language) and a way to specify distributions over (independent) probabilistic inputs, or a syntactic variant of this. We will also assume that there are some observations, and that there are some query proposition for which we want the posterior probability.

In developing the semantics of a probabilistic programming language, we first define the set of possible worlds, and then a probability measure over sets of possible worlds [Halpern 2003].

In probabilistic programming, there are (at least) two sets of possible world that interact semantically. It is easiest to see these in terms of the above example. In the above belief network, there were three random variables $A$, $B$ and $C$, which had complex inter-dependencies amongst them. With three binary random variables, there are 8 possible worlds. These eight possible worlds give a concise characterization of the probability distribution over these variables. I will call this the *concise* space of possible worlds.

In the corresponding probabilistic program, there is an augmented space with five inputs, each of which can be considered a random variable (these are $A$, $Bifa$, $Bifna$, $Cifb$ and $Cifnb$ in the logic representation). With five binary random variables, there are 32 possible worlds. The reason to increase the number of variables, and thus possible worlds, is that in this the *augmented* space, the random variables can be independent.

Note that the variables in the augmented space do not *have* to be independent. For example, $P(bifna|a)$ can be assigned arbitrarily since, when $a$ is true, no other variable depends on $bifna$. In the augmented space, there is enough freedom to make the variables independent. Thus, we can arbitrarily set $P(bifna|a) = P(bifna|\neg a)$, which will be the same as $P(b|\neg a)$. The independence assumption makes the semantics and the computation simpler.

There are three semantics that could be given to a probabilistic program:

- The rejection sampling semantics; running the program with a random number generator, removing those runs that do not predict the observations, the posterior probability of a proposition is the limit, as the number of runs increases, of the proportion of the non-rejected runs that have the proposition true.

- The independent-choice semantics, where a possible world specifies the outcome of all possible draws. Each of these draws is considered to be independent. Given a world, the (deterministic) program would specify what follows. In this semantics, a possible world would select values for all five of the input variables in the example above, and thus gives rise to the augmented space of the above program with 32 worlds.

- The program-trace semantics, where a possible world specifies a sequence of outcomes for the draws encountered in one run of the program. In this semantics, a world would specify the values for three of the draws in the above program, as only three draws are encountered in any run of the program, and thus there would be 8 worlds.

In the logical definition of the belief network (or in the Simula definition if the draws are named), there are 32 worlds in the independent choice semantics:

| World | $A$ | $Bifa$ | $Bifna$ | $Cifb$ | $Cifnb$ | Probability |
|---|---|---|---|---|---|---|
| $w_0$ | false | false | false | false | false | $0.9 \times 0.2 \times 0.7 \times 0.6 \times 0.25$ |
| $w_1$ | false | false | false | false | true | $0.9 \times 0.2 \times 0.7 \times 0.6 \times 0.75$ |
| ... | | | | | | |
| $w_{30}$ | true | true | true | true | false | $0.1 \times 0.8 \times 0.3 \times 0.4 \times 0.75$ |
| $w_{31}$ | true | true | true | true | true | $0.1 \times 0.8 \times 0.3 \times 0.4 \times 0.75$ |

The probability of each world is the product of the probability of each variable (as each of these variables is assumed to be independent). Note that in worlds $w_{30}$ and $w_{31}$, the original variables $A$, $B$ and $C$ are all true; the value of $Cifnb$ is not used when $B$ is true. These variables are also all true in the worlds that only differ in the value of $Bifna$, as again, $Bifna$ is not used when $A$ is true.

In the program-trace semantics there are 8 worlds for this example, but not all of the augmented variables are defined in all worlds.

| World | $A$ | $Bifa$ | $Bifna$ | $Cifb$ | $Cifnb$ | Probability |
|---|---|---|---|---|---|---|
| $w_0$ | false | $\bot$ | false | $\bot$ | false | $0.9 \times 0.7 \times 0.25$ |
| $w_1$ | false | $\bot$ | false | $\bot$ | true | $0.9 \times 0.7 \times 0.75$ |
| ... | | | | | | |
| $w_7$ | true | true | $\bot$ | false | $\bot$ | $0.1 \times 0.8 \times 0.6$ |
| $w_8$ | true | true | $\bot$ | true | $\bot$ | $0.1 \times 0.8 \times 0.4$ |

where $\bot$ means the variable is not defined in this world. These worlds cover all 8 cases of truth values for the original worlds that give values for $A$, $B$ and $C$. The values of $A$, $B$ and $C$ can be obtained from the program. The idea is that a run of the program is never going to encounter an undefined value. The augmented worlds can be obtained from the worlds defined by the program trace by splitting the worlds on each value of the undefined variables. Thus each augmented world corresponds to a set of possible worlds, where the distinctions that are ignored do not make a difference in the probability of the original variables.

While it may seem that we have not made any progress, after all this is just a simple Bayesian network, we can do the same thing for any program with probabilistic inputs. We just need to define the independent inputs (often these are called *noise inputs*), and a deterministic program that gives the consequences of the choices of values for these inputs. It is reasonably easy to see that any belief network can be represented in this way, where the number of independent inputs is equal to the number of free parameters of the belief network. However, we are not restricted to belief networks. The programs can be arbitrarily complex. We also do not need special "original variables", but can define the augmented worlds with respect to any variables of interest. Observations and queries (about which we want the posterior probability) can be propositions about the behavior of the program (e.g., that some assignment of the program variables becomes true).

When the language is Turing-equivalent, the worlds can be countably infinite, and thus there can be uncountably many worlds. A typical assumption is that the program eventually infers the observations and the query, that is, each run of the program will eventually (with probability 1) assign a truth value to any given observation and query. This is not always the case, such as when the query is to determine the fixed point of a Markov chain (see e.g., Pfeffer and Koller [2000]). We could also have non-discrete choices using continuous variables, which complicates but does not invalidate the discussion here.

A probability measure is over sets of possible worlds that form an algebra or a $\sigma$-algebra, depending on whether we want finite additivity or countable additivity [Halpern 2003]. For a programming language, we typically want countable additivity, as this allows us to not have a bound on the number of steps it takes to prove a query. For example, consider a person who plays the lottery until they win. The person will win eventually. This case is easy to represent as a probabilistic program, but requires reasoning about an infinite set of worlds.

The typical $\sigma$-algebra is the set of worlds that can be finitely described, and their (countable) union. Finitely describable means there is a finite set of draws that have their outcomes specified. Thus the probability measure is over sets of worlds that all have the same outcomes for a finite set of draws, and the union of such sets of worlds. We have a measure over such sets by treating the draws to be independent.

## 3 Abductive Characterization

Abduction is a form of reasoning characterized by "reasoning to the best explanation". It is typically characterized by finding a minimal consistent set of assumables that imply some observation. This set of assumables is called an *explanation* of the observation.

Poole [1991, 1993b] gave an abductive characterization of a probabilistic programming language, which gave a mapping between the independent possible world structure, and the descriptions of the worlds produced by abduction. This notion of abduction lets us construct a concise set of sets of possible worlds that is adequate to infer the posterior probability of a query.

The idea is that the the independent inputs become assumables. Given a probabilistic program, a particular observation *obs* and a query $q$, we characterize a (minimal) partition of possible worlds, where

- in each partition either $\neg obs$, $obs \wedge q$ or $obs \wedge \neg q$ can be inferred, and

# Probabilistic Programming Languages

- in each partition the same (finite) set of choices for the values of some of the inputs is made.

This is similar to the program-trace semantics, but will only need to make distinctions relevant to computing $P(q|obs)$. Given a probabilistic program, an observation and a query, the "explanations" of the observation conjoined with the query or its negation, produces such a partition of possible worlds.

In the example above, if the observation was $C = true$, and the query was $B$, we want the minimal set of assignments of values to the independent choices that gives $C = true \land B = true$ or $C = true \land B = false$. There are 4 such explanations:

- $A = true$, $Bifa = true$, $Cifb = true$
- $A = true$, $Bifa = false$, $Cifnb = true$
- $A = false$, $Bifna = true$, $Cifb = true$
- $A = false$, $Bifna = false$, $Cifnb = true$

The probability of each of these explanations is the product of the choices made, as these choices are independent. The posterior probability $P(B|C = true)$ can be easily computed by the weighted sum of the explanations in which $B$ is true. Note also that the same explanations would be true even if $C$ has unobserved descendants. As the number of descendants could be infinite if they were generated by a program, it is better to construct the finite relevant parts than prune the infinite irrelevant parts.

In an analogous way to how the probability of a real-variable is defined as a limit of discretizations, we can compute the posterior probability of a query given a probabilistic programming language. This may seem unremarkable until it is realized that for programs that are guaranteed to halt, there can be countably many possible worlds, and so there are uncountably many sets of possible worlds, over which to place a measure. For programs that are not guaranteed to halt, such as a sequence of lotteries, there are uncountably many possible worlds, and even more sets of possible worlds upon which to place a measure. Abduction gives us the sets of possible worlds in which to answer a conditional probability query. When the programs are not guaranteed to halt, the posterior probability of a query can be defined as the limit of the sets of possible worlds created by abduction, as long as the query can be derived in finite time for all but a set of worlds with measure zero.

In terms of the Simula program, explanations correspond to execution paths. In particular, an explanation corresponds to the outcomes of the draws in one trace of the program that infers the observations and a query or its negation. The set of traces of the program gives a set of possible worlds from which to compute probabilities.

When the program is a logic program, it isn't obvious what the program-trace semantics is. However, the semantics in terms of independent choices and abduction is well-defined. Thus it seems like the semantics in terms of abduction is more general than the program-trace semantics, as it is more generally applicable. It is also possible to define the abductive characterization independently of the details of the programming language, whereas defining a trace or run of a program depends on the details of the programming language.

Note that this abductive characterization is unrelated to MAP or MPE queries; we are defining the marginal posterior probability distribution over the query variables.

## 4 Inference

Earlier algorithms (e.g. Poole [1993a]) extract the minimal explanations and compute conditional probabilities from these. Later algorithms, such as used in IBAL [Pfeffer 2001], use sophisticated variable elimination to carry out inference in this space. IBAL's computation graph corresponds to a graphical representation of the explanations. Problog [De Raedt, Kimmig, and Toivonen 2007] compiles the computation graph into BDDs.

In algorithms that exploit the conditional independent structure, like variable elimination or recursive conditioning, the order that variables are summed out or split on makes a big difference to efficiency. In the independent choice semantics, there are more options available for summing out variables, thus there are more options available for making inference efficient. For example, consider the following fragment of a Simula program:

```
begin
   Boolean x;
   x := draw(0.1);
   if x then
      begin
         Boolean y := draw(0.2);
         ...
      end
   else
      begin
         Boolean z := draw(0.7);
         ...
      end
   ...
end
```

Here $y$ is only defined when $x$ is true and $z$ is only defined when $x$ is false. In the program-trace semantics, $y$ and $z$ are never both defined in any world. In the independent-choice semantics, $y$ and $z$ are defined in all worlds. Efficiency considerations may mean that we want to sum out $X$ first. In the independent-choice semantics, there is no problem, the joint probability on $X$ and $Y$ makes perfect sense. However, in the program trace semantics, it isn't clear what the joint probability of $X$ and $Y$ means. In order to allow for flexible elimination orderings in variable elimination or splitting ordering in recursive conditioning, the independent choice semantics is the natural choice.

Another possible way to implement probabilistic programming is to use MCMC [Milch, Marthi, Russell, Sontag, Ong, and Kolobov 2005; Goodman, Mansinghka, Roy, Bonawitz, and Tenenbaum 2008; McCallum, Schultz, and Singh 2009]. It is possible to do MCMC in either of the spaces of worlds above. The difference arises in conditionals. In the augmented space, for the example above, an MCMC state would include values for all of $X$, $Y$ and $Z$. In the program-trace semantics, it would contain values for $X$ and $Y$ when $X = true$, and values for $X$ and $Z$ when $X = false$, as $Y$ and $Z$ are never simultaneously defined. Suppose $X$'s value changes from $true$ to $false$. In the augmented space, it would just use the remembered values for $Z$. In the program-trace semantics, $Z$ was not defined when $Z$ was true, thus changing $X$ from $true$ to $false$ means re-sampling all of the variables defined in that branch, including $Z$.

BLOG [Milch, Marthi, Russell, Sontag, Ong, and Kolobov 2005] and Church [Goodman, Mansinghka, Roy, Bonawitz, and Tenenbaum 2008] assign values to all of the variables in the augmented space. FACTORIE [McCallum, Schultz, and Singh 2009] works in what we have called the abductive space. Which is these is more efficient is an empirical question.

## 5 Learning Probabilities

The other aspect of modern probabilistic programming languages is the ability to learn the probabilities. As the input variables are rarely observed, the standard way to learn the probabilities is to use EM. Learning probabilities using EM in probabilistic programming languages is described by Sato [1995] and Koller and Pfeffer [1997]. In terms of available programming languages, EM forms the basis for learning in Prism [Sato and Kameya 1997; Sato and Kameya 2001], IBAL [Pfeffer 2001; Pfeffer 2007] and many subsequent languages.

One can do EM learning in either of the semantic structures. The difference is whether some data updates the probabilities of parameters that were not involved in computing the data. By making this choice explicit, it is easy to see that one should use the abductive characterization to only update the probabilities of the choices that were used to derive the data.

Structure learning for probabilistic programming languages has really only been explored in the context of logic programs, where the techniques of inductive logic programming can be applied. De Raedt, Frasconi, Kersting, and Muggleton [2008] overview this active research area.

## 6 Causal Models

It is interesting that the research on causal modelling and probabilistic programming languages have gone on in parallel, with similar foundations, but only recently have researchers started to combine them by adding causal constructs to probabilistic programming languages [Finzi and Lukasiewicz 2003; Baral and Hunsaker 2007; Vennekens, Denecker, and Bruynooghe 2009].

In some sense, the programming languages can be seen as representations for all of the counterfactual situations. A programming language gives a model when some condition is true, but also defines the "else" part of a condition; what happens when the condition is false.

In the future, I expect that programming languages will be the preferred way to specify causal models, and for interventions and counterfactual reasoning to become part of the repertoire of probabilistic programming languages.

## 7 Observation Languages

These languages can be used to compute conditional probabilities by having an "observer" (either humans or sensors) making observations of the world that are conditioned on. One problem that has long gone unrecognized is that it is often not obvious how to condition when the language allows for multiple individuals and relations among them. There are two main problems:

- The observer has to know what to specify and what vocabulary to use. Unfortunately, we can't expect an observer to "tell us everything that is observed". First, there are an unbounded number of true things one can say about a world. Second, the observer does not know what vocabulary to use to describe their observations of the world. As probabilistic models get more integrated into society, the models have to be able to use observations from multiple people and sensors. Often these observations are historic, or are created asynchronously by people who don't even know the model exists and are unknown when the model is being built.

- When there are are unique names, so that the observer knows which object(s) a model is referring to, an observer can provide a value to the random variable corresponding to the property of the individual. However, models often refer to roles [Poole 2007]. The problem is that the observer does not know which individual in the world fills a role referred to in the program (indeed there is often a probability distribution over which individuals fill a role). There needs to be some other mechanism other than asking for the observed value of a random variable or program variable, or the value of a property of an individual.

The first problem can be solved using ontologies. An ontology is a specification of the meaning of the symbols used in an information system. There are two main areas that have spun off from the expert systems work of the 1970's and 1980's. One is the probabilistic revolution pioneered by Pearl. The other is often under the umbrella of knowledge representation and reasoning [Brachman and Levesque 2004]. A major aspect of this work is in the representation of ontologies that specify the meaning of symbols. An ontology language that has come to prominence recently is the language OWL [Hitzler, Krötzsch, Parsia, Patel-Schneider, and Rudolph 2009] which is one of the foundations of the semantic web [Berners-Lee, Hendler, and Lassila 2001]. There has recently been work on representing ontologies to integrate with probabilistic inference [da Costa, Laskey, and Laskey 2005; Lukasiewicz 2008; Poole, Smyth, and Sharma 2009]. This is important for Bayesian reasoning, where we need to condition on all available evidence; potentially applicable evidence is (or should be) published all over the world. Finding and using this evidence is a major challenge. This problem in being investigated under the umbrella of semantic science [Poole, Smyth, and Sharma 2008].

To understand the second problem, suppose we want to build a probabilistic program to model what apartments people will like for an online apartment finding service. This is an example where models of what people want and descriptions of the world are built asynchronously. Rather than modelling people's preferences, suppose we want to model whether they would want to move in and be happy there in 6 months time (this is what the landlord cares about, and presumably what the tenant wants too). Suppose Mary is looking for an apartment for her and her daughter, Sam. Whether Mary likes an apartment depends on the existence and the properties of Mary's bedroom and of Sam's bedroom (and whether they are the same room). Whether Mary likes a room depends on whether it is large. Whether Sam likes a room depends on whether it is green. Figure 1 gives one possible probability model, using a belief network, that follows the above story.

If we observe a particular apartment, such as the one on the right of Figure 1, it isn't obvious how to condition on the observations to determine the posterior probability that the apartment is suitable for Mary. The problem is that apartments don't come labelled with Mary's bedroom and Sam's bedroom. We need some role assignment that specifies which bedroom is Mary's and which bedroom is Sam's. However, which room Sam chooses depends on the colour of the room. We may also like to know the probability that a bachelor's apartment (that contains no bedrooms) would be suitable.

To solve the second problem, we need a representation of observations. These observations and the programs need to refer to interoperating ontologies. The observations need to refer to the existence of objects, and so would seem to need some subset of the first-order predicate calculus. However, we probably don't want to allow arbitrary first-order predicate calculus descriptions of observations. Arguably, people do not observe arbitrary disjunctions. One simple, yet powerful, observation language, based on RDF [Manola and Miller 2004] was proposed by Sharma, Poole, and Smyth [2010]. It is designed to allow for the specification of observations of

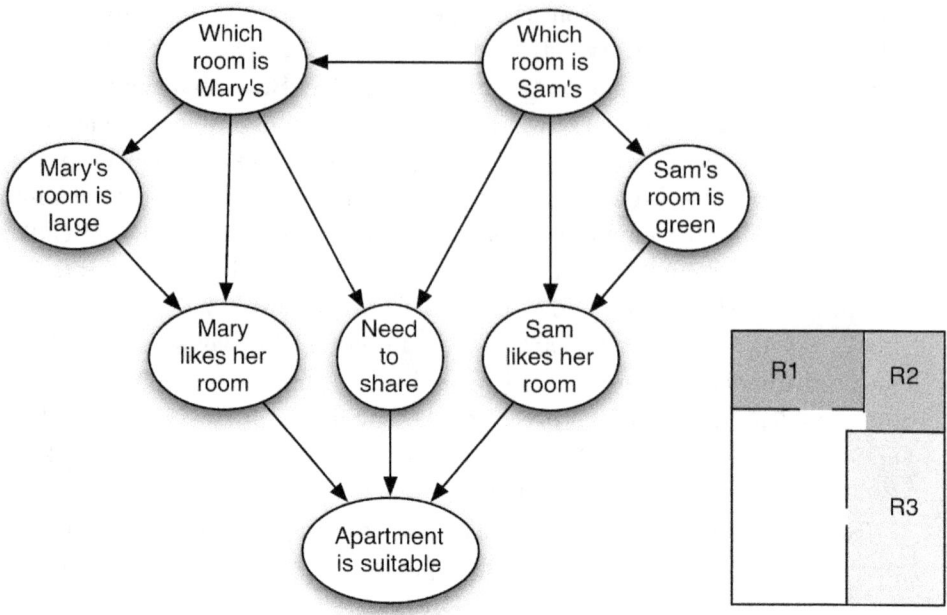

Figure 1. A possible belief network and observed apartment for the apartment example

the existence and non-existence of objects, and the properties and non-properties of objects. Observations are represented in terms of quadruples of the form:

$$\langle object, property, value, truthvalue \rangle$$

When the range of *property* is a fixed set, this quadruple means $property(object, value)$ or $\neg property(object, value)$, depending on the truth value.

When *value* is a world object, this quadruple means $\exists value\ property(object, value)$ or $\neg \exists value\ property(object, value)$, where the other mentions of *value* are in the scope of the quantification. This simple language seems to have the power to represent real observations, without representing arbitrary first-order formulae. It is then part of the program or the programming language to determine the correspondence between the objects in the language and the observed objects.

In the above example, the apartment can be described in terms of the existence of three bedrooms, one medium-sized and red (which we call $R1$), one small and pink (which we call $R2$) one large and green (which we will call $R3$). We also observe that there is not a fourth bedroom. This can be represented as:

$$\langle apr, hasBedroom, R1, true \rangle$$
$$\langle R1, size, medium, true \rangle$$
$$\langle R1, color, red, true \rangle$$
$$\langle apr, hasBedroom, R2, true \rangle$$
$$\ldots$$
$$\langle apr, hasBedroom, R4, false \rangle$$

Thus this language is analogous to observing conjunctions of propositional atoms. However, it also lets us observe the existence and non-existence of objects, without allowing for representing arbitrary disjunctions.

Such observational languages are an important complement to probabilistic programming languages.

## 8  Pivotal Probabilistic Programming Language References

Probabilistic Horn abduction [Poole 1991; Poole 1993b] is the first language with a probabilistic semantics that allows for conditioning. Much of the results of this paper were presented there, in the context of logic programs. Probabilistic Horn abduction was refined into the Independent Choice Logic [Poole 1997] that allowed for choices made by multiple agents, and there is a clean integration with negation as failure [Poole 2000]. Prism introduced learning into essentially the same framework [Sato and Kameya 1997; Sato and Kameya 2001]. More recently, Problog [De Raedt, Kimmig, and Toivonen 2007] has become a focus to implement many logical languages into a common framework.

In parallel to the work on probabilistic logic programming languages, has been work on developing probabilistic functional programming languages starting with Stochastic Lisp [Koller, McAllester, and Pfeffer 1997], including IBAL [Pfeffer 2001; Pfeffer 2007], A-Lisp [Andre and Russell 2002] and Church [Goodman, Mansinghka, Roy, Bonawitz, and Tenenbaum 2008].

Other probabilistic programming languages are based on more imperative languages such as CES [Thrun 2000], based on C, and the languages BLOG [Milch, Marthi, Russell, Sontag, Ong, and Kolobov 2005] and FACTORIE [McCallum, Schultz, and Singh 2009] based on object-oriented languages. BLOG concentrates on number and identity uncertainty, where the probabilistic inputs include the number of objects and whether two names refer to the same object or not.

## 9  Conclusion

This paper has concentrated on similarities, rather than the differences, between the probabilistic programing languages. Much of the research in the area has concentrated on specific languages, and this paper is an attempt to put a unifying structure on this work, in terms of independent choices and abduction.

Unfortunately, it is difficult to implement an efficient learning probabilistic programming language. Most of the languages that exist have just one implementation; the one developed by the designers of the language. As these are typically research code, the various implementations have concentrated on different aspects. For example, the Prism implementation has concentrated on incorporating different learning algorithms, the IBAL implementation has concentrated on efficient inference, my AILog2 implementation of ICL has concentrated on debugging and explanation and use by beginning students[1]. Fortunately, many of the implementations are publicly available and open-source, so that they are available for others to modify.

One of the problems with the current research is that the language and the implementation are often conflated. This means that researchers feel the need to invent a new language in order to investigate a new learning or inference technique. For example, the current IBAL implementation uses exact inference, but it does not need to; different inference procedures could be used with the same language. If we want people to use such languages, they should be able to take advantage of the advances in inference or learning techniques without changing their code. One interesting project is the ProbLog project [De Raedt, Kimmig, and Toivonen 2007], which is building an infrastructure so that many of the different logic programming systems can be combined, and so that the user can use a standard language, and it can incorporate advances in inference and learning.

Probabilistic programming languages have an exciting future. We will want to have rich languages to specify causal mechanisms, processes, and rich models. How to program these models, learn them, and efficiently implement these are challenging research problems.

**Acknowledgments:** Thanks to Peter Carbonetto, Mark Crowley, Jacek Kisyński and Daphne Koller for comments on earlier versions of this paper. Thanks to Judea Pearl for bringing probabilistic reasoning to the forefront of AI research. This work could not have been done without the foundations he lay. This work was supported by an NSERC discovery grant to the author.

# References

Andre, D. and S. Russell (2002). State abstraction for programmable reinforcement learning agents. In *Proc. AAAI-02*.

Baral, C. and M. Hunsaker (2007). Using the probabilistic logic programming language P-log for causal and counterfactual reasoning and non-naive conditioning. In *Proc. IJCAI 2007*, pp. 243–249.

Berners-Lee, T., J. Hendler, and O. Lassila (2001). The semantic web: A new

---

[1] We actually use it to teach logic programming to beginning students. They use it for assignments before they learn that it can also handle probabilities. The language shields the students from the non-declarative aspects of languages such as Prolog, and has many fewer built-in predicates to encourage students to think about the problem they are trying to solve.

form of web content that is meaningful to computers will unleash a revolution of new possibilities. *Scientific American May*, 28–37.

Brachman, R. and H. Levesque (2004). *Knowledge Representation and Reasoning.* Morgan Kaufmann.

da Costa, P. C. G., K. B. Laskey, and K. J. Laskey (2005, Nov). PR-OWL: A Bayesian ontology language for the semantic web. In *Proceedings of the ISWC Workshop on Uncertainty Reasoning for the Semantic Web*, Galway, Ireland.

Dahl, O.-J. and K. Nygaard (1966). Simula : an ALGOL-based simulation language. *Communications of the ACM 9*(9), 671–678.

De Raedt, L., P. Frasconi, K. Kersting, and S. H. Muggleton (Eds.) (2008). *Probabilistic Inductive Logic Programming.* Springer.

De Raedt, L., A. Kimmig, and H. Toivonen (2007). ProbLog: A probabilistic Prolog and its application in link discovery. In *Proceedings of the 20th International Joint Conference on Artificial Intelligence (IJCAI-2007)*, pp. 2462–2467.

Druzdzel, M. and H. Simon (1993). Causality in Bayesian belief networks. In *Proc. Ninth Conf. on Uncertainty in Artificial Intelligence (UAI-93)*, Washington, DC, pp. 3–11.

Finzi, A. and T. Lukasiewicz (2003). Structure-based causes and explanations in the independent choice logic. In *Proceedings of the 19th Conference on Uncertainty in Artificial Intelligence (UAI 2003)*, Acapulco, Mexico, pp. 225–232.

Goodman, N., V. Mansinghka, D. M. Roy, K. Bonawitz, and J. Tenenbaum (2008). Church: a language for generative models. In *Proc. Uncertainty in Artificial Intelligence (UAI)*.

Halpern, J. Y. (2003). *Reasoning about Uncertainty.* Cambridge, MA: MIT Press.

Hitzler, P., M. Krötzsch, B. Parsia, P. F. Patel-Schneider, and S. Rudolph (2009). *OWL 2 Web Ontology Language Primer.* W3C.

Koller, D., D. McAllester, and A. Pfeffer (1997). Effective Bayesian inference for stochastic programs. In *Proceedings of the 14th National Conference on Artificial Intelligence (AAAI)*, Providence, Rhode Island, pp. 740–747.

Koller, D. and A. Pfeffer (1997). Learning probabilities for noisy first-order rules,. In *Proceedings of the 15th International Joint Conference on Artificial Intelligence (IJCAI)*, Nagoya, Japan, pp. 1316–1321.

Laplace, P. S. (1814). *Essai philosophique sur les probabilities.* Paris: Courcier. Reprinted (1812) in English, F.W. Truscott amd F. L. Emory (Trans.) by Wiley, New York.

Lukasiewicz, T. (2008). Expressive probabilistic description logics. *Artificial Intelligence 172*(6-7), 852–883.

Manola, F. and E. Miller (2004). *RDF Primer*. W3C Recommendation 10 February 2004.

McCallum, A., K. Schultz, and S. Singh (2009). Factorie: Probabilistic programming via imperatively defined factor graphs. In *Neural Information Processing Systems Conference (NIPS)*.

Milch, B., B. Marthi, S. Russell, D. Sontag, D. L. Ong, and A. Kolobov (2005). BLOG: Probabilistic models with unknown objects. In *Proc. 19th International Joint Conf. Artificial Intelligence (IJCAI-05)*, Edinburgh.

Pearl, J. (1988). *Probabilistic Reasoning in Intelligent Systems: Networks of Plausible Inference*. San Mateo, CA: Morgan Kaufmann.

Pearl, J. (2000). *Causality: Models, Reasoning and Inference*. Cambridge University Press.

Pfeffer, A. (2001). IBAL: A probabilistic rational programming language. In *Proc. 17th International Joint Conf. Artificial Intelligence (IJCAI-01)*.

Pfeffer, A. (2007). The design and implementation of IBAL: A general-purpose probabilistic language. In L. Getoor and B. Taskar (Eds.), *Statistical Relational Learning*. MIT Press.

Pfeffer, A. and D. Koller (2000). Semantics and inference for recursive probability models,. In *National Conference on Artificial Intelligence (AAAI)*.

Poole, D. (1991, July). Representing Bayesian networks within probabilistic Horn abduction. In *Proc. Seventh Conf. on Uncertainty in Artificial Intelligence (UAI-91)*, Los Angeles, pp. 271–278.

Poole, D. (1993a). Logic programming, abduction and probability: A top-down anytime algorithm for computing prior and posterior probabilities. *New Generation Computing 11*(3–4), 377–400.

Poole, D. (1993b). Probabilistic Horn abduction and Bayesian networks. *Artificial Intelligence 64*(1), 81–129.

Poole, D. (1997). The independent choice logic for modelling multiple agents under uncertainty. *Artificial Intelligence 94*, 7–56. special issue on economic principles of multi-agent systems.

Poole, D. (2000). Abducing through negation as failure: stable models in the Independent Choice Logic. *Journal of Logic Programming 44*(1–3), 5–35.

Poole, D. (2007, July). Logical generative models for probabilistic reasoning about existence, roles and identity. In *22nd AAAI Conference on AI (AAAI-07)*.

Poole, D., C. Smyth, and R. Sharma (2008). Semantic science: Ontologies, data and probabilistic theories. In P. C. da Costa, C. d'Amato, N. Fanizzi, K. B.

Laskey, K. Laskey, T. Lukasiewicz, M. Nickles, and M. Pool (Eds.), *Uncertainty Reasoning for the Semantic Web I*, LNAI/LNCS. Springer.

Poole, D., C. Smyth, and R. Sharma (2009, Jan/Feb). Ontology design for scientific theories that make probabilistic predictions. *IEEE Intelligent Systems 24*(1), 27–36.

Sato, T. (1995). A statistical learning method for logic programs with distribution semantics. In *Proceedings of the 12th International Conference on Logic Programming (ICLP95)*, Tokyo, pp. 715–729.

Sato, T. and Y. Kameya (1997). PRISM: A symbolic-statistical modeling language. In *Proceedings of the 15th International Joint Conference on Artificial Intelligence (IJCAI97)*, pp. 1330–1335.

Sato, T. and Y. Kameya (2001). Parameter learning of logic programs for symbolic-statistical modeling. *Journal of Artificial Intelligence Research (JAIR) 15*, 391–454.

Sharma, R., D. Poole, and C. Smyth (2010). A framework for ontologically-grounded probabilistic matching. *International Journal of Approximate Reasoning In press*.

Thrun, S. (2000). Towards programming tools for robots that integrate probabilistic computation and learning. In *Proceedings of the IEEE International Conference on Robotics and Automation (ICRA)*, San Francisco, CA. IEEE.

Vennekens, J., M. Denecker, and M. Bruynooghe (2009). CP-logic: A language of causal probabilistic events and its relation to logic programming. *Theory and Practice of Logic Programming (TPLP) to appear*.

# 16
# Arguing with a Bayesian Intelligence
INGRID ZUKERMAN

## 1 Introduction

Bayesian Networks (BNs) [Pearl 1988] constitute one of the most influential advances in Artificial Intelligence, with applications in a wide range of domains, e.g., meteorology, agriculture, medicine and environment. To further capitalize on its clear technical advantages, a Bayesian intelligence (a computer system that employs a BN as its knowledge representation and reasoning formalism) should be able to communicate with its users, i.e., users should be able to put forward their views, and the system should be able to generate responses in turn. However, communication between a Bayesian and a human intelligence poses some challenges, as people generally do not engage in normative probabilistic reasoning when faced with uncertainty [Evans, Barston, and Pollard 1983; Lichtenstein, Fischhoff, and Phillips 1982; Tversky and Kahneman 1982]. In addition, human discourse is typically *enthymematic* (i.e., it omits easily inferable information), and usually the beliefs and inference patterns of conversational partners are not perfectly synchronized. As a result, an addressee's understanding may differ from the message intended by his or her conversational partner.

In this chapter, we offer a mechanism that enables a Bayesian intelligence to interpret human arguments for or against a proposition. This mechanism, which is implemented in a system called BIAS (*Bayesian Interactive Argumentation System*), constitutes a building block of a future system that will enable a Bayesian reasoner to communicate with people.[1]

In order to address the above challenges, we adopt the view that discourse interpretation is the process of integrating the contribution of a conversational partner into the addressee's mental model [Kashihara, Hirashima, and Toyoda 1995; Kintsch 1994], which in BIAS's case is a BN. Notice, however, that when performing such an integration, one cannot be sure that one is drawing the intended inferences or reinstating the exact information omitted by the user. All an addressee can do is construct an account of the conversational partner's discourse that makes sense to him or her. An interpretation of an argument that makes sense to BIAS is a subnet of its BN and a set of beliefs.

To illustrate these ideas, consider the argument in Figure 1(a) regarding the guilt

---

[1]The complementary building block, a mechanism that generates arguments from BNs, is described in [Korb, McConachy, and Zukerman 1997; Zukerman, McConachy, and Korb 1998].

Ingrid Zukerman

**Fingerprints being found on the gun**, and **forensics matching the fingerprints with Mr Green** implies that Mr Green probably had the means to murder Mr Body.
**The Bayesian Times reporting that Mr Body seduced Mr Green's girlfriend** implies that Mr Green possibly had a motive to murder Mr Body. Since Mr Green probably had the means to murder Mr Body, and Mr Green possibly had a motive to murder Mr Body, then Mr Green possibly murdered Mr Body.

(a) Sample argument

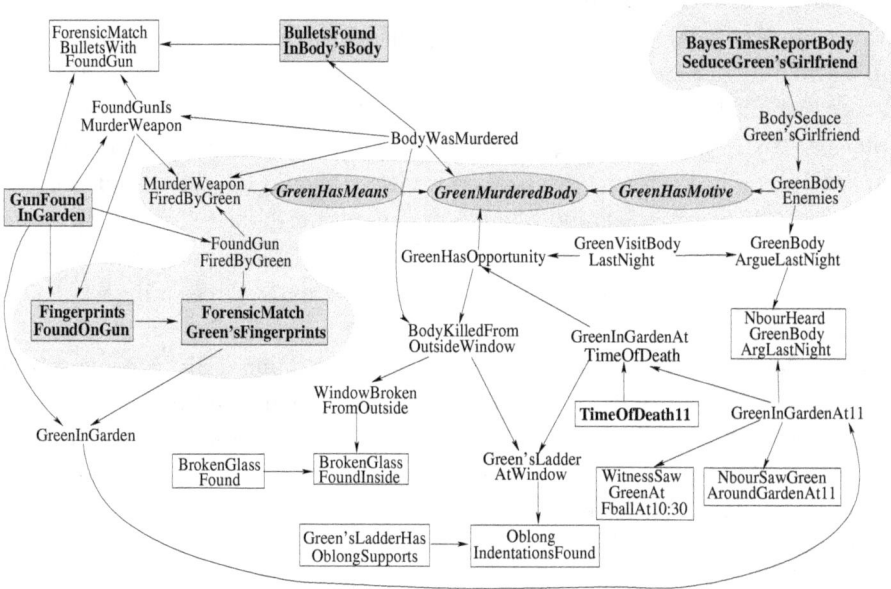

(b) BN and argument interpretation

Figure 1. Argument, domain BN and interpretation of the argument

of Mr Green in the murder of Mr Body (this argument is a gloss of an argument entered through a web interface).[2] The argument is interpreted in the context of the BN in Figure 1(b), which depicts the scenario for this murder mystery — the domain where we tested our ideas [Zukerman and George 2005]. The light-shaded bubble demarcates the BN subnet corresponding to the interpretation preferred by BIAS.[3] This interpretation contains propositions in the BN which bridge the

---

[2] We use the following linguistic terms, which are similar to those used in [Elsaesser 1987], to convey degree of belief: *Very Probable, Probable, Possible* and their negations, and *Even Chance*. According to our surveys, these terms are the most consistently understood by people [George, Zukerman, and Niemann 2007].

[3] The observable evidence nodes are boxed, and the evidence nodes that were actually observed by the user are boldfaced, as are the evidence nodes employed in the argument. The

reasoning gaps in the user's (enthymematic) argument.

Our approach casts the problem of finding a good interpretation of a user's argument as a model selection problem, where the interpretation is the model and the argument is the data. The criterion for selecting an interpretation is inspired by the Minimum Message Length (MML) principle [Wallace 2005] — an operational form of Occam's Razor that balances *model complexity* against *data fit*. That is, we aim to select *the simplest model (interpretation) that explains well the observed data (argument)*. The complexity of a model may be viewed as its probability in light of background knowledge: models that depart from the background knowledge are less probable than models that match the background knowledge, and structurally complex models are less probable than simpler models. Data fit may be viewed as the probability of the data given the model, i.e., the probability that a user who intended a particular interpretation presented the given argument.

The model selection problem is represented as a search problem, where a search procedure generates alternative interpretations, and an evaluation function assesses the merit of each interpretation. Since interpretations must be generated in real time, we use an (almost) anytime algorithm [Dean and Boddy 1988; Horvitz, Suermondt, and Cooper 1989] as our search procedure (Section 3). Our evaluation function is a probabilistic formulation of key aspects of the MML principle (Section 4).

This chapter is organized as follows. In the next section, we define an interpretation. Our algorithm for postulating interpretations is described in Section 3, and our probabilistic formulation for assessing an interpretation in Section 4. In Section 5 we present results of our evaluations, followed by a discussion of the limitations of our system, and advances required to support practical Bayesian argumentation systems.

## 2  What is an Interpretation?

As mentioned in Section 1, we view the interpretation of an argument as a "self explanation" — an account of the argument that makes sense to the addressee. For BIAS, such an account is specified by a tuple $\{IG, SC, EE\}$, where $IG$ is an *interpretation graph*, $SC$ is a *supposition configuration*, and $EE$ are *explanatory extensions*.[4]

- An **interpretation graph** is a subnet of the domain BN that connects the propositions in an argument. This subnet bridges inferential leaps in the argument, but the bridges so constructed may not be those intended by the user.

---

nodes corresponding to the consequents in the user's argument (GreenHasMeans, GreenHasMotive and GreenMurderedBody) are italicized and oval-shaded.

[4] In our initial work, our interpretations contained only interpretation graphs [Zukerman and George 2005]. Subsequent trials with users demonstrated the need for supposition configurations and explanatory extensions [George, Zukerman, and Niemann 2007].

- A **supposition configuration** is a set of beliefs attributed to the user to account for the beliefs expressed in the argument. A supposition may maintain a belief shared with the system (i.e., nothing is supposed), instantiate a node in a BN to TRUE or FALSE, or uninstantiate (forget) a previously instantiated node.

- **Explanatory extensions** consist of nodes and links that are added to an interpretation graph in order to make the inferences in the interpretation more acceptable to people (in early trials of the system, inferences were deemed unacceptable if they contained increases in certainty or large jumps in belief between the antecedents and the consequent). Contrary to suppositions, the beliefs in explanatory extensions are shared by the user and the system.

To illustrate these components, consider the brief argument in Figure 2(a) in relation to our murder mystery, and the three segments under it. Each segment, which highlights one of these components, shows the Bayesian subnet corresponding to the preferred interpretation and its textual rendition. Figure 2(b) shows the interpretation graph alone (the node that connects between the propositions in the argument appears in boldface italics); Figure 2(c) adds a supposition to the interpretation (in a shaded box); and Figure 2(d) adds an explanatory extension (white text in a dark box).

Let us now examine in more detail the three segments in Figure 2. Only one proposition (GreenInGardenAtTimeOfDeath, in boldface italics) needs to be added to connect the argument propositions in the domain BN, and create the interpretation graph in Figure 2(b). Note that the beliefs in this interpretation graph (obtained by Bayesian propagation of the system's beliefs in the domain BN) do not match those in the argument. The argument antecedent GreenInGardenAt11 yields a belief of *PossiblyNot* in GreenInGardenAtTimeOfDeath, which in turn implies that Mr Green *ProbablyNot* had the opportunity to kill Mr Body, and *VeryProbablyNot* committed the murder. To address this problem, the system *supposes* that the user believes that TimeOfDeath11 is TRUE (instead of the system's belief of *Probably*). Figure 2(c) shows how this supposition (in a shaded box) fits in the interpretation graph, and depicts its impact on the beliefs in the interpretation. These beliefs now match those in the argument. However, now the last inference in the argument goes from Mr Green *Possibly* having the opportunity to kill Mr Body to Mr Green *PossiblyNot* murdering Mr Body — a "jump in belief" which people find unacceptable. This problem is addressed by an *explanatory extension* that justifies the consequent on the basis of beliefs presumably shared with the user.[5] In this case, the selected proposition is that Mr Green *ProbablyNot* had the means to murder Mr Body. Figure 2(d) shows how this explanatory extension (white text in a dark box) fits in

---

[5]Note that the interpretation graph in Figure 2(a) also requires explanatory extensions for all the inferences (to overcome the jump in belief in the first inference, and the increases in certainty in the next two inferences). We omitted these explanatory extensions for clarity of exposition.

Arguing with a Bayesian Intelligence

*Mr Green probably being in the garden at 11 implies that he possibly had
the opportunity to kill Mr Body, but he possibly did not murder Mr Body.*

(a) Sample argument

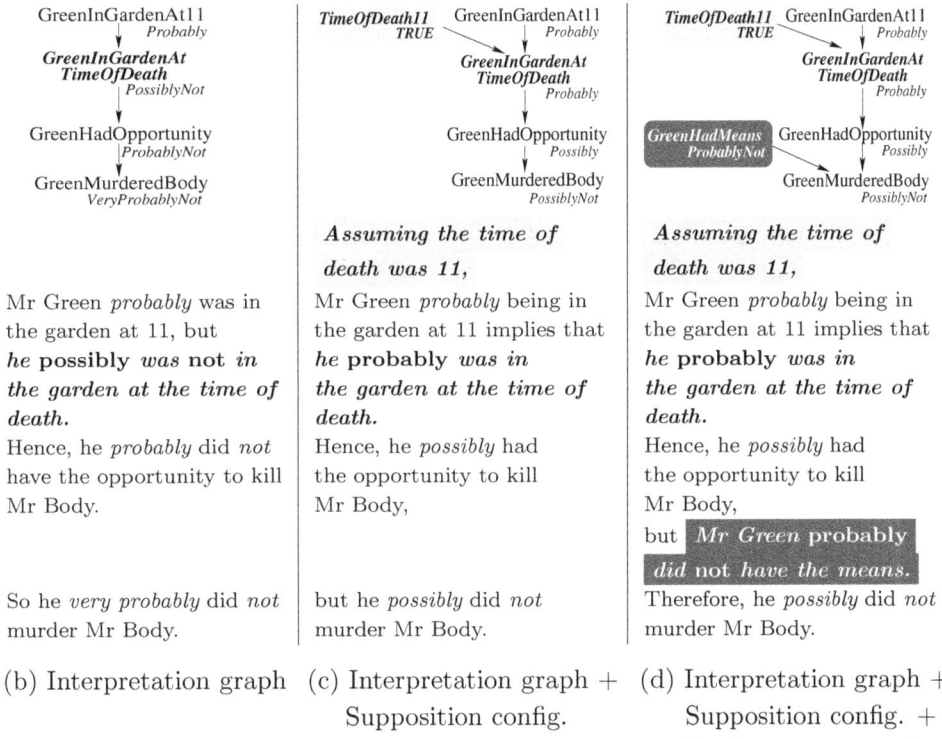

(b) Interpretation graph  (c) Interpretation graph +  (d) Interpretation graph +
                              Supposition config.         Supposition config. +
                                                          Explanatory extension

Figure 2. Interpretation graph, supposition configuration and explanatory extension

the interpretation graph. Note that explanatory extensions do not affect the beliefs in an interpretation, as they simply state previously held beliefs.

## 3 Proposing Interpretations

The problem of finding the best interpretation is exponential, as there are many candidates for each component of an interpretation, and complex interactions between supposition configurations and interpretation graphs. For example, making a supposition could invalidate an otherwise sound line of reasoning.

In order to generate reasonable interpretations in real time, we apply Algorithm 1 — an (almost) anytime algorithm [Dean and Boddy 1988; Horvitz, Suermondt, and Cooper 1989] that iteratively proposes interpretations until time runs out, i.e., until the system has to act upon a preferred interpretation or show the user one or more interpretations for validation [George, Zukerman, and Niemann 2007; Zukerman and George 2005]. At present, our interaction with the user stops when interpretations

| |
|---|
| **Algorithm 1** Argument Interpretation |
| **Require:** User argument, domain knowledge BN |
| 1: **while** there is time **do** |
| 2:   Propose a supposition configuration $SC_i$ — this can be null, an existing supposition configuration or a new one. |
| 3:   Propose a new interpretation graph $IG_{ij}$ under supposition configuration $SC_i$. |
| 4:   Propose explanatory extensions $EE_{ij}$ for interpretation graph $IG_{ij}$ under supposition configuration $SC_i$ as necessary. |
| 5:   Estimate the probability of interpretation $\{SC_i, IG_{ij}, EE_{ij}\}$. |
| 6:   Retain the top $K$ most probable interpretations. |
| 7: **end while** |
| 8: Present the retained interpretations to the user for validation. |

are presented for validation. However, in a complete system, a dialogue module would have to determine a course of action based on the generated interpretations.

In each iteration, the algorithm proposes an interpretation which consists of a supposition configuration, an interpretation graph and explanatory extensions (Steps 2-4). It then estimates the probability of this interpretation (Step 5), and retains the top $K$ most probable interpretations (Step 6). The procedure for building interpretation graphs is described in [Zukerman and George 2005], and the procedures for postulating supposition configurations and generating explanatory extensions are described in [George, Zukerman, and Niemann 2007]. Here we outline the general interpretation process and briefly describe these procedures.

Figure 3(a) depicts a portion of the search tree generated by our algorithm, with multiple supposition configurations considered in the first level, multiple interpretation graphs in the second level, and one set of explanatory extensions per interpretation graph in the third level. A supposition configuration is proposed first because suppositions change the beliefs in the domain and affect the manner in which beliefs influence each other. This happens because suppositions are implemented as instantiations or uninstantiations of nodes, which may block a path in a BN (precluding the propagation of evidence through this path), or unblock a previously blocked path. These interactions, which are difficult to predict until an interpretation graph is complete, motivate the large number of alternatives considered in the first two levels of the search tree. In contrast, explanatory extensions do not seem to have complex interactions with interpretation graphs. Hence, they are generated deterministically in the third level of the search tree — only one set of explanatory extensions is proposed for each interpretation.

Figure 3(b) shows a portion of the search tree instantiated for the short argument at the root node of this tree: "*Mr Green probably being in the garden at 11 implies that Mr Green possibly had the opportunity to kill Mr Body*". In this example, the

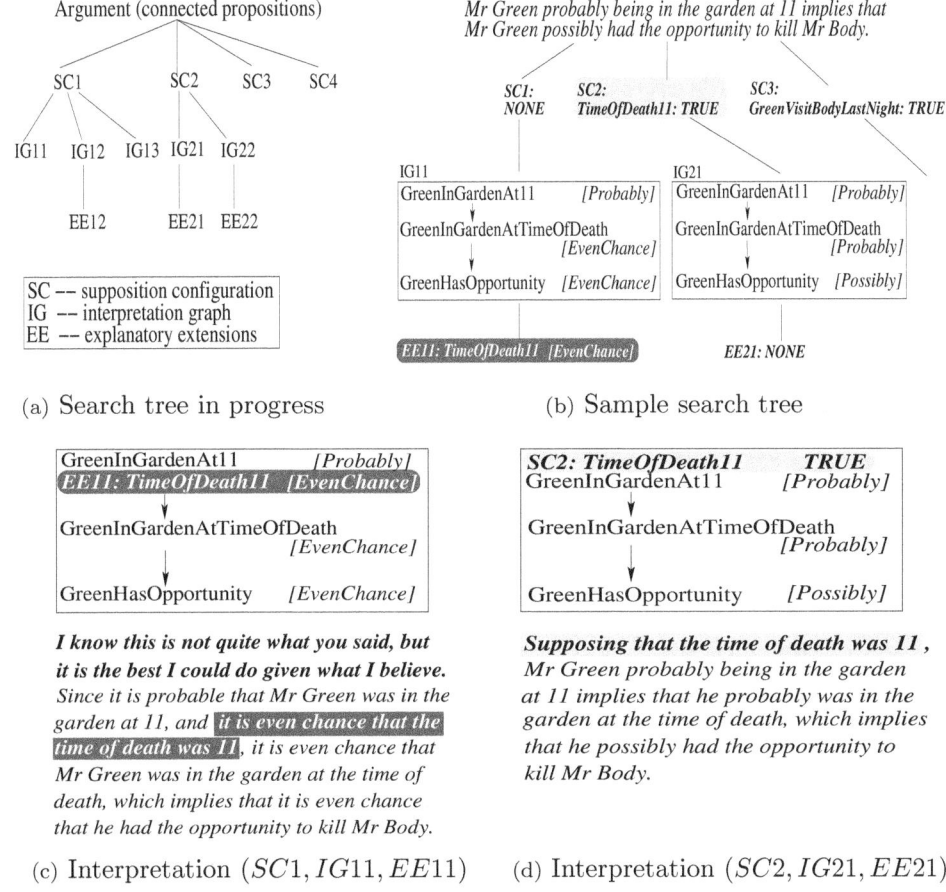

Figure 3. Argument interpretation process

user's belief in the consequent of the argument differs from the belief obtained by BIAS by means of Bayesian propagation from the evidence nodes in the domain BN. As indicated above, BIAS attempts to address this problem by making suppositions about the user's beliefs. The first level of the sample search tree in Figure 3(b) contains three supposition configurations $SC1$, $SC2$ and $SC3$. $SC1$ posits no beliefs that differ from those in the domain BN, thereby retaining the mismatch between the user's belief in the consequent and BIAS's belief; $SC2$ posits that the user believes that the time of death is 11; and $SC3$ posits that the user believes that Mr Green visited Mr Body last night.

The best interpretation graph for $SC1$ is $IG11$ (the evaluation of the goodness of an interpretation is described in Section 4). Here the belief in the consequent differs from that stated by the user, prompting the generation of a preface that acknowledges this fact. In addition, the interpretation graph has a large jump in belief (from *Probably* to *EvenChance*), which causes BIAS to add the mutually believed proposition TimeOfDeath11*[EvenChance]* as an explanatory extension. The resultant interpretation and its gloss appear in Figure 3(c). The best interpretation graph for $SC2$ is $IG21$, which matches the beliefs in the user's argument. The resultant interpretation and its gloss appear in Figure 3(d). Note that both $(SC1, IG11, EE11)$ and $(SC2, IG21, EE21)$ mention TimeOfDeath11, but in the first interpretation this proposition is used as an explanatory extension (with a belief of *EvenChance* obtained by Bayesian propagation), while in the second interpretation it is used as a supposition (with a belief of TRUE). Upon completion of this process, BIAS retains the $K$ most probable interpretations. In this example, the best interpretation is $\{SC2, IG21, EE21\}$.

## 3.1 Generating individual components

Owing to the different ways in which supposition configurations, interpretation graphs and explanatory extensions interact, we employ different techniques to generate each of these components: a dynamic priority queue is used to generate interpretation graphs; supposition configurations are drawn from a static pool based on a dynamic priority queue; and a deterministic algorithm is applied to generate explanatory extensions.

### Generating interpretation graphs

A priority queue is initialized with the smallest BN subnet that connects a user's statements. An iterative process is then followed, where in each iteration, the candidate at the top of the queue is selected, its "children" are generated, and their probability is calculated (Section 4). A child of an interpretation graph is the smallest interpretation graph (BN subnet) that connects the argument propositions in a modified BN where an arc from the parent interpretation graph has been removed. For example, in Figure 1(b), the smallest interpretation graph that connects GreenInGardenAt11 with GreenHasOpportunity goes through GreenInGardenAtTimeOfDeath. If we remove the arc between GreenInGardenAt11 and GreenInGardenAtTimeOfDeath, BIAS generates an interpretation graph that goes from GreenInGardenAt11 to NbourHeardGreenBodyArgLastNight, GreenBodyArgueLastNight and GreenVisitBodyLastNight. The newly generated children are then slotted in the priority queue according to their probability. This process yields good results for interpretation graphs, as the order in which these graphs appear in the queue is indicative of their goodness (graphs that appear earlier are usually better).

### Generating supposition configurations

Supposition configurations are generated by considering the following options for every node in the domain BN: (1) suppose nothing; (2) if the node is uninstantiated then instantiate it to TRUE and to FALSE; and (3) if the node is instantiated, uninstantiate it. The probability of each option is determined by its type (suppose nothing, instantiate or uninstantiate), and by how close a supposed belief is to the belief in the node in question, e.g., the probability of instantiating to TRUE a node with a belief of 0.8 is higher than the probability of instantiating it to FALSE. A supposition configuration is generated by first taking the highest probability option for all the nodes (which is the "suppose nothing" option), then taking the next best option for each node in turn (leaving the others as they are), and so on. For instance, in our domain BN, the second most probable option consists of setting TimeOfDeath11 to TRUE without changing the beliefs in the other nodes, next setting just GreenVisitBodyLastNight to TRUE, and so on.

This process is based on the closeness between the (new) supposed beliefs and the system's beliefs obtained by Bayesian propagation from evidence. However, it does not take into account how well the system's resultant beliefs in the argument propositions match the user's stated beliefs. Hence, we cannot simply rely on supposition configurations that are generated early in the interpretation process, as later configurations may yield a better belief match overall. In order to be able to access these later candidates and still achieve close to anytime performance, we create a static pool of promising candidates at the start of the interpretation process. This pool is populated by calling a priority queue of supposition configurations $M$ times as described above, and retaining the $m$ best candidates ($M \gg m$) on the basis of both (1) how close are the supposed beliefs to the original beliefs in the BN, and (2) how close are the system's resultant beliefs in the argument propositions to those stated by the user (these factors are respectively related to model complexity and data fit, Section 4). During the interpretation process, a new supposition configuration is probabilistically selected from this pool (the priority queue is never recalled).

**Generating explanatory extensions**

We conducted surveys to assess the influence of the beliefs in the antecedents and consequent of probabilistic inferences on the acceptability of these inferences. The main insights from our surveys are that people object to two types of inferences: (1) those which have more certainty regarding the consequent than regarding the antecedents, e.g., *Probably A* ⇒ ***Very*** *Probably C*; and (2) those where there is a large change in certainty from the antecedents to the consequent, e.g., *Probably A* ⇒ *EvenChance C* [Zukerman and George 2005]. In addition, among *acceptable* inferences, people prefer *BothSides* inferences to *SameSide* inferences. *BothSides* inferences have antecedents with beliefs on both "sides" of the belief in the consequent (higher probability and lower), e.g., *A[VeryProbably] & B[PossiblyNot]* ⇒ *C[Possibly]*; while all the antecedents in *SameSide* inferences have beliefs on "one side" of the belief in the consequent, e.g., *A[VeryProbably] & B[Probably]* ⇒ *C[Possibly]* [George, Zukerman, and Niemann 2007].[6]

Explanatory extensions are generated by considering the siblings of the an-

---

[6]Our surveys were restricted to *direct* inferences, where a high/low probability for an antecedent yields a high/low probability for the consequent. We posit similar preferences for *inverse* inferences, where a low/high probability antecedent yields a high/low probability consequent. The work described in [George, Zukerman, and Niemann 2007] contains additional, more fine-grained categories of inferences, but here we restrict our discussion to the main ones.

tecedents of an unacceptable inference in an interpretation graph (these siblings are not in the graph), and assigning each sibling to an inference category according to its effect on the inference. The siblings in the most preferred inference category are then added to the interpretation graph as an explanatory extension, e.g., if there is a set of siblings that turns an unacceptable inference into a *BothSides* inference, then it is chosen. This simple approach yields interpretations that people find acceptable (Section 5). However, further investigation is required to determine whether different combinations of siblings would yield better results.

## 4  Probabilistic Formalism

As mentioned in Section 1, the Minimum Message Length (MML) principle [Wallace 2005] selects the simplest model that explains the observed data. In our case, the data are the argument given by a user, and the candidate models are interpretations of this argument. In addition to the data and the model, the MML principle requires the specification of background knowledge — information shared by the system and the user prior to the argument, e.g., domain knowledge (including shared beliefs) and dialogue history.

We posit that the best interpretation is that with the highest posterior probability.

$$IntBest = \mathrm{argmax}_{i=1,\ldots,q} \Pr(IG_i, SC_i, EE_i | Argument, Background)$$

where $q$ is the number of interpretations.

After assuming conditional independence between the argument and the background given an interpretation, this probability is represented as follows.

$$IntBest = \mathrm{argmax}_{i=1,\ldots,q} \{\Pr(IG_i, SC_i, EE_i | Background) \times \qquad (1)$$
$$\Pr(Argument | IG_i, SC_i, EE_i)\}$$

The first factor, which is also known as *model complexity*, represents the prior probability of the model, and the second factor represents *data fit*.

- The **prior probability** of a model or **model complexity** reflects how "easy" it is to construct the model (interpretation) from background knowledge. For instance, complex models (e.g., interpretations with larger interpretation graphs) usually have a lower prior probability than simpler models.

- **Data fit** measures how similar the data (argument) are to the model (interpretation). The closer the data are to the model, the higher the probability of the data given the model (i.e., the probability that the user uttered the argument when he or she intended the interpretation in question).

Both the argument and its interpretation contain *structural* information and *beliefs*. The beliefs are simply those stated in the argument and in the interpretation, and suppositions made as part of the interpretation. The structural part of the argument comprises the stated propositions and the relationships between them, while the structural part of the interpretation comprises the interpretation graph and explanatory extensions. As stated above, smaller, simpler structures usually have a higher prior probability than larger, more complex ones. However, the simplest structure is not necessarily the best overall. For instance, the simplest possible interpretation for any argument consists of a single proposition, but this interpretation usually yields a poor data fit with most arguments. An increase in structural complexity (and corresponding reduction in probability) may reduce the discrepancy between the argument structure and the structure of the interpretation graph, thereby improving data fit. If this improvement overcomes the reduction in probability due to the higher model complexity, we obtain a higher-probability interpretation overall.

The techniques employed to calculate the prior probability and data fit for both types of information are outlined below (a detailed description appears in [George, Zukerman, and Niemann 2007; Zukerman and George 2005]).

## 4.1 Prior probability of an interpretation: $\Pr(IG_i, SC_i, EE_i | Background)$

As mentioned above, the prior probability of an interpretation reflects how well it fits the background knowledge. In our case, the background knowledge comprises (1) domain knowledge – the evidence in the BN (known to the user and the system); (2) dialogue history – previously mentioned propositions; and (3) presentation preferences – features of acceptable inferences (obtained from user surveys).

To estimate the prior probability of an interpretation, we separate the structure of an interpretation and the beliefs in it (note that $SC_i$ comprises only beliefs, and $EE_i$ comprises only structure).

$$\Pr(IG_i, SC_i, EE_i | Background) = $$
$$\Pr(\text{beliefs in } IG_i, \text{struct of } IG_i, SC_i, EE_i | Background)$$

After applying the chain rule of probability, we obtain

$$\Pr(IG_i, SC_i, EE_i | Background) = \qquad (2)$$
$$\Pr(\text{beliefs in } IG_i | \text{struct of } IG_i, SC_i, EE_i, Background) \times$$
$$\Pr(EE_i | \text{struct of } IG_i, SC_i, Background) \times$$
$$\Pr(\text{struct of } IG_i | SC_i, Background) \times \Pr(SC_i | Background)$$

The factors in Equation 2 are described below (we consider them from last to first for clarity of exposition).

### Supposition configuration: $\Pr(SC_i | Background)$

A supposition configuration addresses mismatches between the beliefs expressed in an argument and those in an interpretation. It comprises beliefs attributed to the user in light of the beliefs shared with the system, which are encoded in the background knowledge. Making suppositions has a lower probability than not making suppositions (which has no discrepancy with the background knowledge). However, as seen in the example in Figure 2(c), making a supposition that reduces or eliminates the discrepancy between the beliefs stated in the argument and those in the interpretation increases the data fit for beliefs.

$\Pr(SC_i|Background)$ reflects how close the suppositions in a supposition configuration are to the current beliefs in the background knowledge. The closer they are, the higher the probability of the supposition configuration. Assuming conditional independence between the supposition for each node given the background knowledge yields

$$\Pr(SC_i|Background) = \prod_{j=1}^{N} \Pr(s_{ji}|Bel_{Bkgrd}(j))$$

where $N$ is the number of nodes in the BN, $s_{ij}$ is the supposition made for node $j$ in supposition configuration $SC_i$, and $Bel_{Bkgrd}(j)$ is the belief in node $j$ according to the background knowledge. $\Pr(s_{ji}|Bel_{Bkgrd}(j))$ is estimated by means of the heuristic function $\mathcal{H}$.

$$\Pr(s_{ji}|Bel_{Bkgrd}(j)) = \mathcal{H}(\mathit{Type}(s_{ji}), Bel(s_{ji})|Bel_{Bkgrd}(j))$$

where $\mathit{Type}(s_{ji})$ is the type of supposition $s_{ji}$ (supposing nothing, supposing evidence, or forgetting evidence), and $Bel(s_{ji})$ is the value of the supposition (TRUE or FALSE when evidence is supposed for node $j$; and the belief in node $j$ obtained from belief propagation in the BN when evidence is forgotten for node $j$). Specifically, we posit that supposing nothing has the highest probability, and supposing the truth or falsehood of an inferred value is more probable than forgetting seen evidence [George, Zukerman, and Niemann 2007]. In addition, strongly believed (high probability) propositions are more likely to be supposed TRUE than weakly believed (lower probability) propositions, and weakly believed propositions are more likely to be supposed FALSE than strongly believed propositions [Lichtenstein, Fischhoff, and Phillips 1982].

**Structure of an interpretation: Pr(struct of IG$_i$|SC$_i$, *Background*)**

$\Pr(\text{struct of } IG_i|SC_i, Background)$ is the probability of selecting the nodes and arcs in $IG_i$ from the domain BN (which is part of the background knowledge). The calculation of this probability is described in detail in [Zukerman and George 2005]. In brief, the prior probability of the structure of an interpretation graph is estimated using the combinatorial notion of selecting the nodes and arcs in the graph from those in the domain BN. To implement this idea, we specify an interpretation graph $IG_i$ by indicating the number of nodes in it ($n_i$), the number of arcs ($a_i$), and the actual nodes and arcs in it ($Nodes_i$ and $Arcs_i$ respectively). Thus, the probability of the structure of $IG_i$ in the context of the domain BN (composed of $A$ arcs and

$N$ nodes) is defined as follows.

$$\Pr(\text{struct } IG_i|Background) = \Pr(Arcs_i, Nodes_i, a_i, n_i|Background)$$

Applying the chain rule of probability yields

$$\Pr(\text{struct } IG_i|Background) = \Pr(Arcs_i|Nodes_i, a_i, n_i, Background) \times \quad (3)$$
$$\Pr(a_i|Nodes_i, n_i, Background) \times$$
$$\Pr(Nodes_i|n_i, Background) \times \Pr(n_i|Background)$$

These probabilities are calculated as follows.

- $\Pr(n_i|Background)$ is the probability of having $n_i$ nodes in an interpretation graph. We model this probability by means of a truncated Poisson distribution, $Poisson(\beta)$, where $\beta$ is the average number of nodes in an interpretation (obtained from user trials).

- $\Pr(Nodes_i|n_i, Background)$ is the probability of selecting the particular $n_i$ nodes in $Nodes_i$ from the $N$ nodes in the domain BN. The simplest calculation assumes that all nodes in an interpretation graph have an equal probability of being selected, i.e., there are $\binom{N}{n_i}$ ways to select these nodes. This calculation generally prefers small models to larger ones.[7] In [Zukerman and George 2005], we considered salience — obtained from dialogue history, which is part of the background knowledge — to moderate the probability of selecting a node. According to this scheme, recently mentioned nodes are more salient (and have a higher probability of being selected) than nodes mentioned less recently.

- $\Pr(a_i|Nodes_i, n_i, Background)$ is the probability of having $a_i$ arcs in an interpretation graph. The number of arcs in an interpretation is between the minimum number of arcs needed to connect $n_i$ nodes ($n_i - 1$), and the actual number of arcs in the domain BN that connect the nodes in $Nodes_i$, denoted $va_i$. We model the probability of $a_i$ by means of a uniform distribution between $n_i - 1$ and $va_i$.

- $\Pr(Arcs_i|Nodes_i, a_i, n_i, Background)$ is the probability of selecting the particular $a_i$ arcs in $Arcs_i$ from the $va_i$ arcs in the domain BN that connect the nodes in $IG_i$. Assuming an equiprobable distribution, there are $\binom{va_i}{a_i}$ ways to select these arcs.

**Structure of an explanatory extension:**
**$\Pr(\mathbf{EE_i}|\text{struct of } \mathbf{IG_i}, \mathbf{SC_i}, Background)$**

Explanatory extensions are added to an interpretation graph to accommodate people's expectations regarding the relationship between the antecedents of an inference

---

[7] In the rare cases where the number of propositions in an interpretation exceeds $N/2$, smaller models do not yield lower probabilities.

and its consequent (rather than to connect between the propositions in an argument). These expectations, which are part of the background knowledge, were obtained from our user studies. Explanatory extensions have no belief component, as the nodes in them do not provide additional evidence, and hence do not affect the beliefs in a BN.

Interpretations with explanatory extensions are more complex, and hence have a lower probability, than interpretations without such extensions. At the same time, as shown in the example in Figure 2(d), an explanatory extension that overcomes an expectation violation regarding the consequent of an inference improves the acceptance of the interpretation, thereby increasing the probability of the model.

According to our surveys, explanatory extensions that yield *BothSides* inferences are preferred to those that yield *SameSide* inferences. In addition, as for interpretation graphs, shorter explanatory extensions are preferred to longer ones. Thus, our estimate of the structural probability of explanatory extensions balances the size of explanatory extensions (*number of propositions*) against their type (*inference category*), as follows.[8]

$$\Pr(\text{struct of } EE_i | \text{struct of } IG_i, SC_i, Background) =$$
$$\prod_{j=1}^{NF_i} \Pr(InfCategory(EE_{ij}), np(EE_{ij}) | \text{struct of } IG_i, SC_i, Background)$$

where $NF_i$ is the number of inferences in $IG_i$, $InfCategory(EE_{ij})$ is the category of the inference obtained by adding explanatory extension $EE_{ij}$ to the $j$th inference in $IG_i$, and $np(EE_{ij})$ is the number of propositions in $EE_{ij}$.

Applying the chain rule of probability yields

$$\Pr(\text{struct of } EE_i | \text{struct of } IG_i, SC_i, Background) = \quad (4)$$
$$\prod_{j=1}^{NF_i} \left\{ \begin{array}{l} \Pr(InfCategory(EE_{ij}) | np(EE_{ij}), \text{struct of } IG_i, SC_i, Background) \times \\ \Pr(np(EE_{ij}) | \text{struct of } IG_i, SC_i, Background) \end{array} \right\}$$

These probabilities are calculated as follows.

- $\Pr(InfCategory(EE_{ij}) | np(EE_{ij}), \text{struct of } IG_i, SC_i, Background)$ is estimated using a heuristic function that represents people's preferences: an explanatory extension that yields a *BothSides* inference has a higher probability than an explanatory extension that yields a *SameSide* inference.

- As for interpretation graphs, $\Pr(np(EE_{ij}) | \text{struct of } IG_i, SC_i, Background)$ is estimated by means of a truncated Poisson distribution, $Poisson(\mu)$, where $\mu$ is the average number of nodes in an explanatory extension.

---

[8] We do not estimate the probability of including particular nodes in an explanatory extension, because the nodes in an explanatory extension are completely determined by their inference category.

**Beliefs in an interpretation:**
**Pr(beliefs in IG$_i$|struct of IG$_i$, SC$_i$, EE$_i$, *Background*)**

The beliefs in an interpretation $IG_i$ are estimated by performing Bayesian propagation from the beliefs in the domain BN and the suppositions. This is an algorithmic process, hence the probability of obtaining the beliefs in $IG_i$ is 1. However, the background knowledge has another aspect, viz users' expectations regarding inferred beliefs. In our preliminary trials, users objected to inferences that had increases in certainty or large changes in belief from their antecedents to their consequent [Zukerman and George 2005].

Thus, interpretations that contain objectionable inferences have a lower probability than interpretations where the beliefs in the consequents of the inferences fall within an "acceptable range" of the beliefs in their antecedents. We use the categories of acceptable inferences obtained from our surveys to estimate the probability of each inference in an interpretation — these categories define an *acceptable range* of beliefs for the consequent of an inference given its antecedents. For example, an inference with antecedents *A[Probably]* & *B[Possibly]* has the acceptable belief range {*Probably, Possibly, EvenChance*} for its consequent. The probability of an inference whose consequent falls within the acceptable range is higher than the probability of an inference whose consequent falls outside this range. In addition, we extrapolate from the results of our surveys, and posit that the probability of an unacceptable inference decreases as the distance of its consequent from the acceptable range increases. We use the Zipf distribution to model the probability of an inference, where the "rank" is the distance between the belief in the consequent and the acceptable belief range.

As mentioned above, explanatory extensions are generated to satisfy people's expectations about the relationship between the beliefs in the antecedents of inferences and the belief in their consequent (i.e., they bring the consequent into the acceptable range of an inference, or at least closer to this range). Thus, they increase the belief probability of an interpretation at the expense of its structural probability.

## 4.2 Data fit between argument and interpretation: Pr(*Argument*|IG$_i$, SC$_i$, EE$_i$)

As mentioned above, data fit reflects the probability that a user who intended a particular interpretation generated the given argument. This probability is a function of the similarity between the argument and the interpretation: the higher this similarity, the higher the probability of the argument given the interpretation. As for prior probabilities, we consider structural similarity and belief similarity.

$$\Pr(Argument|IG_i, SC_i, EE_i) = \Pr(\text{struct of } Argument, \text{beliefs in } Argument | \text{struct of } IG_i, \text{beliefs in } IG_i, SC_i, EE_i)$$

We assume that given an interpretation graph, the argument is independent of the suppositions and explanatory extensions, which yields

$$\Pr(Argument|IG_i, SC_i, EE_i) = \Pr(\text{struct of } Argument|\text{struct of } IG_i) \times \quad (5)$$
$$\Pr(\text{beliefs in } Argument|\text{beliefs in } IG_i)$$

**Structure of the argument given the structure of an interpretation: Pr(struct of *Argument*|struct of IG$_i$)**

The estimation of the structural similarity between an argument and an interpretation is based on the idea that the nodes and arcs in the argument are selected from those in the interpretation graph. This idea is similar to that used to calculate the prior probability of the structure of the interpretation graph, where the nodes and arcs in $IG_i$ were selected from those in the domain BN (Section 4.1). However, in this case there is a complicating factor, since as seen in our examples, a user may mention implications (arcs) which are absent from the interpretation graph (our web interface prevents the inclusion of nodes that do not appear in the domain BN). Hence, the calculation of $\Pr(\text{struct of } Argument|\text{struct of } IG_i)$ is similar to the calculation of $\Pr(\text{struct of } IG_i|Background)$ in Equation 3, but it distinguishes between arcs in *Argument* that are selected from $IG_i$ and arcs that are newly inserted.

**Beliefs in the argument given the beliefs in an interpretation: Pr(beliefs in *Argument*|beliefs in IG$_i$)**

The closer the beliefs stated in an argument are to the beliefs in an interpretation, the higher the probability that the user presented the argument when he or she intended this interpretation. Thus, suppositions that reduce belief discrepancies between the beliefs in an argument and those in an interpretation improve data fit (at the expense of the prior probability of the interpretation, Section 4.1). We employ the Zipf distribution to estimate the probability that the beliefs in an interpretation were intended in the argument, where the "rank" is the difference between the corresponding beliefs.

In a final step, we inspect the interpretation graph to determine whether it has blocked paths. The presence of blocked paths in an interpretation graph suggests that the lines of reasoning in the interpretation do not match those in the argument. Thus, if blocked paths are found, the probability of the belief match between the interpretation graph and the argument is reduced.

## 5 Evaluation

Our evaluation was designed to determine whether our approach to argument interpretation by a Bayesian intelligence yields interpretations that are acceptable to users. However, our target users are not those who constructed the argument, but those who read the argument. Specifically, our evaluation determines whether people reading someone else's argument find BIAS's highest-probability interpretations acceptable (and better than other options).

We evaluated the components of an interpretation separately in the order in which they were developed: first interpretation graphs, next supposition configurations, and finally explanatory extensions. Our evaluations relied on scenarios based on BNs that were similar to that in Figure 1(b). Our participants were staff and students in the Faculty of Information Technology at Monash University and people known to the project team members (the participants exhibited different levels of computer literacy).

**Interpretation graphs.** We constructed four evaluation sets, where each set contained an argument (the argument in Figure 1(a), and three short arguments similar to that in Figure 2(a)) and BIAS's preferred interpretations.[9] Between 17-25 participants read each argument and the interpretations. They were then asked to give each interpretation a score between 1 (Very UNreasonable) and 5 (Very Reasonable), and to comment on aspects of the interpretations that they liked or disliked. People generally found our interpretations acceptable, with average scores between 3.35 and 4. The lower scores were attributed to three main problems: (1) participants' disagreement with the systems' domain-related inferences, (2) discrepancies between the argument's beliefs in the consequents of inferences and the system's beliefs, and (3) unacceptable inferences. The first problem is discussed in Section 6, and the other two problems are addressed by supposition configurations and explanatory extensions respectively.

**Supposition configurations.** We constructed four scenarios, where each scenario had two versions of an argument: (1) an original version, whose beliefs were obtained by Bayesian propagation in the domain BN; and (2) a version given by a hypothetical user, whose conclusion did not match that of the original version (BIAS had to make a supposition in order to account for the beliefs in the user's argument). 34 participants read both arguments, and were then asked whether it was reasonable to make suppositions about the user's beliefs in order to make sense of his or her argument. To answer this question, they could (1) select one of four suppositions we showed them (which included BIAS's top-ranked supposition and other highly ranked suppositions), (2) include an alternative supposition of their choice (from BIAS's knowledge base or of their own devising), or (3) indicate that no suppositions were required. BIAS's preferred supposition was consistently ranked first or second by our trial subjects, with its average rank being the lowest (best) among all the options. In addition, very few respondents felt that no suppositions were warranted.

**Explanatory extensions.** We constructed two evaluation sets, each consisting of a short argument and two alternative interpretations — one with explanatory extensions and one without. These sets were shown to 20 participants. The majority of the participants preferred the interpretations with explanatory extensions. At

---

[9]The arguments were generated by project team members. We also conducted experiments were people not associated with the project entered arguments, but interface problems affected the evaluation (Section 6.4).

the same time, about half of the participants felt that the extended interpretations were too verbose. This problem may be partially attributed to the presentation of the nodes as direct renditions of their propositional content, which makes the interpretations appear repetitive in style. The generation of stylistically diverse text is the subject of active research in Natural Language Generation, e.g., [Gardent and Kow 2005].

## 6 Discussion

This chapter offers a probabilistic approach to argument interpretation by a system that uses a BN as its knowledge representation and reasoning formalism. An interpretation of a user's argument is represented as beliefs in the BN (suppositions) and a Bayesian subnet (interpretation graph and explanatory extensions). Our evaluations show that people found BIAS's interpretations generally acceptable, and its suppositions and explanatory extensions both necessary and reasonable.

Our approach casts the generation of an interpretation as a model selection task, and employs an (almost) anytime algorithm to generate candidate interpretations. Our model selection approach balances the probability of the model in light of background knowledge against its data fit (similarity between the model and the data). In other words, our formalism balances the cost of adding extra elements to an interpretation (e.g., suppositions) against the benefits obtained from these elements. The calculations that implement this idea are based on three main elements: (1) combinatoric principles for extracting an interpretation graph from the domain BN, and an argument from an interpretation; (2) known distributions, such as Poisson for the number of nodes in an interpretation graph or explanatory extension, and Zipf for modeling discrepancies in belief; and (3) manually-generated distributions for suppositions and for preferences regarding different types of inferences. The parameterization of these distributions requires specific information. For instance, the mean of the Poisson distribution, which determines the "penalty" for having too many nodes in an interpretation or explanatory extension, must be empirically determined. Similarly, the hand-tailored distributions for supposition configurations and explanatory extensions require experimental fine-tuning or user studies to gather these probabilities.

The applicability of our approach is mainly affected by our assumption that the nodes in the domain BN are binary. Other factors to be considered when applying our formalism are: the characteristics of the domain, the expressive power of BNs *vis a vis* human reasoning, and the ability of users to interact with the system.

### 6.1 Binary node BNs

The assumption that the nodes in the domain BN are binary simplifies the estimation of the probability of suppositions and explanatory extensions. The relaxation of this assumption to multi-valued nodes would increase the search space for suppositions, and necessitate a generalization of the heuristics used to calculate the probability of a supposition (Section 4.1). The incorporation of multi-valued nodes would also require a generalization of the procedure for generating explanatory extensions and estimating their probability (Sections 3.1 and 4.1 respectively). This in turn would necessitate user studies to determine people's presentation preferences and expectations about inferences involving multi-valued nodes. For instance, people may prefer such inferences to be presented in terms of a particular value of a node or in terms of an aggregate of several values; and different models of expectations may be required for nodes with ordinal values (e.g., low, medium and high) and nodes with scalar values (e.g., colours). Further, as indicated in Section 6.2, these preferences and expectations are domain dependent.

## 6.2 Domain of argumentation

We selected a "commonsense" domain both for ease of design and to be able to conduct trials with non-experts. The nodes and arcs in the domain BN and the values in the Conditional Probability Tables (CPTs) were devised by the project team members. A consequence of working in a commonsense domain is that people, rather than computer systems, are the domain experts. As a result, users may postulate ideas of which the system is unaware (e.g., Mr Green and Ms Scarlet were in cahoots), and their inferences may validly differ from those of the system. For instance, according to BIAS, Mr Green and Mr Body being enemies implies that Mr Green very probably has a motive to kill Mr Body — an inference that several users found objectionable.

To deal with the first of these issues, an argumentation system can (1) restrict the user to use only the propositions known to the system, (2) ignore the user's propositions that are not known to the system, or (3) try to learn the import of new propositions. Our experience with BIAS shows that the first solution is frustrating for users, as people did not like having to shoehorn their reasoning into the propositions known to the system. The second solution leads to only a partial understanding of the user's intentions, and hence potentially to a mis-directed discussion. The third solution, which also applies to the synchronization of inference patterns between the user and the system, is clearly the most sound. However, incorporating new propositions into a BN, and modifying inference patterns, have significant implications with respect to the system's reasoning, and present non-trivial interface design problems. These observations, together with the fact that at present the strength of computer systems is their ability to perform expert reasoning, indicate that a fruitful domain for the incorporation of argumentation capabilities into BNs is an expert domain, where the system's knowledge generally exceeds that of users.

Our procedures for generating interpretation graphs and supposition configurations are domain independent. However, the generation of explanatory extensions is domain dependent. This is because explanatory extensions are generated to explain surprising outcomes, and what is surprising often depends on the domain. Further, in some domains what matters is the increase or reduction in probability, rather than its absolute value, e.g., an increase from 6% to 10% in the probability of a patient having cancer may require an explanatory extension, even though both probabilities belong to the *VeryProbablyNot* belief category.

## 6.3 BN reasoning and human reasoning

Our approach assumes that the underlying BN represents only ground predicates, while human reasoning often involves general statements (quantified predicates). Getoor *et al.* [Getoor, Friedman, Koller, and Taskar 2001] and Taskar *et al.* [Taskar, Abbeel, and Koller 2002] studied probabilistic relational models, which combine advantages of relational logic and BNs, and can generalize over a variety of situations. This is a promising representation for the interpretation of arguments that include quantified predicates.

Belief propagation in BNs differs from human belief propagation when users employ different inference patterns from those in the BN, and when users do not engage in normative probabilistic reasoning. As mentioned above, the synchronization of inference patterns between the user and the system is a challenging task which falls under the purview of probabilistic reasoning and human-computer interfaces.

People's non-normative probabilistic reasoning is partly attributed to *reasoning fallacies* [Evans, Barston, and Pollard 1983; Lichtenstein, Fischhoff, and Phillips 1982; Tversky and Kahneman 1982]. In previous research, we augmented a Bayesian argument generation system with a (rather coarse) model of certain types of human reasoning fallacies [Korb, McConachy, and Zukerman 1997]. An interesting avenue for future research consists of developing finer, domain dependent models of human reasoning fallacies, and incorporating them into our interpretation process.

## 6.4 Argumentation interface

BIAS requires users to construct their arguments using only propositions known to the system, and assumes that the arguments are in premise-to-goal form. As mentioned in Section 6.2, users disliked having to shoehorn their ideas into a restricted set of propositions. An alternative approach, which we considered in [Zukerman, George, and Wen 2003], allowed users to provide Natural Language statements, and then mapped these statements to propositions in the system's knowledge base. However, such a process runs the risk of producing an erroneous mapping. Hence, this process should be able to determine when a mapping is questionable, and handle this situation appropriately.

In addition to a premise-to-goal argumentation strategy, people employ strategies such as reductio-ad-absurdum, inference to best explanation, and reasoning by cases [Zukerman, McConachy, and Korb 2000]. These strategies must be identified prior to rendering an argument into an interpretation graph. An interesting approach for addressing this problem involves using a graphical interface to help users structure an argument [van Gelder 2005], while allowing them to express propositions in Natural Language.

## Acknowledgments

The author thanks her collaborators on the research described in this chapter: Sarah George and Michael Niemann. This research was supported in part by grant DP0878195 from the Australian Research Council (ARC) and by the ARC Centre for Perceptive and Intelligent Machines in Complex Environments.

## References

Dean, T. and M. Boddy (1988). An analysis of time-dependent planning. In *AAAI88 – Proceedings of the 7th National Conference on Artificial Intelligence*, St. Paul, Minnesota, pp. 49–54.

Elsaesser, C. (1987). Explanation of probabilistic inference for decision support systems. In *Proceedings of the AAAI-87 Workshop on Uncertainty in Artificial Intelligence*, Seattle, Washington, pp. 394–403.

Evans, J., J. Barston, and P. Pollard (1983). On the conflict between logic and belief in syllogistic reasoning. *Memory and Cognition 11*, 295–306.

Gardent, C. and E. Kow (2005). Generating and selecting grammatical paraphrases. In *ENLG-05 – Proceedings of the 10th European Workshop on Natural Language Generation*, Aberdeen, Scotland, pp. 49–57.

George, S., I. Zukerman, and M. Niemann (2007). Inferences, suppositions and explanatory extensionsin argument interpretation. *User Modeling and User-Adapted Interaction 17*(5), 439–474.

Getoor, L., N. Friedman, D. Koller, and B. Taskar (2001). Learning probabilistic models of relational structure. In *Proceedings of the 18th International Conference on Machine Learning*, Williamstown, Massachusetts, pp. 170–177.

Horvitz, E., H. Suermondt, and G. Cooper (1989). Bounded conditioning: flexible inference for decision under scarce resources. In *UAI89 – Proceedings of the 1989 Workshop on Uncertainty in Artificial Intelligence*, Windsor, Canada, pp. 182–193.

Kashihara, A., T. Hirashima, and J. Toyoda (1995). A cognitive load application in tutoring. *User Modeling and User-Adapted Interaction 4*(4), 279–303.

Kintsch, W. (1994). Text comprehension, memory and learning. *American Psychologist 49*(4), 294–303.

Korb, K. B., R. McConachy, and I. Zukerman (1997). A cognitive model of argumentation. In *Proceedings of the 19th Annual Conference of the Cognitive Science Society*, Stanford, California, pp. 400–405.

Lichtenstein, S., B. Fischhoff, and L. Phillips (1982). Calibrations of probabilities: The state of the art to 1980. In D. Kahneman, P. Slovic, and A. Tversky (Eds.), *Judgment under Uncertainty: Heuristics and Biases*, pp. 306–334. Cambridge University Press.

Pearl, J. (1988). *Probabilistic Reasoning in Intelligent Systems*. San Mateo, California: Morgan Kaufmann Publishers.

Taskar, B., P. Abbeel, and D. Koller (2002). Discriminative probabilistic models for relational data. In *Proceedings of the 18th Conference on Uncertainty in Artificial Intelligence*, Alberta, Canada, pp. 485–490.

Tversky, A. and D. Kahneman (1982). Evidential impact of base rates. In D. Kahneman, P. Slovic, and A. Tversky (Eds.), *Judgment under Uncertainty: Heuristics and Biases*, pp. 153–160. Cambridge University Press.

van Gelder, T. (2005). Teaching critical thinking: some lessons from cognitive science. *College Teaching 45*(1), 1–6.

Wallace, C. (2005). *Statistical and Inductive Inference by Minimum Message Length*. Berlin, Germany: Springer.

Zukerman, I. and S. George (2005). A probabilistic approach for argument interpretation. *User Modeling and User-Adapted Interaction 15*(1-2), 5–53.

Zukerman, I., S. George, and Y. Wen (2003). Lexical paraphrasing for document retrieval and node identification. In *IWP2003 – Proceedings of the 2nd International Workshop on Paraphrasing: Paraphrase Acquisition and Applications*, Sapporo, Japan, pp. 94–101.

Zukerman, I., R. McConachy, and K. B. Korb (1998). Bayesian reasoning in an abductive mechanism for argument generation and analysis. In *AAAI98 – Proceedings of the 15th National Conference on Artificial Intelligence*, Madison, Wisconsin, pp. 833–838.

Zukerman, I., R. McConachy, and K. B. Korb (2000). Using argumentation strategies in automated argument generation. In *INLG'2000 – Proceedings of the 1st International Conference on Natural Language Generation*, Mitzpe Ramon, Israel, pp. 55–62.

# Part III: Causality

# 17
# Instrumental Sets
CARLOS BRITO

## 1 Introduction

The research of Judea Pearl in the area of causality has been very much acclaimed. Here we highlight his contributions for the use of graphical languages to represent and reason about causal knowledge.[1]

The concept of causation seems to be fundamental to our understanding of the world. Philosophers like J. Carroll put it in these terms: "With regard to our total conceptual apparatus, causation is the center of the center" [Carroll 1994]. Perhaps more dramatically, David Hume states that causation together with resemblance and contiguity are "the only ties of our thoughts, ... for us the cement of the universe" [Hume 1978]. In view of these observations, the need for an adequate language to talk about causation becomes clear and evident.

The use of graphical languages was present in the early times of causal modelling. Already in 1934, Sewall Wright [Wright 1934] represented the causal relation among several variables with diagrams formed by points and arrows (i.e., a directed graph), and noted that the correlations observed between the variables could be associated with the various paths between them in the diagram. From this observation he obtained a method to estimate the strength of the causal connections known as The Method of Path Coefficients, or simply Path Analysis.

With the development of the research in the field, the graphical representation gave way to a mathematical language, in which causal relations are represented by equations of the form $Y = \alpha + \beta X + e$. This movement was probably motivated by an increasing interest in the quantitative aspects of the model, or by the rigorous and formal appearance offered by the mathematical language. However it may be, the consequence was a progressive departure from our basic causal intuitions. Today people ask whether such an equation represents a functional or a causal relation [Reiss 2005]. Sewall Wright and Judea Pearl would presumably answer: "Causal, of course!".

## 2 The Identification Problem

We explore the feasibility of inferring linear cause-effect relationships from various combinations of data and theoretical assumptions. The assumptions are represented

---
[1] This contribution is a simplified version of a joint paper with Judea Pearl in UAI 2002. A great deal of technicality was removed, and new discussion was added, in the hope that the reader will be able to easily follow and enjoy the argument.

Figure 1. (a) a bow-pattern; and (b) a bow-free model.

in the form of an acyclic causal diagram, which contains both arrows and bidirected arcs [Pearl 1995; Pearl 2000a]. The arrows represent the potential existence of direct causal relationships between the corresponding variables, and the bidirected arcs represent spurious correlations due to unmeasured common causes. All interactions among variables are assumed to be linear. Our task is to decide whether the assumptions represented in the diagram are sufficient for assessing the strength of causal effects from non-experimental data, and, if sufficiency is proven, to express the target causal effect in terms of estimable quantities.

This decision problem has been tackled in the past half century, primarily by econometricians and social scientists, under the rubric "The Identification Problem" [Fisher 1966] - it is still unsolved. Certain restricted classes of models are nevertheless known to be identifiable, and these are often assumed by social scientists as a matter of convenience or convention [Duncan 1975]. A hierarchy of three such classes is given in [McDonald 1997]: (1) no bidirected arcs, (2) bidirected arcs restricted to root variables, and (3) bidirected arcs restricted to variables that are not connected through directed paths.

In a further development [Brito and Pearl 2002], we have shown that the identification of the entire model is ensured if variables standing in direct causal relationship (i.e., variables connected by arrows in the diagram) do not have correlated errors; no restrictions need to be imposed on errors associated with indirect causes. This class of models was called "bow-free", since their associated causal diagrams are free of any "bow-pattern" [Pearl 2000a] (see Figure 1).

Most existing conditions for identification in general models are based on the concept of Instrumental Variables (IV) [Pearl 2000b; Bowden and Turkington 1984]. IV methods take advantage of conditional independence relations implied by the model to prove the identification of specific causal-effects. When the model is not rich in conditional independence relations, these methods are not informative. In [Brito and Pearl 2002] we proposed a new graphical criterion for identification which does not make direct use of conditional independence, and thus can be successfully applied to models in which the IV method would fail.

The result presented in this paper is a generalization of the graphical version of the method of instrumental variables, offered by Judea Pearl [Pearl 2000a], to deal with several parameters of the model simultaneously. The traditional method of instrumental variables involves conditions on the independence of the relevant variables and on the rank of a certain matrix of correlations [McFadden ]. The first of these is captured by the notion of d-separation. As for the second, since we know from [Wright 1934] that correlations correspond to paths in the causal diagram, we can investigate which structural properties of the model give rise to the proper conditions of the IV method. The results are graphical criteria that allow us to conclude the identification of some parameters from consideration of the qualitative information represented in the causal diagram.

## 3 Linear Models and Identification

A linear model for the random variables $Y_1, \ldots, Y_n$ is defined by a set of equations of the form:

(1) $\quad Y_j = \sum_i c_{ji} Y_i + e_j, \quad j = 1, \ldots, n$

An equation $Y = cX + e$ encodes two distinct assumptions: (1) the possible existence of (direct) causal influence of $X$ on $Y$; and, (2) the absence of causal influence on $Y$ of any variable that does not appear on the right-hand side of the equation. The parameter $c$ quantifies the (direct) causal effect of $X$ on $Y$. That is, the equation claims that a unit increase in $X$ would result in $c$ units increase of $Y$, assuming that everything else remains the same. The variable $e$ is called an error or disturbance; it represents unobserved background factors that the modeler decides to keep unexplained; this variable is assumed to have a normal distribution with zero mean.

The specification of the equations and the pairs of error-terms $(e_i, e_j)$ with non-zero correlation defines the structure of the model. This structure can be represented by a directed graph, called causal diagram, in which the set of nodes is defined by the variables $Y_1, \ldots, Y_n$, and there is a directed edge from $Y_i$ to $Y_j$ if $Y_i$ appears on the right-hand side of the equation for $Y_j$. Additionally, if error-terms $e_i$ and $e_j$ are assumed to have non-zero correlation, we add a (dashed) bidirected edge between $Y_i$ and $Y_j$. Figure 2 shows a model with the respective causal diagram.

In this work, we consider only recursive models, which are defined by the restriction that $c_{ji} = 0$, for all $i \geq j$. This simply means that the directed edges in the causal diagram do not form cycles.

The set of parameters of the model, denoted by $\Theta$, is formed by the coefficients $c_{ij}$ and the non-zero entries of the error covariance matrix $\Psi$, $[\Psi_{ij}] = cov(e_i, e_j)$.

Fixing the structure and assigning values to the parameters $\Theta$, the model determines a unique covariance matrix $\Sigma$ over the observed variables $Y_1, \ldots, Y_n$, given by (see [Bollen 1989], page 85):

(2) $\quad \Sigma(\Theta) = (I - C)^{-1} \Psi [(I - C)^{-1}]'$

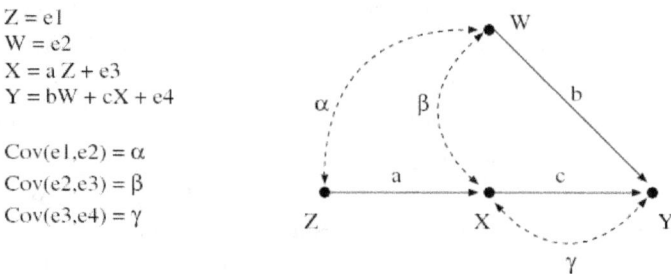

Figure 2. A simple linear model and its causal diagram.

where $C$ is the matrix of coefficients $c_{ji}$.

Conversely, in the Identification Problem, after fixing the structure of the model, one attempts to solve for $\Theta$ in terms of the observed covariance $\Sigma$. This is not always possible. In some cases, no parametrization of the model is compatible with a given $\Sigma$. In other cases, the structure of the model may permit several distinct solutions for the parameters. In these cases, the model is called *non-identified*.

However, even if the model is non-identified, some parameters may still be uniquely determined by the given assumptions and data. Whenever this is the case, the specific parameters are said to be *identified*.

Finally, since the conditions we seek involve the structure of the model alone, and do not depend on the numerical values of the parameters $\Theta$, we insist only on having identification almost everywhere, allowing few pathological exceptions. The concept of identification almost everywhere can be formalized as follows.

Let $h$ denote the total number of parameters in the model. Then, each vector $\Theta \in \Re^h$ defines a parametrization of the model. For each parametrization $\Theta$, the model $G$ generates a unique covariance matrix $\Sigma(\Theta)$. Let $\Theta(\lambda_1, \ldots, \lambda_n)$ denotes the vector of values assigned by $\Theta$ to the parameters $\lambda_1, \ldots, \lambda_n$.

Parameters $\lambda_1, \ldots, \lambda_n$ are identified almost everywhere if

$$\Sigma(\Theta) = \Sigma(\Theta') \quad \text{implies} \quad \Theta(\lambda_1, \ldots, \lambda_n) = \Theta'(\lambda_1, \ldots, \lambda_n)$$

except when $\Theta$ resides on a subset of Lebesgue measure zero of $\Re^h$.

## 4 Graph Background

DEFINITION 1.

1. A *path* in a graph is a sequence of edges such that each pair of consecutive edges share a common node, and each node appears only once along the path.

2. A *directed path* is a path composed only by directed edges, all of them oriented

in the same direction. If there is a directed path going from $X$ to $Y$ we say that $Y$ is a *descendant* of $X$.

3. A path is *closed* if it has a pair of consecutive edges pointing to their common node (e.g., $\ldots \to X \leftarrow \ldots$ or $\ldots \leftrightarrow X \leftarrow \ldots$). In this case, the common node is called a *collider*. A path is *open* if it is not closed.

DEFINITION 2. A path $p$ is blocked by a set of nodes $\mathbf{Z}$ (possibly empty) if either

1. $\mathbf{Z}$ contains some non-collider node of $p$, or

2. at least one collider of $p$ and all of its descendants are outside $\mathbf{Z}$.

The idea is simple. If the path is closed, then it is naturally blocked by its colliders. However, if a collider, or one of its descendants, belongs to $\mathbf{Z}$, then it ceases to be an obstruction. But if a non-collider of $p$ belongs to $\mathbf{Z}$, then the path is definitely blocked.

DEFINITION 3. A set of nodes $\mathbf{Z}$ d-separates $X$ and $Y$ if $\mathbf{Z}$ simultaneously blocks all the paths between $X$ and $Y$. If $\mathbf{Z}$ is empty, then we simply say that $X$ and $Y$ are d-separated.

The significance of this definition comes from a result showing that if $X$ and $Y$ are d-separated by $\mathbf{Z}$ in the causal diagram of a linear model, then the variables $X$ and $Y$ are conditionally independent given $\mathbf{Z}$ [Pearl 2000a]. It is this sort of result that makes the connection between the mathematical and graphical languages, and allows us to express our conditions for identification in graphical terms.

DEFINITION 4. Let $p_1, \ldots, p_n$ be unblocked paths connecting the variables $Z_1, \ldots, Z_n$ and the variables $X_1, \ldots, X_n$, respectively. We say that the set of paths $p_1, \ldots, p_n$ is incompatible if we cannot rearrange their edges to form a different set of unblocked paths $p'_1, \ldots, p'_n$ between the same variables.

A set of disjoint paths (i.e., paths with no common nodes) consists in a simple example of an incompatible set of paths.

## 5 Instrumental Variable Methods

### 5.1 Identification of a Single Parameter

The method of Instrumental Variables (IV) for the identification of causal effects is intended to address the situation where we cannot attribute the entire correlation between two variables, say $X$ and $Y$, to their causal connection. That is, part of the correlation between $X$ and $Y$ is due to common causes and/or correlations between disturbances. Figure 3 shows examples of this situation.

In the simplest cases, like in Figure 3a, we can find a conditioning set $\mathbf{W}$ such that the partial correlation of $X$ and $Y$ given $\mathbf{W}$ can indeed be attributed to the causal relation. In this example, if we take $\mathbf{W} = \{W\}$ we eliminate the source

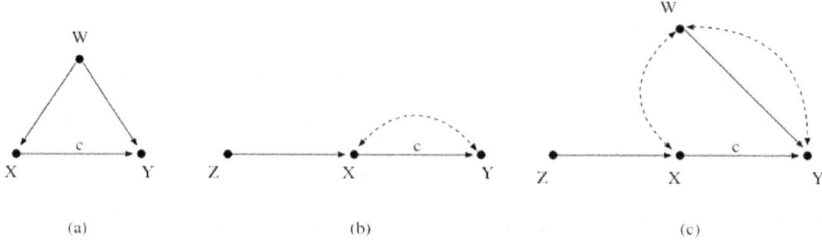

Figure 3. Models with spurious correlation between $X$ and $Y$.

of spurious correlation. The causal effect of $X$ on $Y$ is identified and given by $c = \sigma_{XY.W}$.

There are cases, however, where this idea does not work, either because the spurious correlation is originated by disturbances outside the model (Figure 3b), or else because the conditioning itself introduces spurious correlations (Figure 3c). In situations like these, the IV method asks us to look for a variable $Z$ with the following properties [Bowden and Turkington 1984]:

IV-1. $Z$ is not independent of $X$.

IV-2. $Z$ is independent of all error terms that have an influence on $Y$ that is not mediated by $X$.

The first condition simply states that there is a correlation between $Z$ and $X$. The second condition says that the only source of correlation between $Z$ and $Y$ is due to a covariation bewteen $Z$ and $X$ that subsequently affects $Y$ through the causal connection $X \xrightarrow{c} Y$.

If we can find a variable $Z$ with these properties, then the causal effect of $X$ on $Y$ is identified and given by $c = \sigma_{ZY}/\sigma_{ZX}$.

Using the notion of d-separation we can express the conditions (1) and (2) of the IV method in graphical terms, thus obtaining a criterion for identification that can be applied directly to the causal diagram of the model. Let $G$ be the graph representing the causal diagram of the model, and let $G_c$ be the graph obtained after removing the edge $X \xrightarrow{c} Y$ from $G$ (see Figure 4). Then, $Z$ is an instrumental variable relative to $X \xrightarrow{c} Y$ if:

1. $Z$ is not d-separated from $X$ in $G_c$.

2. $Z$ is d-separated from $Y$ in $G_c$.

Using this criterion, it is easy to verify that $Z$ is an instrumental variable relative to $X \xrightarrow{c} Y$ in the models of Figure 3b and c.

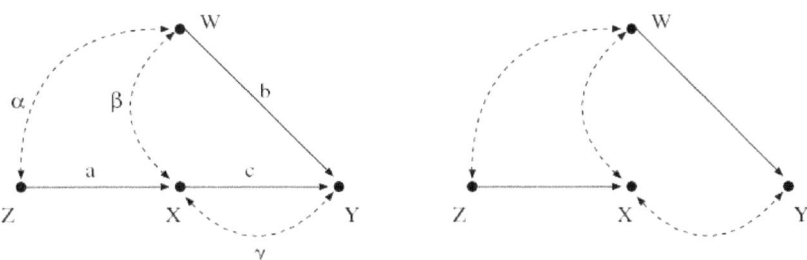

Figure 4. The causal diagram $G$ of a linear model and the graph $G_c$.

## 5.2 Conditional Instrumental Variables

A generalization of the method of instrumental variables is offered through the use of conditioning. A conditional instrumental variable is a variable $Z$ that may not have the properties (IV-1) and (IV-2) above, but after conditioning on a subset $\mathbf{W}$ these properties do hold. When such pair $(Z, \mathbf{W})$ is found, the causal effect of $X$ on $Y$ is identified and given by $c = \sigma_{ZY\cdot\mathbf{W}}/\sigma_{ZX\cdot\mathbf{W}}$.

Again, we obtain a graphical criterion for a conditional IV using the notion of d-separation. Variable $Z$ is a conditional instrumental variable relative to $X \xrightarrow{c} Y$ given $\mathbf{W}$, if

1. $\mathbf{W}$ contains only non-descendants of $Y$.

2. $\mathbf{W}$ does not d-separate $Z$ from $X$ in $G_c$.

3. $\mathbf{W}$ d-separates $Z$ from $Y$ in $G_c$.

## 5.3 Identification of Multiple Parameters

So far we have been concerned with the identification of a single parameter of the model, but in its full version the method of instrumental variables allows to prove simultaneously the identification of several parameters in the same equation (i.e., the causal effects of several variables $X_1, \ldots, X_k$ on the same variable $Y$).

Following [McFadden ], assume that we have the equation

$$Y = c_1 X_1 + \ldots + c_k X_k + e$$

in our linear model. The variables $Z_1, \ldots, Z_j$, with $j \geq k$, are called instruments if

1. The matrix of correlations between the variables $X_1, \ldots, X_k$ and the variables $Z_1, \ldots, Z_j$ is of maximum possible rank (i.e., rank $k$).

2. The variables $Z_1, \ldots, Z_j$ are uncorrelated with the error term $e$.

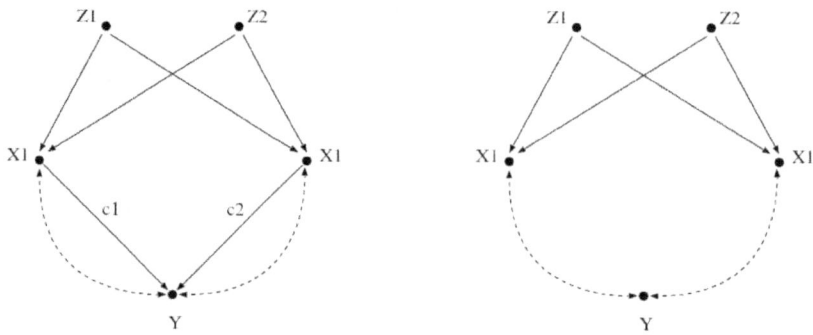

Figure 5. The causal diagram $G$ of a linear model and the graph $\bar{G}$.

Next, we develop our graphical intuition and obtain a graphical criterion for identification that corresponds to the full version of the IV method.

Consider the model in Figure 5a. Here, the variables $Z_1$ and $Z_2$ do not qualify as instrumental variables (or even conditional IVs) with respect to either $X_1 \xrightarrow{c_1} Y$ or $X_2 \xrightarrow{c_2} Y$. But, following ideas similar to the ones developed in the previous sections, in Figure 5b we show the graph obtained by removing edges $X_1 \to Y$ and $X_2 \to Y$ from the causal diagram. Observe that now both d-separation conditions for an instrumental variable hold for $Z_1$ and $Z_2$. This leads to the idea that $Z_1$ and $Z_2$ could be used together as instruments to prove the identification of parameters $c_1$ and $c_2$. Indeed, next we give a graphical criterion that is sufficient to guarantee the identification of a subset of parameters of the model.

Fix a variable $Y$, and consider the edges $X_1 \xrightarrow{c_1} Y, \ldots, X_k \xrightarrow{c_k} Y$ in the causal diagram $G$ of the model. Let $\bar{G}$ be the graph obtained after removing the edges $X_1 \to Y, \ldots, X_k \to Y$ from $G$. The variables $Z_1, \ldots, Z_k$ are instruments relative to $X_1 \xrightarrow{c_1} Y, \ldots, X_k \xrightarrow{c_k} Y$ if

1. There exists an incompatible set of unblocked paths $p_1, \ldots, p_k$ connecting the variables $Z_1, \ldots, Z_k$ to the variables $X_1, \ldots, X_k$.

2. The variables $Z_i$ are d-separated from $Y$ in $\bar{G}$.

3. Each variable $Z_i$ is not d-separated from the corresponding variable $X_i$ in $\bar{G}$.[2]

THEOREM 5. *If we can find variables $Z_1, \ldots, Z_k$ satisfying the conditions above, then the parameters $c_1, \ldots, c_k$ are identified almost everywhere, and can be computed by solving a system of linear equations.*

---

[2]Notice that this condition is redundant, since it follows from the first condition.

Instrumental Sets

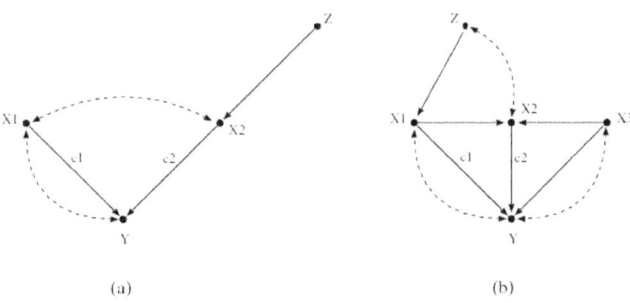

(a)  (b)

Figure 6. More examples of the new criterion.

Figure 6 shows more examples of application of the new graphical criterion. Model (a) illustrates an interesting case, in which variable $X_2$ is used as the instrumental variable for $X_1 \to Y$, while $Z$ is the instrumental variable for $X_2 \to Y$. Finally, in model (b) we have an example in which the parameter of edge $X_3 \to Y$ is non-identified, and still the graphical criterion allows to show the identfication of $c_1$ and $c_2$.

## 6 Wright's Method of Path Coefficients

Here, we describe an important result introduced by Sewall Wright [Wright 1934], which is extensively explored in our proofs.

Given variables $X$ and $Y$ in a recursive linear model, the correlation coefficient of $X$ and $Y$, denoted by $\rho_{XY}$, can be expressed as a polynomial on the parameters of the model. More precisely,

$$(3) \quad \sigma_{XY} = \sum_p T(p)$$

where the summation ranges over all unblocked paths $p$ between $X$ and $Y$, and each term $T(p)$ represents the contribution of the path $p$ to the total correlation between $X$ and $Y$. The term $T(p)$ is given by the product of the parameters of the edges along the path $p$. We refer to Equation 3 as Wright's equation for $X$ and $Y$.

Wright's method of path coefficients for identification consists in forming Wright's equations for each pair of variables in the model, and then solving for the parameters in terms of the observed correlations. Whenever there is a unique solution for a parameter $c$, this parameter is identified.

## 7 Proof of Theorem 1

### 7.1 Notation

Fix a variable $Y$ in the model. Let $\mathbf{X} = \{X_1, \ldots, X_n\}$ be the set of all non-descendants of $Y$ which are connected to $Y$ by an edge. Define the following set of edges incoming $Y$:

(4) $Inc(Y) = \{(X_i, Y) : X_i \in \mathbf{X}\}$

Note that for some $X_i \in \mathbf{X}$ there may be more than one edge between $X_i$ and $Y$ (one directed and one bidirected). Thus, $|Inc(Y)| \geq |\mathbf{X}|$. Let $\lambda_1, \ldots, \lambda_m$, $m \geq k$, denote the parameters of the edges in $Inc(Y)$.

It follows that edges $X_1 \xrightarrow{c_1} Y, \ldots, X_k \xrightarrow{c_k} Y$ all belong to $Inc(Y)$, because $X_1, \ldots, X_k$ are clearly non-descendants of $Y$. We assume that $\lambda_i = c_i$, for $i = 1, \ldots, k$, while $\lambda_{k+1}, \ldots, \lambda_m$ are the parameters of the remaining edges of $Inc(Y)$.

Let $Z$ be any non-descendant of $Y$. Wright's equation for the pair $(Z, Y)$ is given by:

$$(5) \quad \sigma_{ZY} = \sum_p T(p)$$

where each term $T(p)$ corresponds to an unblocked path $p$ between $Z$ and $Y$. The next lemma proves a property of such paths.

LEMMA 6. *Any unblocked path between $Y$ and one of its non-descendants $Z$ must include exactly one edge from $Inc(Y)$.*

Lemma 6 allows us to write equation 4 as:

$$(6) \quad \sigma_{ZY} = \sum_{j=1}^{m} a_j \cdot \lambda_j$$

Thus, the correlation between $Z$ and $Y$ can be expressed as a linear function of the parameters $\lambda_1, \ldots, \lambda_m$, with no constant term. In addition, we can say something about the coefficients $a_j$. Each term in Equation 5 corresponds to an unblocked path that reaches $Y$ through some egge, say $X_j \xrightarrow{\lambda_j} Y$. When we group the terms together according to the parameter $\lambda_j$ and factor it out, we are, in a sense, removing the edge $X_j \to Y$ from those paths. Thus, each coefficient $a_j$ in Equation 6 is a sum of terms associated with unblocked paths between $Z$ and $X_j$.

### 7.2 Basic Linear Equations

We have just seen that the correlations between the instrumental variables $Z_i$ and $Y$ can be written as a linear function of the parameters $\lambda_1, \ldots, \lambda_m$:

$$(7) \quad \rho_{Z_i Y} = \sum_{j=1}^{m} a_{ij} \cdot \lambda_j$$

Next, we prove an important result

LEMMA 7. *The coefficients $a_{i,k+1}, \ldots, a_{im}$ in Equation 7 are all identically zero.*

**Proof.** The fact that $Z_i$ is d-separated from $Y$ in $\bar{G}$ implies that $\rho_{Z_i Y} = 0$ in any probability distribution compatible with $\bar{G}$. Hence, the expression for $\rho_{Z_i Y}$ must vanish when evaluated in the causal diagram $\bar{G}$. But this implies that each coefficient $a_{ij}$ in Equation 7 is identically zero, when the expression is evaluated in $\bar{G}$.

Next, we show that the only difference between the expression for $\rho_{Z_iY}$ on the causal diagrams $G$ and $\bar{G}$ are the coefficients of the parameters $\lambda_1, \ldots, \lambda_k$.

Recall from the previous section that each coefficient $a_{ij}$ is a sum of terms associated with paths which can be extended by the edge $\xrightarrow{\lambda_j} Y$ to form an unblocked path between $Z$ and $Y$.

Fixing $j > k$, we observe that the insertion of edges $x_1 \to Y, \ldots, X_k \to Y$ in $\bar{G}$ does not create any new such path (and clearly does not eliminate any existing one). Hence, for $j > k$, the coefficients $a_{ij}$ in the expression for $\rho_{Z_iY}$ in the causal diagrams $G$ and $\bar{G}$ are exactly the same, namely, identically zero. □

The conclusion from Lemma 7 is that the expression for $\rho_{Z_iY}$ is a linear function only of parameters $\lambda_1, \ldots, \lambda_k$:

$$(8) \quad \rho_{Z_iY} = \sum_{j=1}^{k} a_{ij} \cdot \lambda_j$$

## 7.3 System of Equations $\Phi$

Writing Equation 8 for each instrumental variable $Z_i$, we obtain the following system of linear equations on the parameters $\lambda_1, \ldots, \lambda_k$:

$$(9) \quad \Phi = \begin{cases} \rho_{Z_1Y} = a_{11}\lambda_1 + \ldots, a_{1k}\lambda_k \\ \ldots \\ \rho_{Z_kY} = a_{k1}\lambda_1 + \ldots, a_{kk}\lambda_k \end{cases}$$

Our goal now is to show that $\Phi$ can be solved uniquely for the parameters $\lambda_i$, and so prove the identification of $\lambda_1, \ldots, \lambda_k$. Next lemma proves an important result in this direction.

Let $A$ denote the matrix of coefficients of $\Phi$.

LEMMA 8. *Det(A) is a non-trivial polynomial on the parameters of the model.*

**Proof.** The determinant of $A$ is defined as the weighted sum, for all permutations $\pi$ of $\langle 1, \ldots, k \rangle$, of the product of the entries selected by $\pi$. Entry $a_{ij}$ is selected by a permutation $\pi$ if the $i^{th}$ element of $\pi$ is $j$. The weights are either 1 or -1, depending on the parity of the permutation.

Now, observe that each diagonal entry $a_{ii}$ is a sum of terms associated with unblocked paths between $Z_i$ and $X_i$. Since $p_i$ is one such path, we can write $a_{ii} = T(p_i) + \hat{a}_{ii}$. From this, it is easy to see that the term

$$(10) \quad T^* = \prod_{j=1}^{k} T(p_j)$$

appears in the product of permutation $\pi = \langle 1, \ldots, n \rangle$, which selects all the diagonal entries of $A$.

We prove that $det(A)$ does not vanish by showing that $T^*$ is not cancelled out by any other term in the expression for $det(A)$.

Let $\tau$ be any other term appearing in the summation that defines the determinant of $A$. This term appears in the product of some permutation $\pi$, and has as factors exactly one term from each entry $a_{ij}$ selected by $\pi$. Thus, associated with such factor there is an unblocked path between $Z_i$ and $X_j$. Let $p'_1, \ldots, p'_k$ be the unblocked paths associated with the factors of $\tau$.

We conclude the proof observing that, since $p_1, \ldots, p_k$ is an incompatible set, its edges cannot be rearranged to form a different set of unblocked paths between the same variables, and so $\tau \neq T^*$. Hence, the term $T^*$ is not cancelled out in the summation, and the expression for $det(A)$ does not vanish. □

## 7.4 Identification of $\lambda_1, \ldots, \lambda_k$

Lemma 8 gives that $det(Q)$ is a non-trivial polynomial on the parameters of the model. Thus, $det(Q)$ only vanishes on the roots of this polynomial. However, [Okamoto 1973] has shown that the set of roots of a polynomial has Lebesgue measure zero. Thus, the system $\Phi$ has unique solution almost everywhere.

It just remains to show that we can estimate the entries of the matrix of coefficients $A$ from the data. But this is implied by the following observation.

Once again, coefficient $a_{ij}$ is given by a sum of terms associated with unblocked paths between $Z_i$ and $X_j$. But, in principle, not every unblocked path between $Z_i$ and $X_j$ would contribute with a term to the sum; just those which can be extended by the edge $X_j \to Y$ to form an unblocked path between $Z_i$ and $Y$. However, since the edge $X_j \to Y$ does not point to $X_j$, every unblocked path between $Z_i$ and $X_j$ can be extended by the edge $X_j \to Y$ without creating a collider. Hence, the terms of all unblocked paths between $Z_i$ and $X_j$ appear in the expression for $a_{ij}$, and by the method of path coefficients, we have $a_{ij} = \rho_{Z_i X_j}$.

We conclude that each entry of matrix $A$ can be estimated from data, and we can solve the system of linear equations $\Phi$ to obtain the parameters $\lambda_1, \ldots, \lambda_k$.

## References

Bollen, K. (1989). *Structural Equations with Latent Variables*. John Wiley, New York.

Bowden, R. and D. Turkington (1984). *Instrumental Variables*. Cambridge Univ. Press.

Brito, C. and J. Pearl (2002). A graphical criterion for the identification of causal effects in linear models. In *Proc. of the AAAI Conference, Edmonton, Canada.*.

Carroll, J. (1994). *Laws of Nature*. Cambridge University Press.

Duncan, O. (1975). *Introduction to Structural Equation Models.* Academic Press.

Fisher, F. (1966). *The Identification Problem in Econometrics.* McGraw-Hill.

Hume, D. (1978). *A Treatise of Human Nature.* Oxford University Press.

McDonald, R. (1997). Haldane's lungs: A case study in path analysis. *Mult. Beh. Res.*, 1–38.

McFadden, D. *Lecture Notes for Econ 240b.* Dept of Economics, UC Berkeley.

Okamoto, M. (1973). Distinctness of the eigenvalues of a quadratic form in a multivariate sample. *Annals of Statistics*, 763–765.

Pearl, J. (1995). Causal diagrams for empirical research. *Biometrika*, 669–710.

Pearl, J. (2000a). *Causality: Models, Reasoning and Inference.* Cambridge Press.

Pearl, J. (2000b). Parameter identification: A new perspective. *Technical Report R-276, UCLA.*

Reiss, J. (2005). Causal instrumental variables and interventions. *Philosophy of Science.* 72, 964–976.

Wright, S. (1934). The method of path coefficients. *Ann. Math. Statistics.*, 161–215.

# 18
# Seeing and Doing: The Pearlian Synthesis

PHILIP DAWID

## 1 Introduction

It is relatively recently that much attention has focused on what, for want of a better term, we might call "statistical causality", and the subject has developed in a somewhat haphazard way, without a very clear logical basis. There is in fact a variety of current conceptions and approaches [Campaner and Galavotti 2007; Hitchcock 2007; Galavotti 2008]—here we shall distinguish in particular *agency*, *graphical*, *probabilistic* and *modular* conceptions of causality—that tend to be mixed together in an informal and half-baked way, based on "definitions" that often do not withstand detailed scrutiny. In this article I try to unpick this tangle and expose the various different strands that contribute to it. Related points, with a somewhat different emphasis, are made in a companion paper [Dawid 2009].

The approach of Judea Pearl [2009] cuts through this Gordian knot like the sword of Alexander. Whereas other conceptions of causality may be philosophically questionable, definitionally unclear, pragmatically unhelpful, theoretically skimpy, or simply confused, Pearl's theory is none of these. It provides a valuable framework, founded on a rich and fruitful formal theory, by means of which causal assumptions about the world can be meaningfully represented, and their implications developed. Here we will examine both the relationships of Pearl's theory with the other conceptions considered, and its differences from them. We extract the essence of Pearl's approach as an assumption of "modularity", the transferability of certain probabilistic properties between observational and interventional regimes: so, in particular, forging a synthesis between the very different activities of "seeing" and "doing". And we describe a generalisation of this framework that releases it from any necessary connexion to graphical models.

The plan of the paper is as follows. In § 2, I describe the agency, graphical and probabilistic conceptions of causality, and their connexions and distinctions. Section 3 introduces Pearl's approach, showing its connexions with, and differences from, the other theories. Finally, in § 4, I present the generalisation of that approach, emphasising the modularity assumptions that underlie it, and the usefulness of the theory of "extended conditional independence" for describing and manipulating these.

**Disclaimer** I have argued elsewhere [Dawid 2000, 2007a, 2010] that it is important to distinguish arguments about "Effects of Causes" (EoC, otherwise termed "type", or "generic" causality"), from those about "Causes of Effects" (CoE, also termed "token", or "individual" causality); and that these demand different formal frameworks and analyses. My concern here will be entirely focused on problems of generic causality, EoC. A number of the current frameworks for statistical causality, such as Rubin's "potential response models" [Rubin 1974, 1978], or Pearl's "probabilistic causal models" [Pearl 2009, Chapter 7], are more especially suited for handling CoE type problems, and will not be discussed further here. There are also numerous other conceptions of causality, such as *mechanistic causality* [Salmon 1984; Dowe 2000], that I shall not be considering here.

## 2 Some conceptions of causality

There is no generally agreed understanding of what "causality" is or how it should behave. There are two conceptions in particular that are especially relevant for "statistical causality": *Agency Causality* and *Probabilistic Causality*. The latter in turn is closely related to what we might term *Graphical Causality*.

### 2.1 Agency causality

The "agency" or "manipulability" interpretation of causality [Price 1991; Hausman 1998; Woodward 2003] depends on an assumed notion of external "manipulation" (or "intervention"), that might itself be taken as a primitive—at any rate we shall not try and explicate it further here. The basic idea is that causality is all about how an external manipulation that sets the value of some variable (or set of variables) $X$ will affect some other (unmanipulated) "response variable" (or set of variables) $Y$. The emphasis is usually on comparison of the responses ensuing from different settings $x$ for $X$: a version of the "contrastive" or "difference-making" understanding of causality. Much of Statistical Science—for example, the whole subfield of Experimental Design—aims to address exactly these kinds of questions about the comparative effects of interventions on a system, which are indeed a major object of all scientific enquiry.

We can define certain causal terms quite naturally within the agency theory [Woodward 2003]. Thus we could interpret the statement

"$X$ *has no effect on* $Y$"[1]

as holding whenever, considering regimes that manipulate only $X$, the resulting value of $Y$ (or some suitable codification of uncertainty about $Y$, such as its probability distribution) does not depend on the value $x$ assigned to $X$. When this fails, $X$ *has an effect on* $Y$; we might then go on to quantify this dependence in various ways.

We could likewise interpret

"$X$ *has no (direct) effect on* $Y$, *after controlling for* $W$"

as the property that, considering regimes where we manipulate both $W$ and $X$, when we manipulate $W$ to some value $w$ and $X$ to some value $x$, the ensuing value (or uncertainty) for $Y$ will depend only on $w$, and not further on $x$.

Now suppose that, explicitly or implicitly, we restrict consideration to some collection $\mathcal{V}$ of manipulable variables. Then we might interpret the statement

---

[1] Just as "zero" is fundamental to arithmetic and "independence" is fundamental to probability, so the concept of "no effect" is fundamental to causality.

"*X is a direct cause of Y* (relative to $\mathcal{V}$)"

(where $\mathcal{V}$ might be left unmentioned, but must be clearly understood) as the negation of "*X has no direct effect on Y*, after controlling for $\mathcal{V} \setminus \{X, Y\}$".[2]

It is important to bear in mind that all these assertions relate to properties of the real world under the various regimes considered: in particular, they can not be given purely mathematical definitions. And in real world problems there are typically various ways of manipulating variables, so we must be very clear as to exactly what is intended.

EXAMPLE 1. **Ideal gas law**

Consider the "ideal gas law":

(1)  $PV = kNT$

where $P$ is the absolute pressure of the gas, $V$ is its volume, $N$ is the number of molecules of gas present, $k$ is Boltzmann's constant, and $T$ is the absolute temperature. For our current purposes this will be supposed to be universally valid, no matter how the values of the variables in (1) may have come to arise.

Taking a fixed quantity $N$ of gas in an impermeable container, we might consider interventions on any of $P$, $V$ and $T$. (Note however that, because of the constraint (1), we can not simultaneously and arbitrarily manipulate all three variables.)

An intervention that sets $V$ to $v$ and $T$ to $t$ will lead to the unique value $p = kNt/v$ for $P$. Because this depends on both $v$ and $t$, we can say that there is a *direct effect* of each of $V$ and $T$ on $P$ (relative to $\mathcal{V} = \{V, P, T\}$). Similarly, $P$ has a direct effect on each of $V$ and $T$.

What if we wish to quantify, say, "the causal effect of $V$ on $P$"? Any attempt to do this must take account of the fact that the problem requires additional specification to be well-defined. Suppose the volume of the container can be altered by applying a force to a piston. Initially the gas has $V = v_0$, $P = p_0$, $T = t_0$. We wish to manipulate $V$ to a new value $v_1$. If we do this *isothermally*, *i.e.* by sufficiently slow movement of the piston that, through flow of heat through the walls of the container, the temperature of the gas always remains the same as that of the surrounding heat bath, we will end up with $V = v_1$, $P = p_1 = v_0 p_0 / v_1$, $T = t_1 = t_0$. But if we move the piston *adiabatically*, *i.e.* so fast that no heat can pass through the walls of the container, the relevant law is $PV^\gamma =$ constant, where $\gamma = 5/3$ for a monatomic gas. Then we get $V = v_1$, $P = p_1^* = p_0(v_0/v_1)^\gamma$, $T = t_1^* = p_1^* v_1 / kN$.

## 2.2 Graphical causality

By *graphical causality* we shall refer to an interpretation of causality in terms of an underlying *directed acyclic graph* (DAG) (noting in passing that other graphical representations are also possible). As a basis for this, we suppose that there is a suitable "causal ambit"[3] $\mathcal{A}$ of variables (not all necessarily observable) that we regard as relevant, and a "causal DAG"

---

[2] Neapolitan [2003, p. 56] has a different and more complex interpretation of "direct cause".
[3] The importance of the causal ambit will become apparent later.

$\mathcal{D}$ over a collection $\mathcal{V} \subseteq \mathcal{A}$. These ingredients are "known to Nature", though not necessarily to us: $\mathcal{D}$ is "Nature's DAG". Given such a causal DAG $\mathcal{D}$, for $X, Y \in \mathcal{V}$ we interpret "$X$ is a *direct cause* of $Y$" as synonymous with "$X$ is a parent of $Y$ in $\mathcal{D}$", and similarly equate "*cause*" with "ancestor in $\mathcal{D}$". One can also use the causal DAG to introduce further graphically defined causal terms, such as "causal chain", "intermediate variable", ...

The concepts of causal ambit and causal DAG might be regarded as primitive notions, or attempts might be made to define them in terms of pre-existing understandings of causal concepts. In either case, it would be good to have criteria to distinguish a putative causal ambit from a non-causal ambit, and a causal DAG from a non-causal DAG.

For example, we typically read [Hernán and Robins 2006]:

"A *causal DAG* $\mathcal{D}$ is a DAG in which:

(i). the lack of an arrow from $V_j$ to $V_m$ can be interpreted as the absence of a direct causal effect of $V_j$ on $V_m$ (relative to the other variables on the graph)

(ii). all common causes, even if unmeasured, of any pair of variables on the graph are themselves on the graph.[4]

If we start with a DAG $\mathcal{D}$ over $\mathcal{V}$ that we accept as being a causal DAG, and interpret "direct cause" *etc.* in terms of that, then conditions (i) and (ii) will be satisfied by definition. However, this begs the question of how we are to tell a causal from a non-causal DAG.

More constructively, suppose we start with a prior understanding of the term "direct cause" (relative to $\mathcal{V}$)—for example, though by no means necessarily,[5] based on the agency interpretation described in § 2.1 above. It appears that we could then use the above definition to check whether a proposed DAG $\mathcal{D}$ is indeed "causal". But while this is essentially straightforward so far as condition (i) is concerned (except that there is no obvious reason to require a DAG representation), interpretation and implementation of condition (ii) is more problematic. First, what is a "common cause"? Spirtes et al. [2000, p. 44] say that a variable $X$ is a common cause of variables $Y$ and $Z$ if and only if $X$ is both a direct cause of $Y$ and a direct cause of $Z$ — but in each case relative to the set $\{X, Y, Z\}$, so that this definition is not dependent on the causal ambit $\mathcal{V}$. Neapolitan [2003, p. 57] has a different interpretation, which apparently is relative to an essentially arbitrary set $\mathcal{V}$ — but then states that that problems can arise when at least one common cause is not in $\mathcal{V}$, a possibility that seems to be precluded by his definition.

As another attempt at clarification, Spirtes and Scheines [2004] require "that the set of variables in the causal graph be *causally sufficient*, *i.e.* if $\mathcal{V}$ is the set of variables in the causal graph, that there is no variable $L$ not in $\mathcal{V}$ that is a direct cause (relative to $\mathcal{V} \cup \{L\}$) of two variables in $\mathcal{V}$". If "$L \notin \mathcal{V}$ is not a direct cause of $V \in \mathcal{V}$" is interpreted in agency terms, it would mean that $V$ would not respond to manipulations of $L$, when holding fixed all the other variables in $\mathcal{V}$. But whatever the interpretation of direct cause, such a "definition" of causal sufficiency is ineffective when the range of possible choices

---

[4]The motivation for this requirement is not immediately obvious, but is related to the defensibility of the *causal Markov* property described in § 2.3 below.

[5]See § 2.2 below.

for the additional variable $L$ is entirely unrestricted—for then how could we ever be sure that it holds, without conducting an infinite search over all unmentioned variables $L$? That is why we posit an appropriate clearly-defined "causal ambit" $\mathcal{A}$: we can then restrict the search to $L \in \mathcal{A}$.

It seems to me that we should, realistically, allow that "causality" can operate, in parallel, at several different levels of granularity. Thus while it may or may not be possible to describe the medical effects of aspirin treatment in terms of quantum theory, even if we could, it would be a category error to try and do so in the context of a clinical trial. So there may be various different causal descriptions of the world, all operating at different levels, each with its associated causal ambit $\mathcal{A}$ of variables and various causal DAGs $\mathcal{D}$ over sets $\mathcal{V} \subseteq \mathcal{A}$. The meaning of any causal terms used should then be understood in relation to the appropriate level of description.

The obvious questions to ask about graphical causality, which are however not at all easy to answer, are: "When can a collection $\mathcal{A}$ of variables be regarded as a causal ambit?", and "When can a DAG be regarded as a causal DAG?".

In summary, so long as we *start* with a DAG $\mathcal{D}$ over $\mathcal{V}$ that we are willing to accept as a *causal* DAG (taken as a primitive concept), we can take $\mathcal{V}$ itself as our causal ambit, and use the structure of $\mathcal{D}$ to *define* causal terms. Without having a prior primitive notion of what constitutes a "causal DAG", however, conditions such as (i) and (ii) are unsatisfactory as a definition. At the very least, they require that we have specified (but how?) an appropriate causal ambit $\mathcal{A}$, relevant to our desired level of description, and have a clear pre-existing understanding (*i.e.* not based on the structure of $\mathcal{D}$, since that would be logically circular) of the terms "direct causal effect", "common cause" (perhaps relative to a set $\mathcal{V}$).

**Agency causality and graphical causality**

It is tempting to use the agency theory as a basis for such prior causal understanding. However, graphical causality does not really sit well with agency causality. For, as seen clearly in Example 1, in the agency intepretation it is perfectly possible for two variables each to have a direct effect on the other—which could not hold under any DAG representation. Similarly [Halpern and Pearl 2005; Hall 2000] there is no obvious reason to expect agency causality to be a transitive relation, which would again be a requirement under the graphical conception. For better or worse, the agency theory does not currently seem to be endowed with a sufficiently rich axiomatic structure to guide manipulations of its causal properties; and however such a general axiomatic structure might look, it would seem unduly restrictive to relate it closely to DAG models.

## 2.3 Probabilistic causality

*Probabilistic Causality* [Reichenbach 1956; Suppes 1970; Spohn 2001] depends on the existence and properties of a probability distribution $P$ over quantities of interest. At its (over-)simplest, it equates *causality* with *probability raising*: "$A$ is a *cause* of $B$" (where $A$ and $B$ are events) if $P(B \mid A) > P(B)$. This is more usefully re-expressed in its null form, and referred to random variables $X$ and $Y$: $X$ is not a cause of $Y$ if the distribution of $Y$ given $X$ is the same as the marginal distribution of $Y$; and this is equivalent to *probabilistic independence* of $Y$ from $X$: $Y \perp\!\!\!\perp X$. But this is clearly unsatisfactory as it stands, since we could have dependence between $X$ and $Y$, $Y \not\!\perp\!\!\!\perp X$, with, at the same time, conditional independence given some other variable (or set of variables) $Z$: $Y \perp\!\!\!\perp X \mid Z$. If $Z$ can be regarded as delimiting the context in which we are considering the relationship between $X$ and $Y$, we might still regard $X$ and $Y$ as "causally unrelated". Thus probabilistic causality is based on *conditional* (in)dependence properties of probability distributions. However there remain obvious problems in simply equating the non-symmetrical relation of cause-and-effect with the symmetrical relation of probabilistic (in)dependence, and with clarifying what counts as an appropriate conditioning "context" variable $Z$, so that additional structure and assumptions (*e.g.* related to an assumed "causal order", possibly but not necessarily temporal) are required to complete the theory.

Most modern accounts locate probabilistic causality firmly within the graphical conception — so inheriting all the features and difficulties of that approach. It is *assumed* that there is a DAG $\mathcal{D}$, over a suitable collection $\mathcal{V}$ of variables, such that

(i). $\mathcal{D}$ can be interpreted as a *causal* DAG; and, in addition,

(ii). the joint probability distribution $P$ of the variables in $\mathcal{V}$ is *Markov* over $\mathcal{D}$, *i.e.* its probabilistic conditional independence (CI) properties are represented by the same DAG $\mathcal{D}$, according to the "$d$-separation" semantics described by Pearl [1986], Verma and Pearl [1990], Lauritzen et al. [1990].

In particular, from (ii), for any $V \in \mathcal{V}$, $V$ is independent of its non-descendants, $\mathrm{nd}(V)$, in $\mathcal{D}$, given its parents, $\mathrm{pa}(V)$, in $\mathcal{D}$. Given the further interpretation (i) of $\mathcal{D}$ as a causal DAG, this can be expressed as "$V$ is independent of its non-effects, given its direct causes in $\mathcal{V}$"— the so-called *causal Markov* assumption. Also, (ii) implies that, for any sets of variables $X$ and $Y$ in $\mathcal{D}$, $X \perp\!\!\!\perp Y \mid \mathrm{an}(X) \cap \mathrm{an}(Y)$ (where $\mathrm{an}(X)$ denotes the set of ancestors of $X$ in $\mathcal{D}$, including $X$ itself): again with $\mathcal{D}$ interpreted as causal, this can be read as saying "$X$ and $Y$ are conditionally independent, given their common causes in $\mathcal{V}$". In particular, marginal independence (where $X \perp\!\!\!\perp Y$ is represented in $\mathcal{D}$) holds if and only if $\mathrm{an}(X) \cap \mathrm{an}(Y) = \emptyset$, *i.e.* (using (i)) "$X$ and $Y$ have no common cause" (including each other) in $\mathcal{V}$; in the "if" direction, this has been termed the *weak causal Markov* assumption [Scheines and Spirtes 2008]. Many workers regard the causal and weak causal Markov assumptions as compelling—but this must depend on making the "right" choice for $\mathcal{V}$ (essentially, through appropriate delineation of the causal ambit.)

Note that this conception of causality involves, simultaneously, two very different ways of interpreting the DAG $\mathcal{D}$ (see Dawid [2009] for more on this). The $d$-separation semantics by means of which we relate $\mathcal{D}$ to conditional independence properties of the joint distribution $P$, while clearly defined, are somewhat subtle: in particular, the arrows in $\mathcal{D}$ are somewhat incidental "construction lines", that only play a small rôle in the semantics. But as soon as we also give $\mathcal{D}$ an interpretation as a "causal DAG" we are into a completely different way of interpreting it, where the arrows themselves are regarded as directly carrying causal meaning. Probabilistic causality can thus be thought of as the progeny of a shotgun wedding between two ill-matched parties.

**Causal discovery**

The enterprise of *Causal Discovery* [Spirtes et al. 2000; Glymour and Cooper 1999; Neapolitan 2003] is grounded in this probabilistic-cum-graphical conception of causality. There are many variations, but all share the same basic philosophy. Essentially, one analyses observational data in an attempt to identify conditional independencies (possibly involving unobserved variables) in the distribution from which they arise. Some of these might be discarded as "accidental" (perhaps because they are inconsistent with an *a priori* causal order); those that remain might be represented by a DAG. The hope is that this discovered conditional independence DAG can also be interpreted as a causal DAG. When, as is often the case, there are several Markov equivalent DAG representations of the discovered CI relationships, which, moreover, cannot be causally distinguished on *a priori* grounds (*e.g.* in terms of an assumed causal order), this hope can not be fully realised; but if we can assume that one of these, at least, is a causal DAG, then at least an arrow common to all of them can be interpreted causally.

## 2.4 A spot of bother

Spirtes et al. [2000] and Pearl [2009], among others, have stressed the fundamental importance of distinguishing between the activities of *Seeing* and *Doing*. *Seeing* involves passive observation of a system in its natural state. *Doing*, on the other hand, relates to the behaviour of the system in a disturbed state brought about by external intervention. As a simple point of pure logic, there is no reason for there to be any relationship between these two types of behaviour of a system.

The probabilistic interpretation of causality relates solely to the *seeing* regime, whereas the agency account focuses entirely on what happens in *doing* regimes. As such these two interpretations inhabit totally unrelated universes. There are non-trivial foundational difficulties with the probabilistic (or other graphical) interpretations of causality (what exactly is a causal DAG? how will we know when we have got one?); on the other hand agency causality, while less obviously problematic and perhaps more naturally appealing, does not currently appear to offer a rich enough theory to be very useful. Even at a purely technical level, agency and probabilistic causality have very little in common. Probabilistic causality, through its close ties with conditional independence, has at its disposal the well-developed theoretical machinery of that concept, while the associated graphical structure allows for ready interpretation of concepts such as "causal pathway". Such considerations are however of marginal relevance to agency causality, which need not involve any probabilistic or graphical connexions.

From the point of view of a statistician, this almost total disconnect between the causal theories relating to the regimes of seeing and doing is particularly worrying. For one of the major purposes of "causal inference" is to draw conclusions, from purely observational "seeing" data on a system, about "doing": how would the system behave were we to intervene in it in certain ways? But not only is there no necessary logical connexion between the behaviours in the different regimes, the very concepts and representations by which we try to understand causality in the different regimes are worlds apart.

## 3   The Pearlian Synthesis

Building on ideas introduced by Spirtes et al. [2000], Pearl's approach to causality, as laid out for example in his book [Pearl 2009],[6] attempts to square this circle: it combines the two apparently incommensurable approaches of agency causality and probabilistic causality[7] in a way that tries to bring together the best features of both, while avoiding many of their individual problems and pitfalls.

Pearl considers a type of stochastic model, described by a DAG $\mathcal{D}$ over a collection $\mathcal{V}$ of variables, that can be simultaneously interpreted in terms of both agency and probabilistic causality. We could, if we wished, think of $\mathcal{V}$ as a "causal ambit", and $\mathcal{D}$ as a "causal DAG", but little is gained (or lost) by doing so, since the interpretations of any causal terms we may employ are provided internally by the model, rather than built on any pre-existing causal conceptions.

In its probabilistic interpretation, such a DAG $\mathcal{D}$ represents the conditional independence properties of the undisturbed system, which is supposed Markov with respect to $\mathcal{D}$. In its agency interpretation, the same DAG $\mathcal{D}$ is used to describe precisely how the system responds, probabilistically, to external interventions that set the values of (an arbitrary collection of) its variables. Specifically, such a disturbed probability distribution is supposed still Markov with respect to $\mathcal{D}$, and the conditional distribution of any variable $V$ in $\mathcal{V}$, given its parents in $\mathcal{D}$, is supposed the same in all regimes, seeing or doing (except of course those that directly set the value of $V$ itself, say at $v$, for which that distribution is replaced by the 1-point distribution at $v$). The "parent-child" conditional distributions thus constitute invariant "modular components" that (with the noted exception) can be transferred unchanged from one regime to another.

We term such a causal DAG model "Pearlian". Whether or not a certain DAG $\mathcal{D}$ indeed supplies a Pearlian DAG model for a given system can never be a purely syntactical question about its graphical structure, but is, rather, a semantic question about its relationship with the real world: do the various regimes actually have the probabilistic properties and relationships asserted? This may be true or false, but at least it is a meaningful question, and it is clear in principle how it can be addressed in purely empirical fashion: by observing and comparing the behaviours of the system under the various regimes.[8] A Pearlian DAG

---

[6]We in fact shall deal only with Pearl's earlier, fully stochastic, theory. More recently (see the second-half of Pearl [2009], starting with Chapter 7), he has moved to an interpretation of DAG models based on deterministic functional relationships, with stochasticity deriving solely from unobserved exogenous variables. That interpretation does however imply all the properties of the stochastic theory, and can be regarded as a specialisation of it. We shall not here be considering any features (such as the possibility of counterfactual analysis) dependent on the additional structure of Pearl's deterministic approach, since these only become relevant when analysing "causes of effects"—see Dawid [2000, 2002] for more on this.

[7]We have already remarked that probabilistic causality is itself the issue of an uneasy alliance between two quite different ways of interpreting graphs. Further miscegenation with the agency conception of causality looks like a eugenically risky endeavour!

[8]For this to be effective, the variables in $\mathcal{V}$ should have clearly-defined meanings and be observable in the real-world. Some Pearlian models incorporate unobservable latent variables without clearly identified external referents, in which case only the implications of such a model for the behaviour of observables can be put to empirical test.

model thus has the great virtue, all too rare in treatments of causality, of being totally clear and explicit about what is being said—allowing one to accord it, in a principled way, acceptance or rejection, as deemed appropriate, in any given application. And when a system can indeed be described by a Pearlian DAG, it is straightforward to learn (not merely qualitatively, but quantitatively too), from purely observational data, about the (probabilistic) effects of any interventions on variables in the system.

## 3.1 Justification

The falsifiability of the property of being a Pearlian DAG (unlike, for example, the somewhat ill-defined property of being a "causal DAG") is at once a great strength of the theory (especially for those with a penchant for Karl Popper's "falsificationist" Philosophy of Science), and something of an Achilles' heel. For all too often it will be impossible, for a variety of pragmatic, ethical or financial reasons, to conduct the experiments that would be needed to falsify the Pearlian assumptions. A lazy reaction might then simply be to assume that a DAG found, perhaps by "causal discovery", to represent observational conditional independencies, but without any interventions having been applied, is indeed Pearlian—and so also describes what would happen under interventions. While this may well be an interesting working hypothesis to guide further experimental investigations, it would be an illogical and dangerous point at which to conclude our studies. In particular, further experimental investigations could well result in rejection of our assumed Pearlian model.

Nevertheless, if forced to make a tentative judgment on the Pearlian nature, or otherwise, of a putative DAG model[9] of a system, there are a number of more or less reasonable, more or less intuitive, arguments that can be brought to bear. As a very simple example, we would immediately reject any putative "Pearlian DAG" in which an arrow goes backwards in time,[10] or otherwise conflicts with an accepted causal order. As another, if an "observational" regime itself involves an imposed physical randomisation to generate the value of some variable $X$, in a way that might possibly take account of variables $Z$ temporally prior to $X$, we might reasonably regard the conditional distribution of some later variable $Y$, given $X$ and $Z$, as a modular component, that would be the same in a regime that intervenes to set the value of $X$ as it is in the (observational) randomisation regime.[11] Such arguments can be further extended to "natural experiments", where it is Nature that imposed the external randomisation. This is the case for "Mendelian randomisation" [Didelez and Sheehan 2007], which capitalises on the random assortment of genes under Mendelian genetics. Other natural experiments rely on other causal assumptions about Nature: thus the "discontinuity design" [Trochim 1984] assumes that Nature supplies continuous dose-response cause-effect relationships. But all such justifications are, and must be, based on (what we think are) properties of the real world, and not solely on the internal structure of

---

[9]Assumed, for the sake of non-triviality, already to be a Markov model of its observational probabilistic properties.

[10]Assuming, as most would accept, that an intervention in a variable at some time can not affect any variable whose value is determined at an earlier time.

[11]See Dawid [2009] for an attempted argument for this, as well as caveats as to its general applicability.

the putative Pearlian DAG. In particular, they are founded on pre-existing ideas we have about causal and non-causal processes in the world, even though these ideas may remain unformalised and woolly: the important point is that we have enough, perhaps tacit, shared understanding of such processes to convince both ourselves and others that they can serve as external justification for a suggested Pearlian model. Unless we have sufficient justification of this kind, all the beautiful analysis (*e.g.* in Pearl [2009]) that develops the implications of a Pearlian model will be simply irrelevant. To echo Cartwright [1994, Chapter 2], "No causes in, no causes out".

## 4 Modularity, extended conditional independence and decision-theoretic causality

Although Pearlian causality as described above appears to be closely tied to graphical representation, this is really an irrelevance. We can strip it of its graphical clothing, laying bare its core ingredient: the property that certain conditional distributions[12] are the same across several different regimes. This *modular* conception provides us with yet another interpretation of causality. When, as here, the regimes considered encompass both observation (seeing) and intervention (doing), it has the great advantage over other theories of linking those disparate universes, thus supporting *causal inference*.

The modularity assumption can be conveniently expressed formally in the algebraic language of conditional independence, suitably interpreted [Dawid 1979, 2002, 2009], making no reference to graphs. Thus let $F$ be a "regime indicator", a non-stochastic parameter variable, whose value indicates the regime whose probabilistic properties are under consideration. If $X$ and $Y$ are stochastic variables, the "extended conditional independence" (ECI) property

(2) $\quad Y \perp\!\!\!\perp F \mid X$

can be interpreted as asserting that the conditional distribution of $Y$, for specified regime $F = f$ and given observed value $X = x$, depends only on $x$ and not further on the regime $f$ that is operating: in terms of densities we could write $p(y \mid f, x) = p(y \mid x)$. If $F$ had been a stochastic variable this would be entirely equivalent to stochastic conditional independence of $Y$ and $F$ given $X$; but it remains meaningful, with the above interpretation, even when $F$ is a non-stochastic regime indicator: Indeed, it asserts exactly the modular nature of the conditional distribution $p(y \mid x)$, as being the same across all the regimes indicated by values of $F$. Such modularity properties, when expressed in terms of ECI, can be formally manipulated—and, in those special cases where this is possible and appropriate, represented and manipulated graphically—in essentially the same fashion as for regular probabilistic conditional independence.

For applications of ECI to causal inference, we would typically want one or more of the regimes indicated by $F$ to represent the behaviour of the system when subjected to an intervention of a specified kind—thus linking up nicely with the agency interpretation; and one

---

[12]More generally, we could usefully identify features of the different regimes other than conditional distributions—for example, conditional expectations, or odds ratios—as modular components.

regime to describe the undisturbed system on which observations are made—thus allowing the possibility of "causal inference" and making links with probabilistic causality, but in a non-graphical setting. Modularity/ECI assumptions can now be introduced, as considered appropriate, and their implications extracted by algebraic or graphical manipulations, using the established theory of conditional independence. We emphasise that, although the notation and technical machinery of conditional independence is being used here, this is applied in a way that is very different from the approach of probabilistic causality: no assumptions need be made connecting causal relationships with ordinary probabilistic conditional independence.

Because it concerns the probabilistic behaviour of a system under interventions—a particular interpretation of agency causality—this general approach can be termed "decision-theoretic" causality. With the emphasis now on modularity, intuitive or graphically motivated causal terms such as "direct effect" or "causal pathway" are best dispensed with (and with them such assumptions as the causal Markov property). The decision-theoretic approach should not be regarded as providing a philosophical foundation for "causality", or even as a way of interpreting causal terms, but rather as very useful machinery for expressing and manipulating whatever modularity assertions one might regard as appropriate in a given problem.

## 4.1 Intervention DAGs

The assumptions that are implicit in a Pearlian model can be displayed very explicitly in the decision-theoretic framework, by associating a non-stochastic "intervention variable" $F_X$ with each "domain variable" $X \in \mathcal{V}$. The assumed ECI properties are conveniently displayed by means of a DAG, $\mathcal{D}^*$, which extends the Pearlian DAG $\mathcal{D}$ by adding extra nodes for these regime indicators, and extra arrows, from $F_X$ to $X$ for each $X \in \mathcal{V}$ [Spohn 1976; Spirtes et al. 2000; Pearl 2009; Dawid 2002; Dawid 2009]. If $\mathcal{X}$ is the set of values for $X$, then that for $F_X$ is $\mathcal{X} \cup \{\emptyset\}$: the intended interpretation is that $F_X = \emptyset$ (the "idle" regime) corresponds to the purely observational regime, while $F_X = x \in \mathcal{X}$ corresponds to "setting" $X$ at $x$.

To be precise, we specify the distribution of $X \in \mathcal{V}$ given its parents $(\text{pa}(X), F_X)$ in $\mathcal{D}^*$ (where $\text{pa}(X)$ denotes the "domain" parents of $X$, in $\mathcal{D}$) as follows. When $F_X = \emptyset$, this is the same as the observational conditional distribution of $X$, given $\text{pa}(X)$; and when $F_X = x$ it is just a 1-point distribution on $x$, irrespective of the values of $\text{pa}(X)$. The extended DAG $\mathcal{D}^*$, supplied with these parent-child specifications, is the *intervention DAG* representation of the problem.

With this construction, for any settings of all the regime indicators, some to idle and some to fixed values, the implied joint distribution of all the domain variables in that regime is exactly as required for the Pearlian DAG interpretation. But a valuable added bonus of the intervention DAG representation is that the Pearlian assumptions are explicitly represented. For example, the standard $d$-separation semantics applied to $\mathcal{D}^*$ allows us to read off the ECI property $X \perp\!\!\!\perp \{F_Y : Y \neq X\} \mid (\text{pa}(X), F_X)$, which asserts the modular property of the conditional distribution of $X$ given $\text{pa}(X)$: when $F_X = \emptyset$ (the only non-trivial case) the conditional distribution of $X$ given $\text{pa}(X)$ is the same, no matter how the other variables are set (or left idle).

### 4.2 More general causal models

It is implicit in the Pearlian conception that every variable in $\mathcal{V}$ should be manipulable (the causal Markov property then follows). But there is no real reason to require this. We can instead introduce intervention variables for just those variables that we genuinely wish to consider as "settable". The advantage of this is that fewer assumptions need be made and justified, but useful conclusions can often still be drawn.

EXAMPLE 2. (**Instrumental variable**)

Suppose we are interested in the "causal effect" of a binary exposure variable $X$ on some response $Y$. However we can not directly manipulate $X$. Moreover the observational relationship between $X$ and $Y$ may be distorted because of an unobserved "confounder" variable, $U$, associated with both $X$ and $Y$. In an attempt to evade this difficulty, we also measure an "instrumental variable" $Z$.

To express our interest in the *causal* effect of $X$ on $Y$, we introduce an intervention variable $F_X$ associated with $X$, defined and interpreted exactly as in §4.1 above. The aim of our causal inference is to make some kind of comparison between the distributions of the response $Y$ in the interventional regimes, $F_X = 0$ and $F_X = 1$, corresponding to manipulating the value of $X$. The available data, however, are values of $(X, Y, Z)$ generated under the observational regime, $F_X = \emptyset$. We must make some assumptions if we are to be able to use features of that observational joint distribution to address our causal question, and clearly these must involve some kind of transference of information across regimes.

A useful (when valid!) set of assumptions about the relationships between all the variables in the problem is embodied in the following set of ECI properties (the "core conditions")[13] for basing causal inferences on an instrumental variable):

$$(U, Z) \perp\!\!\!\perp F_X \tag{3}$$

$$U \perp\!\!\!\perp Z \mid F_X \tag{4}$$

$$Y \perp\!\!\!\perp F_X \mid (X, U) \tag{5}$$

$$Y \perp\!\!\!\perp Z \mid (X, U; F_X) \tag{6}$$

$$X \not\perp\!\!\!\perp Z \mid F_X = \emptyset \tag{7}$$

Property (3) is to be interpreted as saying that the joint distribution of $(U, Z)$ is independent of the regime $F_X$: *i.e.*, it is the same in all three regimes. That is to say, it is entirely unaffected by whether, and if so how, we intervene to set the value of $X$. The identity of this joint distribution across the two interventional regimes, $F_X = 0$ and $F_X = 1$, can be interpreted as expressing a causal property: manipulating $X$ has no (probabilistic) effect

---

[13]In addition to these core conditions, precise identification of a causal effect by means of an instrumental variable requires further modelling assumptions, such as linear regressions [Didelez and Sheehan 2007].

on the pair of variables $(U, Z)$. Moreover, since this common joint distribution is also supposed the same in the idle regime, $F_X = \emptyset$, we could in principle use observational data to estimate it—thus opening up the possibility of causal inference.

Property (4) asserts that, in their (common) joint distribution in any regime, $U$ and $Z$ are independent (this however is a purely probabilistic, not a causal, property).

Property (5) says that the conditional distribution of $Y$ given $(X, U)$ is the same in both interventional regimes, as well as in the observational regime, and can thus be considered as a modular component, fully transferable between the three regimes—again, I regard this as expressing a causal property.

Property (6) asserts that this common conditional distribution is unaffected by further conditioning on $Z$ (not in itself a causal property).

Finally, property (7) requires that $Z$ be genuinely associated with $X$ in the observational regime.

Of course, these ECI properties should not simply be assumed without some attempt at justification: for example, Mendelian randomisation attempts this in the case that $Z$ is an inherited gene. But because we have no need to consider interventions at any node other than $X$, less by way of justification is required than if we were to do so.

Once expressed in terms of ECI, these core conditions can be manipulated algebraically using the general theory of conditional independence [Dawid 1979]. Depending on what further modelling assumptions are made, it may then be possible to identify, or to bound, the desired causal effect in terms of properties of the observational joint distribution of $(X, Y, Z)$ [Dawid 2007b, Chapter 11].

In this particular case, although the required ECI conditions are expressed without reference to any graphical representation, it is possible (though not obligatory!) to give them one. This is shown in Figure 1. Properties (3)–(6) can be read off this DAG directly using the standard $d$-separation semantics. (Property (7) is only represented under a further assumption that the graphical representation is faithful.) We term such a DAG an *augmented DAG*: it differs from a Pearlian DAG in that some, but not necessarily all, variables have associated intervention indicators.

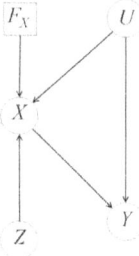

Figure 1. Instrumental variable: Augmented DAG representation

Just as for regular CI, it is possible for a collection of ECI properties, constituting a

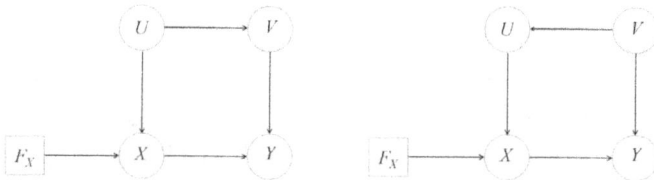

Figure 2. Two Markov-equivalent augmented DAGs

decision-theoretic causal model, to have no (augmented) DAG representation, or more than one. This latter is the case for Figure 2, where the direction of the arrow between $U$ and $V$ is not determined. This emphasises that, even when we do have an augmented DAG representation, we can not necessarily interpret the direction of an arrow in it as directly related to the direction of causality. Even in Figure 1 (and in spite of the natural connotation of the term "instrument"), the arrow pointing from $Z$ to $X$ is not be interpreted as necessarily causal, since the dependence between $Z$ and $X$ could be due to a "common cause" $U^*$ without affecting the ECI properties (3)–(6) [Dawid 2009], and Figure 1 is merely a graphical representation of these properties, based on $d$-separation semantics. In particular, one should be cautious of using an augmented DAG, which is nothing but a way of representing certain ECI statements, to introduce graphically motivated concepts such as "causal pathway". The general decision-theoretic description of causality *via* modularity, expressed in terms of ECI properties, where there is no requirement that the assumptions be representable by means of an augmented DAG at all, allows us to evade some of the restrictions of graphical causality, while still retaining a useful "agency-cum-probabilistic" causal theory.

The concept of an "interventional regime" can be made much more general, and in particular we need not require that it have the properties assumed above for an intervention variable associated with a domain variable. We could, for example, incorporate "fat hand" interventions that do not totally succeed in their aim of setting a variable to a fixed value, or interventions (such as kicking the system) that simultaneously affect several domain variables [Duvenaud et al. 2009]. So long as we understand what such regimes refer to in the real world, and can make and justify assumptions of modularity of appropriate conditional distributions as we move across regimes, we can apply the decision-theoretic ECI machinery. And at this very general level we can even apply a variant of "causal discovery" algorithms—so long as we can make observations under all the regimes considered.[14] For example, if we can observe $(X, Y)$ under the different regimes described by $F$, we can readily investigate the validity of the ECI property $X \perp\!\!\!\perp F \mid Y$ using standard tests (*e.g.*

---

[14]Or we might make parametric modelling assumptions about the relationships across regimes, to fill in for regimes we are not able to observe. This would be required for example when want to consider the effect of setting the value of a continuous "dose" variable. At this very general level we can even dispense entirely with the assumption of modular conditional distributions [Duvenaud et al. 2009].

the $\chi^2$-test) for conditional independence. Such discovered ECI properties (whether or not they can be expressed graphically) can then be used to model the "causal structure" of the problem.

## 5 Conclusion

Over many years, Judea Pearl's original and insightful approach to understanding uncertainty and causality have had an enormous influence on these fields. They have certainly had a major influence on my own research directions: I have often—as evidenced by this paper—found myself following in his footsteps, picking up a few crumbs here and there for further digestion.

Pearl's ideas do not however exist in a vacuum, and I believe it is valuable both to relate them to their precursors and to assess the ways in which they may develop. In attempting this task I fully acknowledge the leadership of a peerless researcher, whom I feel honoured to count as a friend.

## References

Campaner, R. and M. C. Galavotti (2007). Plurality in causality. In P. K. Machamer and G. Wolters (Eds.), *Thinking About Causes: From Greek Philosophy to Modern Physics*, pp. 178–199. Pittsburgh: University of Pittsburgh Press.

Cartwright, N. (1994). *Nature's Capacities and Their Measurement*. Oxford: Clarendon Press.

Dawid, A. P. (1979). Conditional independence in statistical theory (with Discussion). *Journal of the Royal Statistical Society, Series B 41*, 1–31.

Dawid, A. P. (2000). Causal inference without counterfactuals (with Discussion). *Journal of the American Statistical Association 95*, 407–448.

Dawid, A. P. (2002). Influence diagrams for causal modelling and inference. *International Statistical Review 70*, 161–189. Corrigenda, *ibid.*, 437.

Dawid, A. P. (2007a). Counterfactuals, hypotheticals and potential responses: A philosophical examination of statistical causality. In F. Russo and J. Williamson (Eds.), *Causality and Probability in the Sciences*, Volume 5 of *Texts in Philosophy*, pp. 503–32. London: College Publications.

Dawid, A. P. (2007b). Fundamentals of statistical causality. Research Report 279, Department of Statistical Science, University College London. <http://www.ucl.ac.uk/Stats/research/reports/psfiles/rr279.pdf>

Dawid, A. P. (2010). Beware of the DAG! *Journal of Machine Learning Research*. To appear.

Dawid, A. P. (2010). The rôle of scientific and statistical evidence in assessing causality. In R. Goldberg, J. Paterson, and G. Gordon (Eds.), *Perspectives on Causation*, Oxford. Hart Publishing. To appear.

Didelez, V. and N. A. Sheehan (2007). Mendelian randomisation as an instrumental variable approach to causal inference. *Statistical Methods in Medical Research 16*, 309–330.

Dowe, P. (2000). *Physical Causation*. Cambridge: Cambridge University Press.

Duvenaud, D., D. Eaton, K. Murphy, and M. Schmidt (2010). Causal learning without DAGs. *Journal of Machine Learning Research*. To appear.

Galavotti, M. C. (2008). Causal pluralism and context. In M. C. Galavotti, R. Scazzieri, and P. Suppes (Eds.), *Reasoning, Rationality and Probability*, Chapter 11, pp. 233–252. Chicago: The University of Chicago Press.

Glymour, C. and G. F. Cooper (Eds.) (1999). *Computation, Causation and Discovery*. Menlo Park, CA: AAAI Press.

Hall, N. (2000). Causation and the price of transitivity. *Journal of Philosophy XCVII*, 198–222.

Halpern, J. Y. and J. Pearl (2005). Causes and explanations: A structural-model approach. Part I: Causes. *British Journal for the Philosophy of Science 56*, 843–887.

Hausman, D. (1998). *Causal Asymmetries*. Cambridge: Cambridge University Press.

Hernán, M. A. and J. M. Robins (2006). Instruments for causal inference: An epidemiologist's dream? *Epidemiology 17*, 360–372.

Hitchcock, C. (2007). How to be a causal pluralist. In P. K. Machamer and G. Wolters (Eds.), *Thinking About Causes: From Greek Philosophy to Modern Physics*, pp. 200–221. Pittsburgh: University of Pittsburgh Press.

Lauritzen, S. L., A. P. Dawid, B. N. Larsen, and H.-G. Leimer (1990). Independence properties of directed Markov fields. *Networks 20*, 491–505.

Neapolitan, R. E. (2003). *Learning Bayesian Networks*. Upper Saddle River, New Jersey: Prentice Hall.

Pearl, J. (1986). A constraint–propagation approach to probabilistic reasoning. In L. N. Kanal and J. F. Lemmer (Eds.), *Uncertainty in Artificial Intelligence*, Amsterdam, pp. 357–370. North-Holland.

Pearl, J. (2009). *Causality: Models, Reasoning and Inference* (Second ed.). Cambridge: Cambridge University Press.

Price, H. (1991). Agency and probabilistic causality. *British Journal for the Philosophy of Science 42*, 157–176.

Reichenbach, H. (1956). *The Direction of Time*. Berkeley: University of Los Angeles Press.

Rubin, D. B. (1974). Estimating causal effects of treatments in randomized and nonrandomized studies. *Journal of Educational Psychology 66*, 688–701.

Rubin, D. B. (1978). Bayesian inference for causal effects: the role of randomization. *Annals of Statistics 6*, 34–68.

Salmon, W. C. (1984). *Scientific Explanation and the Causal Structure of the World*. Princeton: Princeton University Press.

Scheines, R. and P. Spirtes (2008). Causal structure search: Philosophical foundations and future problems. Paper presented at NIPS 2008 Workshop "Causality: Objectives and Assessment", Whistler, Canada.

Spirtes, P., C. Glymour, and R. Scheines (2000). *Causation, Prediction and Search* (Second ed.). New York: Springer-Verlag.

Spirtes, P. and R. Scheines (2004). Causal inference of ambiguous manipulations. *Philosophy of Science 71*, 833–845.

Spohn, W. (1976). *Grundlagen der Entscheidungstheorie*. Ph.D. thesis, University of Munich. (Published: Kronberg/Ts.: Scriptor, 1978).

Spohn, W. (2001). Bayesian nets are all there is to causal dependence. In M. C. Galavotti, P. Suppes, and D. Costantini (Eds.), *Stochastic Dependence and Causality*, Chapter 9, pp. 157–172. Chicago: University of Chicago Press.

Suppes, P. (1970). *A Probabilistic Theory of Causality*. Amsterdam: North Holland.

Trochim, W. M. K. (1984). *Research Design for Program Evaluation: The Regression-Discontinuity Approach*. SAGE Publications.

Verma, T. and J. Pearl (1990). Causal networks: Semantics and expressiveness. In R. D. Shachter, T. S. Levitt, L. N. Kanal, and J. F. Lemmer (Eds.), *Uncertainty in Artificial Intelligence 4*, Amsterdam, pp. 69–76. North-Holland.

Woodward, J. (2003). *Making Things Happen: A Theory of Causal Explanation*. Oxford: Oxford University Press.

# 19

# Effect Heterogeneity and Bias in Main-Effects-Only Regression Models

FELIX ELWERT AND CHRISTOPHER WINSHIP

## 1 Introduction

The overwhelming majority of OLS regression models estimated in the social sciences, and in sociology in particular, enter all independent variables as main effects. Few regression models contain many, if any, interaction terms. Most social scientists would probably agree that the assumption of constant effects that is embedded in main-effects-only regression models is theoretically implausible. Instead, they would maintain that regression effects are historically and contextually contingent; that effects vary across individuals, between groups, over time, and across space. In other words, social scientists doubt constant effects and believe in effect heterogeneity.

But why, if social scientists believe in effect heterogeneity, are they willing to substantively interpret main-effects-only regression models? The answer—not that it's been discussed explicitly—lies in the implicit assumption that the main-effects coefficients in linear regression represent straightforward averages of heterogeneous individual-level causal effects.

The belief in the averaging property of linear regression has previously been challenged. Angrist [1998] investigated OLS regression models that were correctly specified in all conventional respects except that effect heterogeneity in the main treatment of interest remained unmodeled. Angrist showed that the regression coefficient for this treatment variable gives a rather peculiar type of average—a conditional variance weighted average of the heterogeneous individual-level treatment effects in the sample. If the weights differ greatly across sample members, the coefficient on the treatment variable in an otherwise well-specified model may differ considerably from the arithmetic mean of the individual-level effects among sample members.

In this paper, we raise a new concern about main-effects-only regression models. Instead of considering models in which heterogeneity remains unmodeled in only one effect, we consider standard linear path models in which unmodeled heterogeneity is potentially pervasive.

Using simple examples, we show that unmodeled effect heterogeneity in more than one structural parameter may mask confounding and selection bias, and thus lead to biased estimates. In our simulations, this heterogeneity is indexed by latent (unobserved) group membership. We believe that this setup represents a fairly realistic scenario—one in which the analyst has no choice but to resort to a main-effects-only regression model because she cannot include the desired interaction terms since group-membership is unobserved. Drawing on Judea Pearl's theory of directed acyclic graphs (DAG) [1995, 2009] and VanderWeele and Robins [2007], we then show that the specific biases we report can be predicted from an analysis of the appropriate DAG. This paper is intended as a serious warning to applied regression modelers to beware of unmodeled effect heterogeneity, as it may lead to gross misinterpretation of conventional path models.

We start with a brief discussion of conventional attitudes toward effect heterogeneity in the social sciences and in sociology in particular, formalize the notion of effect heterogeneity, and briefly review results of related work. In the core sections of the paper, we use simulations to demonstrate the failure of main-effects-only regression models to recover average causal effects in certain very basic three-variable path models where unmodeled effect heterogeneity is present in more than one structural parameter. Using DAGs, we explain which constellations of unmodeled effect heterogeneity will bias conventional regression estimates. We conclude with a summary of findings.

## 2 A Presumed Averaging Property of Main-Effects-Only Regression

### 2.1 Social Science Practice

The great majority of empirical work in the social sciences relies on the assumption of constant coefficients to estimate OLS regression models that contain nothing but main effect terms for all variables considered.[1] Of course, most researchers do not believe that real-life social processes follow the constant-coefficient ideal of conventional regression. For example, they aver that the effect of marital conflict on children's self-esteem is larger for boys than for girls [Amato and Booth 1997]; or that the death of a spouse increases mortality more for white widows than for African American widows [Elwert and Christakis 2006]. When pressed, social scientists would probably agree that the causal effect of almost any treatment on almost any outcome likely varies from group to group, and from person to person.

But if researchers are such firm believers in effect heterogeneity, why is the constant-coefficients regression model so firmly entrenched in empirical practice? The answer lies in the widespread belief that the coefficients of linear regression models estimate averages of heterogeneous parameters—average causal effects—representing the average of the individual-level causal effects across sample members. This (presumed) averaging property of standard regression models is important for empirical practice for at least three reasons. First, sample sizes in the social sciences are often too small to investigate effect heterogeneity by including interaction terms between the treatment and more than a few common effect modifiers (such as sex, race, education, income, or place of residence); second, the variables needed to explicitly model heterogeneity may well not have been measured; third, and most importantly, the complete list of effect modifiers along which the causal effect of treatment on the outcome varies is typically unknown (indeed, unknowable) to the analyst in any specific application. Analysts thus rely on faith that

---

[1]Whether a model requires an interaction depends on the functional form of the dependent and/or independent variables. For example, a model with no interactions in which the independent variables are entered in log form, would require a whole series of interactions in order to approximate this function if the independent variables where entered in nonlog form.

their failure to anticipate and incorporate all dimensions of effect heterogeneity into regression analysis simply shifts the interpretation of regression coefficients from individual-level causal effects to average causal effects, without imperiling the causal nature of the estimate.

## 2.2 Defining Effect Heterogeneity

We start by developing our analysis of the consequences of causal heterogeneity within the counterfactual (potential outcomes) model. For a continuous treatment $T \in (-\infty, \infty)$, let $T = t$ denote some specific treatment value and $T = 0$ the control condition. $Y(t)_i$ is the potential outcome of individual i for treatment $T = t$, and $Y(0)_i$ is the potential outcome of individual i for the control condition. For a particular individual, generally only one value of $Y(t)_i$ will be observed. The *individual-level causal effect* (ICE) of treatment level $T = t$ compared to $T = 0$ is then defined as: $\delta_{i,t} = Y(t)_i - Y(0)_i$ (or $\delta_i$, for short, if T is binary).

Since $\delta_{i,t}$ is generally not directly estimable, researchers typically attempt estimating the *average causal effect* (ACE) for some sample or population:

$$\overline{\delta}_t = \sum_{i=1}^{N} \delta_{i,t} / N$$

We say that the effect of treatment T is *heterogeneous* if: $\delta_{i,t} \neq \overline{\delta}_t$ for at least one i.

In other words, effect heterogeneity exists if the causal effect of the treatment differs across individuals. The basic question of this paper is whether a regression estimate for the causal effect of the treatment can be interpreted as an average causal effect if effect heterogeneity is present.

## 2.3 Regression Estimates as Conditional Variance Weighted Average Causal Effects

The ability of regression to recover average causal effects under effect heterogeneity has previously been challenged by Angrist [1998].[2] Here, we briefly sketch the main result. For a binary treatment, T=0,1, Angrist assumed a model where treatment was ignorable given covariates X and the effect of treatment varied across strata defined by the values of X. He then analyzed the performance of an OLS regression model that properly controlled for confounding in X but was misspecified to include only a main effect term for T and no interactions between T and X. Angrist showed that the regression estimate for the main effect of treatment can be expressed as a weighted average of stratum-specific treatment effects, albeit one that is difficult to interpret. For each stratum defined by fixed values of X, the numerator of the OLS estimator has the form $\delta_x W_x P(X=x)$,[3] where $\delta_x$ is the stratum-specific causal effect and $P(X=x)$ is the relative size of the stratum in the sample. The weight, $W_x$, is a function of the propensity score, $P_x = P(T=1 \mid X)$, associated with the stratum, $W_x = P_x (1 - P_x)$, which equals the stratum-specific variance of treatment. This variance, and hence the weight, is largest if $P_x = .5$ and smaller as $P_x$ goes to 0 or 1.

---

[2]This presentation follows Angrist [1998] and Angrist and Pischke [2009].
[3]The denominator of the OLS estimator is just a normalizing constant that does not aid intuition.

If the treatment effect is constant across strata, these weights make good sense. OLS gives the minimum variance linear unbiased estimator of the model parameters under homoscedasticity assuming correct specification of the model. Thus in a model without interactions between treatment and covariates X the OLS estimator gives the most weight to strata with the smallest variance for the estimated within-stratum treatment effect, which, not considering the size of the strata, are those strata with the largest treatment variance, i.e. with the $P_x$ that are closest to .5. However, if effects are heterogeneous across strata, this weighting scheme makes little substantive sense: in order to compute the average causal effect, $\bar{\delta}$, as defined above, we would want to give the same weight to every individual in the sample. As a variance-weighted estimator, however, regression estimates under conditions of unmodeled effect heterogeneity do not give the same weight to every individual in the sample and thus do not converge to the (unweighted) average treatment effect.

## 3 Path Models with Pervasive Effect Heterogeneity

Whereas Angrist analyzed a misspecified regression equation that incorrectly assumed no treatment-covariate interaction for a *single* treatment variable, we investigate the ability of a main-effects-only regression model to recover unbiased average causal effects in simple path models with unmodeled effect heterogeneity across *multiple* parameters.

*Setup:* To illustrate how misleading the belief in the averaging power of the constant-coefficient model can be in practice, we present simulations of basic linear path models, shown in summary in Figure 1 (where we have repressed the usual uncorrelated error terms).

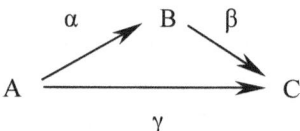

**Figure 1.** A simple linear path model

To introduce effect heterogeneity, let $G = 0, 1$ index membership in a latent group and permit the possibility that the three structural parameters $\alpha$, $\beta$, and $\gamma$ vary across (but not within) levels of G. The above path model can then be represented by two linear equations: $B = A\alpha_G + \varepsilon_B$ and $C = A\gamma_G + B\beta_G + \varepsilon_C$. In our simulations, we assume that $A \sim N(0,1)$ and $\varepsilon_B$, and $\varepsilon_C$ are iid $N(0,1)$, and hence all variables are normally distributed. From these equations, we next simulate populations of N=100,000 observations, with $P(G=1) = P(G=0) = 1/2$. We start with a population in which all three parameters are constant across the two subgroups defined by G, and then systematically introduce effect heterogeneity by successively permitting the structural parameters to vary by group, yielding one population for each of the $2^3 = 8$ possible combinations of constant/varying parameters. To fix ideas, we choose the group-specific parameter values shown in Table

1. For simulations in which one or more parameters do not vary by group, we set the constant parameter(s) to the average of the group specific parameters, e.g. $\alpha = (\alpha_0 + \alpha_1)/2$.

**Table 1:** Group-specific structural parameters for simulations

|  | $\alpha_G$ | $\beta_G$ | $\gamma_G$ |
|---|---|---|---|
| Group: |  |  |  |
| G=0 | 0.4 | 0.5 | 0.6 |
| G=1 | 1.2 | 2.5 | 1.4 |
| Average | 0.8 | 1.5 | 1.0 |

Finally, we estimate a conventional linear regression model for the effects of A and B on C using the conventional default specification, in which all variables enter as main effects only, $C = A\gamma + B\beta + \varepsilon$. (Note that G is latent and therefore cannot be included in the model.) The parameter, $\gamma$ refers to the direct effect of A on C holding B constant, and $\beta$ refers to the total effect of B on C.[4] In much sociological and social science research, this main-effects regression model is intended to recover average structural (causal) effects, and is commonly believed to be well suited for the purpose.

*Results:* Table 2 shows the regression estimates for the main effect parameters across the eight scenarios of effect heterogeneity. We see that the main effects regression model correctly recovers the desired (average) parameters, $\gamma=1$ and $\beta=1.5$ if none of the parameters vary across groups (column 1), or if only one of the three parameters varies (columns 2-4).

Other constellations of effect heterogeneity, however, produce biased estimates. If $\alpha_G$ and $\beta_G$ (column 5); or $\alpha_G$ and $\gamma_G$ (column 6); or $\alpha_G$, $\beta_G$, and $\gamma_G$ (column 8) vary across groups, the main-effects-only regression model fails to recover the true (average) parameter values known to underlie the simulations. For our specific parameter values, the estimated (average) effect of B on C in these troubled scenarios is always too high, and the estimated average direct effect of A on C is either too high or too low. Indeed, if we set $\gamma=0$ but let $\alpha_G$ and $\beta_G$ vary across groups, the estimate for $\gamma$ in the main-effects-only regression model would suggest the presence of a direct effect of A on C even though it is known by design that no such direct effect exists (not shown).

Failure of the regression model to recover the known path parameters is not merely a function of the number of paths that vary. Although none of the scenarios in which fewer than two parameters vary yield incorrect estimates, and the scenario in which all three parameters vary is clearly biased, results differ for the three scenarios in which exactly two parameters vary. In two of these scenarios (columns 5 and 6), regression fails to recover the desired (average) parameters, while regression does recover the correct average parameters in the third scenario (column 7).

---

[4]The notion of direct and indirect effects is receiving deserved scrutiny in important recent work by Robins and Greenland [1992]; Pearl [2001]; Robins [2003]; Frangakis and Rubin [2002]; Sobel [2008]; and VanderWeele [2008].

**Table 2:** OLS regression estimates for the main effects of A and B on C across eight different combinations of effect heterogeneity in α, β, and/or γ

|  | (1) | (2) | (3) | (4) | (5) | (6) | (7) | (8) |
|---|---|---|---|---|---|---|---|---|
| Heterogeneity in: | - | α | β | γ | α, β | α, γ | β, γ | α, β, γ |
| Group: | G0  G1 | G0  G1 | G0  G1 | G0  G1 | G0  G1 | G0  G1 | G0  G1 | G0  G1 |
| α | 0.8 | 0.4  1.2 | 0.8 | 0.8 | 0.4  1.2 | 0.4  1.2 | 0.8 | 0.4  1.2 |
| β | 1.5 | 1.5 | 0.5  2.5 | 1.5 | 0.5  2.5 | 1.5 | 0.5  2.5 | 0.5  2.5 |
| γ | 1.0 | 1.0 | 1.0 | 0.6  1.4 | 1.0 | 0.6  1.4 | 0.6  1.4 | 0.6  1.4 |
| Pooled OLS estimate: | | | | | | | | |
| β | 1.50 | 1.50 | 1.50 | 1.50 | **1.77** | **1.64** | 1.50 | **1.91** |
| γ | 1.00 | 1.00 | 1.00 | 1.00 | **1.17** | **0.89** | 1.00 | **1.07** |

*Note:* Bold estimates are biased for the true (average) parameters. Results from independent simulations of N=100,000 for each scenario using (group-specific) parameters listed above. See text for details.

In sum, the naïve main-effects-only linear regression model recovers the correct (average) parameter values only under certain conditions of limited effect heterogeneity, and it fails to recover the true average effects in certain other scenarios, including the scenario we consider most plausible in the majority of sociological applications, i.e., where all three parameters vary across groups. If group membership is latent—because group membership is unknown to or unmeasured by the analyst— and thus unmodeled, linear regression generally will fail to recover the true average effects.

## 4 DAGs to the Rescue

These results spell trouble for empirical practice in sociology. Judea Pearl's work on causality and directed acyclic graphs (DAGs) [1995, 2009] offers an elegant and powerful approach to understanding the problem. Focusing on the appropriate DAGs conveys the critical insight for the present discussion that effect heterogeneity, rather than being a nuisance that is easily averaged away, encodes structural information that analysts ignore at their peril.

Pearl's DAGs are nonparametric path models that encode causal dependence between variables: an arrow between two variables indicates that the second variable is causally dependent on the first (for detailed formal expositions of DAGs, see Pearl [1995, 2009]; for less technical introductions see Robins [2001]; Greenland, Pearl and Robins [1999] in epidemiology, and Morgan and Winship [2007] in sociology). For example, the DAG in Figure 2 indicates that Z is a function of X and Y, $Z= f(X,Y,\varepsilon_Z)$, where $\varepsilon_Z$ is an unobserved error term independent of $(X,Y)$.

In a non-parametric DAG—as opposed to a conventional social science path model—the term $f(\ )$ can be any function. Thus, the DAG in Figure 2 is consistent with a linear structural equation in which X only modifies (i.e. introduces heterogeneity into) the effect of Y on Z, $Z = Y\xi + YX\psi + \varepsilon_Z$.[5] In the language of VanderWeele and Robins [2007], who provide the most extensive treatment of effect heterogeneity using DAGs to date, one may call X a "direct effect modifier" of the effect of Y on Z. The point is that a variable that modifies the effect of Y on Z is causally associated with Z, as represented by the arrow from X to Z.

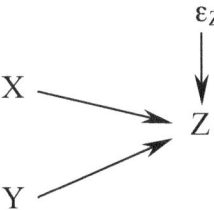

**Figure 2.** DAG illustrating direct effect modification of the effect of Y on Z in X

Returning to our simulation, one realizes that the social science path model of Figure 1, although a useful tool for informally illustrating the data generation process, does not, generally, provide a sufficiently rigorous description of the causal structure underlying the simulations. Figure 1, although truthfully representing the separate data generating mechanism for each group and each individual in the simulated population, is not the correct DAG for the pooled population containing groups $G = 0$ and $G = 1$ for all of the heterogeneity scenarios considered above. Specifically, in order to turn the informal social science path model of Figure 1 into a DAG, one would have to integrate the source of heterogeneity, G, into the picture. How this is to be done depends on the structure of heterogeneity. If only $\beta_G$ (the effect of B on C) and/or $\gamma_G$ (the direct effect of A on C holding B constant) varied with G, then one would add an arrow from G into C. If $\alpha_G$ (the effect of A on B) varied with G, then one would add an arrow from G into B. The DAG in Figure 3 thus represents those scenarios in which $\alpha_G$ as well as either $\beta_G$ or $\gamma_G$, or both, vary with G (columns 5, 6, and 8). Interpreted in terms of a linear path model, this DAG is consistent with the following two structural equations: $B = A\alpha_0 + AG\alpha_1 + \varepsilon_B$ and $C = A\gamma_0 + AG\gamma_1 + B\beta_0 + BG\beta_1 + \varepsilon_C$ (where the iid errors, $\varepsilon$, have been omitted from the DAG and are assumed to be uncorrelated).[6]

In our analysis, mimicking the reality of limited observational data with weak substantive theory, we have assumed that A, B, and C are observed, but that G is not observed. It is immediately apparent that the presence of G in Figure 3 means that, first, G is a confounder for the effect of B on C; and, second, that B is a "collider" [Pearl 2009] on

---

[5]It is also consistent with an equation that adds a main effect of X. For the purposes of this paper it does not matter whether the main effect is present.

[6]By construction of the example, we assume that A is randomized and thus marginally independent of G. Note, however, that even though G is mean independent of B and C (no main effect of G on either B or C), G is not marginally independent of B or C because var(B|G=1)≠var(B|G=0) and var(C|G=1)≠var(C|G=0), which explains the arrows from G into B and C. Adding main effects of G on B and C would not change the arguments presented here.

the path from A to C via B and G. Together, these two facts explain the failure of the main-effects-only regression model to recover the true parameters in panels 5, 6, and 8: First, in order to recover the effect of B on C, β, one would need to condition on the confounders A and G. But G is latent so it cannot be conditioned on. Second, conditioning on the collider B in the regression opens a "backdoor path" from A to C via B and G (when G is not conditioned on), i.e. it induces a non-causal association between A and C, creating selection bias in the estimate for the direct effect of A on C, γ [Pearl 1995, 2009; Hernán et al 2004]. Hence, both coefficients in the main-effects-only regression model will be biased for the true (average) parameters.

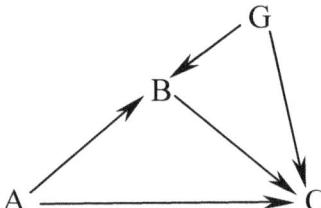

**Figure 3.** DAG consistent with effect modification of the effects of A on B, and B on C and/or A on C, in G

By contrast, if G modifies neither β nor γ, then the DAG would not contain an arrow from G into C; and if G does not modify α then the DAG would not contain an arrow from G into B. Either way, if either one (or both) of the arrows emanating from G are missing, then G is not a confounder for the effect of B on C, and conditioning on B will not induce selection bias by opening a backdoor path from A to C. Only then would the main effects regression model be unbiased and recover the true (average) parameters, as seen in panels 1-4 and 7.

In sum, Pearl's DAGs neatly display the structural information encoded in effect heterogeneity [VanderWeele and Robins 2007]. Consequently, Pearl's DAGs immediately draw attention to problems of confounding and selection bias that can occur when more than one effect in a causal system varies across sample members. Analyzing the appropriate DAG, the failure of main-effects-only regression models to recover average structural parameters in certain constellations of effect heterogeneity becomes predictable.

## 5 Conclusion

This paper considered a conventional structural model of a kind commonly used in the social sciences and explored its performance under various basic scenarios of effect heterogeneity. Simulations show that the standard social science strategy of dealing with effect heterogeneity—by ignoring it—is prone to failure. In certain situations, the main-effects-only regression model will recover the desired quantities, but in others it will not. We believe that effect heterogeneity in all arrows of a path model is plausible in many, if not most, substantive applications. Since the sources of heterogeneity are often not theorized, known, or measured, social scientists continue routinely to estimate main-effects-only regression models in hopes of recovering average causal effects. Our examples demonstrate that the belief in the averaging powers of main-effects-only regression models may be misplaced if heterogeneity is pervasive, as estimates can be mildly or wildly off the mark. Judea Pearl's DAGs provide a straightforward explanation for these difficulties—DAGs remind analysts that effect heterogeneity may encode structural information about confounding and selection bias that requires consideration when designing statistical strategies for recovering the desired average causal effects.

**Acknowledgments:** We thank Jamie Robins for detailed comments on a draft version of this paper, and Michael Sobel, Stephen Morgan, Hyun Sik Kim, and Elizabeth Wrigley-Field for advice. Genevieve Butler provided editorial assistance.

# References

Amato, Paul R., and Alan Booth. (1997). *A Generation at Risk: Growing Up in an Era of Family Upheaval*. Cambridge, MA: Harvard University Press.

Angrist, Joshua D. (1998). "Estimating the Labor Market Impact on Voluntary Military Service Using Social Security Date on Military Applicants." *Econometrica 66*: 249-88.

Angrist, Joshua D. and Jörn-Steffen Pischke. (2009). *Mostly Harmless Econometrics: An Empiricist's Companion.* Princeton, NJ: Princeton University Press.

Elwert, Felix, and Nicholas A. Christakis. (2006). "Widowhood and Race." *American Sociological Review 71*: 16-41.

Frangakis, Constantine E., and Donald B. Rubin. (2002). "Principal Stratification in Causal Inference." *Biometrics 58*: 21–29.

Greenland, Sander, Judea Pearl, and James M. Robbins. (1999). "Causal Diagrams for Epidemiologic Research." *Epidemiology 10*: 37-48.

Hernán, Miguel A., Sonia Hernández-Diaz, and James M. Robins. (2004). "A Structural Approach to Section Bias." *Epidemiology 155* (2): 174-184.

Morgan, Stephen L. and Christopher Winship. (2007). *Counterfactuals and Causal Inference: Methods and Principles of Social Research.* Cambridge: Cambridge University Press.

Pearl, Judea. (1995). "Causal Diagrams for Empirical Research." *Biometrika 82* (4): 669-710.

Pearl, Judea. (2001). "Direct and Indirect Effects." In *Proceedings of the Seventeenth Conference on Uncertainty in Artificial Intelligence*. San Francisco, CA: Morgan Kaufmann, 411-420.

Pearl, Judea. (2009). *Causality: Models, Reasoning, and Inference*. Second Edition. Cambridge: Cambridge University Press.

Robins, James M. (2001). "Data, Design, and Background Knowledge in Etiologic Inference," *Epidemiology 11* (3): 313-320.

Robins, James M. (2003). "Semantics of Causal DAG Models and the Identification of Direct and Indirect Effects." In: *Highly Structured Stochastic Systems*, P. Green, N. Hjort and S. Richardson, Eds. Oxford: Oxford University Press.

Robins, James M, and Sander Greenland. (1992). "Identifiability and Exchangeability for Direct and Indirect Effects." *Epidemiology 3*:143-155.

Sobel, Michael. (2008). "Identification of Causal Parameters in Randomized Studies with Mediating Variables," *Journal of Educational and Behavioral Statistics 33* (2): 230-251.

VanderWeele, Tyler J. (2008). "Simple Relations Between Principal Stratification and Direct and Indirect Effects." *Statistics and Probability Letters 78*: 2957-2962.

VanderWeele, Tyler J. and James M. Robins. (2007). "Four Types of Effect Modification: A Classification Based on Directed Acyclic Graphs." *Epidemiology 18* (5): 561-568.

# 20

# Causal and Probabilistic Reasoning in P-log

MICHAEL GELFOND AND NELSON RUSHTON

## 1 Introduction

In this paper we give an overview of the knowledge representation (KR) language P-log [Baral, Gelfond, and Rushton 2009] whose design was greatly influenced by work of Judea Pearl. We introduce the syntax and semantics of P-log, give a number of examples of its use for knowledge representation, and discuss the role Pearl's ideas played in the design of the language. Most of the technical material presented in the paper is not new. There are however two novel technical contributions which could be of interest. First we expand P-log semantics to allow domains with infinite Herbrand bases. This allows us to represent infinite sequences of random variables and (indirectly) continuous random variables. Second we generalize the logical base of P-log which improves the degree of elaboration tolerance of the language.

The goal of the P-log designers was to create a KR-language allowing natural and elaboration tolerant representation of commonsense knowledge involving logic and probabilities. The logical framework of P-log is Answer Set Prolog (ASP) — a language for knowledge representation and reasoning based on the answer set semantics (*aka* stable model semantics) of logic programs [Gelfond and Lifschitz 1988; Gelfond and Lifschitz 1991]. ASP has roots in declarative programing, the syntax and semantics of standard Prolog, disjunctive databases, and non-monotonic logic. The semantics of ASP captures the notion of possible beliefs of a reasoner who adheres to the *rationality principle* which says that "One shall not believe anything one is not forced to believe". The entailment relation of ASP is non-monotonic[1], which facilitates a high degree of elaboration tolerance in ASP theories. ASP allows natural representation of defaults and their exceptions, causal relations (including effects of actions), agents' intentions and obligations, and other constructs of natural language. ASP has a number of efficient reasoning systems, a well developed mathematical theory, and a well tested methodology of representing and using knowledge for computational tasks (see, for instance, [Baral 2003]). This, together with the fact that some of the designers of P-log came from the ASP community made the choice of a logical foundation for P-log comparatively easy.

---

[1] Roughly speaking, a language $L$ is *monotonic* if whenever $\Pi_1$ and $\Pi_2$ are collections of statements of $L$ with $\Pi_1 \subset \Pi_2$, and $W$ is a model of $\Pi_2$, then $W$ is a model of $\Pi_1$. A language which is not monotonic is said to be *nonmonotonic*.

The choice of a probabilistic framework was more problematic and that is where Judea's ideas played a major role. Our first problem was to choose from among various conceptualizations of probability: classical, frequentist, subjective, etc. Understanding the intuitive readings of basic language constructs is crucial for a software/knowledge engineer — probably more so than for a mathematician who may be primarily interested in their mathematical properties. Judea Pearl in [Pearl 1988] introduced the authors to the subjective view of probability — i.e. understanding of probabilities as degrees of belief of a rational agent — and to the use of subjective probability in AI. This matched well with the ASP-based logic side of the language. The ASP part of a P-log program can be used for describing possible beliefs, while the probabilistic part would allow knowledge engineers to quantify the degrees of these beliefs.

After deciding on an intuitive reading of probabilities, the next question was *which sorts of probabilistic statements to allow*. Fortunately, the question of concise and transparent representation of probability distributions was already addressed by Judea in [Pearl 1988], where he showed how Bayesian nets can be successfully used for this purpose. The concept was extended in [Pearl 2000] where Pearl introduced the notion of Causal Bayesian Nets (CBN's). Pearl's definition of CBN's is pioneering in three respects. First, he gives a framework where nondeterministic causal relations are the primitive relations among random variables. Second, he shows how relationships of correlation and (classical) independence *emerge* from these causal relationships in a natural way; and third he shows how this emergence is faithful to our intuitions about the difference between causality and (mere) correlation.

As we mentioned above, one of the primary desired features in the design of P-log was elaboration tolerance — defined as the ability of a representation to incorporate new knowledge with minimal revision [McCarthy 1999]. P-log inherited from ASP the ability to naturally incorporate many forms of new logical knowledge. An extension of ASP, called CR-Prolog, further improved this ability [Balduccini and Gelfond 2003]. The term "elaboration tolerance" is less well known in the field of probabilistic reasoning, but one of the primary strengths of Bayes nets as a representation is the ability to systematically and smoothly incorporate new knowledge through conditioning, using Bayes Theorem as well as algorithms given by Pearl [Pearl 1988] and others. Causal Bayesian Nets carry this a step further, by allowing us to formalize interventions in addition to (and as distinct from) observations, and smoothly incorporate either kind of new knowledge in the form of updates. Thus from the standpoint of elaboration tolerance, CBN's were a natural choice as a probabilistic foundation for P-log.

Another reason for choosing CBN's is that we simply believe Pearl's distinction between observations and interventions to be central to commonsense probabilistic reasoning. It gives a precise mathematical basis for distinguishing between the following questions: (1) what can I expect to happen given that I observe $X = x$, and (2) what can I expect to happen if I *intervene in the normal operation of a probabilistic system* by fixing value of variable $X$ to $x$? These questions could in theory be answered using classical methods, but only by creating a separate probabilistic model for each question. In a CBN these two questions may be treated as conditional probabilities (one conditioned on an observation and the other on an action) of a single probabilistic model.

P-log carries things another step. There are many actions one could take to manipulate a system besides fixing the values of (otherwise random) variables — and the effects of such actions are well studied under headings associated with ASP. Moreover, besides actions, there are many sorts of information one might gain besides those which simply eliminate possible worlds: one may gain knowledge which introduces new possible worlds, alters the probabilities of possible worlds, introduces new logical rules, etc. ASP has been shown to be a good candidate for handling such updates in non-probabilistic settings, and our hypothesis was that it would serve as well when combined with a probabilistic representation. Thus some of the key advantages of Bayesian nets, which are amplified by CBN's, show plausible promise of being even further amplified by their combination with ASP. This is the methodology of P-log: to combine a well studied method for elaboration tolerant probabilistic representations (CBN's) with a well studied method for elaboration tolerant logical representations (ASP).

Finally let us say a few words about the current status of the language. It is comparatively new. The first publication on the subject appeared in [Baral, Gelfond, and Rushton 2004], and the full journal paper describing the language appeared only recently in [Baral, Gelfond, and Rushton 2009]. The use of P-log for knowledge representation was also explored in [Baral and Hunsaker 2007] and [Gelfond, Rushton, and Zhu 2006]. A prototype reasoning system based on ASP computation allowed the use of the language for a number of applications (see, for instance, [Baral, Gelfond, and Rushton 2009; Pereira and Ramli 2009]). We are currently working on the development and implementation of a more efficient system, and on expanding it to allow rules of CR-Prolog. Finding ways for effectively combining ASP-based computational methods of P-log with recent advanced algorithms for Bayesian nets is probably one of the most interesting open questions in this area.

The paper is organized as follows. Section 2 contains short introduction to ASP and CR-Prolog. Section 3 describes the syntax and informal semantics of P-log, illustrating both through a nontrivial example. Section 4 gives another example, similar in nature to Simpson's Paradox. Section 5 states a new theorem which extends the semantics of P-log from that given in [Baral, Gelfond, and Rushton 2009] to cover programs with infinitely many random variables. The basic idea of Section 5 is accessible to a general audience, but its technical details require an understanding of the material presented in [Baral, Gelfond, and Rushton 2009].

## 2  Preliminaries

This section contains a description of syntax and semantics of both ASP and CR-Prolog. In what follows we use a standard notion of a sorted signature from classical logic. Terms and atoms are defined as usual. An atom $p(\bar{t})$ and its negation $\neg p(\bar{t})$ are referred to as *literals*. Literals of the form $p(\bar{t})$ and $\neg p(\bar{t})$ are called *contrary*. ASP and CR-Prolog also contain connectives *not* and *or* which are called *default negation* and *epistemic disjunction* respectively. Literals possibly preceded by default negation are called *extended literals*.

An ASP program is a pair consisting of a signature $\sigma$ and a collection of rules of the form

$$l_0 \text{ or } \ldots \text{ or } l_m \leftarrow l_{m+1}, \ldots, l_k, \text{not } l_{k+1}, \ldots, \text{not } l_n \qquad (1)$$

where $l$'s are literals. The right-hand side of of the rule is often referred to as the rule's *body*, the left-hand side as the rule's head.

The answer set semantics of a logic program $\Pi$ assigns to $\Pi$ a collection of *answer sets* – partial interpretations[2] corresponding to possible sets of beliefs which can be built by a rational reasoner on the basis of rules of $\Pi$. In the construction of such a set $S$, the reasoner is assumed to be guided by the following informal principles:

- $S$ must satisfy the rules of $\Pi$;

- the reasoner should adhere to the *rationality principle*, which says that *one shall not believe anything one is not forced to believe*.

To understand the former let us consider a partial interpretation $S$ viewed as a possible set of beliefs of our reasoner. A ground atom $p$ is satisfied by $S$ if $p \in S$, i.e., the reasoner believes $p$ to be true. According to the semantics of our connectives $\neg p$ means that $p$ is false. Consequently, $\neg p$ is satisfied by $S$ iff $\neg p \in S$, i.e., the reasoner believes $p$ to be false. Unlike $\neg p$, *not* $p$ has an epistemic character and is read as *there is no reason to believe that $p$ is true*. Accordingly, $S$ satisfies *not* $l$ if $l \notin S$. (Note that it is possible for the reasoner to believe neither $p$ nor $\neg p$). An epistemic disjunction $l_1$ *or* $l_2$ is satisfied by $S$ if $l_1 \in S$ or $l_2 \in S$, i.e., the reasoner believes at least one of the disjuncts to be true. Finally, $S$ satisfies the body (resp., head) of rule (1) if $S$ satisfies all of the extended literals occurring in its body (resp., head); and $S$ satisfies rule (1) if $S$ satisfies its head or does not satisfy its body.

What is left is to capture the intuition behind the rationality principle. This will be done in two steps.

DEFINITION 1 (Answer Sets, Part I). Let program $\Pi$ consist of rules of the form:

$$l_0 \text{ or } \ldots \text{ or } l_i \leftarrow l_{i+1}, \ldots, l_m.$$

An answer set of $\Pi$ is a consistent set $S$ of ground literals such that:

---

[2] By partial interpretation we mean a consistent set of ground literals of $\sigma(\Pi)$.

- $S$ satisfies the rules of $\Pi$.

- $S$ is minimal; i.e., no proper subset of $S$ satisfies the rules of $\Pi$.

The rationality principle here is captured by the minimality condition. For example, it is easy to see that $\{\ \}$ is the only answer set of program consisting of the single rule $p \leftarrow p$, and hence the reasoner associated with it knows nothing about the truth or falsity of $p$. The program consisting of rules

$p(a)$.
$q(a)$ or $q(b) \leftarrow p(a)$.

has two answer sets: $\{p(a), q(a)\}$ and $\{p(a), q(b)\}$. Note that no rule requires the reasoner to believe in both $q(a)$ and $q(b)$. Hence he believes that the two formulas $p(a)$ and $(q(a)$ or $q(b))$ are true, and that $\neg p(a)$ is false. He remains undecided, however, about, say, the two formulas $p(b)$ and $(\neg q(a)$ or $\neg q(b))$. Now let us consider an arbitrary program:

DEFINITION 2 (Answer Sets, Part II). Let $\Pi$ be an arbitrary collection of rules (1) and $S$ a set of literals. By $\Pi^S$ we denote the program obtained from $\Pi$ by

1. removing all rules containing *not l* such that $l \in S$;

2. removing all other premises containing *not* .

$S$ is an answer set of $\Pi$ iff $S$ is an answer set of $\Pi^S$.

To illustrate the definition let us consider a program

$p(a)$.
$p(b)$.
$\neg p(X) \leftarrow$ *not* $p(X)$.

where $p$ is a unary predicate whose domain is the set $\{a, b, c\}$. The last rule, which says that if $X$ is not believed to satisfy $p$ then $p(X)$ is false, is the ASP formalization of a Closed World Assumption for a relation $p$ [Reiter 1978]. It is easy to see that $\{p(a), p(b), \neg p(c)\}$ is the only answer set of this program. If we later learn that $c$ satisfies $p$, this information can be simply added to the program as $p(c)$. The default for $c$ will be defeated and the only answer set of the new program will be $\{p(a), p(b), p(c)\}$.

The next example illustrates the ASP formalization of a more general default. Consider a statement: "Normally, computer science courses are taught only by computer science professors. The logic course is an exception to this rule. It may be taught by faculty from the math department." This is a typical *default* with a *weak exception*[3] which can be represented in ASP by the rules:

---

[3] An exception to a default is called *weak* if it stops application of the default without defeating its conclusion. Otherwise it is called *strong*.

$$\neg may\_teach(P,C) \leftarrow \neg member(P,cs),$$
$$course(C,cs),$$
$$not\ ab(d_1(P,C)),$$
$$not\ may\_teach(P,C).$$
$$ab(d_1(P,logic)) \leftarrow not\ \neg member(P,math).$$

Here $d_1(P,C)$ is the name of the default rule and $ab(d_1(P,C))$ says that default $d_1(P,C)$ is not applicable to the pair $\langle P,C \rangle$. The second rule above stops the application of the default in cases where the class is *logic* and $P$ may be a math professor. Used in conjunction with rules:

$member(john,cs).$
$member(mary,math).$
$member(bob,ee).$
$\neg member(P,D) \leftarrow not\ member(P,D).$
$course(logic,cs).$
$course(data\_structures,cs).$

the program will entail that Mary does not teach data structures while she may teach logic; Bob teaches neither logic nor data structures, and John may teach both classes.

The previous examples illustrate the representation of defaults and their strong and weak exceptions. There is another type of possible exception to defaults, sometimes referred to as an **indirect exception**. Intuitively, these are rare exceptions that come into play only as a last resort, to restore the consistency of the agent's world view when all else fails. The representation of indirect exceptions seems to be beyond the power of ASP. This observation led to the development of a simple but powerful extension of ASP called **CR-Prolog** (or ASP with consistency-restoring rules). To illustrate the problem let us consider the following example.

Consider an ASP representation of the default "elements of class $c$ normally have property $p$":

$$p(X) \leftarrow c(X),$$
$$not\ ab(d(X)),$$
$$not\ \neg p(X).$$

together with the rule

$$q(X) \leftarrow p(X).$$

and the facts $c(a)$ and $\neg q(a)$. Let us denote this program by $E$, where $E$ stands for "exception".

It is not difficult to check that $E$ is inconsistent. No rules allow the reasoner to prove that the default is not applicable to $a$ (i.e. to prove $ab(d(a))$) or that $a$ does not have property $p$. Hence the default must conclude $p(a)$. The second rule implies $q(a)$ which contradicts one of the facts. However, there seems to exists a commonsense argument which may allow a reasoner to avoid inconsistency, and to conclude that $a$ is an indirect exception to the default. The argument is based on the **Contingency Axiom** for default $d(X)$ which says that *any element of class $c$ can be an exception to the default $d(X)$ above, but such a possibility is very rare, and, whenever possible, should be ignored*. One may informally argue that since the application of the default to $a$ leads to a contradiction, the possibility of $a$ being an exception to $d(a)$ cannot be ignored and hence $a$ must satisfy this rare property.

In what follows we give a brief description of CR-Prolog — an extension of ASP capable of encoding and reasoning about such rare events.

A program of CR-Prolog is a four-tuple consisting of

1. A (possibly sorted) signature.

2. A collection of regular rules of ASP.

3. A collection of rules of the form

$$l_0 \stackrel{+}{\leftarrow} l_1, \ldots, l_k, not\ l_{k+1}, \ldots, not\ l_n \qquad (2)$$

   where $l$'s are literals. Rules of this type are called *consistency restoring* rules (CR-rules).

4. A partial order, $\leq$, defined on sets of CR-rules. This partial order is often referred to as a **preference relation**.

Intuitively, rule (2) says that if the reasoner associated with the program believes the body of the rule, then he "may possibly" believe its head. However, this possibility may be used only if there is no way to obtain a consistent set of beliefs by using only regular rules of the program. The partial order over sets of CR-rules will be used to select preferred possible resolutions of the conflict. Currently the inference engine of CR-Prolog [Balduccini 2007] supports two such relations, denoted $\leq_1$ and $\leq_2$. One is based on the set-theoretic inclusion ($R_1 \leq_1 R_2$ holds iff $R_1 \subseteq R_2$). The other is defined by the cardinality of the corresponding sets ($R_1 \leq_2 R_2$ holds iff $|R_1| \leq |R_2|$). To give the precise semantics we will need some terminology and notation.

The set of regular rules of a CR-Prolog program $\Pi$ will be denoted by $\Pi^r$, and the set of CR-rules of $\Pi$ will be denoted by $\Pi^{cr}$. By $\alpha(r)$ we denote a regular rule obtained from a consistency restoring rule $r$ by replacing $\stackrel{+}{\leftarrow}$ by $\leftarrow$. If $R$ is a set of CR-rules then $\alpha(R) = \{\alpha(r) : r \in R\}$. As in the case of ASP, the semantics of CR-Prolog will be given for ground programs. A rule with variables will be viewed as a shorthand for a set of ground rules.

DEFINITION 3. (Abductive Support)
A minimal (with respect to the preference relation of the program) collection $R$ of

CR-rules of $\Pi$ such that $\Pi^r \cup \alpha(R)$ is consistent (i.e. has an answer set) is called an **abductive support** of $\Pi$.

DEFINITION 4. (Answer Sets of CR-Prolog)
A set $A$ is called an *answer set* of $\Pi$ if it is an answer set of a regular program $\Pi^r \cup \alpha(R)$ for some abductive support $R$ of $\Pi$.

Now let us show how CR-Prolog can be used to represent defaults and their indirect exceptions. The CR-Prolog representation of the default $d(X)$, which we attempted to represent in ASP program $E$, may look as follows

$$\begin{aligned} p(X) &\leftarrow c(X), \\ &\quad not\ ab(d(X)), \\ &\quad not\ \neg p(X). \\ \neg p(X) &\stackrel{+}{\leftarrow} c(X). \end{aligned}$$

The first rule is the standard ASP representation of the default, while the second rule expresses the Contingency Axiom for the default $d(X)$[4]. Consider now a program obtained by combining these two rules with an atom $c(a)$.

Assuming that $a$ is the only constant in the signature of this program, the program's unique answer set will be $\{c(a), p(a)\}$. Of course this is also the answer set of the regular part of our program. (Since the regular part is consistent, the Contingency Axiom is ignored.) Let us now expand this program by the rules

$q(X) \leftarrow p(X).$
$\neg q(a).$

The regular part of the new program is inconsistent. To save the day we need to use the Contingency Axiom for $d(a)$ to form the abductive support of the program. As a result the new program has the answer set $\{\neg q(a), c(a), \neg p(a))\}$. The new information does not produce inconsistency, as it did in ASP program $E$. Instead the program withdraws its previous conclusion and recognizes $a$ as a (strong) exception to default $d(a)$.

## 3 The Language

A P-log program consists of its *declarations, logical rules, random selection rules, probability atoms, observations,* and *actions*. We will begin this section with a brief description of the syntax and informal readings of these components of the programs, and then proceed to an illustrative example.

The declarations of a P-log program give the types of objects and functions in the program. Logical rules are "ordinary" rules of the underlying logical language

---

[4]In this form of Contingency Axiom, we treat $X$ as a strong exception to the default. Sometimes it may be useful to also allow weak indirect exceptions; this can be achieved by adding the rule $ab(d(X)) \stackrel{+}{\leftarrow} c(X)$.

written using light syntactic sugar. For purposes of this paper, the underlying logical language is CR-Prolog.

P-log uses *random selection rules* to declare random attributes (essentially random variables) of the form $a(\bar{t})$, where $a$ is the name of the attribute and $\bar{t}$ is a vector of zero or more parameters. In this paper we consider random selection rules of the form

$$[\,r\,]\ random(a(\bar{t})) \leftarrow B. \qquad (3)$$

where $r$ is a term used to name the random causal process associated with the rule and $B$ is a conjunction of zero or more extended literals. The name $[\,r\,]$ is optional and can be omitted if the program contains exactly one random selection rule for $a(\bar{t})$. Statement (3) says that *if $B$ were to hold, the value of $a(\bar{t})$ would be selected at random from its range by process $r$, unless this value is fixed by a deliberate action.* More general forms of random selection rules, where the values may be selected from a range which depends on context, are discussed in [Baral, Gelfond, and Rushton 2009].

Knowledge of the numeric probabilities of possible values of random attributes is expressed through *causal probability atoms*, or *pr-atoms*. A *pr*-atom takes the form

$$pr_r(a(\bar{t}) = y|_c B) = v$$

where $a(\bar{t})$ is a random attribute, $B$ a conjunction of literals, $r$ is a causal process, $v \in [0,1]$, and $y$ is a possible value of $a(\bar{t})$. The statement says that *if the value of $a(\bar{t})$ is fixed by process $r$, and $B$ holds, then the probability that $r$ causes $a(\bar{t}) = y$ is $v$.* If $r$ is uniquely determined by the program then it can be omitted. The "causal stroke" '$|_c$' and the "rule body" $B$ may also be omitted in case $B$ is empty.

Observations and actions of a P-log program are written, respectively, as

$$obs(l). \qquad do(a(\bar{t}) = y)).$$

where $l$ is a literal, $a(\bar{t})$ a random attribute, and $y$ a possible value of $a(\bar{t})$. $obs(l)$ is read *$l$ is observed to be true*. The action $do(a(\bar{t}) = y)$ is read *the value of $a(\bar{t})$, instead of being random, is set to $y$ by a deliberate action.*

This completes a general introductory description of P-log. Next we give an example to illustrate this description. The example shows how certain forms of knowledge may be represented, including deterministic causal knowledge, probabilistic causal knowledge, and strict and defeasible logical rules (a rule is *defeasible* if it states an overridable presumption; otherwise it is *strict*). We will use this example to illustrate the syntax of P-log, and, afterward, to provide an indication of the formal semantics. Complete syntax and semantics are given in [Baral, Gelfond, and Rushton 2009], and the reader is invited to refer there for more details.

EXAMPLE 5. [Circuit]

A circuit has a motor, a breaker, and a switch. The switch may be open or closed. The breaker may be tripped or not; and the motor may be turning or not. The operator may toggle the switch or reset the breaker. If the switch is closed and the system is functioning normally, the motor turns. The motor never turns when the switch is open, the breaker is tripped, or the motor is burned out. The system may break and if so the break could consist of a tripped breaker, a burned out motor, or both, with respective probabilities .9, .09, and .01. Breaking, however, is rare, and should be considered only in the absence of other explanations.

Let us show how to represent this knowledge in P-log. First we give declarations of sorts and functions relevant to the domain. As typical for representation of dynamic domains we will have sorts for actions, fluents (properties of the domain which can be changed by actions), and time steps. Fluents will be partitioned into inertial fluents and defined fluents. The former are subject to the law of inertia [Hayes and McCarthy 1969] (which says that things stay the same by default), while the latter are specified by explicit definitions in terms of already defined fluents. We will also have a sort for possible types of breaks which may occur in the system. In addition to declared sorts P-log contains a number of predefined sorts, e.g. a sort *boolean*. Here are the sorts of the domain for the circuit example:

$action = \{toggle, reset, break\}$.

$inertial\_fluent = \{closed, tripped, burned\}$.

$defined\_fluent = \{turning, faulty\}$.

$fluent = inertial\_fluent \cup defined\_fluent$.

$step = \{0, 1\}$.

$breaks = \{trip, burn, both\}$.

In addition to sorts we need to declare functions (referred in P-log as *attributes*) relevant to our domain.

$holds \; : \; fluent \times step \rightarrow boolean$.

$occurs \; : \; action \times step \rightarrow boolean$.

Here $holds(f, T)$ says that fluent $f$ is true at time step $T$ and $occurs(a, T)$ indicates that action $a$ was executed at $T$.

The last function we need to declare is a random attribute $type\_of\_break(T)$ which denotes the type of an occurrence of action *break* at step $T$.

$type\_of\_break \; : \; step \rightarrow breaks$.

The first two logical rules of the program define the direct effects of action *toggle*.

$$holds(closed, T+1) \leftarrow occurs(toggle, T),$$
$$\neg holds(closed, T).$$
$$\neg holds(closed, T+1) \leftarrow occurs(toggle, T),$$
$$holds(closed, T).$$

They simply say that toggling opens and closes the switch. The next rule says that resetting the breaker untrips it.

$$\neg holds(tripped, T+1) \leftarrow occurs(reset, T).$$

The effects of action *break* are described by the rules

$$holds(tripped, T+1) \leftarrow occurs(break, T),$$
$$type\_of\_break(T) = trip.$$
$$holds(burned, T+1) \leftarrow occurs(break, T),$$
$$type\_of\_break(T) = burn.$$
$$holds(tripped, T+1) \leftarrow occurs(break, T),$$
$$type\_of\_break(T) = both.$$
$$holds(burned, T+1) \leftarrow occurs(break, T),$$
$$type\_of\_break(T) = both.$$

The next two rules express the inertia axiom which says that *by default, things stay as they are*. They use default negation *not* — the main nonmonotonic connective of ASP —, and can be viewed as typical representations of defaults in ASP and its extensions.

$$holds(F, T+1) \leftarrow inertial\_fluent(F),$$
$$holds(F, T),$$
$$not \neg holds(F, T+1).$$
$$\neg holds(F, T+1) \leftarrow inertial\_fluent(F),$$
$$\neg holds(F, T),$$
$$not\ holds(F, T+1).$$

Next we explicitly define fluents *faulty* and *turning*.

$$holds(faulty, T) \leftarrow holds(tripped, T).$$
$$holds(faulty, T) \leftarrow holds(burned, T).$$
$$\neg holds(faulty, T) \leftarrow not\ holds(faulty, T).$$

The rules above say that the system is functioning abnormally if and only if the breaker is tripped or the motor is burned out. Similarly the next definition says that the motor turns if and only if the switch is closed and the system is functioning normally.

$$holds(turning, T) \leftarrow holds(closed, T),$$
$$\neg holds(faulty, T).$$
$$\neg holds(turning, T) \leftarrow not\ holds(turning, T).$$

The above rules are sufficient to define causal effects of actions. For instance if we assume that at Step 0 the motor is turning and the breaker is tripped, i.e.

action *break* of the type *trip* occurred at 0, then in the resulting state we will have $holds(tripped, 1)$ as the direct effect of this action; while $\neg holds(turning, 1)$ will be its indirect effect[5].

We will next have a default saying that for each action $A$ and time step $T$, in the absence of a reason to believe otherwise we assume $A$ does not occur at $T$.

$\neg occurs(A, T) \leftarrow action(A), not\ occurs(A, T).$

We next state a CR-rule representing possible exceptions to this default. The rule says that a break to the system may be considered if necessary (that is, necessary in order to reach a consistent set of beliefs).

$occurs(break, 0) \stackrel{+}{\leftarrow} .$

The next collection of facts describes the initial situation of our story.

$\neg holds(closed, 0). \quad \neg holds(burned, 0). \quad \neg holds(tripped, 0). \quad occurs(toggle, 0).$

Next, we state a random selection rule which captures the non-determinism in the description of our circuit.

$random(type\_of\_break(T)) \leftarrow occurs(break, T).$

The rule says that if action *break* occurs at step $T$ then the type of break will be selected at random from the range of possible types of breaks, unless this type is fixed by a deliberate action. Intuitively, *break* can be viewed as a non-deterministic action, with non-determinism coming from the lack of knowledge about the precise type of *break*.

Let $\pi_0$ be the circuit program given so far. Next we will give a sketch of the formal semantics of P-log, using $\pi_0$ as an illustrative example.

The *logical part* of a P-log program $\Pi$ consists of its declarations, logical rules, random selection rules, observations, and actions; while its *probabilistic part* consists of its *pr*-atoms (though the above program does not have any). The semantics of P-log describes a translation of the logical part of $\Pi$ into an "ordinary" CR-Prolog program $\tau(\Pi)$. The semantics of $\Pi$ is then given by

---

[5] It is worth noticing that, though short, our formalization of the circuit is non-trivial. It is obtained using the general methodology of representing dynamic systems modeled by transition diagrams whose nodes correspond to physically possible states of the system and whose arcs are labeled by actions. A transition $\langle \sigma_0, a, \sigma_1 \rangle$ indicates that state $\sigma_1$ may be a result of execution of $a$ in $\sigma_0$. The problem of finding concise and mathematically accurate description of such diagrams has been a subject of research for over 30 years. Its solution requires a good understanding of the nature of causal effects of actions in the presence of complex interrelations between fluents. An additional level of complexity is added by the need to specify what is not changed by actions. As noticed by John McCarthy, the latter, known as the Frame Problem, can be reduced to finding a representation of the Inertia Axiom which requires the ability to represent defaults and to do non-monotonic reasoning. The representation of this axiom as well as that of the interrelations between fluents we used in this example is a simple special case of general theory of action and change based on logic programming under the answer set semantics.

1. a collection of answer sets of $\tau(\Pi)$ viewed as the set of possible worlds of a rational agent associated with $\Pi$, along with

2. a probability measure over these possible worlds, determined by the collection of the probability atoms of $\Pi$.

To obtain $\tau(\pi_0)$ we represent sorts as collections of facts. For instance, sort *step* would be represented in CR-Prolog as

$step(0).\quad step(1).$

For a non-boolean function *type_of_break* the occurrences of atoms of the form $type\_of\_break(T) = trip$ in $\pi_0$ are replaced by $type\_of\_break(T, trip)$. Similarly for *burn* and *both*. The translation also contains the axiom

$$\neg type\_of\_break(T, V_1) \leftarrow breaks(V_1), breaks(V_2), V_1 \neq V_2,$$
$$type\_of\_break(T, V_2).$$

to guarantee that *type_of_break* is a function. In general, the same transformation is performed for all non-boolean functions.

Logical rules of $\pi_0$ are simply inserted into $\tau(\pi_0)$. Finally, the random selection rule is transformed into

$$type\_of\_break(T, trip) \text{ or } type\_of\_break(T, burn) \text{ or } type\_of\_break(T, both) \leftarrow$$
$$occurs(break, T),$$
$$not\ intervene(type\_of\_break(T)).$$

It is worth pointing out here that while CBN's represent the notion of intervention in terms of transformations on graphs, P-log axiomatizes the semantics of intervention by including *not intervene*(...) in the body of the translation of each random selection rule. This amounts to a *default presumption* of randomness, overridable by intervention. We will see next how actions using *do* can defeat this presumption.

Observations and actions are translated as follows. For each literal $l$ in $\pi_0$, $\tau(\pi_0)$ contains the rule

$\leftarrow obs(l), not\ l.$

For each atom $a(\bar{t}) = y$, $\tau(\pi)$ contains the rules

$a(\bar{t}, y) \leftarrow do(a(\bar{t}, y)).$

and

$intervene(a(\bar{t})) \leftarrow do(a(\bar{t}, Y)).$

The first rule eliminates possible worlds of the program failing to satisfy $l$. The second rule makes sure that interventions affect their intervened-upon variables in the expected way. The third rule defines the relation *intervene* which, for each action, cancels the randomness of the corresponding attribute.

It is not difficult to check that under the semantics of CR-Prolog, $\tau(\pi_0)$ has a unique possible world $W$ containing $holds(closed, 1)$ and $holds(turning, 1)$, the direct and indirect effects, respectively, of the action *close*. Note that the collection of regular ASP rules of $\tau(\pi_0)$ is consistent, i.e., has an answer set. This means that CR-rule $occurs(break, 0) \stackrel{+}{\leftarrow}$ is not activated, break does not occur, and the program contains no randomness.

Now we will discuss how probabilities are computed in P-log. Let $\Pi$ be a P-log program containing the random selection rule $[r]\ random(a(\bar{t})) \leftarrow B_1$ and the pr-atom $pr_r(a(\bar{t}) = y \mid_c B_2) = v$. Then if $W$ is a possible world of $\Pi$ satisfying $B_1$ and $B_2$, the *assigned probability* of $a(\bar{t}) = y$ in $W$ is defined [6] to be $v$. In case $W$ satisfies $B_1$ and $a(\bar{t}) = y$, but there is no pr-atom $pr_r(a(\bar{t}) = y \mid_c B_2) = v$ of $\Pi$ such that $W$ satisfies $B_2$, then the *default probability* of $a(\bar{t}) = y$ in $W$ is computed using the "indifference principle", which says that two possible values of a random selection are equally likely if we have no reason to prefer one to the other (see [Baral, Gelfond, and Rushton 2009] for details). The *probability* of each random atom $a(\bar{t}) = y$ occurring in each possible world $W$ of program $\Pi$, written $P_\Pi(W, a(\bar{t}) = y)$, is now defined to be the assigned probability or the default probability, as appropriate.

Let $W$ be a possible world of $\Pi$. The *unnormalized probability*, $\hat{\mu}_\Pi(W)$, of a possible world $W$ *induced by* $\Pi$ is

$$\hat{\mu}_\Pi(W) =_{def} \prod_{a(\bar{t},y) \in\ W} P_\Pi(W, a(\bar{t}) = y)$$

where the product is taken only over atoms for which $P(W, a(\bar{t}) = y)$ is defined.

Suppose $\Pi$ is a P-log program having at least one possible world with nonzero unnormalized probability, and let $\Omega$ be the set of possible worlds of $\Pi$. The *measure*, $\mu_\Pi(W)$, of a possible world $W$ *induced by* $\Pi$ is the unnormalized probability of $W$ divided by the sum of the unnormalized probabilities of all possible worlds of $\Pi$, i.e.,

$$\mu_\Pi(W) =_{def} \frac{\hat{\mu}_\Pi(W)}{\sum_{W_i \in \Omega} \hat{\mu}_\Pi(W_i)}$$

When the program $\Pi$ is clear from context we may simply write $\hat{\mu}$ and $\mu$ instead of $\hat{\mu}_\Pi$ and $\mu_\Pi$ respectively.

This completes the discussion of how probabilities of possible worlds are defined in P-log. Now let us return to the circuit example. Let program $\pi_1$ be the union of $\pi_0$ with the single observation

$obs(\neg holds(turning, 1))$

The observation contradicts our previous conclusion $holds(turning, 1)$ reached by using the effect axiom for *toggle*, the definitions of *faulty* and *turning*, and the

---

[6] For the sake of well definiteness, we consider only programs in which at most one $v$ satisfies this definition.

inertia axiom for *tripped* and *burned*. The program $\tau(\pi_1)$ will resolve this contradiction by using the CR-rule $occurs(break, 0) \stackrel{+}{\leftarrow}$ to conclude that the action *break* occurred at Step 0. Now *type_of_break* randomly takes one of its possible values. Accordingly, $\tau(\pi_1)$ has three answer sets: $W_1$, $W_2$, and $W_3$. All of them contain $occurs(break, 0)$, $holds(faulty, 1)$, $\neg holds(turning, 1)$. One, say $W_1$ will contain

$type\_of\_break(1, trip)$, $holds(tripped, 1)$, $\neg holds(burned, 1)$

$W_2$ and $W_3$ will respectively contain

$type\_of\_break(1, burn)$, $\neg holds(tripped, 1)$, $holds(burned, 1)$

and

$type\_of\_break(1, both)$, $holds(tripped, 1)$, $holds(burned, 1)$

In accordance with our general definition, $\pi_1$ will have three possible worlds, $W_1$, $W_2$, and $W_3$. The probabilities of each of these three possible worlds can be computed as 1/3, using the indifference principle.

Now let us add some quantitative probabilities to our program. If $\pi_2$ is the union of $\pi_1$ with the following three *pr*-atoms

$pr(type\_of\_break(T) = trip \mid_c break(T)) = 0.9$
$pr(type\_of\_break(T) = burned \mid_c break(T)) = 0.09$
$pr(type\_of\_break(T) = both \mid_c break(T)) = 0.01$

then program $\pi_2$ has the same possible worlds as $\Pi_1$. Not surprisingly, $P_{\pi_2}(W_1) = 0.9$. Similarly $P_{\pi_2}(W_2) = 0.09$ and $P_{\pi_2}(W_3) = 0.01$. This demonstrates how a P-log program may be written in stages, with quantitative probabilities added as they are needed or become available.

Typically we are interested not just in the probabilities of individual possible worlds, but in the probabilities of certain interesting sets of possible worlds described, e.g., those described by formulae. For current purposes a rather simple definition suffices. Viz., recalling that possible worlds are sets of literals, for an arbitrary set $C$ of literals we define

$$P_\pi(C) =_{def} P_\pi(\{W : C \subseteq W\}).$$

For example, $P_{\pi_1}(holds(turning, 1)) = 0$, $P_{\pi_1}(holds(tripped, 1)) = 1/3$, and $P_{\pi_2}(holds(tripped, 1)) = 0.91$.

Our example is in some respects rather simple. For instance, every possible world of our program contains at most one atom of the form $a(\bar{t}) = y$ where $a(\bar{t})$ is a random attribute. We hope, however, that this example gives a reader some insight in the syntax and semantics of P-log. It is worth noticing that the example shows the ability of P-log to mix logical and probabilistic reasoning, including reasoning about causal effects of actions and explanations of observations. In addition it demonstrates the non-monotonic character of P-log, i.e. its ability to react to new knowledge by changing probabilistic models of the domain and creating new possible worlds.

The ability to introduce new possible worlds as a result of conditioning is of interest from two standpoints. First, it reflects the common sense semantics of utterances such as "the motor might be burned out." Such a sentence does not eliminate existing possible beliefs, and so there is no classical (i.e., monotonic) semantics in which the statement would be informative. If it is informative, as common sense suggests, then its content seems to introduce new possibilities into the listener's thought process.

Second, nonmonotonicity can improve performance. Possible worlds tend to proliferate exponentially with the size of a program, quickly making computations intractable. The ability to consider only those random selections which may explain our abnormal observations may make computations tractable for larger programs. Even though our current solver is in its early stages of development, it is based on well researched answer set solvers which efficiently eliminate impossible worlds from consideration based on logical reasoning. Thus even our early prototype has shown promising performance on problems where logic may be used to exclude possible worlds from consideration in the computation of probabilities [Gelfond, Rushton, and Zhu 2006].

## 4 Spider Example

In this section, we consider a variant of Simpson's paradox, to illustrate the formalization of interventions in P-log. The story we would like to formalize is as follows:

In Stan's home town there are two kinds of poisonous spider, the creeper and the spinner. Bites from the two are equally common in Stan's area — though spinner bites are more common on a worldwide basis. An experimental anti-venom has been developed to treat bites from either kind of spider, but its effectiveness is questionable.

One morning Stan wakes to find he has a bite on his ankle, and drives to the emergency room. A doctor examines the bite, and concludes it is a bite from either a creeper or a spinner. In deciding whether to administer the anti-venom, the doctor examines the data he has on bites from the two kinds of spiders: out of 416 people bitten by the creeper worldwide, 312 received the anti-venom and 104 did not. Among those who received the anti-venom, 187 survived; while 73 survived who did not receive anti-venom. The spinner is more deadly and tends to inhabit areas where the treatment is less available. Of 924 people bitten by the spinner, 168 received the anti-venom, 34 of whom survived. Of the 756 spinner bite victims who did not receive the experimental treatment, only 227 survived.

For a random individual bitten by a creeper or spinner, let $s$, $a$, and $c$ denote the events of *survival, administering anti-venom,* and *creeper bite.* Based on the fact that the two sorts of bites are equally common in Stan's region, the doctor assigns a 0.5 probability to either kind of bite. He also computes a probability of survival, with and without treatment, from each kind of bite, based on the sampling distribution of the available data. He similarly computes the probabilities that victims of each kind of bite received the anti-venom. We may now imagine the doctor uses Bayes' Theorem to compute $P(s \mid a) = 0.522$ and $P(s \mid \neg a) = 0.394$.

Thus we see that if we choose a historical victim, in such a way that he has a 50/50 chance of either kind of bite, those who received anti-venom would have a substantially higher chance of survival. Stan is in the situation of having a 50/50 chance of either sort of bite; however, he is *not* a historical victim. Since we are intervening in the decision of whether he receives anti-venom, the computation above is not germane (as readers of [Pearl 2000] already know) — though we can easily imagine the doctor making such a mistake. A correct solution is as follows. Formalizing the relevant parts of the story in a P-log program $\Pi$ gives

*survive, antivenom* : *boolean.*
*spider* : {*creeper, spinner*}.

*random*(*spider*).
*random*(*survive*).
*random*(*antivenom*).

$pr(spider = creeper) = 0.5$.

$pr(survive \mid_c spider = creeper, antivenom) = 0.6$.
$pr(survive \mid_c spider = creeper, \neg antivenom) = 0.7$.
$pr(survive \mid_c spider = spinner, antivenom) = 0.2$.
$pr(survive \mid_c spider = spinner, \neg antivenom) = 0.3$.

and so, according to our semantics,

$P_{\Pi \cup \{do(antivenom)\}}(survive) = 0.4$
$P_{\Pi \cup \{do(\neg antivenom)\}}(survive) = 0.5$

Thus, the correct decision, assuming we want to intervene to maximize Stan's chance of survival, is to not administer antivenom.

In order to reach this conclusion by classical probability, we would need to consider separate probability measures $P_1$ and $P_2$, on the sets of patients who received or did not receive antivenom, respectively. If this is done correctly, we obtain $P_1(s) = 0.4$ and $P_2(s) = 0.5$, as in the P-log program.

Thus we can get a correct classical solution using separate probability measures. Note however, that we could also get an *incorrect* classical solution using separate measures, since there exist probability measures $\hat{P}_1$ and $\hat{P}_2$ on the sets of historical bite victims which capture classical conditional probabilities given $a$ and $\neg a$ respectively. We may define

$\hat{P}_1(E) =_{def} \frac{P(E \cap a)}{0.3582}$

$\hat{P}_2(E) =_{def} \frac{P(E \cap \neg a)}{0.6418}$

It is well known that each of these is a probability measure. They are seldom seen only because classical conditional probability gives us simple notations for them *in terms of a single measure capturing common background knowledge*. This allows us to refer to probabilities conditioned on observations without defining a new measure for each such observation. What we do not have, classically, is a similar mechanism for probabilities conditioned on intervention — which is sometimes of interest as the example shows. The ability to condition on interventions in this way has been a fundamental contribution of Pearl; and the inclusion in P-log of such conditioning-on-intervention is a direct result of the authors' reading of his book.

## 5 Infinite Programs

The definitions given so far for P-log apply only to programs with finite numbers of random selection rules. In this section we state a theorem which allows us to extend these semantics to programs which may contain infinitely many random selection rules. No changes are required from the syntax given in [Baral, Gelfond, and Rushton 2009], and the probability measure described here agrees with the one in [Baral, Gelfond, and Rushton 2009] whenever the former is defined.

We begin by defining the class of programs for which the new semantics are applicable. The reader is referred to [Baral, Gelfond, and Rushton 2009] for the definitions of *causally ordered, unitary*, and *strict probabilistic levelling*.

DEFINITION 6. [Admissible Program]
A P-log program is *admissible* if it is causally ordered and unitary, and if there exists a strict probabilistic levelling $||$ on $\Pi$ such that no ground literal occurs in the heads of rules in infinitely many $\Pi_i$ with respect to $||$.

The condition of admissibility, and the definitions it relies on, are all rather involved to state precisely, but the intuition is as follows. Basically, a program is unitary if the probabilities assigned to the possible outcomes of each selection rule are either all assigned and sum to 1, or are not all assigned and their sum does not exceed 1. The program is causally ordered if its causal dependencies are acyclic and if the only nondeterminism in it is a result of random selection rules. A strict probabilistic levelling is a well ordering of the selection rules of a program which witnesses the fact that it is causally ordered. Finally, a program which meets these conditions is admissible if every ground literal in the program logically depends on only finitely many random experiments. For example, the following program is not unitary:

$random(a) : boolean.$
$pr(a) = 1/2.$
$pr(\neg a) = 2/3.$

The following program is not causally ordered:

$random(a) : boolean.$
$random(b) : boolean.$
$pr_r(a|_c \, b) = 1/3.$
$pr_r(a|_c \, \neg b) = 2/3.$
$pr_r(b|_c \, a) = 1/5.$

and neither is the following:

$p \leftarrow not \; q.$
$q \leftarrow not \; p.$

since it has two answer sets which arise from circularity of defaults, rather than random selections. The following program is both unitary and causally ordered, but not admissible, because $atLeastOneTail$ depends on infinitely many coin tosses.

$coin\_toss : positive\_integer \rightarrow \{head, tail\}.$
$atLeastOneTail : boolean.$
$random(coin\_toss(N)).$
$atLeastOneTail \leftarrow coin\_toss(N) = tail.$

We need one more definition before stating the main theorem:

DEFINITION 7. [Cylinder algebra of $\Pi$]
Let $\Pi$ be a countably infinite P-log program with random attributes $a_i(t)$, $i > 0$, and let $C$ be the collection of sets of the form $\{\omega \; : \; a_i(t) = y \in \omega\}$ for arbitrary $t$, $i$, and $y$. The sigma algebra generated by $C$ will be called the *cylinder algebra* of program $\Pi$.

Intuitively, the cylinder algebra of a program $\Pi$ is the collection of sets which can be formed by performing countably many set operations (union, intersection, and complement) upon sets whose probabilities are defined by finite subprograms. We are now ready to state the main proposition of this section.

PROPOSITION 8. [Admissible programs]
Let $\Pi$ be an admissible P-log program with at most countably infinitely many ground rules, and let $A$ be the cylinder algebra of $\Pi$. Then there exists a unique probability measure $P_\Pi$ defined on $A$ such that whenever $[r] \; random(a(\bar{t})) \leftarrow B_1$ and $pr_r(a(\bar{t}) = y \mid B_2) = v$ occur in $\Pi$, and $P_\Pi(B_1 \wedge B_2) > 0$, we have $P_\Pi(a(\bar{t}) = y \mid B_1 \wedge B_2) = v$.

Recall that the semantic value of a P-log program $\Pi$ consists of (1) a set of possible worlds of $\Pi$ and (2) a probability measure on those possible worlds. The proposition now puts us in position to give semantics for programs with infinitely many random selection rules. The possible worlds of the program are the answer sets of the associated (infinite) CR-Prolog program, as determined by the usual definition — while the probability measure is $P_\Pi$, as defined in Proposition 8.

We next give an example which exercises the proposition, in a form of a novel paradox. Imagine a casino which offers an infinite sequence of games, of which our agent may decide to play as many or as few as he wishes. For the $n^{th}$ game, a fair coin is tossed $n$ times. If the agent chooses to play the $n^{th}$ game, then the agent wins $2^{n+1} + 1$ dollars if all tosses made in the $n^{th}$ game are heads and otherwise loses one dollar.

We can formalize this game as an infinite P-log program $\Pi$. First, we declare a countable sequence of games and an integer valued variable, representing the player's net winnings after each game.

*game* : *positive_integer*.
*winnings* : *game* $\to$ *integer*.
*play* : *game* $\to$ *boolean*.
*coin* : $\{\langle M, N \rangle \mid 1 \leq M \leq N\} \to \{head, tail\}$.

Note that the declaration for *coin* is not written in the current syntax of P-log; but to save space we use set-builder notation here as a shorthand for the more lengthy formal declaration. Similarly, the notation $\langle M, N \rangle$ is also a shorthand. From this point on we will write $coin(M, N)$ instead of $coin(\langle M, N \rangle)$.

$\Pi$ also contains a declaration to say that the throws are random and the coin is known to be fair:

*random*($coin(M, N)$).
$pr(coin(M, N) = head) = 1/2$.

The conditions of winning the $N^{th}$ game are described as follows:

$lose(N) \leftarrow play(N), coin(N, M) = tail$.
$win(N) \leftarrow play(N), not\ lose(N)$.

The amount the agent wins or loses on each game is given by

$winnings(0) = 0$.
$winnings(N + 1) = winnings(N) + 1 + 2^{N+1} \leftarrow win(N)$.
$winnings(N + 1) = winnings(N) - 1 \leftarrow lose(N)$.
$winnings(N + 1) = winnings(N) \leftarrow \neg play(N)$.

Finally the program contains rules which describe the agent's strategy in choosing which games to play. Note that the agent's expected winnings in the $N^{th}$ game are given by $(1/2^N)(1 + 2^{N+1}) - (1 - 1/2^N) = 1$, so each game has positive expectation for the player. Thus a reasonable strategy might be to play every game, represented as

$play(N)$.

This completes program $\Pi$. It can be shown to be admissible, and hence there is a unique probability measure $P_\Pi$ satisfying the conclusion of Proposition 1. Thus, for example, $P_\Pi(coin(3,2) = head) = 1/2$, and $P_\Pi(win(10)) = 1/2^{10}$. Each of these probabilities can be computed from finite sub-programs. As more interesting example, let $S$ be the set of possible worlds in which the agent wins infinitely many games. The probability of this event cannot be computed from any finite sub-program of $\Pi$. However, $S$ is a countable intersection of countable unions of sets whose probabilities are defined by finite subprograms. In particular,

$$S = \bigcap_{N=1}^{\infty} \bigcup_{J=N}^{\infty} \{W \mid win(J) \in W\}$$

and therefore, $S$ is in the cylinder algebra of $\Pi$ and so its probability is given by the measure defined in Proposition 1.

So where is the Paradox? To see this, let us compute the probability of $S$. Since $P_\Pi$ is a probability measure, it is monotonic in the sense that no set has greater probability than any of its subsets. $P_\Pi$ must also be *countably subadditive*, meaning that the probability of a countable union of sets cannot exceed the sum of their probabilities. Thus, from the above we get for every $N$,

$$\begin{aligned} P_\Pi(S) &< P_\Pi(\bigcup_{J=N}^{\infty} \{W \mid win(J) \in W\} \\ &\leq \sum_{J=N}^{\infty} P_\Pi(\{W \mid win(J) \in W\}) \\ &= \sum_{J=N}^{\infty} 1/2^J \\ &= 1/2^N \end{aligned}$$

Now since right hand side can be made arbitrarily small by choosing a sufficiently large $N$, it follows that $P_\Pi(S) = 0$. Consequently, with probability 1, our agent will *lose* all but finitely many of the games he plays. Since he loses one dollar per play indefinitely after his final win, his winnings converge to $-\infty$ with probability 1, even though each of his wagers has positive expectation!

**Acknowledgement**

The first author was partially supported in this research by iARPA.

# References

Balduccini, M. (2007). CR-MODELS: An inference engine for CR-Prolog. In C. Baral, G. Brewka, and J. Schlipf (Eds.), *Proceedings of the 9th Inter-*

national Conference on Logic Programming and Non-Monotonic Reasoning (LPNMR'07), Volume 3662 of Lecture Notes in Artificial Intelligence, pp. 18–30. Springer.

Balduccini, M. and M. Gelfond (2003, Mar). Logic Programs with Consistency-Restoring Rules. In P. Doherty, J. McCarthy, and M.-A. Williams (Eds.), *International Symposium on Logical Formalization of Commonsense Reasoning*, AAAI 2003 Spring Symposium Series, pp. 9–18.

Baral, C. (2003). *Knowledge representation, reasoning and declarative problem solving with answer sets*. Cambridge University Press.

Baral, C., M. Gelfond, and N. Rushton (2004, Jan). Probabilistic Reasoning with Answer Sets. In *Proceedings of LPNMR-7*.

Baral, C., M. Gelfond, and N. Rushton (2009). Probabilistic reasoning with answer sets. *Journal of Theory and Practice of Logic Programming (TPLP)* 9(1), 57–144.

Baral, C. and M. Hunsaker (2007). Using the probabilistic logic programming language p-log for causal and counterfactual reasoning and non-naive conditioning. In *Proceedings of IJCAI-2007*, pp. 243–249.

Gelfond, M. and V. Lifschitz (1988). The stable model semantics for logic programming. In *Proceedings of ICLP-88*, pp. 1070–1080.

Gelfond, M. and V. Lifschitz (1991). Classical negation in logic programs and disjunctive databases. *New Generation Computing* 9(3/4), 365–386.

Gelfond, M., N. Rushton, and W. Zhu (2006). Combining logical and probabilistic reasoning. AAAI 2006 Spring Symposium Series, pp. 50–55.

Hayes, P. J. and J. McCarthy (1969). Some Philosophical Problems from the Standpoint of Artificial Intelligence. In B. Meltzer and D. Michie (Eds.), *Machine Intelligence 4*, pp. 463–502. Edinburgh University Press.

McCarthy, J. (1999). Elaboration tolerance. In progress.

Pearl, J. (1988). *Probabistic reasoning in intelligent systems: networks of plausable inference*. Morgan Kaufmann.

Pearl, J. (2000). *Causality*. Cambridge University Press.

Pereira, L. M. and C. Ramli (2009). Modelling decision making with probabilistic causation. *Intelligent Decision Technologies (IDT)*. to appear.

Reiter, R. (1978). *On Closed World Data Bases*, pp. 119–140. Logic and Data Bases. Plenum Press.

# 21
# On Computers Diagnosing Computers

MOISES GOLDSZMIDT

## 1  Introduction

I came to UCLA in the fall of 1987 and immediately enrolled in the course titled "Probabilistic Reasoning in Intelligent Systems" where we, as a class, went over the draft of Judea's book of the same title [Pearl 1988]. The class meetings were fun and intense. Everybody came prepared, having read the draft of the appropriate chapter and having struggled through the list of homework exercises that were due that day. There was a high degree of discussion and participation, and I was very impressed by Judea's attentiveness and interest in our suggestions. He was fully engaged in these discussions and was ready to incorporate our comments and change the text accordingly. The following year, I was a teaching assistant (TA) for that class. The tasks involved with being a TA gave me a chance to rethink and really digest the contents of the course. It dawned on me then what a terrific insight Judea had to focus on formalizing the notion of conditional independence: All the "juice" he got in terms of making "reasoning under uncertainty" computationally effective came from that formalization. Shortly thereafter, I had a chance to chat with Judea about these and related thoughts. I was in need of formalizing a notion of "relevance" for my own research and thought that I could adapt some ideas from the graphoid models [Pearl 1988]. In that opportunity Judea shared another of his great insights with me. After hearing me out, Judea said one word: "causality". I don't remember the exact words he used to elaborate, but the gist of what he said to me was: "we as humans perform extraordinarily complex reasoning tasks, being able to select the relevant variables, circumscribe the appropriate context, and reduce the number of factors that we should manipulate. I believe that our intuitive notions of causality enable us to do so. Causality is the holly grail [for Artificial Intelligence]".

In this short note, I would like to pay tribute to Judea's scientific work by speculating on the very realistic possibility of computers using his formalization of causality for automatically performing a nontrivial reasoning task commonly reserved for humans. Namely designing, generating, and executing experiments in order to conduct a proper diagnosis and identify the causes of performance problems on code being executed in large clusters of computers. What follows in the next two sections is not a philosophical exposition on the meaning of "causality" or on the reasoning powers of automatons. It is rather a brief description of the current state of the art in programming large clusters of computers and then, a brief account argumenting that the conditions are ripe for embarking on this research path.

## 2 Programming large clusters of computers made easy

There has been a recent research surge in systems directed at providing programmers with the ability to write efficient parallel and distributed applications [Hadoop 2008; Dean and Ghemawat 2004; Isard et al. 2007]. Programs written in these environments are automatically parallelized and executed on large clusters of commodity machines. The tasks of enabling programmers to effectively write and deploy parallel and distributed application has of course been a long-standing problem. Yet, the relatively recent emergence of large-scale internet services, which depend on clusters of hundreds of thousands of general purpose servers, have given the area a forceful push. Indeed, this is not merely an academic exercise; code written in these environments has been deployed and is very much in everyday use at companies such as Google, Microsoft, and Yahoo (and many others). These programs process web pages in order to feed the appropriate data to the search and news summarization engines; render maps for route planning services; and update usage and other statistics from these services. Year old figures estimate that Dryad, the specific such environment created at Microsoft [Isard et al. 2007], is used to crunch on the order of a petabyte a day at Microsoft. In addition, in our lab at Microsoft Research, a cluster of 256 machines controlled by Dryad runs daily at a 100% utilization. This cluster mostly runs tests and experiments on research algorithms in machine learning, privacy, and security that process very large amounts of data.

The intended model in Dryad is for the programmer to build code as if she were programming one computer. The system then takes care of a) distributing the code to the actual cluster and b) managing the execution of the code in the cluster. All aspects of execution, including data partition, communications, and fault tolerance, are the responsibility of Dryad.

With these new capabilities comes the need for new tools for debugging code, profiling execution performance, and diagnosing system faults. By the mere fact that clusters of large numbers of computers are being employed, rare bugs will manifest themselves more often, and devices will fail in more runs (due to both software and hardware problems). In addition, as the code will be executed in a networked environment and the data will be partitioned (usually according to some hash function), communication bandwidth, data location, contention for shared disks, and data skewness will impact the performance of the programs. Most of the times the impact of these factors will be hard to reproduce in a single machine, making it an imperative that the diagnosis, profiling, and debugging be performed in the same environment and conditions as those in which the code is running.

## 3 Computers diagnosing computers

The good news is that the same infrastructure that enables the programming and control of these clusters can be used for debugging and diagnosis. Normally the computation proceeds in stages where the different nodes in the cluster perform the same computation in parallel on different portions of the data. For purposes of fault tolerance, there are mechanisms in Dryad to monitor the execution time of each node at any computation stage. It is therefore possible to gather robust statistics about the expected execution time of any particular node at a given stage and identify especially slow nodes. Currently, this information is used to restart those nodes or to migrate the computation to other nodes.

We can take this further and collect the copious amount of data that is generated by the various built-in monitors looking at things such as cpu utilization, memory utilization, garbage collection, disk utilization, and statistics on I/O.[1] The statistical analysis of these signals may provide clues pointing at the probable causes of poor performance and even of failures. Indeed we have built a system called Artemis [Creţu-Ciocârlie et al. 2008], that takes advantage of the Dryad infrastructure to collect and preprocess the data from these signals in a distributed and opportunistic fashion. Once the data is gathered, Artemis will run a set of statistical and machine learning algorithms ranging from summarizations to regression and pattern classification. In this paper we propose one more step. We can imagine a system that guided with the information from these analyses, performs active experiments on the execution of the code. The objective will be to causally diagnose problems, and properly profile dependencies between the various factors affecting the performance of the computations.

Let us ground this idea in a realistic example. Suppose that through the analysis of the execution logs of some large task we identify that, on a computationally intensive stage, a small number of machines performed significantly worse that the average/median (in terms of overall processing speed). Through further analysis, for example logistic regression with L1 regularization, we are able to identify the factors that differentiate the slower machines. Thus, we narrow down the possibilities and determine that the main difference between these machines and the machines that performed well is the speed at which the data is read by the slower machines.[2] Further factors influencing this speed are whether the data resides on a local disk and whether there are other computational nodes that share that disk (and introduce contention), and on the speed of the network. Figure 1 shows a (simplified) causal model of this scenario depicting two processing nodes. The dark nodes represent factors/variables that can be controlled or where intervention is possible. Conducting controlled experiments guided by this graph would enable the

---

[1] The number of counters and other signals that these monitors yield can easily reach on the order of hundreds per machine.

[2] This particular case was encountered by the author while running a benchmark based on Terasort on a cluster with hundreds of machines [Creţu-Ciocârlie et al. 2008].

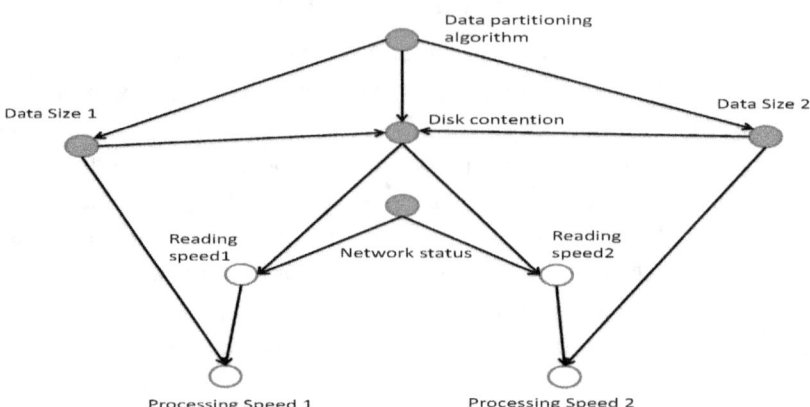

Figure 1. Simplified causal network depicting the processing speed scenario. Dark nodes represent the factor/variables than can be controlled or where intervention is possible.

precise characterization of the relationship between processing speed, data skewness, and disk contention, so that we can figure out how to partition and locate the data more efficiently and avoid having slow processing nodes. As the causal graph clearly exposes, controlling the data sizes for the computing nodes is not enough: if they reside on the same disk, contention may still cause slowdowns. This is obvious from the representation and algebra proposed by Judea in [Pearl 2000], as applied to this graph. This model also makes clear that intervening directly on the level of contention in the disk will indeed eliminate the dependency between the reading speed and the size of the data.

The idea of using graphical models for diagnosing computer systems goes back at least to [Breese and Heckerman 1996; Blake and Breese 1995]. It took close to 10 years after those papers for the first publication reporting the use of Bayesian networks for diagnosis in a nontrivial system in production to appear in a top tier systems conference [Cohen et al. 2004]. The methods in [Cohen et al. 2004] involve passive observation, and the authors make very clear that inferences concern correlation and not necessarily causation. However, hinting at root cause through correlation may not be enough in the very near future. Complexity and scale in current networked distributed systems keeps on increasing at a rapid pace. Because of service availability and reliability requirements, root cause analysis pointing at effective repair actions and accurate empirical characterization of dependencies between the different factors affecting computation are rapidly becoming a must.

Systems such as Dryad[Isard et al. 2007] enable the effective programming of large cluster of computers. In addition, they provide effective mechanisms for controlling the "variables" of interest and setting up experiments in these clusters. Systems such as Artemis [Creţu-Ciocârlie et al. 2008] enable efficient collection and processing of extensive monitoring data, including the recording of the system state for recreating particular troublesome scenarios. The final ingredient for having machines automatically set up and conduct experiments is a language to describe these experiments and an algebra to reason about them in order to guarantee that the right variables are being controlled, and that we are intervening in the right spots in order to get to the correct conclusions. Through his seminal work in [Pearl 2000] and follow up papers, Judea Pearl has already given us that ingredient.

**Acknowledgments:** The author wishes to thank Mihai Budiu for numerous technical discussions on the topics of this paper, Joe Halpern for his help with the presentation, and very especially Judea Pearl for his continuous inspiration in the relentless and honest search for scientific truth.

# References

Blake, R. and J. Breese (1995). Automatic bottleneck detection. In *UAI'95: Proceedings of the Conference on Uncertainty in Artificial Intelligence*.

Breese, J. and D. Heckerman (1996). Decision theoretic troubleshooting. In *UAI'96: Proceedings of the Conference on Uncertainty in Artificial Intelligence*.

Cohen, I., M. Goldszmidt, T. Kelly, J. Symons, and J. Chase (2004). Correlating instrumentation data to systems states: A building block for automated diagnosis and control. In *OSDI'04: Proceedings of the 6th conference on Symposium on Opearting Systems Design & Implementation*. USENIX Association.

Creţu-Ciocârlie, G. F., M. Budiu, and M. Goldszmidt (2008). Hunting for problems with Artemis. In *USENIX Workshop on the Analysis of System Logs (WASL)*.

Dean, J. and S. Ghemawat (2004). Mapreduce: simplified data processing on large clusters. In *OSDI'04: Proceedings of the 6th conference on Symposium on Opearting Systems Design & Implementation*. USENIX Association.

Hadoop (2008). The hadoop project. http://hadoop.apache.org.

Isard, M., M. Budiu, Y. Yu, A. Birrell, and D. Fetterly (2007). Dryad: distributed data-parallel programs from sequential building blocks. In *EuroSys '07: Proceedings of the 2nd ACM SIGOPS/EuroSys European Conference on Computer Systems 2007*. ACM.

Pearl, J. (1988). *Probabilistic Reasoning in Intelligent Systems: Networks of Plausible Inference*. Morgan Kaufmann.

Pearl, J. (2000). *Causality: Models, Reasoning, and Inference*. Cambridge Univ. Press.

# 22
# Overthrowing the Tyranny of Null Hypotheses Hidden in Causal Diagrams

SANDER GREENLAND

## 1 Introduction

Graphical models have a long history before and outside of causal modeling. Mathematical graph theory extends back to the 1700s and was used for circuit analysis in the 19$^{th}$ century. Its application in probability and computer science dates back at least to the 1960s (Biggs et al., 1986), and by the 1980s graphical models had become fully developed tools for these fields (e.g., Pearl, 1988; Hajek et al., 1992; Lauritzen, 1996).

As *Bayesian networks*, graphical models are carriers of direct conditional independence judgments, and thus represent a collection of assumptions that confine prior support to a lower dimensional manifold of the space of prior distributions over the nodes. Such dimensionality reduction was recognized as essential in formulating explicit and computable algorithms for digital-machine inference, an essential task of artificial-intelligence (AI) research. By the 1990s, these models had been merged with causal path diagrams long used in observational health and social science (OHSS) (Wright, 1934; Duncan, 1975), resulting in a formal theory of causal diagrams (Spirtes et al., 1993; Pearl, 1995, 2000).

It should be no surprise that some of the most valuable and profound contributions to these *developments* were from Judea Pearl, a renowned AI theorist. He motivated causal diagrams as *causal* Bayesian networks (Pearl, 2000), in which the basis for the dimensionality reduction is grounded in judgments of causal independence (and especially, autonomy) rather than mere probabilistic independence. Beyond his extensive technical and philosophical contributions, Pearl fought steadfastly to roll back prejudice against causal modeling and causal graphs in statistics. Today, only a few statisticians still regard causality as a metaphysical notion to be banned from formal modeling (Lad, 1999). While a larger minority still reject some aspects of causal-diagram or potential-outcome theory (e.g., Dawid, 2000, 2008; Shafer, 2002), the spreading wake of applications display the practical value of these theories, and formal causal diagrams have advanced into applied journals and books (e.g., Greenland et al., 1999; Cole and Hernán, 2002; Hernán et al., 2002; Jewell, 2004; Morgan and Winship, 2007; Glymour and Greenland, 2008) – although their rapid acceptance in OHSS may well have been facilitated by the longstanding informal use of path diagrams to represent qualities of causal systems (e.g., Susser, 1973; Duncan, 1975).

Graphs are unsurpassed tools for illustrating certain mathematical results that hold in functional systems (whether stochastic or not, or causal or not). Nonetheless, it is essential to recognize that many if not most causal judgments in OHSS are based on observational (purely associational) data, with little or nothing in the way of manipulative (or "surgical") experiment to test these judgments. Time order is usually known, which insures that the chosen arrow directions are correct; but rarely is there a sound basis for deleting an arrow, leaving autonomy in question. When all empirical constraints encoded by the causal network come from passive frequency observations rather than experiments, the primacy of causal independence judgments has to be questioned. In these situations (which characterize observational research), we should not neglect associational models (including graphs) that encode frequency-based judgments, for these models may be all that are identified by available data. Indeed, a deep philosophical commitment to statistically identified quantities seems to drive the arguments of certain critics of potential outcomes and causal diagrams (Dawid, 2000, 2008). Even if we reject this philosophy, however, we should retain the distinction between levels of identification provided by our data, for even experimental data will not identify everything we would like to know.

I will argue that, in some ways, the distinction of nonidentification from identification is as fundamental to modeling and statistical inference about causal effects as is the distinction of causation from association (Gustafson, 2005; Greenland, 2005a, 2009a, 2009b). Indeed, I believe that some of the controversy and confusion over causation versus association stems from the inability of statistical observations to point identify (consistently estimate) many of the causal parameters that astute scientists legitimately ask about. Furthermore, if we consider strategies that force identification from available data (such as node or arrow deletions from graphical models) we will find that identification may arise only by declaring some types of joint frequencies as justifying the corresponding conditional independence assumptions. This leads directly into the complex topic of pruning algorithms, including the choice of target or loss function.

I will outline these problems in their most basic forms, for I think that in the rush to adopt causal diagrams some realism has been lost by neglecting problems of nonidentification and pruning. My exposition will take the form of a series of vignettes that illustrate some basic points of concern. I will not address equally important concerns that many of the nodes offered as "treatments" may be ill-defined or nonmanipulable, or may correspond poorly to the treatments they ostensibly represent (Greenland, 2005b; Hernán, 2005; Cole and Frangakis, 2009; VanderWeele, 2009).

## 2 Nonidentification from Unfaithfulness in a Randomized Trial

Nonidentification can be seen and has caused controversy in the simplest causal-inference settings. Consider an experiment that randomizes a node R. Inferences on causal effects of R from subsequent associations of R with later events would then be justified, since R would be an exogenous node. R would also be an instrumental variable for certain descendants under further conditional-independence assumptions.

A key problem is how one could justify removing arrows along the line of descent from R to another node Y, even if R is exogenous. The overwhelmingly dominant approach licenses such removal if the observed R-Y association fails to meet some criterion for departure from pure randomness. This schematic for a causal-graph pruning algorithm was employed by Spirtes et al. (1993), unfortunately with a very naïve Neyman-Pearsonian criterion (basically, allowing removal of arrows when a *P*-value exceeds an $\alpha$ level). These and related graphical algorithms (Pearl and Verma, 1991) produce what appear to be results in conflict with practical intuitions, namely causal "discovery" algorithms for single observational data sets, with no need for experimental evidence. These algorithms have been criticized philosophically on grounds related to the identification problem (Freedman and Humphreys, 1999; Robins and Wasserman, 1999ab), and there are also objections based on statistical theory (Robins et al., 2003).

One controversial assumption in these algorithms is *faithfulness* (or stability) that all connected nodes are associated. Although arguments have been put forward in its favor (e.g., Spirtes et al., 1993; Pearl, 2000, p. 63), this assumption coheres poorly with prior beliefs of some experienced researchers. Without faithfulness, two nodes may be independent even if there is an arrow linking them directly, if that arrow represents the presence of causal effects among units in a target population. A classic example of such unfaithfulness appeared in the debates between Fisher and Neyman in the 1930s, in which they disagreed on how to formulate the causal null hypothesis (Senn, 2004). The framework of their debate would be recognized today as the *potential-outcome* or counterfactual model, although in that era the model (when named) was called the randomization model. This model illustrates the benefit of randomization as a means of detecting a signal by injecting white noise into a system to drown out uncontrolled influences.

To describe the model, suppose we are to study the effect of a treatment X on an outcome $Y_{obs}$ observable on units in a specific target population. Suppose further we can fully randomize X, so X will equal the randomized node R. In the potential-outcome formulation, the outcome becomes a vector **Y** indexed by X. Specifically, X determines which component $Y_x$ of **Y** is observable conditional on X=x: $Y_{obs} = Y_x$ given X=x. To say X can causally affect a unit makes no reference to observation, however; it merely means that some components of **Y** are unequal. With a binary treatment and outcome, there are four types of units in the target population about a binary treatment X which indexes a binary potential-outcome vector **Y** (Copas, 1973):

1) Noncausal units with outcomes **Y**=(1,1) under X=1,0 ("doomed' to $Y_{obs}$=1);
2) Causal units with outcomes **Y**=(1,0) under X=1,0 (X=1 causes $Y_{obs}$=1);
3) Causal units with outcomes **Y**=(0,1) under X=1,0 (X=1 prevents $Y_{obs}$=1); and
4) Noncausal units with outcomes **Y**=(0,0) under X=1,0 ("immune" to $Y_{obs}$=1).

Suppose the proportion of type *i* in the trial population is $p_i$. There are now two null hypotheses:

$H_s$: There are no causal units: $p_2=p_3=0$ (sharp or strong null),
$H_w$: There is no net effect of treatment on the distribution of $Y_{obs}$: $p_2=p_3$ (weak null).

Under the randomization distribution we have

$E(Y_{obs}|X=1) = Pr(Y_{obs}=1|do[X=1]) = Pr(Y_1=1) = p_1+p_2$ and
$E(Y_{obs}|X=0) = Pr(Y_{obs}=1|do[X=0]) = Pr(Y_0=1) = p_1+p_3$;

hence $H_w$: $p_2=p_3$ is equivalent to the hypothesis that the expected outcome is the same for both treatment groups, and that the proportions with $Y_{obs}$=1 under the extreme population intervention do[X=1] to every unit and do[X=0] to every unit are equal. Note however that only $H_s$ entails that the proportion with $Y_{obs}=1$ would be the same under *every* possible allocation of treatment X among the units; this property implies that the Y margin is fixed under $H_s$, and thus provides a direct causal rationale for Fisher's exact test of $H_s$ (Greenland, 1991).

$H_s$ also entails $H_w$ (or, in terms of parameter subspaces, $H_s \subset H_w$). The converse is false; but, under any of the "optimal" statistical tests that can be formulated from data on X and $Y_{obs}$ only, power is identical to the test size on all alternatives to the sharp null with $p_2=p_3$, i.e., $H_s$ is not identifiable within $H_w$, so within $H_w$ the power of any valid test of $H_s$ will not exceed its nominal alpha level. Thus, following Neyman, it is only relevant to think in terms of $H_w$, because $H_w$ could be rejected whenever $H_s$ could be rejected. Furthermore, some later authors would disallow $H_w - H_s$: $p_2 = p_3 \neq 0$ because it violates faithfulness (Spirtes et al., 2001) or because it represents an extreme treatment-by-unit interaction with no main effect (Senn, 2004).

There is also a Bayesian argument for focusing exclusively on $H_w$. $H_w$ is of Lebesgue measure zero, so under the randomization model, distinctions within $H_w$ can be ignored by inferences based on an absolutely continuous prior on $\mathbf{p} = (p_1, p_2, p_3)$ (Spirtes et al., 1993). More generally, any distinction that remains *a posteriori* can be traced to the prior. A more radical stance would dismiss both $H_s$ and the model defined by 1-4 above as "metaphysical," because it invokes constraints on the joint distribution of the components $Y_1$ and $Y_0$, and that joint distribution is not identified by randomization of X if only X and $Y_{obs}$ are observed (Dawid, 2000).

On the other hand, following Fisher one can argue that the null of key scientific and practical interest is $H_s$, and that $H_w - H_s$: $p_2 = p_3 \neq 0$ is a scientifically important and distinct hypothesis. For instance, $p_2>0$, $p_3>0$ entails the existence of units who should be treated quite differently, and provides an imperative to seek covariates that discriminate between the two causal types, even if $p_2=p_3$. Furthermore, rejection of the stronger $H_s$ is a *weaker* inference than rejection of the weaker $H_w$, and thus rejecting only $H_s$ would be a conservative interpretation of a "significant" test statistic. Thus, focusing on $H_s$ is compatible with a strictly falsificationist view of testing in which acceptance of the null is disallowed. Finally, there are real examples in which X=1 causes Y=1 in some units and causes Y=0 in others; in some of these cases there may be near-perfect balance of causation and prevention, as predicted by certain physical explanations for the observations (e.g., as in Neutra et al., 1980).

To summarize, identification problems arose in the earliest days of formal causal modeling, even when considering only the simplest of trials. Those problems pivoted not on whether one should attempt formal modeling of causation as distinct from association, but rather on what could be identified by standard experimental designs. In the face of limited (and limiting) design strategies, these problems initiated a long history of attempts to banish identification problems based on idealized inference systems and absolute philosophical assertions. But a counter-tradition of arguments, both practical and philosophical, has regarded identification problems as carriers of valuable scientific information: They are signs of study limitations which need to be recognized and can only be dealt with effectively by innovative data collection (e.g., measuring more covariates or deploying new study designs), instead of by increasing sample sizes and defining the problems away so that "identical replications" are sufficient to narrow inferences.

## 3 Causal Diagrams Encode Numerous Uncertain Null Hypotheses

To move to the observational setting that is my main concern, consider figure 1, a typical causal diagram used to illustrate assumptions used by methods for estimating "the effect of X on Y" from observational data.

Figure 1: Naïve causal diagram

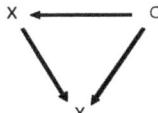

The first point to note is that this diagram is woefully incomplete relative to the epidemiologic reality, because it ignores
   a) unmodeled confounders (variables not in the graph that affect more than one node in the graph),
   b) selection effects (effects of factors in the graph on selection), and
   c) measurement errors (which require addition of measurement nodes for each imperfectly measured node).

Put another way, typical causal DAGs like that in figure 1 are full of hidden, assumed null hypotheses, in the form of assumptions that imply problems a, b, and c are absent. For example, a causal DAG assumes that for **every** node pair (A,B) in the DAG,
   1) there is **no** shared ancestor not in graph (not A↔B),
   2) there is **no** unmarked conditioning event that has opened a path between A and B (not A—B),
   3) if A and B are nonadjacent (neither A→B nor A←B), there is **no** mechanism that leads directly from one node to another (thus bypassing other nodes in the graph).

Not every study will seriously violate all of these assumptions. But in most studies in OHSS, none of the nulls 1-3 will have convincing support, and any purported test of a causal effect will really be a test of these 3 nulls as well as the specified causal null. This fact is just a special case of longstanding observations that statistical tests are really tests of all assumptions used in the test, not just the particular null of interest (Fisher, 1943; Box, 1980). In this regard, note that absence of arrows between nodes (3) encodes particularly strong nulls that are routinely presumed but rarely have supporting data. More often in OHSS, we observe only a conditional temporal sequence such as "A precedes B," which may be due to A→B, A↔B, A—B or some combination.

While sensitivity analysis is often recommended to examine the impact of deviations from assumptions, it becomes unintelligible if not infeasible as the number of assumptions (or corresponding parameters) increase. Then too, some causal inferences will display unlimited sensitivity to certain assumptions, requiring the introduction of priors on the corresponding parameters in order to salvage any inference (Greenland, 1998, 2005a; Gustafson, 2005). This problem arises in the model given below.

## 4 Eliminating Unsupported Nulls (graphical realism)

Let conditioning be denoted with square brackets around the conditioned event node. Then, in contrast to Figure 1, realistic causal graphs for OHSS will have

1) numerous unobserved (latent) nodes, often more of them than observed nodes,
2) few node pairs without an arc between them,
3) no **observed** set of variables sufficient for bias control, and
4) a selection node S that is bracketed and potentially affected by most other nodes.

In particular, when all variables are subject to measurement error, a realistic causal model for a single exposure-disease analysis will have at least:

X = Exposure, X*: measured X
Y = Outcome, Y*: measured Y
C = Known antecedents, C*: measured C
U = Other antecedents (unmeasured and possibly unknown)
S = Selection into the analysis (from selection into the study plus exclusions).

Because analysis is always conditioned on S=1, we should always show this conditioning event on the graph with a circle or brackets around it, e.g., as [S=1].

As an example, fig. 2 shows what I'd consider a **minimal** realistic causal graph for a typical case-control study of a life history and a degenerative disease outcome (e.g, nicotine intake X and Alzheimer's disease Y), which has 25 of 28 possible adjacencies.

### Figure 2: Realistic causal diagram

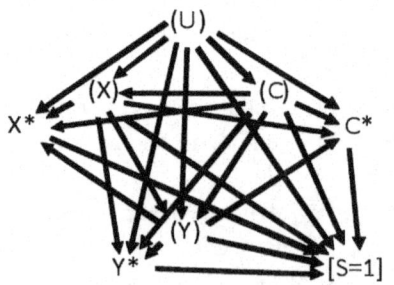

What can fig. 2 provide if further assumptions cannot be justified? The only observed distribution is p(c*,x*,y*|S=1), which is not a factor in the causal Markov decomposition entailed by the graph,

$p(u,c,x,y,c^*,x^*,y^*,s) =$
$p(u)p(c|u)p(x|u,c)p(y|u,c,x)p(c^*|u,c,x,y)p(x^*|u,c,x,y)p(y^*|u,c,x,y)p(s|u,c,x,y,c^*,x^*,y^*)$,
which involves both S=0 events (not selected) and S=1 events (selected), i.e., the lowercase "s" is used when S can be either 0 or 1.

The marginal (total-population) potential-outcome distribution for Y after intervention on X, $p(y_x)$, equals $p(y|do[X=x])$, which under fig. 2 equals the standardized (mixing) distribution of Y given X standardized to (weighted by or mixed over) $p(c,u) = p(c|u)p(u)$:

$p(y_x) = p(y|do[x]) = \sum_{u,c} p(y|u,c,x)p(c|u)p(u)$.

This estimand involves only three factors in the decomposition, but none of them are identified if U is unobserved and no further assumptions are made. Analysis of the causal estimand $p(y_x)$ must somehow relate it to the observed distribution $p(c^*,x^*,y^*|S=1)$ using known or estimable quantities, or else remain purely speculative (i.e., a sensitivity analysis).

It is a long, hard road from $p(c^*,x^*,y^*|S=1)$ to $p(y_x)$, much longer than the current "causal inference" literature often makes it look. To appreciate the distance, rewrite the summand of the standardization formula for $p(y_x)$ as an inverse-probability-weighted (IPW) term derived from an observation $(c^*,x^*,y^*|S=1)$: From fig. 2,

$p(y|u,c,x)p(c|u)p(u) =$
$p(c^*,x^*,y^*|S=1)p(S=1)p(u,c,x,y|c^*,x^*,y^*,S=1)/$
$p(x|u,c)p(c^*|u,c,x,y)p(x^*|u,c,x,y)p(y^*|u,c,x,y)p(S=1|u,c,x,y,c^*,x^*,y^*)$.

The latter expression includes
1) the exposure dependence on its parents, $p(x|u,c)$;
2) the measurement distributions $p(c^*|u,c,x,y)$, $p(x^*|u,c,x,y)$, $p(y^*|u,c,x,y)$; and
3) the fully conditioned selection probability $p(S=1|u,c,x,y,c^*,x^*,y^*)$.

The absence of effects corresponding to 1–3 from graphs offered as "causal" suggests that "causal inference" from observational data using formal causal models remains a theoretical and largely speculative exercise (albeit often presented without explicit acknowledgement of that fact).

When adjustments for these effects are attempted, we are usually forced to use crude empirical counterparts of terms like those in 1–3, with each substitution demanding nonidentified assumptions. Consider that, for valid inference under figure 2,

1) Propensity scoring and IPW for treatment need $p(x|u,c)$, but all we get from data is $p(x^*|c^*)$. Absence of u and c is usually glossed over by assuming "no unmeasured confounders" or "no residual confounding." These are not credible assumptions in OHSS.

2) IPW for selection and censoring needs $p(S=1|u,c,x,y,c^*,x^*,y^*)$, but usually the most we get from a cohort study or nested study is $p(S=1|c^*,x^*)$. We do not even get that much in a case-control study.

3) Measurement-error correction needs conditional distributions from $p(c^*,x^*,y^*,u,c,x,y|S=1)$, but even when a "validation" study is done, we obtain only alternative measurements $c^\dagger,x^\dagger,y^\dagger$ (which are rarely error-free) on a tiny and

biased subset. So we end up with observations from $p(c^\dagger,x^\dagger,y^\dagger,c^*,x^*,y^*|S=1,V=1)$ where V is the validation indicator.

4) Consistency between the observed X and the intervention variable, in the sense that $P(Y|X=x) = P(Y|do[X=x],X=x)$. This can be hard to believe for common variables such as smoking, body-mass index, and blood pressure, even if $do[X=x]$ is well-defined (which is not usually the case).

In the face of these realities, standard practice seems to be: Present wildly hypothetical analyses that pretend the observed distribution $p(c^*,x^*,y^*|S=1)$, perhaps along with $p(c^\dagger,x^\dagger,y^\dagger,c^*,x^*,y^*|S=1,V=1)$ or $p(S=1|c^*,x^*)$, is sufficient for causal inference. The massive gaps are filled in with models or assumptions, which are priors that reduce dimensionality of the problem to something within computing bounds. For example, use of IPW with $p(S=1|c^*,x^*)$ to adjust for selection bias (as when 1−S is a censoring indicator) depends crucially on a nonidentified ignorability assumption that $S \perp (U,C,X,Y)|(C^*,X^*)$, i.e., that selection S is independent of the latent variables U,C,X,Y given the observed variables $C^*,X^*$. We should expect this condition to be violated whenever a latent variable affects selection directly or shares unobserved causes with selection. If such effects are exist but are missing from the analysis graph, then by some definitions the graph (and hence the resulting analysis) isn't causal, no matter how much propensity scoring (PS), marginal structural modeling (MSM), inverse-probability weighting (IPW), or other causal-modeling procedures we apply to the observations $(c^*,x^*,y^*|S=1)$.

Of course, the overwhelming dimensionality of typical OHSS problems virtually guarantees that arbitrary constraints will enter at some point, and forces even the best scientists to rely on a tiny subset of all the models or explanations consistent with available facts. Personal bias in determining this subset may be unavoidable due to strong cultural influences (such as adherence to received theories, as well as moral strictures and financial incentives), which can also lead to biased censoring of observations (Greenland, 2009c). One means of coping with such bias is by being aware of it, then trying to test it against the facts one can muster (which are often few).

The remaining sections sketch some alternatives to pretending we can identify unbiased or assuredly valid estimators of causal effects in observational data, as opposed to within hypothetical models for data generation (Greenland, 1990; Robins, 2001). In these approaches, both frequentist and Bayesian analyses are viewed as hypotheticals conditioned on a data-generation model of unknown validity. Frequentist analysis provides only inferences of the form "if the data-generation process behaves like this, here is how the proposed decision rule would perform," while Bayesian analysis provides only inferences of the form "if I knew that its data-generation process behaves like this, here is how this study would alter my bets."[1] If we aren't sure how the data-generation

---

[1]This statement describes Bayes factors (Good, 1983) conditioned on the model. That model may include an unknown parameter that indexes a finite number of submodels scattered over some high-dimensional subspace, in which case the Bayesian analysis is called "model averaging," usually with an implicit uniform prior over the models. Model averaging may also operate over continuous parameters via priors on those parameters.

process behaves, no statistical analysis can provide more, no matter how much causal modeling is done.

## 5  Predictive Analysis

If current models for observed-data generators (whether logistic, structural, or propensity models) can't be taken seriously as "causal", what can we make of their outputs? It is hard to believe the usual excuses offered for regression outputs (e.g., that they are "descriptive") when the fitted model is asserted to be causal or "structural." Are we to consider the outputs of (say) and IPW-fitted MSM to be some sort of data summary? Or will it function as some kind of optimal predictor of outcomes in a purely predictive context? No serious case has been made for causal models in either role, and it seems that some important technical improvements are needed before causal modeling methods become credible predictive tools.

Nonetheless, graphical models remain useful (and might be less misleading) even when they are not "causal," serving instead as mere carriers of conditional independence assumptions within a time-ordered framework. In this usage, one may still employ presumed causal independencies as prior judgments for specification. In particular, for predictive purposes, some or all of the arrows in the graph may retain informal causal interpretations; but they may be causally wrong, and yet the graph can still be correct for predictive purposes.

In this regard, most of the graphical modeling literature in statistics imposes little in the way of causal burden on the graph, as when graphs are used as influence diagrams, belief and information networks, and so on without formal causal interpretation (that is, without representing a formal causal model, e.g., Pearl, 1988; Hajek et al., 1992; Cox and Wermuth, 1996; Lauritzen, 1996). DAG rules remain valid for prediction if the absence of an open path from X to Y is interpreted as entailing $X \perp Y$, or equivalently if the absence of a directed path from X to Y (in causal terms, X is not a cause of Y; equivalently, Y is not affected by X) is interpreted as entailing $X \perp Y | \mathbf{pa}_X$, the noncausal Markov condition (where $\mathbf{pa}_X$ is the set of parents of X). In that case, X→Y can be used in the graph even if X has no effect on Y, or vice-versa.

As an example, suppose X and Y are never observed without them affecting selection S, as when X is affects miscarriage S and Y is congenital malformation. If the target population is births, X predicts malformations Y among births (which have S=1). As another example, suppose X and Y are never observed without an uncontrolled, ungraphed confounder U, as when X is diet and Y is health status. If one wishes to target those at high risk for screening or actuarial purposes it does not matter if X→Y represents a causally confounded relation. Lack of a directed path from X to Y now corresponds to lack of additional predictive value for Y from X given $\mathbf{pa}_X$. Arrow directions in temporal (time-ordered) predictive graphs correspond to point priors about time order, just as they do in causal graphs.

Of course, if misinterpreted as causal, predictive inferences from graphs (or any predictive modeling) may be potentially disastrous for judging interventions on X. But, in OHSS, the causality represented by a directed path in a so-called causal diagram rarely corresponds to more than a hypothesis, plausible perhaps but only one among a myriad of others. If most arrows shown in a graph encode no real data other than an observed conditional temporal sequencing, then labeling the graph as a "causal diagram" sets the stage for the disaster.

Figure 3 is the temporal predictive diagram for the observables in the earlier example, assuming those events occur in the order C*, X*, Y*, [S=1].

### Figure 3: Temporally predictive diagram

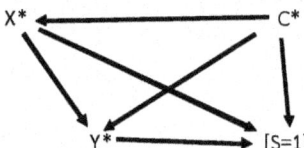

Comparison to the causal diagram in figure 2 illustrates how a temporal predictive diagram for an observable frequency distribution may be derived from an underlying causal diagram for a nonidentified theory. Figure 3 is saturated in the sense that all nodes are connected by an edge, but this need not be so for a predictive diagram derived from a causal one. If there is temporal ambiguity among the observables, there may be multiple predictive diagrams compatible with the causal diagram (which will form a subset of the multiple probability graphs compatible with the causal diagram).

If we treat causal models as carriers of prior information about conditional independencies, they appear as legitimate candidates to consider as predictive models. For example, MSMs can be evaluated as devices for prediction from fixed sequences and structural nested models can be evaluated as devices for prediction from stochastic processes. I would thus offer this challenge to the current "longitudinal causal modeling" literature: If we know our observations are just a dim and distant projection of the causal structure and we can only identify predictive links among observed quantities, are there predictive advantages of structural modeling (modeling potential outcomes as well as observed outcomes)? If not, what precisely is the advantage of fitting such models (compared to noncausal models) when effects are not identified?

I believe there *are* advantages of causal models, precisely as described by Pearl (2000): They provide an encoding for qualitative (structural) prior information expressed in terms of "cause" and "effect." But in current practice, fitting methods for complex causal models are quite primitive, and need to incorporate properly smoothness and other information that can be freely assumed in purely predictive-modeling approaches. This is a general problem of semi-parametric theory: It necessarily focuses sharp constraints in some dimensions and none in "most" dimensions (represented by the infinite-dimensional time component in standard Cox models). When relevant dimensions for constraint (those where much background information is available) are not well represented by the dimensions constrained by the model, considerably efficiency can be lost for estimating parameters of interest. A striking example given by Whittemore and Keller (1986) displayed the poor small-sample performance for estimating a survival curve when using an unsmoothed nonparametric hazard estimator (Kaplan-Meier or Nelson-Altschuler estimation), relative to spline smoothing of the hazard.

## 6 Pruning the Identified Portion of the Model

Over recent decades, great strides have been made in creating predictive algorithms; the question remains however, what role should these algorithms play in causal inference? It would seem that these algorithms can be beneficially applied to fitting the marginal distribution identified by the observations. Nonetheless, the targets of causal inference in observational studies lie beyond the identified margin, and thus beyond the reach of these algorithms. At best, then, the algorithms can provide the identified foundation for building into unobserved dimensions of the phenomena under study.

Even if we focus only on the identified margin, however, there may be far more nodes and edges than seem practical to allow in the final model. A prominent feature of modern predictive algorithms is that they start with an impractically large number of terms and then aggressively prune the model, and may re-grow and re-prune repeatedly (Hastie et al., 2009). These strategies coincide with the intuition that omitting a term is justified when its contribution is too small to stand out against bias and background noise; e.g., we do not include variables like patient identification number because we know that are usually pure noise.

Nonetheless, automated algorithms often delete variables or connections that prior information instead suggests are relevant or related; thus shields from pruning are often warranted. Furthermore, a deleted node or arrow may indeed be important from a contextual perspective even if does not meet algorithmic retention criteria. Thus, model simplification strategies such as pruning may be justified by a need for dimensionality reduction, but should be recognized as part of algorithmic compression or computational prediction, not as a mode of inference about structural models.

Apart from these vague cautions, it has long been recognized that if our goal is to evaluate causal effects, different loss functions are needed from those in the pruning algorithms commonly applied by researchers. Specifically, the loss or benefit entailed by pruning needs to be evaluated in reference to the target effect under study, and not simply successful prediction of identified quantities. Operationalizing this imperative requires building out into the nonidentified (latent) realm of the target effects, which is the focus of *bias modeling*.

## 7 Modeling Latent Causal Structures (Bias Modeling)

The target effects in causal inference are functions of unobserved dimensions of the data-generating process, which consist primarily of bias sources (Greenland, 2005a). Once we recognize the nonidentification this structure entails, the major analysis task shifts away from mathematical statistics to prior specification, because with nonidentification only proper priors on nonidentified parameters can lead to proper posteriors.

Even the simplest point-exposure case can involve complexities that transform simple and precise-looking conventional results into complex and utterly ambiguous posteriors (Greenland, 2009a, 2009b). In a model complex enough to reflect Figure 2, there are far too many elements of specification to contextually justify them all in detail. For example, one could only rarely justify fewer than two free structural parameters per arrow, and the distributional form for each parameter prior would call for at least two hyperparameters per parameter (e.g., a mean and a variance), leading to at least 50 parameters and 100 hyperparameters in a graph with 25 arrows. Allowing but one prior association parameter (e.g., a correlation) per parameter pair adds over 1,000 (50 choose 2) more hyperparameters.

As a consequence of the exponential complexity of realistic models, prior specification is difficult, ugly, ad hoc, highly subjective, and tentative in the extreme. In addition, the hard-won model will lack generalizability and elegance, making it distasteful to both the applied scientist and the theoretical statistician. Nor will it please the applied statistician concerned with "data analysis," for the analysis will instead revolve around numerous contextual judgments that enlist diverse external sources of information. In contrast to the experimental setting (in which the data-generation model may be dictated entirely by the design), the usually sharp distinction between prior and data information will be blurred by the dependence of the data-generation model on external information.

These facts raise another challenge to the current "causal modeling" literature: If we know our observations are just a dim and distant projection of the causal structure and we can only identify predictive links among observed quantities, how can we incorporate simultaneously all error sources (systematic as well as random) known to be important into a complex longitudinal framework involving mismeasurement of entire sequences of exposures and confounders? Some progress on this front has been made, but primarily in contexts with validation data available (Cole et al., 2010), which is not the usual case.

## 8 The Descriptive Alternative

In the face of the extraordinary complexity of realistic models for OHSS, it should be an option of each study to focus on describing the study and its data thoroughly, sparing us attempts at inference about nonidentified quantities such as "causal effects." This option will likely never be popular, but should be allowed and even encouraged (Greenland et al., 2004). After all, why should I care about your causal inferences, especially if they are based on or grossly over-weighted by the one or few studies that you happened to be involved in? If I am interested in forming my own inferences, I do want to see your data and get an accurate narrative of the physical processes that produced them. In this regard, statistics may supply data summaries. Nonetheless, it must be made clear exactly how the statistics offered reflect the data as opposed to some hypothesis about the population from which they came; $P$-values do not satisfy this requirement (Greenland, 1993; Poole, 2001).

Here then is a final challenge to the "causal modeling" literature: If we know our observations are just a dim and distant projection of the causal structure and we can only identify associations among observed quantities, how can we interpret the outputs of "structural modeling" (such as confidence limits for ostensibly causal estimands which are not in fact identified) as data summaries? We should want to see answers that are sensible when the targets are effects in a context at least as complex as in fig. 2.

## 9 What is a Causal Diagram?

The above considerations call into question some epidemiological accounts of causal diagrams. Pearl (2000) describes a causal model $M$ as a formal functional system giving relations among a set of variables. $M$ defines a joint probability distribution p() and an intervention operator do[] on the variables. A causal diagram is then a directed graph $G$ that implies the usual Markov decomposition for p() and displays additional properties relating p() and do[]. In particular, each child-parent family {X, $\mathbf{pa}_X$} in $G$ satisfies

1) p(x|do[$\mathbf{pa}_X$=a]) = p(x|$\mathbf{pa}_X$=a), and
2) if Z is not in {X, $\mathbf{pa}_X$}, p(x|do[Z=z],$\mathbf{pa}_X$=a) = p(x|$\mathbf{pa}_X$=a).

(e.g., see Pearl, 2000, p. 24). These properties stake out $G$ as an illustration (mapping) of structure within $M$.

Condition 1 is often described as stating that the association of each node X with its parent vector $\mathbf{pa}_X$ is unconfounded given $M$. Condition 2 says that, given $M$, the only variables in $G$ that affect a node X are its parents, and is often called the causal Markov condition (CMC). Nonetheless, as seems to happen often as time passes and methods become widely adopted, details have gotten lost. In the more applied literature, causal diagrams have come to be described as "unconfounded graphs" without reference to an underlying causal model (e.g., Hernán et al., 2004; VanderWeele and Robins, 2007; Glymour and Greenland, 2008). This description not only misses the CMC (2) but, taken literally, means that all shared causes are in the graph.

Condition 1 is a property relating two mathematical objects, $G$ and $M$. To claim a diagram is unconfounded is to instead make a claim about the relation of $G$ the real world, thus inviting confusion between a *model* for causal processes and the actual processes. For many experts in OHSS, the claim of unconfoundedness has zero probability of being correct because of its highly restrictive empirical content (e.g., see Robins and Wasserman, 1999ab). At best, we can only hope that the diagram provides a useful computing aid for predicting the outcomes of intervention strategies.

As with regression models, causal models in OHSS are always false. Because we can never know we have a correct model (and in fact in OHSS we can't even know if we are very close), to say $G$ is causal if unconfounded is a scientifically vacuous definition: It is saying the graph is causal if the causal model it represents is correct. This is akin to saying a monotone increasing function from the range of X to [0,1] is not a probability distribution if it is not in fact how X is distributed; thus a normal($\mu,\sigma^2$) cumulative function wouldn't be a probability distribution unless it is *the* actual probability distribution for X (whether that distribution is an objective event generator or a subjective betting schedule).

So, to repeat: To describe a causal diagram as an "unconfounded graph" blurs the distinction between models and reality. Model-based deductions are logical conditionals of the form "model $M$ deductively yields these conclusions," and have complete certainty *given* the model $M$. But the model, and hence reality, is never known with certainty, and in OHSS cannot be claimed as known except in the most crude fashion. The point is brought home above by appreciating just how unrealistic all causal models and diagrams in OHSS must be. Thus I would encourage the description of causal diagrams as graphical causal models (or more precisely, graphical representations of certain equivalence classes of causal models), rather than as "unconfounded graphs" (or similar phrases). This usage might even be acceptable to some critics of the current causal-modeling literature (Dawid, 2008).

## 10 Summary and Conclusions

I would be among the last to deny the utility of causal diagrams; but I argue that their practical utility in OHSS is limited to (i) compact and visually immediate representation of assumptions, and (ii) illustration of sources of nonidentification and bias given realistic assumptions. Converse claims about their utility for identification seem only the latest in a long line of promises to "solve" the problem of causal inference. These promises are akin to claims of preventing and curing all cancers; while progress is possible, the enormous complexity of real systems should leave us skeptical about claims of "solutions" to the real problem.

Many authors have recognized that the problem of effect identification is unsolvable in principle. Although this logical impossibility led some to deny the scientific merit of causal thinking, it has not prevented development of useful tools that have causal-modeling components. Nonetheless, the most precision we can realistically hope for estimating effects in OHSS is about one-digit accuracy, and in many problems even that seems too optimistic. Thus some practical sense is needed to determine what is and isn't important to include as model components. Yet, despite the crudeness of OHSS, good sense seems to lead almost inevitably to including more components than can be identified by available data.

My main point is that effect identification (in the frequentist sense of identification by the observed data) should be abandoned as a primary goal in causal modeling in OHSS. My reasons are practical: Identification will often demand dropping too much of importance from the model, thus imposing null hypotheses that have no justification in either past frequency observations or in priors about mechanisms generating the observations, thus leading to overconfident and biased inferences. In particular, defining a graph as "causal" if it is unconfounded assumes a possibly large set of causal null hypotheses (at least two for every pair of nodes in the graph: no shared causes or conditioned descendants not in the graph). In OHSS, the only graphs that satisfy such a definition will need many latent nodes to be "causal" in this sense, and as a consequence will reveal the nonidentified nature of target effects. Inference may then proceed by imposing contextually defensible priors or penalties (Greenland, 2005a, 2009a, 2009b, 2010).

Despite my view and similar ones (e.g., Gustafson, 2005), I suspect the bulk of causal-inference statistics will trundle on relying exclusively on artificially identified models. It will thus be particularly important to remember that just because a method is labeled a "causal modeling" method does not mean it gives us estimates and tests of actual causal effects. For those who find identification too hard to abandon in formal analysis, the only honest recourse is to separate identified and nonidentified components of the model, focus technique on the identified portion, and leave the latent residual as a topic for sensitivity analysis, speculative modeling, and further study. In this task, graphs can be used without the burden of causality if we allow them a role as pure prediction tools, and they can also be used as causal diagrams of the largely latent structure that generates the data.

**Acknowledgments:** I am most grateful to Tyler VanderWeele, Jay Kaufman, and Onyebuchi Arah for their extensive and useful comments on this chapter.

# References

Biggs, N., Lloyd, E. and Wilson, R. (1986). *Graph Theory, 1736-1936*. Oxford University Press.

Box, G.E.P. (1980). Sampling and Bayes inference in scientific modeling and robustness. *Journal of the Royal Statistical Society, Series A* **143**, 383–430.

Cole S.R. and M.A. Hernán (2002). Fallibility in estimating direct effects. *International Journal of Epidemiology* **31**, 163–165.

Cole, S.R. and C.E. Frangakis (2009). The consistency assumption in causal inference: a definition or an assumption? *Epidemiology* **20**, 3–5.

Cole, S.R., L.P. Jacobson, P.C. Tien, L. Kingsley, J.S. Chmiel and K. Anastos (2010). Using marginal structural measurement-error models to estimate the long-term effect of antiretroviral therapy on incident AIDS or death. *American Journal of Epidemiology* **171**, 113-122.

Copas, J.G. (1973). Randomization models for matched and unmatched 2x2 tables. *Biometrika* **60**, 267-276.

Cox, D.R. and N. Wermuth. (1996). *Multivariate Dependencies: Models, Analysis and Interpretation*. Boca Raton, FL: CRC/Chapman and Hall.

Dawid, A.P. (2000). Causal inference without counterfactuals (with discussion). *Journal of the American Statistical Association* **95**, 407-448.

Dawid, A.P. (2008). Beware of the DAG! In: *NIPS 2008 Workshop Causality: Objectives and Assessment*. JMLR Workshop and Conference Proceedings.

Duncan, O.D. (1975). *Introduction to Structural Equation Models*. New York: Academic Press.

Fisher, R.A. (1943; reprinted 2003). Note on Dr. Berkson's criticism of tests of significance. *Journal of the American Statistical Association* **38**, 103–104. Reprinted in the *International Journal of Epidemiology* **32**, 692.

Freedman, D.A. and Humphreys, P. (1999). Are there algorithms that discover causal structure? *Synthese* **121**, 29–54.

Glymour, M.M. and S. Greenland (2008). Causal diagrams. Ch. 12 in: Rothman, K.J., S. Greenland and T.L. Lash, eds. *Modern Epidemiology*, 3rd ed. Philadelphia: Lippincott.

Good, I.J. (1983). *Good thinking*. Minneapolis: U. Minnesota Press.

Greenland, S. (1990). Randomization, statistics, and causal inference. *Epidemiology* **1**, 421-429.

Greenland, S. (1991). On the logical justification of conditional tests for two-by-two-contingency tables. *The American Statistician* **45**, 248-251.

Greenland, S. (1993). Summarization, smoothing, and inference. *Scandinavian Journal of Social Medicine* **21**, 227-232.

Greenland, S. (1998). The sensitivity of a sensitivity analysis. In: 1997 Proceedings of the Biometrics Section. Alexandria, VA: American Statistical Association, 19-21.

Greenland, S. (2005a). Epidemiologic measures and policy formulation: Lessons from potential outcomes (with discussion). *Emerging Themes in Epidemiology* (online journal) 2:1–4. (Originally published as "Causality theory for policy uses of epidemiologic measures," Chapter 6.2 in: Murray, C.J.L., J.A. Salomon, C.D. Mathers and A.D. Lopez, eds. (2002) *Summary Measures of Population Health*. Cambridge, MA: Harvard University Press/WHO, 291-302.)

Greenland, S. (2005b). Multiple-bias modeling for analysis of observational data (with discussion). *Journal of the Royal Statistical Society, Series A* **168**, 267–308.

Greenland, S. (2009a). Bayesian perspectives for epidemiologic research. III. Bias analysis via missing-data methods. *International Journal of Epidemiology* **38**, 1662–1673.

Greenland, S. (2009b). Relaxation penalties and priors for plausible modeling of nonidentified bias sources. *Statistical Science* **24**, 195-210.

Greenland, S. (2009c). Dealing with uncertainty about investigator bias: disclosure is informative. *Journal of Epidemiology and Community Health* **63**, 593-598.

Greenland, S. (2010). The need for syncretism in applied statistics (comment on "The future of indirect evidence" by Bradley Efron). *Statistical Science* **25**, in press.

Greenland, S., J. Pearl, and J.M. Robins (1999). Causal diagrams for epidemiologic research. *Epidemiology* **10**, 37-48.

Greenland, S., M. Gago-Dominguez, and J.E. Castellao (2004). The value of risk-factor ("black-box") epidemiology (with discussion). *Epidemiology* **15**, 519-535.

Gustafson, P. (2005). On model expansion, model contraction, identifiability, and prior information: two illustrative scenarios involving mismeasured variables (with discussion). *Statistical Science* **20**, 111-140.

Hajek, P., T. Havranek and R. Jirousek (1992). *Uncertain Information Processing in Expert Systems*. Boca Raton, FL: CRC Press.

Hastie, T., R. Tibshirani and J. Friedman (2009). *The elements of statistical learning: Data mining, inference, and prediction*, 2$^{nd}$ ed. New York: Springer.

Hernán, M.A. (2005). Hypothetical interventions to define causal effects—afterthought or prerequisite? *American Journal of Epidemiology* **162**, 618–620.

Hernán M.A., S. Hernandez-Diaz, M.M. Werler and A.A. Mitchell. (2002). Causal knowledge as a prerequisite for confounding evaluation: An application to birth defects epidemiology. *American Journal of Epidemiology* **155**, 176–184.

Hernán M.A., S. Hernandez-Diaz and J.M. Robins (2004). A structural approach to selection bias. *Epidemiology* **15**, 615-625.

Jewell, N. (2004). *Statistics for Epidemiology*. Boca Raton, FL: Chapman and Hall/CRC.

Lad, F. (1999). Assessing the foundations of Bayesian networks: A challenge to the principles and the practice. *Soft Computing* **3**, 174-180.

Lauritzen, S. (1996). *Graphical Models*. Oxford: Clarendon Press.

Leamer, E.E. (1978). *Specification Searches: Ad Hoc Inference with Nonexperimental Data*. New York: Wiley.

Morgan, S.L. and C. Winship. (2007). *Counterfactuals and Causal Inference: Methods and Principles for Social Research*. New York: Cambridge University Press.

Neutra, R.R., S. Greenland, and E.A. Friedman (1980). The effect of fetal monitoring on cesarean section rates. *Obstetrics and Gynecology* **55**, 175-180.

Pearl, J. (1988). *Probabilistic Reasoning in Intelligent Systems*. Morgan Kaufmann, San Mateo, CA.

Pearl, J. (1995). Causal diagrams for empirical research (with discussion). *Biometrika* **82**, 669-710.

Pearl, J. (2000; 2$^{nd}$ ed. 2009). *Causality*. New York: Cambridge University Press.

Pearl, J. and P. Verma (1991). A theory of inferred causation. In: *Principles of Knowledge Representation and Reasoning: Proceedings of the Second International Conference*, Ed. J.A. Allen, R. Filkes and E. Sandewall. San Francisco: Morgan Kaufmann, 441-452.

Poole, C. (2001). Poole C. Low P-values or narrow confidence intervals: Which are more durable? *Epidemiology* **12**, 291–294.

Robins, J.M. (2001). Data, design, and background knowledge in etiologic inference. *Epidemiology* **12**, 313–320.

Robins, J.M. and L. Wasserman (1999a). On the impossibility of inferring causation from association without background knowledge. In: *Computation, Causation, and Discovery*. Glymour, C. and Cooper, G., eds. Menlo Park, CA, Cambridge, MA: AAAI Press/The MIT Press, pp. 305-321.

Robins, J.M. and L. Wasserman (1999b). Rejoinder to Glymour and Spirtes. In: *Computation, Causation, and Discovery*. Glymour, C. and Cooper, G., eds. Menlo Park, CA, Cambridge, MA: AAAI Press/The MIT Press, pp. 333-342.

Robins, J.M., R. Scheines, P. Spirtes and L. Wasserman (2003). Uniform consistency in causal inference. *Biometrika* **90**, 491-515.

Senn, S. (2004). Controversies concerning randomization and additivity in clinical trials. *Statistics in Medicine* **23**, 3729–3753.

Shafer, G. (2002). Comment on "Estimating causal effects," by George Maldonado and Sander Greenland. *International Journal of Epidemiology* **31**, 434-435.

Spirtes, P., C. Glymour and R. Scheines (1993; 2$^{nd}$ ed. 2001). *Causation, Prediction, and Search*. Cambridge, MA: MIT Press.

Susser, M. (1973). *Causal Thinking in the Health Sciences*. New York: Oxford University Press.

VanderWeele, T.J. (2009). Concerning the consistency assumption in causal inference. *Epidemiology* **20**, 880-883.

VanderWeele, T.J. and J.M. Robins (2007). Directed acyclic graphs, sufficient causes and the properties of conditioning on a common effect. *American Journal of Epidemiology* **166**, 1096-1104.

Whittemore, A.S. and J.B. Keller (1986). Survival estimation using splines. *Biometrics* **42**, 495-506.

Wright, S., (1934). The method of path coefficients. *Annals of Mathematical Statistics* **5**, 161-215.

# 23
# Actual Causation and the Art of Modeling
JOSEPH Y. HALPERN AND CHRISTOPHER HITCHCOCK

## 1 Introduction

In *The Graduate*, Benjamin Braddock (Dustin Hoffman) is told that the future can be summed up in one word: "Plastics". One of us (Halpern) recalls that in roughly 1990, Judea Pearl told him that the future was in causality. Pearl's own research was largely focused on causality in the years after that; his seminal contributions are widely known. We were among the many influenced by his work. We discuss one aspect of it, *actual causation*, in this article, although a number of our comments apply to causal modeling more generally.

Pearl introduced a novel account of actual causation in Chapter 10 of *Causality*, which was later revised in collaboration with one of us [Halpern and Pearl 2005]. In some ways, Pearl's approach to actual causation can be seen as a contribution to the philosophical project of trying to analyze actual causation in terms of counterfactuals, a project associated most strongly with David Lewis [1973a]. But Pearl's account was novel in at least two important ways. The first was his use of *structural equations* as a tool for modeling causality. In the philosophical literature, causal structures were often represented using so-called *neuron diagrams*, but these are not (and were never intended to be) all-purpose representational tools. (See [Hitchcock 2007b] for a detailed discussion of the limitations of neuron diagrams.) We believe that the lack of a more adequate representational tool had been a serious obstacle to progress. Second, while the philosophical literature on causality has focused almost exclusively on actual causality, for Pearl, actual causation was a rather specialized topic within the study of causation, peripheral to many issues involving causal reasoning and inference. Thus, Pearl's work placed the study of actual causation within a much broader context.

The use of structural equations as a model for causal relationships was well known long before Pearl came on the scene; it seems to go back to the work of Sewall Wright in the 1920s (see [Goldberger 1972] for a discussion). However, the details of the framework that have proved so influential are due to Pearl. Besides the Halpern-Pearl approach mentioned above, there have been a number of other closely-related approaches for using structural equations to model actual causation; see, for example, [Glymour and Wimberly 2007; Hall 2007; Hitchcock 2001; Hitchcock 2007a; Woodward 2003]. The goal of this paper is to look more carefully at the modeling of causality using structural equations. For definiteness, we use the

Halpern-Pearl (HP) version [Halpern and Pearl 2005] here, but our comments apply equally well to the other variants.

It is clear that the structural equations can have a major impact on the conclusions we draw about causality—it is the equations that allow us to conclude that lower air pressure is the cause of the lower barometer reading, and not the other way around; increasing the barometer reading will not result in higher air pressure. The structural equations express the effects of *interventions*: what happens to the bottle if it is hit with a hammer; what happens to a patient if she is treated with a high dose of the drug, and so on. These effects are, in principle, objective; the structural equations can be viewed as describing objective features of the world. However, as pointed out by Halpern and Pearl [2005] and reiterated by others [Hall 2007; Hitchcock 2001; Hitchcock 2007a], the choice of variables and their values can also have a significant impact on causality. Moreover, these choices are, to some extent, subjective. This, in turn, means that judgments of actual causation are subjective.

Our view of actual causation being at least partly subjective stands in contrast to the prevailing view in the philosophy literature, where the assumption is that the job of the philosopher is to analyze the (objective) notion of causation, rather like that of a chemist analyzing the structure of a molecule. This may stem, at least in part, from failing to appreciate one of Pearl's lessons: actual causality is only part of the bigger picture of causality. There can be an element of subjectivity in ascriptions of actual causality without causation itself being completely subjective. In any case, the experimental evidence certainly suggests that people's views of causality are subjective, even when there is no disagreement about the relevant structural equations. For example, a number of experiments show that broadly normative considerations, including the subject's own moral beliefs, affect causal judgment. (See, for example, [Alicke 1992; Cushman 2009; Cushman, Knobe, and Sinnott-Armstrong 2008; Hitchcock and Knobe 2009; Knobe and Fraser 2008].) Even in relatively non-controversial cases, people may want to focus on different aspects of a problem, and thus give different answers to questions about causality. For example, suppose that we ask for the cause of a serious traffic accident. A traffic engineer might say that the bad road design was the cause; an educator might focus on poor driver's education; a sociologist might point to the pub near the highway where the driver got drunk; a psychologist might say that the cause is the driver's recent breakup with his girlfriend.[1] Each of these answers is reasonable. By appropriately choosing the variables, the structural equations framework can accommodate them all.

Note that we said above "by appropriately choosing the variables". An obvious question is "What counts as an appropriate choice?". More generally, what makes a model an appropriate model? While we do want to allow for subjectivity, we need

---

[1] This is a variant of an example originally due to Hanson [1958].

to be able to justify the modeling choices made. A lawyer in court trying to argue that faulty brakes were the cause of the accident needs to be able to justify his model; similarly, his opponent will need to understand what counts as a legitimate attack on the model. In this paper we discuss what we believe are reasonable bases for such justifications. Issues such as model stability and interactions between the events corresponding to variables turn out to be important.

Another focus of the paper is the use of defaults in causal reasoning. As we hinted above, the basic structural equations model does not seem to suffice to completely capture all aspects of causal reasoning. To explain why, we need to briefly outline how actual causality is defined in the structural equations framework. Like many other definitions of causality (see, for example, [Hume 1739; Lewis 1973b]), the HP definition is based on counterfactual dependence. Roughly speaking, $A$ is a cause of $B$ if, had $A$ not happened (this is the counterfactual condition, since $A$ did in fact happen) then $B$ would not have happened. As is well known, this naive definition does not capture all the subtleties involved with causality. Consider the following example (due to Hall [2004]): Suzy and Billy both pick up rocks and throw them at a bottle. Suzy's rock gets there first, shattering the bottle. Since both throws are perfectly accurate, Billy's would have shattered the bottle had Suzy not thrown. Thus, according to the naive counterfactual definition, Suzy's throw is not a cause of the bottle shattering. This certainly seems counterintuitive.

The HP definition deals with this problem by taking $A$ to be a cause of $B$ if $B$ counterfactually depends on $A$ *under some contingency*. For example, Suzy's throw is the cause of the bottle shattering because the bottle shattering counterfactually depends on Suzy's throw, under the contingency that Billy doesn't throw. (As we will see below, there are further subtleties in the definition that guarantee that, if things are modeled appropriately, Billy's throw is not also a cause.)

While the definition of actual causation in terms of structural equations has been successful at dealing with many of the problems of causality, examples of Hall [2007], Hiddleston [2005], and Hitchcock [2007a] show that it gives inappropriate answers in cases that have structural equations isomorphic to ones where it arguably gives the appropriate answer. This means that, no matter how we define actual causality in the structural-equations framework, the definition must involve more than just the structural equations. Recently, Hall [2007], Halpern [2008], and Hitchcock [2007a] have suggested that using defaults might be a way of dealing with the problem. As the psychologists Kahneman and Miller [1986, p. 143] observe, "an event is more likely to be undone by altering exceptional than routine aspects of the causal chain that led to it". This intuition is also present in the legal literature. Hart and Honoré [1985] observe that the statement "It was the presence of oxygen that caused the fire" makes sense only if there were reasons to view the presence of oxygen as abnormal.

As shown by Halpern [2008], we can model this intuition formally by combining a well-known approach to modeling defaults and normality, due to Kraus, Lehmann, and Magidor [1990] with the structural-equation model. Moreover, doing this leads to a straightforward solution to the problem above. The idea is that, when showing that if $A$ hadn't happened then $B$ would not have happened, we consider only contingencies that are at least as normal as the actual world. For example, if someone typically leaves work at 5:30 PM and arrives home at 6, but, due to unusually bad traffic, arrives home at 6:10, the bad traffic is typically viewed as the cause of his being late, not the fact that he left at 5:30 (rather than 5:20).

But once we add defaults to the model, the problem of justifying the model becomes even more acute. We not only have to justify the structural equations and the choice of variables, but also the default theory. The problem is exacerbated by the fact that default and "normality" have a number of interpretations. Among other things, they can represent moral obligations, societal conventions, prototypicality information, and statistical information. All of these interpretations are relevant to understanding causality; this makes justifying default choices somewhat subtle.

The rest of this paper is organized as follows. In Sections 2 and 3, we review the notion of causal model and the HP definition of actual cause; most of this material is taken from [Halpern and Pearl 2005]. In Section 4, we discuss some issues involved in the choice of variables in a model. In Section 5, we review the approach of [Halpern 2008] for adding considerations of normality to the HP framework, and discuss some modeling issues that arise when we do so. We conclude in Section 6.

## 2 Causal Models

In this section, we briefly review the HP definition of causality. The description of causal models given here is taken from [Halpern 2008], which in turn is based on that of [Halpern and Pearl 2005].

The HP approach assumes that the world is described in terms of random variables and their values. For example, if we are trying to determine whether a forest fire was caused by lightning or an arsonist, we can take the world to be described by three random variables:

- $F$ for forest fire, where $F = 1$ if there is a forest fire and $F = 0$ otherwise;

- $L$ for lightning, where $L = 1$ if lightning occurred and $L = 0$ otherwise;

- $ML$ for match (dropped by arsonist), where $ML = 1$ if the arsonist drops a lit match, and $ML = 0$ otherwise.

Some random variables may have a causal influence on others. This influence is modeled by a set of *structural equations*. For example, to model the fact that if either a match is lit or lightning strikes, then a fire starts, we could use the random variables $ML$, $F$, and $L$ as above, with the equation $F = \max(L, ML)$. (Alternately, if a fire requires both causes to be present, the equation for $F$ becomes $F = \min(L, ML)$.) The equality sign in this equation should be thought of more like an assignment statement in programming languages; once we set the values of $F$

and $L$, then the value of $F$ is set to their maximum. However, despite the equality, if a forest fire starts some other way, that does not force the value of either $ML$ or $L$ to be 1.

It is conceptually useful to split the random variables into two sets: the *exogenous* variables, whose values are determined by factors outside the model, and the *endogenous* variables, whose values are ultimately determined by the exogenous variables. For example, in the forest-fire example, the variables $ML$, $L$, and $F$ are endogenous. However, we want to take as given that there is enough oxygen for the fire and that the wood is sufficiently dry to burn. In addition, we do not want to concern ourselves with the factors that make the arsonist drop the match or the factors that cause lightning. These factors are all determined by the exogenous variables.

Formally, a *causal model* $M$ is a pair $(\mathcal{S}, \mathcal{F})$, where $\mathcal{S}$ is a *signature*, which explicitly lists the endogenous and exogenous variables and characterizes their possible values, and $\mathcal{F}$ defines a set of *modifiable structural equations*, relating the values of the variables. A signature $\mathcal{S}$ is a tuple $(\mathcal{U}, \mathcal{V}, \mathcal{R})$, where $\mathcal{U}$ is a set of exogenous variables, $\mathcal{V}$ is a set of endogenous variables, and $\mathcal{R}$ associates with every variable $Y \in \mathcal{U} \cup \mathcal{V}$ a nonempty set $\mathcal{R}(Y)$ of possible values for $Y$ (that is, the set of values over which $Y$ *ranges*). $\mathcal{F}$ associates with each endogenous variable $X \in \mathcal{V}$ a function denoted $F_X$ such that $F_X : (\times_{U \in \mathcal{U}} \mathcal{R}(U)) \times (\times_{Y \in \mathcal{V} - \{X\}} \mathcal{R}(Y)) \to \mathcal{R}(X)$. This mathematical notation just makes precise the fact that $F_X$ determines the value of $X$, given the values of all the other variables in $\mathcal{U} \cup \mathcal{V}$. If there is one exogenous variable $U$ and three endogenous variables, $X$, $Y$, and $Z$, then $F_X$ defines the values of $X$ in terms of the values of $Y$, $Z$, and $U$. For example, we might have $F_X(u, y, z) = u + y$, which is usually written as $X \leftarrow U + Y$.[2] Thus, if $Y = 3$ and $U = 2$, then $X = 5$, regardless of how $Z$ is set.

In the running forest-fire example, suppose that we have an exogenous random variable $U$ that determines the values of $L$ and $ML$. Thus, $U$ has four possible values of the form $(i, j)$, where both of $i$ and $j$ are either 0 or 1. The $i$ value determines the value of $L$ and the $j$ value determines the value of $ML$. Although $F_L$ gets as arguments the vale of $U$, $ML$, and $F$, in fact, it depends only on the (first component of) the value of $U$; that is, $F_L((i,j), m, f) = i$. Similarly, $F_{ML}((i,j), l, f) = j$. The value of $F$ depends only on the value of $L$ and $ML$. How it depends on them depends on whether either cause by itself is sufficient for the forest fire or whether both are necessary. If either one suffices, then $F_F((i,j), l, m) = \max(l, m)$, or, perhaps more comprehensibly, $F = \max(L, ML)$; if both are needed, then $F = \min(L, ML)$. For future reference, call the former model the *disjunctive* model, and the latter the *conjunctive* model.

The key role of the structural equations is to define what happens in the presence of external interventions. For example, we can explain what happens if the arsonist

---

[2]The fact that $X$ is assigned $U + Y$ (i.e., the value of $X$ is the sum of the values of $U$ and $Y$) does not imply that $Y$ is assigned $X - U$; that is, $F_Y(U, X, Z) = X - U$ does not necessarily hold.

does *not* drop the match. In the disjunctive model, there is a forest fire exactly if there is lightning; in the conjunctive model, there is definitely no fire. Setting the value of some variable $X$ to $x$ in a causal model $M = (\mathcal{S}, \mathcal{F})$ results in a new causal model denoted $M_{X \leftarrow x}$. In the new causal model, the equation for $X$ is very simple: $X$ is just set to $x$; the remaining equations are unchanged. More formally, $M_{X \leftarrow x} = (\mathcal{S}, \mathcal{F}^{X \leftarrow x})$, where $\mathcal{F}^{X \leftarrow x}$ is the result of replacing the equation for $X$ in $\mathcal{F}$ by $X = x$.

The structural equations describe *objective* information about the results of interventions, that can, in principle, be checked. Once the modeler has selected a set of variables to include in the model, *the world* determines which equations among those variables correctly represent the effects of interventions.[3] By contrast, the *choice* of variables is subjective; in general, there need be no objectively "right" set of exogenous and endogenous variables to use in modeling a problem. We return to this issue in Section 4.

It may seem somewhat circular to use causal models, which clearly already encode causal information, to define actual causation. Nevertheless, as we shall see, there is no circularity. The equations of a causal model do not represent relations of *actual causation*, the very concept that we are using them to define. Rather, the equations characterize the results of *all possible* interventions (or at any rate, all of the interventions that can be represented in the model) without regard to what actually happened. Specifically, the equations do not depend upon the actual values realized by the variables. For example, the equation $F = \max(L, ML)$, by itself, does not say anything about whether the forest fire was actually caused by lightning or by an arsonist, or, for that matter, whether a fire even occurred. By contrast, relations of actual causation depend crucially on how things actually play out.

A sequence of endogenous $X_1, \ldots, X_n$ of is a *directed path* from $X_1$ to $X_n$ if the value of $X_{i+1}$ (as given by $F_{X_{i+1}}$) depends on the value of $X_i$, for $1 = 1, \ldots, n-1$. In this paper, following HP, we restrict our discussion to *acyclic* causal models, where causal influence can be represented by an acyclic Bayesian network. That is, there is no cycle $X_1, \ldots, X_n, X_1$ of endogenous variables that forms a directed path from $X_1$ to itself. If $M$ is an acyclic causal model, then given a *context*, that is, a setting $\vec{u}$ for the exogenous variables in $\mathcal{U}$, there is a unique solution for all the equations.

---

[3] In general, there may be uncertainty about the causal model, as well as about the true setting of the exogenous variables in a causal model. Thus, we may be uncertain about whether smoking causes cancer (this represents uncertainty about the causal model) and uncertain about whether a particular patient actually smoked (this is uncertainty about the value of the exogenous variable that determines whether the patient smokes). This uncertainty can be described by putting a probability on causal models and on the values of the exogenous variables. We can then talk about the probability that $A$ is a cause of $B$.

## 3 The HP Definition of Actual Cause

### 3.1 A language for describing causes

Given a signature $\mathcal{S} = (\mathcal{U}, \mathcal{V}, \mathcal{R})$, a *primitive event* is a formula of the form $X = x$, for $X \in \mathcal{V}$ and $x \in \mathcal{R}(X)$. A *causal formula (over $\mathcal{S}$)* is one of the form $[Y_1 \leftarrow y_1, \ldots, Y_k \leftarrow y_k]\phi$, where $\phi$ is a Boolean combination of primitive events, $Y_1, \ldots, Y_k$ are distinct variables in $\mathcal{V}$, and $y_i \in \mathcal{R}(Y_i)$. Such a formula is abbreviated as $[\vec{Y} \leftarrow \vec{y}]\phi$. The special case where $k = 0$ is abbreviated as $\phi$. Intuitively, $[Y_1 \leftarrow y_1, \ldots, Y_k \leftarrow y_k]\phi$ says that $\phi$ would hold if $Y_i$ were set to $y_i$, for $i = 1, \ldots, k$.

A causal formula $\psi$ is true or false in a causal model, given a context. As usual, we write $(M, \vec{u}) \models \psi$ if the causal formula $\psi$ is true in causal model $M$ given context $\vec{u}$. The $\models$ relation is defined inductively. $(M, \vec{u}) \models X = x$ if the variable $X$ has value $x$ in the unique (since we are dealing with acyclic models) solution to the equations in $M$ in context $\vec{u}$ (that is, the unique vector of values for the endogenous variables that simultaneously satisfies all equations in $M$ with the variables in $\mathcal{U}$ set to $\vec{u}$). The truth of conjunctions and negations is defined in the standard way. Finally, $(M, \vec{u}) \models [\vec{Y} \leftarrow \vec{y}]\phi$ if $(M_{\vec{Y} \leftarrow \vec{y}}, \vec{u}) \models \phi$. We write $M \models \phi$ if $(M, \vec{u}) \models \phi$ for all contexts $\vec{u}$.

For example, if $M$ is the disjunctive causal model for the forest fire, and $u$ is the context where there is lightning and the arsonist drops the lit match, then $(M, u) \models [ML \leftarrow 0](F = 1)$, since even if the arsonist is somehow prevented from dropping the match, the forest burns (thanks to the lightning); similarly, $(M, u) \models [L \leftarrow 0](F = 1)$. However, $(M, u) \models [L \leftarrow 0; ML \leftarrow 0](F = 0)$: if the arsonist does not drop the lit match and the lightning does not strike, then the forest does not burn.

### 3.2 A preliminary definition of causality

The HP definition of causality, like many others, is based on counterfactuals. The idea is that if $A$ and $B$ both occur, then $A$ is a cause of $B$ if, if $A$ hadn't occurred, then $B$ would not have occurred. This idea goes back to at least Hume [1748, Section VIII], who said:

> We may define a cause to be an object followed by another, ..., where, if the first object had not been, the second never had existed.

This is essentially the *but-for* test, perhaps the most widely used test of actual causation in tort adjudication. The but-for test states that an act is a cause of injury if and only if, but for the act (i.e., had the the act not occurred), the injury would not have occurred.

There are two well-known problems with this definition. The first can be seen by considering the disjunctive causal model for the forest fire again. Suppose that the arsonist drops a match and lightning strikes. Which is the cause? According to a naive interpretation of the counterfactual definition, neither is. If the match hadn't dropped, then the lightning would still have struck, so there would have been a forest fire anyway. Similarly, if the lightning had not occurred, there still would have been a forest fire. As we shall see, the HP definition declares both lightning and the arsonist causes of the fire. (In general, there may be more than one actual cause of an outcome.)

A more subtle problem is what philosophers have called *preemption*, which is illustrated by the rock-throwing example from the introduction. As we observed, according to a naive counterfactual definition of causality, Suzy's throw would not be a cause.

The HP definition deals with the first problem by defining causality as counterfactual dependency *under certain contingencies*. In the forest-fire example, the forest fire does counterfactually depend on the lightning under the contingency that the arsonist does not drop the match; similarly, the forest fire depends counterfactually on the dropping of the match under the contingency that the lightning does not strike.

Unfortunately, we cannot use this simple solution to treat the case of preemption. We do not want to make Billy's throw the cause of the bottle shattering by considering the contingency that Suzy does not throw. So if our account is to yield the correct verdict in this case, it will be necessary to limit the contingencies that can be considered. The reason that we consider Suzy's throw to be the cause and Billy's throw not to be the cause is that Suzy's rock hit the bottle, while Billy's did not. Somehow the definition of actual cause must capture this obvious intuition.

With this background, we now give the preliminary version of the HP definition of causality. Although the definition is labeled "preliminary", it is quite close to the final definition, which is given in Section 5. The definition is relative to a causal model (and a context); $A$ may be a cause of $B$ in one causal model but not in another. The definition consists of three clauses. The first and third are quite simple; all the work is going on in the second clause.

The types of events that the HP definition allows as actual causes are ones of the form $X_1 = x_1 \wedge \ldots \wedge X_k = x_k$—that is, conjunctions of primitive events; this is often abbreviated as $\vec{X} = \vec{x}$. The events that can be caused are arbitrary Boolean combinations of primitive events. The definition does not allow statements of the form "$A$ or $A'$ is a cause of $B$", although this could be treated as being equivalent to "either $A$ is a cause of $B$ or $A'$ is a cause of $B$". On the other hand, statements such as "$A$ is a cause of $B$ or $B'$" are allowed; this is not equivalent to "either $A$ is a cause of $B$ or $A$ is a cause of $B'$".

DEFINITION 1. (Actual cause; preliminary version) [Halpern and Pearl 2005] $\vec{X} = \vec{x}$ is an *actual cause of* $\phi$ *in* $(M, \vec{u})$ if the following three conditions hold:

AC1. $(M, \vec{u}) \models (\vec{X} = \vec{x})$ and $(M, \vec{u}) \models \phi$.

AC2. There is a partition of $\mathcal{V}$ (the set of endogenous variables) into two subsets $\vec{Z}$ and $\vec{W}$ with $\vec{X} \subseteq \vec{Z}$, and a setting $\vec{x}'$ and $\vec{w}$ of the variables in $\vec{X}$ and $\vec{W}$, respectively, such that if $(M, \vec{u}) \models Z = z^*$ for all $Z \in \vec{Z}$, then both of the following conditions hold:

(a) $(M, \vec{u}) \models [\vec{X} \leftarrow \vec{x}', \vec{W} \leftarrow \vec{w}]\neg \phi$.

(b) $(M, \vec{u}) \models [\vec{X} \leftarrow \vec{x}, \vec{W}' \leftarrow \vec{w}, \vec{Z}' \leftarrow \vec{z}^*]\phi$ for all subsets $\vec{W}'$ of $\vec{W}$ and all subsets $\vec{Z}'$ of $\vec{Z}$, where we abuse notation and write $\vec{W}' \leftarrow \vec{w}$ to denote the assignment where the variables in $\vec{W}'$ get the same values as they would in the assignment $\vec{W} \leftarrow \vec{w}$.

AC3. $\vec{X}$ is minimal; no subset of $\vec{X}$ satisfies conditions AC1 and AC2.

AC1 just says that $\vec{X} = \vec{x}$ cannot be considered a cause of $\phi$ unless both $\vec{X} = \vec{x}$ and $\phi$ actually happen. AC3 is a minimality condition, which ensures that only those elements of the conjunction $\vec{X} = \vec{x}$ that are essential for changing $\phi$ in AC2(a) are considered part of a cause; inessential elements are pruned. Without AC3, if dropping a lit match qualified as a cause of the forest fire, then dropping a match and sneezing would also pass the tests of AC1 and AC2. AC3 serves here to strip "sneezing" and other irrelevant, over-specific details from the cause. Clearly, all the "action" in the definition occurs in AC2. We can think of the variables in $\vec{Z}$ as making up the "causal path" from $\vec{X}$ to $\phi$, consisting of one or more directed paths from variables in $\vec{X}$ to variables in $\phi$. Intuitively, changing the value(s) of some variable(s) in $\vec{X}$ results in changing the value(s) of some variable(s) in $\vec{Z}$, which results in the value(s) of some other variable(s) in $\vec{Z}$ being changed, which finally results in the truth value of $\phi$ changing. The remaining endogenous variables, the ones in $\vec{W}$, are off to the side, so to speak, but may still have an indirect effect on what happens. AC2(a) is essentially the standard counterfactual definition of causality, but with a twist. If we want to show that $\vec{X} = \vec{x}$ is a cause of $\phi$, we must show (in part) that if $\vec{X}$ had a different value, then $\phi$ would have been false. However, this effect of the value of $\vec{X}$ on the truth value of $\phi$ may not hold in the actual context; the value of $\vec{W}$ may have to be different to allow this effect to manifest itself. For example, consider the context where both the lightning strikes and the arsonist drops a match in the disjunctive model of the forest fire. Stopping the arsonist from dropping the match will not prevent the forest fire. The counterfactual effect of the arsonist on the forest fire manifests itself only in a situation where the lightning does not strike (i.e., where $L$ is set to 0). AC2(a) is what allows us to call both the lightning and the arsonist causes of the forest fire. Essentially, it ensures that $\vec{X}$ alone suffices to bring about the change from $\phi$ to $\neg \phi$; setting $\vec{W}$ to $\vec{w}$ merely eliminates possibly spurious side effects that may mask the effect of changing the value of $\vec{X}$. Moreover, when $\vec{X} = \vec{x}$, although the values of variables on the causal path (i.e., the variables $\vec{Z}$) may be perturbed by the change to $\vec{W}$, this perturbation has no impact on the value of $\phi$. If $(M, \vec{u}) \models \vec{Z} = \vec{z}^*$, then $z^*$ is the value of the variable $Z$ in the context $\vec{u}$. We capture the fact that the perturbation has no impact on the value of $\phi$ by saying that if some variables $Z$ on the causal path were set to their original values in the context $\vec{u}$, $\phi$ would still be true, as long as $\vec{X} = \vec{x}$.

EXAMPLE 2. For the forest-fire example, let $M$ be the disjunctive model for the forest fire sketched earlier, with endogenous variables $L$, $ML$, and $F$. We want to show that $L = 1$ is an actual cause of $F = 1$. Clearly $(M,(1,1)) \models F = 1$ and $(M,(1,1)) \models L = 1$; in the context $(1,1)$, the lightning strikes and the forest burns down. Thus, AC1 is satisfied. AC3 is trivially satisfied, since $\vec{X}$ consists of only one element, $L$, so must be minimal. For AC2, take $\vec{Z} = \{L, F\}$ and take $\vec{W} = \{ML\}$, let $x' = 0$, and let $w = 0$. Clearly, $(M,(1,1)) \models [L \leftarrow 0, ML \leftarrow 0](F \neq 1)$; if the lightning does not strike and the match is not dropped, the forest does not burn down, so AC2(a) is satisfied. To see the effect of the lightning, we must consider the contingency where the match is not dropped; the definition allows us to do that by setting $ML$ to 0. (Note that here setting $L$ and $ML$ to 0 overrides the effects of $U$; this is critical.) Moreover, $(M,(1,1)) \models [L \leftarrow 1, ML \leftarrow 0](F = 1)$; if the lightning strikes, then the forest burns down even if the lit match is not dropped, so AC2(b) is satisfied. (Note that since $\vec{Z} = \{L, F\}$, the only subsets of $\vec{Z} - \vec{X}$ are the empty set and the singleton set consisting of just $F$.)

It is also straightforward to show that the lightning and the dropped match are also causes of the forest fire in the context where $U = (1,1)$ in the conjunctive model. Again, AC1 and AC3 are trivially satisfied and, again, to show that AC2 holds in the case of lightning we can take $\vec{Z} = \{L, F\}$, $\vec{W} = \{ML\}$, and $x' = 0$, but now we let $w = 1$. In the conjunctive scenario, if there is no lightning, there is no forest fire, while if there is lightning (and the match is dropped) there is a forest fire, so AC2(a) and AC2(b) are satisfied; similarly for the dropped match.

EXAMPLE 3. Now consider the Suzy-Billy example.[4] We get the desired result—that Suzy's throw is a cause, but Billy's is not—but only if we model the story appropriately. Consider first a coarse causal model, with three endogenous variables:

- $ST$ for "Suzy throws", with values 0 (Suzy does not throw) and 1 (she does);

- $BT$ for "Billy throws", with values 0 (he doesn't) and 1 (he does);

- $BS$ for "bottle shatters", with values 0 (it doesn't shatter) and 1 (it does).

(We omit the exogenous variable here; it determines whether Billy and Suzy throw.) Take the formula for $BS$ to be such that the bottle shatters if either Billy or Suzy throw; that is $BS = \max(BT, ST)$. (We assume that Suzy and Billy will not miss if they throw.) $BT$ and $ST$ play symmetric roles in this model; there is nothing to distinguish them. Not surprisingly, both Billy's throw and Suzy's throw are classified as causes of the bottle shattering in this model. The argument is essentially identical to that in the disjunctive model of the forest-fire example in

---

[4]The discussion of this example is taken almost verbatim from HP.

the context $U = (1,1)$, where both the lightning and the dropped match are causes of the fire.

The trouble with this model is that it cannot distinguish the case where both rocks hit the bottle simultaneously (in which case it would be reasonable to say that both $ST = 1$ and $BT = 1$ are causes of $BS = 1$) from the case where Suzy's rock hits first. To allow the model to express this distinction, we add two new variables to the model:

- $BH$ for "Billy's rock hits the (intact) bottle", with values 0 (it doesn't) and 1 (it does); and

- $SH$ for "Suzy's rock hits the bottle", again with values 0 and 1.

Now our equations will include:

- $SH = ST$;

- $BH = \min(BT, 1 - SH)$; and

- $BS = \max(SH, BH)$.

Now it is the case that, in the context where both Billy and Suzy throw, $ST = 1$ is a cause of $BS = 1$, but $BT = 1$ is not. To see that $ST = 1$ is a cause, note that, as usual, it is immediate that AC1 and AC3 hold. For AC2, choose $\vec{Z} = \{ST, SH, BH, BS\}$, $\vec{W} = \{BT\}$, and $w = 0$. When $BT$ is set to 0, $BS$ tracks $ST$: if Suzy throws, the bottle shatters and if she doesn't throw, the bottle does not shatter. To see that $BT = 1$ is *not* a cause of $BS = 1$, we must check that there is no partition $\vec{Z} \cup \vec{W}$ of the endogenous variables that satisfies AC2. Attempting the symmetric choice with $\vec{Z} = \{BT, BH, SH, BS\}$, $\vec{W} = \{ST\}$, and $w = 0$ violates AC2(b). To see this, take $\vec{Z}' = \{BH\}$. In the context where Suzy and Billy both throw, $BH = 0$. If $BH$ is set to 0, the bottle does not shatter if Billy throws and Suzy does not. It is precisely because, in this context, Suzy's throw hits the bottle and Billy's does not that we declare Suzy's throw to be the cause of the bottle shattering. AC2(b) captures that intuition by allowing us to consider the contingency where $BH = 0$, despite the fact that Billy throws. We leave it to the reader to check that no other partition of the endogenous variables satisfies AC2 either.

This example emphasizes an important moral. If we want to argue in a case of preemption that $X = x$ is the cause of $\phi$ rather than $Y = y$, then there must be a random variable ($BH$ in this case) that takes on different values depending on whether $X = x$ or $Y = y$ is the actual cause. If the model does not contain such a variable, then it will not be possible to determine which one is in fact the cause. This is certainly consistent with intuition and the way we present evidence. If we want to argue (say, in a court of law) that it was $A$'s shot that killed $C$ rather than $B$'s, then we present evidence such as the bullet entering $C$ from the left side (rather than the right side, which is how it would have entered had $B$'s shot been the lethal one). The side from which the shot entered is the relevant random variable in this case. Note that the random variable may involve temporal evidence (if $Y$'s shot had been the lethal one, the death would have occurred a few seconds later), but it certainly does not have to.

## 4 The Choice of Variables

A modeler has considerable leeway in choosing which variables to include in a model. Nature does not provide a uniquely correct set of variables. Nonetheless, there are a number of considerations that guide variable selection. While these will not usually suffice to single out one choice of variables, they can provide a framework for the rational evaluation of models, including resources for motivating and defending certain choices of variables, and criticizing others.

The problem of choosing a set of variables for inclusion in a model has many dimensions. One set of issues concerns the question of how many variables to include in a model. If the modeler begins with a set of variables, how can she know whether she should add additional variables to the model? Given that it is always possible to add additional variables, is there a point at which the model contains "enough" variables? Is it ever possible for a model to have "too many" variables? Can the addition of further variables ever do positive harm to a model?

Another set of issues concerns the values of variables. Say that variable $X'$ is a *refinement* of $X$ if, for each value $x$ in the range of $X$, there is some subset $S$ of the range of $X'$ such that $X = x$ just in case $X'$ is in $S$. When is it appropriate or desirable to replace a variable with a refinement? Can it ever lead to problems if a variable is too fine-grained? Similarly, are there considerations that would lead us to prefer a model that replaced $X$ with a new variable $X''$, whose range is a proper subset or superset of the range of $X$?

Finally, are there constraints on the set of variables in a model over and above those we might impose on individual variables? For instance, can the choice to include a particular variable $X$ within a model require us to include another variable $Y$, or to exclude a particular variable $Z$?

While we cannot provide complete answers to all of these questions, we believe a good deal can be said to reduce the arbitrariness of the choice of variables. The most plausible way to motivate guidelines for the selection of variables is to show how inappropriate choices give rise to systems of equations that are inaccurate, misleading, or incomplete in their predictions of observations and interventions. In the next three subsections, we present several examples to show how such considerations can be brought to bear on the problem of variable choice.

### 4.1 The Number of Variables

We already saw in Example 3 that it is important to choose the variables correctly. Adding more variables can clearly affect whether $A$ is a cause of $B$. When is it appropriate or necessary to add further variables to a model?[5] Suppose that we have an infinite sequence of models $M^1, M^2, \ldots$ such that the variables in $M^i$ are $X_0, \ldots, X_{i+1}, Y$, and $M^{i+1}_{X_{i+1} \leftarrow 1} = M_i$ (so that $M^{i+1}$ can be viewed as an extension of $M^i$). Is it possible that whether $X_0 = 1$ is a cause of $Y = 1$ can alternate as we go through this sequence? This would indicate a certain "instability" in the causality. In this circumstance, a lawyer should certainly be able to argue against using, say, $M^7$ as a model to show that $X_0 = 1$ is cause of $Y = 1$. On the other hand, if the sequence stabilizes, that is, if there is some $k$ such that for all $i \geq k$, $M^i$ delivers the same verdict on some causal claim of interest, that would provide a strong reason to accept $M^k$ as sufficient.

Compare Example 2 with Example 3. In Example 2, we were able to adequately model the scenario using only three endogenous variables: $L$, $ML$, and $F$. By contrast, in Example 3, the model containing only three endogenous variables, $BT$, $ST$, and $BS$, was inadequate. What is the difference between the two scenarios? One difference we have already mentioned is that there seems to be an important feature of the second scenario that cannot be captured in the three-variable model: Suzy's rock hit the bottle before Billy's did. There is also a significant "topological" difference between the two scenarios. In the forest-fire example, there are two directed paths into the variable $F$. We could interpolate additional variables along these two paths. We could, for instance, interpolate a variable representing the occurrence of a small brush fire. But doing so would not fundamentally change the causal structure: there would still be just two directed paths into $F$. In the case of preemption, however, adding the additional variables $SH$ and $BH$ created an additional directed path that was not there before. The three-variable model contained just two directed paths: one from $ST$ to $BS$, and one from $BT$ to $BS$. However, once the variables $SH$ and $BH$ were added, there were three directed paths: $\{ST, SH, BS\}$, $\{BT, BH, BS\}$, and $\{ST, SH, BH, BS\}$. The intuition, then, is that adding additional variables to a model will not affect the relations of actual causation that hold in the model unless the addition of those variables changes the "topology" of the model. A more complete mathematical characterization of the conditions under which the verdicts of actual causality remain stable under the addition of further variables strikes us as a worthwhile research project that has not yet been undertaken.

## 4.2 The Ranges of Variables

Not surprisingly, the set of possible values of a variable must also be chosen appropriately. Consider, for example, a case of "trumping", introduced by Schaffer [2000]. Suppose that a group of soldiers is very well trained, so that they will obey any order given by a superior officer; in the case of conflicting orders, they obey the

---

[5]Although his model of causality is quite different from ours, Spohn [2003] also considers the effect of adding or removing variables, and discusses how a model with fewer variables should be related to one with more variables.

highest-ranking officer. Both a sergeant and a major issue the order to march, and the soldiers march. Let us put aside the morals that Schaffer attempts to draw from this example (with which we disagree; see [Halpern and Pearl 2005] and [Hitchcock 2010]), and consider only the modeling problem. We will presumably want variables $S$, $M$, and $A$, corresponding to the sergeant's order, the major's order, and the soldiers' action. We might let $S = 1$ represent the sergeant's giving the order to march and $S = 0$ represent the sergeant's giving no order; likewise for $M$ and $A$. But this would not be adequate. If the only possible order is the order to march, then there is no way to capture the principle that in the case of conflicting orders, the soldiers obey the major. One way to do this is to replace the variables $M$, $S$, and $A$ by variables $M'$, $S'$ and $A'$ that take on three possible values. Like $M$, $M' = 0$ if the major gives no order and $M' = 1$ if the major gives the order to march. But now we allow $M' = 2$, which corresponds to the major giving some other order. $S'$ and $A'$ are defined similarly. We can now write an equation to capture the fact that if $M' = 1$ and $S' = 2$, then the soldiers march, while if $M' = 2$ and $S' = 1$, then the soldiers do not march.

The appropriate set of values of a variable will depend on the other variables in the picture, and the relationship between them. Suppose, for example, that a hapless homeowner comes home from a trip to find that his front door is stuck. If he pushes on it with a normal force then the door will not open. However, if he leans his shoulder against it and gives a solid push, then the door will open. To model this, it suffices to have a variable $O$ with values either 0 or 1, depending on whether the door opens, and a variable $P$, with values 0 or 1 depending on whether or not the homeowner gives a solid push.

On the other hand, suppose that the homeowner also forgot to disarm the security system, and that the system is very sensitive, so that it will be tripped by any push on the door, regardless of whether the door opens. Let $A = 1$ if the alarm goes off, $A = 0$ otherwise. Now if we try to model the situation with the same variable $P$, we will not be able to express the dependence of the alarm on the homeowner's push. To deal with both $O$ and $A$, we need to extend $P$ to a 3-valued variable $P'$, with values 0 if the homeowner does not push the door, 1 if he pushes it with normal force, and 2 if he gives it a solid push.

These considerations parallel issues that arise in philosophical discussions about the metaphysics of "events".[6] Suppose that our homeowner pushed on the door with enough force to open it. Is there just one event, the push, that can be described at various levels of detail, such as a "push" or a "hard push"? This is the view of Davidson [1967]. Or are there rather many different events corresponding to these different descriptions, as argued by Kim [1973] and Lewis [1986b]? And if we take the latter view, which of the many events that occur should be counted as causes of the door's opening? These strike us as pseudoproblems. We believe that questions

---

[6]This philosophical usage of the word "event" is different from the typical usage of the word in computer science and probability, where an event is just a subset of the state space.

about causality are best addressed by dealing with the methodological problem of constructing a model that correctly describes the effects of interventions in a way that is not misleading or ambiguous.

A slightly different way in which one variable may constrain the values that another may take is by its implicit presuppositions. For example, a counterfactual theory of causation seems to have the somewhat counterintuitive consequence that one's birth is a cause of one's death. This sounds a little odd. If Jones dies suddenly one night, shortly before his 80th birthday, the coroner's inquest is unlikely to list "birth" as among the causes of his death. Typically, when we investigate the causes of death, we are interested in what makes the difference between a person's dying and his surviving. So our model might include a variable $D$ such $D = 1$ holds if Jones dies shortly before his 80th birthday, and $D = 0$ holds if he continues to live. If our model also includes a variable $B$, taking the value 1 if Jones is born, 0 otherwise, then there simply is no value that $D$ would take if $B = 0$. Both $D = 0$ and $D = 1$ implicitly presuppose that Jones was born (i.e., $B = 1$). Our conclusion is that if we have chosen to include a variable such as $D$ in our model, then we cannot conclude that Jones' birth is a cause of his death!

### 4.3 Dependence and Independence

Lewis [1986a] added a constraint to his counterfactual theory of causation. In order for event $c$ to be a cause of event $e$, the two events cannot be logically related. Suppose for instance, that Martha says "hello" loudly. If she had not said "hello", then she certainly could not have said "hello" loudly. But her saying "hello" is not a cause of her saying "hello" loudly. The counterfactual dependence results from a logical, rather than a causal, relationship between the two events.

We must impose a similar constraint upon causal models. Values of different variables should not correspond to events that are logically related. But now, rather than being an *ad hoc* restriction, it has a clear rationale. For suppose that we had a model with variable $H_1$ and $H_2$, where $H_1$ represents "Martha says 'hello'" (i.e., $H_1 = 1$ if Martha says "hello" and $H_1 = 0$ otherwise), and $H_2$ represents "Martha says 'hello' loudly". The intervention $H_1 = 0 \land H_2 = 1$ is meaningless; it is logically impossible for Martha not to say "hello" and to say "hello" loudly.

We doubt that any careful modeler would choose variables that have logically related values. However, the converse of this principle, that the different values of any particular variable *should* be logically related (in fact, mutually exclusive), is less obvious and equally important. Consider Example 3. While, in the actual context, Billy's rock will hit the bottle just in case Suzy's doesn't, this is not a necessary relationship. Suppose that, instead of using two variables $SH$ and $BH$, we try to model the scenario with a variable $H$ that takes the value 1 if Suzy's rock hits, and and 0 if Billy's rock hits. The reader can verify that, in this model, there is no contingency such that the bottle's shattering depends upon Suzy's throw. The problem, as we said, is that $H = 0$ and $H = 1$ are *not* mutually exclusive; there are possible situations in which both rocks hit or neither rock hits the bottle. In particular, this representation does not allow us to consider independent interventions on the rocks hitting the bottle. As the discussion in Example 3 shows, it is precisely such an intervention that is needed to establish that Suzy's throw (and not Billy's) is the actual cause of the bottle shattering.

While these rules are simple in principle, their application is not always transparent.

EXAMPLE 4. Consider cases of "switching", which have been much discussed in the philosophical literature. A train is heading toward the station. An engineer throws a switch, directing the train down the left track, rather than the right track. The tracks re-converge before the station, and the train arrives as scheduled. Was throwing the switch a cause of the train's arrival? HP consider two causal models of this scenario. In the first, there is a random variable $S$ which is 1 if the switch is thrown (so the train goes down the left track) and 0 otherwise. In the second, in addition to $S$, there are variables $LT$ and $RT$, indicating whether or not the train goes down the left track and right track, respectively. Note that with the first representation, there is no way to model the train not making it to the arrival point. With the second representation, we have the problem that $LT = 1$ and $RT = 1$ are arguably not independent; the train cannot be on both tracks at once. If we want to model the possibility of one track or another being blocked, we should use, instead of $LT$ and $RT$, variables $LB$ and $RB$, which indicate whether the left track or right track, respectively, are blocked. This allows us to represent all the relevant possibilities without running into independence problems. Note that if we have only $S$ as a random variable, then $S = 1$ cannot be a cause of the train arriving; it would have arrived no matter what. With $RB$ in the picture, the preliminary HP definition of actual cause rules that $S = 1$ can be an actual cause of the train's arrival; for example, under the contingency that $RB = 1$, the train does not arrive if $S = 0$. (However, once we extend the definition to include defaults, as we will in the next section, it becomes possible once again to block this conclusion.)

These rules will have particular consequences for how we should represent events that might occur at different times. Consider the following simplification of an example introduced by Bennett [1987], and also considered in HP.

EXAMPLE 5. Suppose that the Careless Camper (CC for short) has plans to go camping on the first weekend in June. He will go camping unless there is a fire in the forest in May. If he goes camping, he will leave a campfire unattended, and there will be a forest fire. Let the variable $C$ take the value 1 if CC goes camping, and 0 otherwise. How should we represent the state of the forest?

There appear to be at least three alternatives. The simplest proposal would be to use a variable $F$ that takes the value 1 if there is a forest fire at some time, and 0 otherwise.[7] But now how are we to represent the dependency relations between $F$

---

[7] This is, in effect, how effects have been represented using "neuron diagrams" in late preemption

and $C$? Since CC will go camping only if there is no fire (in May), we would want to have an equation such as $C = 1 - F$. On the other hand, since there will be a fire (in June) just in case CC goes camping, we will also need $F = C$. This representation is clearly not rich enough, since it does not let us make the clearly relevant distinction between whether the forest fire occurs in May or June. The problem is manifested in the fact that the equations are cyclic, and have no consistent solution.[8]

A second alternative, adopted by Halpern and Pearl [2005, p. 860], would be to use a variable $F'$ that takes the value 0 if there is no fire, 1 if there is a fire in May, and 2 if there is a fire in June. Now how should we write our equations? Since CC will go camping unless there is a fire in May, the equation for $C$ should say that $C = 0$ iff $F' = 1$. And since there will be a fire in June if CC goes camping, the equation for $F'$ should say that $F' = 2$ if $C = 1$ and $F' = 0$ otherwise. These equations are cyclic. Moreover, while they do have a consistent solution, they are highly misleading in what they predict about the effects of interventions. For example, the first equation tells us that intervening to create a forest fire in June would cause CC to go camping in the beginning of June. But this seems to get the causal order backwards!

The third way to model the scenario is to use two separate variables, $F_1$ and $F_2$, to represent the state of the forest at separate times. $F_1 = 1$ will represent a fire in May, and $F_1 = 0$ represents no fire in May; $F_2 = 1$ represents a fire in June and $F_2 = 0$ represents no fire in June. Now we can write our equations as $C = 1 - F_1$ and $F_2 = C \times (1 - F_1)$. This representation is free from the defects that plague the other two representations. We have no cycles, and hence there will be a consistent solution for any value of the exogenous variables. Moreover, this model correctly tells us that only an intervention on the state of the forest in May will affect CC's camping plans.

Once again, our discussion of the methodology of modeling parallels certain metaphysical discussions in the philosophy literature. If heavy rains delay the onset of a fire, is it the same fire that would have occurred without the rains, or a different fire? It is hard to see how to gain traction on such an issue by direct metaphysical speculation. By contrast, when we recast the issue as one about what kinds of variables to include in causal models, it is possible to say exactly how the models will mislead you if you make the wrong choice.

---

cases. See Hitchcock [2007b, pp. 85–88] for discussion.

[8]Careful readers will note the the preemption case of Example 3 is modeled in this way. In that model, $BH$ is a cause of $BS$, even though it is the earlier shattering of the bottle that prevents Billy's rock from hitting. Halpern and Pearl [2005] note this problem and offer a dynamic model akin to the one recommended below. As it turns out, this does not affect the analysis of the example offered above.

## 5 Dealing with normality and typicality

While the definition of causality given in Definition 1 works well in many cases, it does not always deliver answers that agree with (most people's) intuition. Consider the following example, taken from Hitchcock [2007a], based on an example due to Hiddleston [2005].

EXAMPLE 6. Assassin is in possession of a lethal poison, but has a last-minute change of heart and refrains from putting it in Victim's coffee. Bodyguard puts antidote in the coffee, which would have neutralized the poison had there been any. Victim drinks the coffee and survives. Is Bodyguard's putting in the antidote a cause of Victim surviving? Most people would say no, but according to the preliminary HP definition, it is. For in the contingency where Assassin puts in the poison, Victim survives iff Bodyguard puts in the antidote.

Example 6 illustrates an even deeper problem with Definition 1. The structural equations for Example 6 are *isomorphic* to those in the forest-fire example, provided that we interpret the variables appropriately. Specifically, take the endogenous variables in Example 6 to be $A$ (for "assassin does not put in poison"), $B$ (for "bodyguard puts in antidote"), and $VS$ (for "victim survives"). Then $A$, $B$, and $VS$ satisfy exactly the same equations as $L$, $ML$, and $F$, respectively. In the context where there is lightning and the arsonists drops a lit match, both the lightning and the match are causes of the forest fire, which seems reasonable. But here it does not seem reasonable that Bodyguard's putting in the antidote is a cause. Nevertheless, any definition that just depends on the structural equations is bound to give the same answers in these two examples. (An example illustrating the same phenomenon is given by Hall [2007].) This suggests that there must be more to causality than just the structural equations. And, indeed, the final HP definition of causality allows certain contingencies to be labeled as "unreasonable" or "too farfetched"; these contingencies are then not considered in AC2(a) or AC2(b). As discussed by Halpern [2008], there are problems with the HP account; we present here the approach used in [Halpern 2008] for dealing with these problems, which involves assuming that an agent has, in addition to a theory of causality (as modeled by the structural equations), a theory of "normality" or "typicality". (The need to consider normality was also stressed by Hitchcock [2007a] and Hall [2007], and further explored by Hitchcock and Knobe [2009].) This theory would include statements like "typically, people do not put poison in coffee" and "typically doctors do not treat patients to whom they are not assigned". There are many ways of giving semantics to such typicality statements (e.g., [Adams 1975; Kraus, Lehmann, and Magidor 1990; Spohn 2009]). For definiteness, we use *ranking functions* [Spohn 2009] here.

Take a *world* to be a complete description of the values of all the random variables. we assume that each world has associated with it a *rank*, which is just a natural number or $\infty$. Intuitively, the higher the rank, the less "normal" or "typical" the world. A world with a rank of 0 is reasonably normal, one with a rank of 1 is somewhat normal, one with a rank of 2 is quite abnormal, and so on. Given a ranking on worlds, the statement "if $p$ then typically $q$" is true if in all the worlds of least rank where $p$ is true, $q$ is also true. Thus, in one model where people do not typically put either poison or antidote in coffee, the worlds where neither poison nor antidote is put in the coffee have rank 0, worlds where either poison or antidote is put in the coffee have rank 1, and worlds where both poison and antidote are put in the coffee have rank 2.

Take an *extended causal model* to be a tuple $M = (\mathcal{S}, \mathcal{F}, \kappa)$, where $(\mathcal{S}, \mathcal{F})$ is a causal model, and $\kappa$ is a *ranking function* that associates with each world a rank. In an acyclic extended causal model, a context $\vec{u}$ determines a world, denoted $s_{\vec{u}}$. $\vec{X} = \vec{x}$ is a *cause of $\phi$ in an extended model $M$ and context $\vec{u}$* if $\vec{X} = \vec{x}$ is a cause of $\phi$ according to Definition 1, except that in AC2(a), there must be a world $s$ such that $\kappa(s) \leq \kappa(s_{\vec{u}})$ and $\vec{X} = \vec{x}' \wedge \vec{W} = \vec{w}$ is true at $s$. This can be viewed as a formalization of Kahneman and Miller's [1986] observation that "an event is more likely to be undone by altering exceptional than routine aspects of the causal chain that led to it".

This definition deals well with all the problematic examples in the literature. Consider Example 6. Using the ranking described above, Bodyguard is not a cause of Victim's survival because the world that would need to be considered in AC2(a), where Assassin poisons the coffee, is less normal than the actual world, where he does not. We consider just one other example here (see [Halpern 2008] for further discussion).

EXAMPLE 7. Consider the following story, taken from (an early version of) [Hall 2004]: Suppose that Billy is hospitalized with a mild illness on Monday; he is treated and recovers. In the obvious causal model, the doctor's treatment is a cause of Billy's recovery. Moreover, if the doctor does *not* treat Billy on Monday, then the doctor's omission to treat Billy is a cause of Billy's being sick on Tuesday. But now suppose that there are 100 doctors in the hospital. Although only doctor 1 is assigned to Billy (and he forgot to give medication), in principle, any of the other 99 doctors could have given Billy his medication. Is the nontreatment by doctors 2–100 also a cause of Billy's being sick on Tuesday?

Suppose that in fact the hospital has 100 doctors and there are variables $A_1, \ldots, A_{100}$ and $T_1, \ldots, T_{100}$ in the causal model, where $A_i = 1$ if doctor $i$ is assigned to treat Billy and $A_i = 0$ if he is not, and $T_i = 1$ if doctor $i$ actually treats Billy on Monday, and $T_i = 0$ if he does not. Doctor 1 is assigned to treat Billy; the others are not. However, in fact, no doctor treats Billy. Further assume that, typically, no doctor is assigned to a given patient; if doctor $i$ is not assigned to treat Billy, then typically doctor $i$ does not treat Billy; and if doctor $i$ *is* assigned to Billy, then typically doctor $i$ treats Billy. We can capture this in an extended causal model where the world where no doctor is assigned to Billy and no doctor treats him has rank 0; the 100 worlds where exactly one doctor is assigned to Billy, and that doctor treats him, have rank 1; the 100 worlds where exactly one doctor is assigned to Billy and no one treats him have rank 2; and the $100 \times 99$ worlds where exactly one doctor is assigned to Billy but some other doctor treats him have rank 3. (The ranking given to other worlds is irrelevant.) In this extended model, in the context where doctor $i$ is assigned to Billy but no one treats him, $i$ is the cause of Billy's sickness (the world where $i$ treats Billy has lower rank than the world where $i$ is assigned to Billy but no one treats him), but no other doctor is a cause of Billy's sickness. Moreover, in the context where $i$ is assigned to Billy and treats him, then $i$ is the cause of Billy's recovery (for AC2(a), consider the world where no doctor is assigned to Billy and none treat him).

Adding a normality theory to the model gives the HP account of actual causation greater flexibility to deal with these kinds of cases. This raises the worry, however, that this gives the modeler too much flexibility. After all, the modeler can now render any claim that $A$ is an actual cause of $B$ false, simply by choosing a normality order that assigns the actual world $s_{\vec{u}}$ a lower rank than any world $s$ needed to satisfy AC2. Thus, the introduction of normality exacerbates the problem of motivating and defending a particular choice of model. Fortunately, the literature on the psychology of counterfactual reasoning and causal judgment goes some way toward enumerating the sorts of factors that constitute normality. (See, for example, [Alicke 1992; Cushman 2009; Cushman, Knobe, and Sinnott-Armstrong 2008; Hitchcock and Knobe 2009; Kahneman and Miller 1986; Knobe and Fraser 2008; Kahneman and Tversky 1982; Mandel, Hilton, and Catellani 1985; Roese 1997].) These factors include the following:

- Statistical norms concern what happens most often, or with the greatest frequency. Kahneman and Tversky [1982] gave subjects a story in which Mr. Jones usually leaves work at 5:30, but occasionally leaves early to run errands. Thus, a 5:30 departure is (statistically) "normal", and an earlier departure "abnormal". This difference affected which alternate possibilities subjects were willing to consider when reflecting on the causes of an accident in which Mr. Jones was involved.

- Norms can involve moral judgments. Cushman, Knobe, and Sinnott-Armstrong [2008] showed that people with different views about the morality of abortion have different views about the abnormality of insufficient care for a fetus, and this can lead them to make different judgments about the cause of a miscarriage.

- Policies adopted by social institutions can also be norms. For instance, Knobe and Fraser [2008] presented subjects with a hypothetical situation in which a department had implemented a policy allowing administrative assistants to take pens from the department office, but prohibiting faculty from doing so. Subjects were more likely to attribute causality to a professor's taking a pen than to an assistant's taking one, even when the situation was otherwise similar.

- There can also be norms of "proper functioning" governing the operations of biological organs or mechanical parts: there are certain ways that hearts and spark plugs are "supposed" to operate. Hitchcock and Knobe [2009] show that these kinds of norms can also affect causal judgments.

The law suggests a variety of principles for determining the norms that are used in the evaluation of actual causation. In criminal law, norms are determined by direct legislation. For example, if there are legal standards for the strength of seat belts in an automobile, a seat belt that did not meet this standard could be judged a cause of a traffic fatality. By contrast, if a seat belt complied with the legal standard, but nonetheless broke because of the extreme forces it was subjected to during a particular accident, the fatality would be blamed on the circumstances of the accident, rather than the seat belt. In such a case, the manufacturers of the seat belt would not be guilty of criminal negligence. In contract law, compliance with the terms of a contract has the force of a norm. In tort law, actions are often judged against the standard of "the reasonable person". For instance, if a bystander was harmed when a pedestrian who was legally crossing the street suddenly jumped out of the way of an oncoming car, the pedestrian would not be held liable for damages to the bystander, since he acted as the hypothetical "reasonable person" would have done in similar circumstances. (See, for example, [Hart and Honoré 1985, pp. 142ff.] for discussion.) There are also a number of circumstances in which deliberate malicious acts of third parties are considered to be "abnormal" interventions, and affect the assessment of causation. (See, for example, [Hart and Honoré 1985, pp. 68ff.].)

As with the choice of variables, we do not expect that these considerations will always suffice to pick out a uniquely correct theory of normality for a causal model. They do, however, provide resources for a rational critique of models.

## 6 Conclusion

As HP stress, causality is relative to a model. That makes it particularly important to justify whatever model is chosen, and to enunciate principles for what makes a reasonable causal model. We have taken some preliminary steps in investigating this issue with regard to the choice of variables and the choice of defaults. However, we hope that we have convinced the reader that far more needs to be done if causal models are actually going to be used in applications.

**Acknowledgments:** We thank Wolfgang Spohn for useful comments. Joseph Halpern was supported in part by NSF grants IIS-0534064 and IIS-0812045, and by AFOSR grants FA9550-08-1-0438 and FA9550-05-1-0055.

# References

Adams, E. (1975). *The Logic of Conditionals.* Dordrecht, Netherlands: Reidel.

Alicke, M. (1992). Culpable causation. *Journal of Personality and Social Psychology 63*, 368–378.

Bennett, J. (1987). Event causation: the counterfactual analysis. In *Philosophical Perspectives, Vol. 1, Metaphysics*, pp. 367–386. Atascadero, CA: Ridgeview Publishing Company.

Cushman, F. (2009). The role of moral judgment in causal and intentional attribution: What we say or how we think?". Unpublished manuscript.

Cushman, F., J. Knobe, and W. Sinnott-Armstrong (2008). Moral appraisals affect doing/allowing judgments. *Cognition 108*(1), 281–289.

Davidson, D. (1967). Causal relations. *Journal of Philosophy LXIV*(21), 691–703.

Glymour, C. and F. Wimberly (2007). Actual causes and thought experiments. In J. Campbell, M. O'Rourke, and H. Silverstein (Eds.), *Causation and Explanation*, pp. 43–67. Cambridge, MA: MIT Press.

Goldberger, A. S. (1972). Structural equation methods in the social sciences. *Econometrica 40*(6), 979–1001.

Hall, N. (2004). Two concepts of causation. In J. Collins, N. Hall, and L. A. Paul (Eds.), *Causation and Counterfactuals.* Cambridge, Mass.: MIT Press.

Hall, N. (2007). Structural equations and causation. *Philosophical Studies 132*, 109–136.

Halpern, J. Y. (2008). Defaults and normality in causal structures. In *Principles of Knowledge Representation and Reasoning: Proc. Eleventh International Conference (KR '08)*, pp. 198–208.

Halpern, J. Y. and J. Pearl (2005). Causes and explanations: A structural-model approach. Part I: Causes. *British Journal for Philosophy of Science 56*(4), 843–887.

Hansson, R. N. (1958). *Patterns of Discovery.* Cambridge, U.K.: Cambridge University Press.

Hart, H. L. A. and T. Honoré (1985). *Causation in the Law* (second ed.). Oxford, U.K.: Oxford University Press.

Hiddleston, E. (2005). Causal powers. *British Journal for Philosophy of Science 56*, 27–59.

Hitchcock, C. (2001). The intransitivity of causation revealed in equations and graphs. *Journal of Philosophy XCVIII*(6), 273–299.

Hitchcock, C. (2007a). Prevention, preemption, and the principle of sufficient reason. *Philosophical Review 116*, 495–532.

Hitchcock, C. (2007b). What's wrong with neuron diagrams? In J. Campbell, M. O'Rourke, and H. Silverstein (Eds.), *Causation and Explanation*, pp. 69–92. Cambridge, MA: MIT Press.

Hitchcock, C. (2010). Trumping and contrastive causation. *Synthese*. To appear.

Hitchcock, C. and J. Knobe (2009). Cause and norm. *Journal of Philosophy*. To appear.

Hume, D. (1739). *A Treatise of Human Nature*. London: John Noon.

Hume, D. (1748). *An Enquiry Concerning Human Understanding*. Reprinted by Open Court Press, LaSalle, IL, 1958.

Kahneman, D. and D. T. Miller (1986). Norm theory: comparing reality to its alternatives. *Psychological Review 94*(2), 136–153.

Kahneman, D. and A. Tversky (1982). The simulation heuristic. In D. Kahneman, P. Slovic, and A. Tversky (Eds.), *Judgment Under Incertainty: Heuristics and Biases*, pp. 201–210. Cambridge/New York: Cambridge University Press.

Kim, J. (1973). Causes, nomic subsumption, and the concept of event. *Journal of Philosophy LXX*, 217–236.

Knobe, J. and B. Fraser (2008). Causal judgment and moral judgment: two experiments. In W. Sinnott-Armstrong (Ed.), *Moral Psychology, Volume 2: The Cognitive Science of Morality*, pp. 441–447. Cambridge, MA: MIT Press.

Kraus, S., D. Lehmann, and M. Magidor (1990). Nonmonotonic reasoning, preferential models and cumulative logics. *Artificial Intelligence 44*, 167–207.

Lewis, D. (1973a). Causation. *Journal of Philosophy 70*, 113–126. Reprinted with added "Postscripts" in D. Lewis, *Philosophical Papers*, Volume II, Oxford University Press, 1986, pp. 159–213.

Lewis, D. (1986a). Causation. In *Philosophical Papers*, Volume II, pp. 159–213. New York: Oxford University Press. The original version of this paper, without numerous postscripts, appeared in the *Journal of Philosophy* **70**, 1973, pp. 113–126.

Lewis, D. (1986b). Events. In *Philosophical Papers*, Volume II, pp. 241–270. New York: Oxford University Press.

Lewis, D. K. (1973b). *Counterfactuals*. Cambridge, Mass.: Harvard University Press.

Mandel, D. R., D. J. Hilton, and P. Catellani (Eds.) (1985). *The Psychology of Counterfactual Thinking*. New York: Routledge.

Pearl, J. (2000). *Causality: Models, Reasoning, and Inference*. New York: Cambridge University Press.

Roese, N. (1997). Counterfactual thinking. *Psychological Bulletin CXXI*, 133–148.

Schaffer, J. (2000). Trumping preemption. *Journal of Philosophy XCVII*(4), 165–181. Reprinted in J. Collins and N. Hall and L. A. Paul (eds.), *Causation and Counterfactuals*, MIT Press, 2002.

Spohn, W. (2003). Dependency equilibria and the causal structure of decision and game situations. In *Homo Oeconomicus XX*, pp. 195–255.

Spohn, W. (2009). A survey of ranking theory. In F. Huber and C. Schmidt-Petri (Eds.), *Degrees of Belief. An Anthology*, pp. 185–228. Dordrecht, Netherlands: Springer.

Woodward, J. (2003). *Making Things Happen: A Theory of Causal Explanation*. Oxford, U.K.: Oxford University Press.

# 24
# From C-Believed Propositions to the Causal Calculator

VLADIMIR LIFSCHITZ

## 1 Introduction

Default rules, unlike inference rules of classical logic, allow us to derive a new conclusion only when it does not conflict with the other available information. The best known example is the so-called commonsense law of inertia: in the absence of information to the contrary, properties of the world can be presumed to be the same as they were in the past. Making the idea of commonsense inertia precise is known as the frame problem [Shanahan 1997]. Default reasoning is nonmonotonic, in the sense that we may be forced to retract a conclusion derived using a default when additional information becomes available.

The idea of a default first attracted the attention of AI researchers in the 1970s. Developing a formal semantics of defaults turned out to be a difficult task. For instance, the attempt to describe commonsense inertia in terms of circumscription outlined in [McCarthy 1986] was unsatisfactory, as we learned from the Yale Shooting example [Hanks and McDermott 1987].

In this note, we trace the line of work on the semantics of defaults that started with Judea Pearl's 1988 paper on the difference between "E-believed" and "C-believed" propositions. That paper has led other researchers first to the invention of several theories of nonmonotonic causal reasoning, then to designing action languages $\mathcal{C}$ and $\mathcal{C}+$, and then to the creation of the Causal Calculator—a software system for automated reasoning about action and change.

## 2 Starting Point: Labels E and C

The paper *Embracing Causality in Default Reasoning* [Pearl 1988] begins with the observation that

> almost every default rule falls into one of two categories: expectation-evoking or explanation-evoking. The former describes association among events in the outside world (e.g., fire is typically accompanied by smoke); the latter describes how we reason about the world (e.g., smoke normally suggests fire).

Thus the rule fire ⇒ smoke is an expectation-evoking, or "causal" default; the rule smoke ⇒ fire is explanation-evoking, or "evidential." To take another example,

(1) rained ⇒ grass_wet

is a causal default;

(2) grass_wet ⇒ sprinkler_on

is an evidential default.

To discuss the distinction between properties of causal and evidential defaults, Pearl labels believed propositions by distinguishing symbols $C$ and $E$. A proposition $P$ is $E$-believed, written $E(P)$, if it is a direct consequence of some evidential rule. Otherwise, if $P$ can be established as a direct consequence of only causal rules, it is said to be $C$-believed, written $C(P)$. The labels are used to prevent certain types of inference chains; in particular, $C$-believed propositions are prevented in Pearl's paper from triggering evidential defaults. For example, both causal rule (1) and evidential rule (2) are reasonable, but using them to infer sprinkler_on from rained is not.

We will see that the idea of using the distinguishing symbols $C$ and $E$ had a significant effect on the study of commonsense reasoning over the next twenty years.

## 3 "Explained" as a Modal Operator

The story continues with Hector Geffner's proposal to turn the label $C$ into a modal operator and to treat Pearl's causal rules as formulas of modal logic. A formula $F$ is considered "explained" if the formula $CF$ holds.

> A rule such as "rain causes the grass to be wet" may thus be expressed as a sentence
>
> $$\text{rain} \rightarrow C\,\text{grass\_wet},$$
>
> which can then be read as saying that if rain is true, grass_wet is explained [Geffner 1990].

The paper defined, for a set of axioms of this kind, which propositions are "causally entailed" by it.

Geffner showed how this modal language can be used to describe effects of actions. We can express that $e(x)$ is an effect of an action $a(x)$ with precondition $p(x)$ by the axiom

(3) $p(x)_t \wedge a(x)_t \rightarrow Ce(x)_{t+1},$

where $p(x)_t$ expresses that fluent $p(x)$ holds at time $t$, and $e(x)_{t+1}$ is understood in a similar way; $a(x)_t$ expresses that action $a(x)$ is executed between times $t$ and $t+1$.

Such axioms explain the value of a fluent at some point in time ($t+1$ in the consequent of the implication) in terms of the past ($t$ in the antecedent). Geffner gives also an example of explaining the value of a fluent in terms of the values of other fluents at the same point in time: if all ducts are blocked at time $t$, that causes the room to be stuffy at time $t$. Such "static" causal dependencies are instrumental when actions with indirect effects are involved. For instance, blocking a duct can indirectly cause the room to become stuffy. We will see another example of this kind in the next section.

## 4  Predicate "Caused"

Fangzhen Lin showed a few years later that the intuitions explored by Pearl and Geffner can be made precise without introducing a new nonmonotonic semantics. Circumscription [McCarthy 1986] will do if we employ, instead of the modal operator C, a new predicate.

> Technically, we introduce a new ternary predicate *Caused* into the situation calculus: $Caused(p, v, s)$ if the proposition $p$ is caused (by something unspecified) to have the truth value $v$ in the state $s$ [Lin 1995].

The counterpart of formula (3) in this language is

(4)  $p(x, s) \to Caused(e(x), true, do(a(x), s))$.

Lin acknowledges his intellectual debt to [Pearl 1988] by noting that his approach echoes the theme of Pearl's paper—the need for a primitive notion of causality in default reasoning.

The proposal to circumscribe *Caused* was a major event in the history of research on the use of circumscription for solving the frame problem. As we mentioned before, the original method [McCarthy 1986] turned out to be unsatisfactory; the improvement described in [Haugh 1987; Lifschitz 1987] is only applicable when actions have no indirect effects. The method of [Lin 1995] is free of this limitation. The main example of that paper is a suitcase with two locks and a spring loaded mechanism that opens the suitcase instantaneously when both locks are in the up position; opening the suitcase may thus become an indirect effect of toggling a switch. The static causal relationship between the fluents $up(l)$ and $open$ is expressed in Lin's language by the axiom

(5)  $up(L1, s) \land up(L2, s) \to Caused(open, true, s)$.

## 5  Principle of Universal Causation

Yet another important modification of Geffner's theory was proposed in [McCain and Turner 1997]. That approach was originally limited to formulas of the form

$$F \to \mathsf{C}G,$$

where $F$ and $G$ do not contain C. (Such formulas are particularly useful; for instance, (3) has this form.) The authors wrote such a formula as

(6)  $F \Rightarrow G$,

so that the thick arrow $\Rightarrow$ represented in their paper a combination of material implication $\rightarrow$ with the modal operator C. In [Turner 1999], that method was extended to the full language of [Geffner 1990].

The key idea of this theory of causal knowledge is described in [McCain and Turner 1997] as follows:

> Intuitively, in a causally possible world history every fact that is caused obtains. We assume in addition the *principle of universal causation*, according to which—in a causally possible world history—every fact that obtains is caused. In sum, we say that a world history is causally possible if exactly the facts that obtain in it are caused in it.

The authors note that the principle of universal causation represents a strong philosophical commitment that is rewarded by the mathematical simplicity of the nonmonotonic semantics that it leads to. The definition of their semantics is indeed surprisingly simple, or at least short. They note also that in applications this strong commitment can be easily relaxed.

The extension of [McCain and Turner 1997] described in [Giunchiglia, Lee, Lifschitz, McCain, and Turner 2004] allows $F$ and $G$ in (6) to be slightly more general than propositional formulas, which is convenient when non-Boolean fluents are involved. In the language of that paper we can write, for instance,

(7) $\quad a_t \Rightarrow f_{t+1} = v$

to express that executing action $a$ causes fluent $f$ to take value $v$.

## 6 Action Descriptions

An action description is a formal expression representing a transition system—a directed graph such that its vertices can be interpreted as states of the world, with edges corresponding to the transitions caused by the execution of actions. In [Giunchiglia and Lifschitz 1998], the nonmonotonic causal logic from [McCain and Turner 1997] was used to define an action description language, called $\mathcal{C}$. The language $\mathcal{C}+$ [Giunchiglia, Lee, Lifschitz, McCain, and Turner 2004] is an extension of $\mathcal{C}$ that accomodates non-Boolean fluents and is also more expressive in some other ways.

The distinguishing syntactic feature of action description languages is that they do not involve symbols for time instants. For example, the counterpart of (7) in $\mathcal{C}+$ is

$$a \textbf{ causes } f = v.$$

The $\mathcal{C}+$ keyword **causes** implicitly indicates a shift from the time instant $t$ when the execution of action $a$ begins to the next time instant $t+1$ when fluent $f$ is evaluated. This keyword represents a combination of three elements: material implication, the Pearl-Geffner causal operator, and time shift.

## 7 The Causal Calculator

Literal completion, defined in [McCain and Turner 1997], is a modification of the completion process familiar from logic programming [Clark 1978]. It is applicable to any finite set $T$ of causal laws (6) whose heads $G$ are literals, and produces a set of propositional formulas such that its models in the sense of propositional logic are identical to the models of $T$ in the sense of the McCain-Turner causal logic. Literal completion can be used to reduce some computational problems involving $\mathcal{C}$ action descriptions to the propositional satisfiability problem.

This idea is used in the design of the Causal Calculator (CCALC)—a software system that reasons about actions in domains described in a subset of $\mathcal{C}$ [McCain 1997]. CCALC performs search by invoking a SAT solver in the spirit of the "planning as satisfiability" method of [Kautz and Selman 1992]. Version 2 of CCALC [Lee 2005] extends it to $\mathcal{C}+$ action descriptions.

The Causal Calculator has been succesfully applied to several challenge problems in the theory of commonsense reasoning [Lifschitz, McCain, Remolina, and Tacchella 2000], [Lifschitz 2000], [Akman, Erdoğan, Lee, Lifschitz, and Turner 2004]. More recently, it was used for the executable specification of norm-governed computational societies [Artikis, Sergot, and Pitt 2009] and for the automatic analysis of business processes under authorization constraints [Armando, Giunchiglia, and Ponta 2009].

## 8 Conclusion

As we have seen, Judea Pearl's idea of labeling the propositions that are derived using causal rules has suggested to Geffner, Lin and others that the condition

$$G \text{ is caused (by something unspecified) if } F \text{ holds}$$

can be sometimes used as an approximation to

$$G \text{ is caused by } F.$$

Eliminating the binary "is caused by" in favor of the unary "is caused" turned out to be a remarkably useful technical device.

## 9 Acknowledgements

Thanks to Selim Erdoğan, Hector Geffner, and Joohyung Lee for comments on a draft of this note. This work was partially supported by the National Science Foundation under Grant IIS-0712113.

## References

Akman, V., S. Erdoğan, J. Lee, V. Lifschitz, and H. Turner (2004). Representing the Zoo World and the Traffic World in the language of the Causal Calculator. *Artificial Intelligence 153(1-2)*, 105–140.

Armando, A., E. Giunchiglia, and S. E. Ponta (2009). Formal specification and automatic analysis of business processes under authorization constraints: an action-based approach. In *Proceedings of the 6th International Conference on Trust, Privacy and Security in Digital Business (TrustBus'09)*.

Artikis, A., M. Sergot, and J. Pitt (2009). Specifying norm-governed computational societies. *ACM Transactions on Computational Logic 9*(1).

Clark, K. (1978). Negation as failure. In H. Gallaire and J. Minker (Eds.), *Logic and Data Bases*, pp. 293–322. New York: Plenum Press.

Geffner, H. (1990). Causal theories for nonmonotonic reasoning. In *Proceedings of National Conference on Artificial Intelligence (AAAI)*, pp. 524–530. AAAI Press.

Giunchiglia, E., J. Lee, V. Lifschitz, N. McCain, and H. Turner (2004). Nonmonotonic causal theories. *Artificial Intelligence 153(1–2)*, 49–104.

Giunchiglia, E. and V. Lifschitz (1998). An action language based on causal explanation: Preliminary report. In *Proceedings of National Conference on Artificial Intelligence (AAAI)*, pp. 623–630. AAAI Press.

Hanks, S. and D. McDermott (1987). Nonmonotonic logic and temporal projection. *Artificial Intelligence 33*(3), 379–412.

Haugh, B. (1987). Simple causal minimizations for temporal persistence and projection. In *Proceedings of National Conference on Artificial Intelligence (AAAI)*, pp. 218–223.

Kautz, H. and B. Selman (1992). Planning as satisfiability. In *Proceedings of European Conference on Artificial Intelligence (ECAI)*, pp. 359–363.

Lee, J. (2005). *Automated Reasoning about Actions*.[1] Ph.D. thesis, University of Texas at Austin.

Lifschitz, V. (1987). Formal theories of action (preliminary report). In *Proceedings of International Joint Conference on Artificial Intelligence (IJCAI)*, pp. 966–972.

Lifschitz, V. (2000). Missionaries and cannibals in the Causal Calculator. In *Proceedings of International Conference on Principles of Knowledge Representation and Reasoning (KR)*, pp. 85–96.

Lifschitz, V., N. McCain, E. Remolina, and A. Tacchella (2000). Getting to the airport: The oldest planning problem in AI. In J. Minker (Ed.), *Logic-Based Artificial Intelligence*, pp. 147–165. Kluwer.

Lin, F. (1995). Embracing causality in specifying the indirect effects of actions. In *Proceedings of International Joint Conference on Artificial Intelligence (IJCAI)*, pp. 1985–1991.

---

[1] http://peace.eas.asu.edu/joolee/papers/dissertation.pdf .

McCain, N. (1997). *Causality in Commonsense Reasoning about Actions.*[2] Ph.D. thesis, University of Texas at Austin.

McCain, N. and H. Turner (1997). Causal theories of action and change. In *Proceedings of National Conference on Artificial Intelligence (AAAI)*, pp. 460–465.

McCarthy, J. (1986). Applications of circumscription to formalizing common sense knowledge. *Artificial Intelligence 26*(3), 89–116.

Pearl, J. (1988). Embracing causality in default reasoning (research note). *Artificial Intelligence 35(2)*, 259–271.

Shanahan, M. (1997). *Solving the Frame Problem: A Mathematical Investigation of the Common Sense Law of Inertia.* MIT Press.

Turner, H. (1999). A logic of universal causation. *Artificial Intelligence 113*, 87–123.

---

[2] ftp://ftp.cs.utexas.edu/pub/techreports/tr97-25.ps.gz .

# 25
# Analysis of the Binary Instrumental Variable Model

THOMAS S. RICHARDSON AND JAMES M. ROBINS

## 1 Introduction

Pearl's seminal work on instrumental variables [Chickering and Pearl 1996; Balke and Pearl 1997] for discrete data represented a leap forwards in terms of understanding: Pearl showed that, contrary to what many had supposed based on linear models, in the discrete case the assumption that a variable was an instrument could be subjected to empirical test. In addition, Pearl improved on earlier bounds [Robins 1989] for the average causal effect (ACE) in the absence of any monotonicity assumptions. Pearl's approach was also innovative insofar as he employed a computer algebra system to derive analytic expressions for the upper and lower bounds.

In this paper we build on and extend Pearl's work in two ways. First we show the geometry underlying Pearl's bounds. As a consequence we are able to derive bounds on the average causal effect for all four compliance types. Our analysis also makes it possible to perform a sensitivity analysis using the distribution over compliance types. Second our analysis provides a clear geometric picture of the instrumental inequalities, and allows us to isolate the counterfactual assumptions necessary for deriving these tests. This may be seen as analogous to the geometric study of models for two-way tables [Fienberg and Gilbert 1970; Erosheva 2005]. Among other things this allows us to clarify which are the alternative hypotheses against which Pearl's test has power. We also relate these tests to recent work of Pearl's on bounding direct effects [Cai, Kuroki, Pearl, and Tian 2008].

## 2 Background

We consider three binary variables, $X$, $Y$ and $Z$. Where:

$Z$ is the instrument, presumed to be randomized e.g. the assigned treatment;

$X$ is the treatment received;

$Y$ is the response.

For $X$ and $Z$, we will use 0 to indicate placebo, and 1 to indicate drug. For $Y$ we take 1 to indicate a desirable outcome, such as survival. $X_z$ is the treatment a

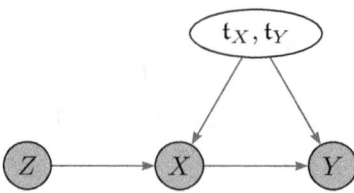

Figure 1. Graphical representation of the IV model given by assumptions (1) and (2). The shaded nodes are observed.

patient would receive if assigned to $Z = z$. We follow convention by referring to the four *compliance* types:

| $X_{z=0}$ | $X_{z=1}$ | Compliance Type | |
|---|---|---|---|
| 0 | 0 | Never Taker | NT |
| 0 | 1 | Complier | CO |
| 1 | 0 | Defier | DE |
| 1 | 1 | Always Taker | AT |

Since we suppose the counterfactuals are well-defined, if $Z = z$ then $X = X_z$. Similarly we consider counterfactuals $Y_{xz}$ for $Y$. Except where explicitly noted we will make the exclusion restrictions:

$$Y_{x=0,z=0} = Y_{x=0,z=1} \qquad Y_{x=1,z=0} = Y_{x=1,z=1} \qquad (1)$$

for each patient, so that a patient's outcome only depends on treatment assigned via the treatment received. One consequence of the analysis below is that these equations may be tested separately. We may thus similarly enumerate four types of patient in terms of their *response* to received treatment:

| $Y_{x=0}$ | $Y_{x=1}$ | Response Type | |
|---|---|---|---|
| 0 | 0 | Never Recover | *NR* |
| 0 | 1 | Helped | *HE* |
| 1 | 0 | Hurt | *HU* |
| 1 | 1 | Always Recover | *AR* |

As before, it is implicit in our notation that if $X = x$, then $Y_x = Y$; this is referred to as the 'consistency assumption' (or axiom) by Pearl among others. In what follows we will use $t_X$ to denote a generic compliance type in the set $\mathbb{D}_X$, and $t_Y$ to denote a generic response type in the set $\mathbb{D}_Y$. We thus have 16 patient types:

$$\langle t_X, t_Y \rangle \in \{NT, CO, DE, AT\} \times \{NR, HE, HU, AR\} \equiv \mathbb{D}_X \times \mathbb{D}_Y \equiv \mathbb{D}.$$

(Here and elsewhere we use angle brackets $\langle t_X, t_Y \rangle$ to indicate an ordered pair.) Let $\pi_{t_X} \equiv p(t_X)$ denote the marginal probability of a given compliance type $t_X \in \mathbb{D}_X$,

and let
$$\pi_X \equiv \{\pi_{t_X} \mid t_X \in \mathbb{D}_X\}$$
denote a marginal distribution on $\mathbb{D}_X$. Similarly we use $\pi_{t_Y|t_X} \equiv p(t_Y \mid t_X)$ to denote the probability of a given response type within the sub-population of individuals of compliance type $t_X$, and $\pi_{Y|X}$ to indicate a specification of all these conditional probabilities:
$$\pi_{Y|X} \equiv \{\pi_{t_Y|t_X} \mid t_X \in \mathbb{D}_X, t_Y \in \mathbb{D}_Y\}.$$
We will use $\pi$ to indicate a joint distribution $p(t_X, t_Y)$ on $\mathbb{D}$.

Except where explicitly noted we will make the randomization assumption that the distribution of types $\langle t_X, t_Y \rangle$ is the same in both arms:
$$Z \perp\!\!\!\perp \{X_{z=0}, X_{z=1}, Y_{x=0}, Y_{x=1}\}. \tag{2}$$
A graph corresponding to the model given by (1) and (2) is shown in Figure 1.

**Notation**

In places we will make use of the following compact notation for probability distributions:
$$\begin{aligned} p_{y_k|x_j z_i} &\equiv p(Y = k \mid X = j, Z = i), \\ p_{x_j|z_i} &\equiv p(X = j \mid Z = i), \\ p_{y_k x_j|z_i} &\equiv p(Y = k, X = j \mid Z = i). \end{aligned}$$

There are several simple geometric constructions that we will use repeatedly. In consequence we introduce these in a generic setting.

## 2.1 Joints compatible with fixed margins

Consider a bivariate random variable $U = \langle U_1, U_2 \rangle \in \{0,1\} \times \{0,1\}$. Now for fixed $c_1, c_2 \in [0,1]$ consider the set
$$\mathcal{P}_{c_1,c_2} = \left\{ p \; \middle| \; \sum_{u_2} p(1, u_2) = c_1 \; ; \; \sum_{u_1} p(u_1, 1) = c_2 \right\}$$
in other words, $\mathcal{P}_{c_1,c_2}$ is the set of joint distributions on $U$ compatible with fixed margins $p(U_i = 1) = c_i$, $i = 1, 2$.

It is not hard to see that $\mathcal{P}_{c_1,c_2}$ is a one-dimensional subset (line segment) of the 3-dimensional simplex of distributions for $U$. We may describe it explicitly as follows:
$$\left\{ \begin{array}{rl} p(1,1) &= t \\ p(1,0) &= c_1 - t \\ p(0,1) &= c_2 - t \\ p(0,0) &= 1 - c_1 - c_2 + t \end{array} \quad t \in \left[\max\{0, (c_1+c_2)-1\}, \min\{c_1, c_2\}\right] \right\}. \tag{3}$$

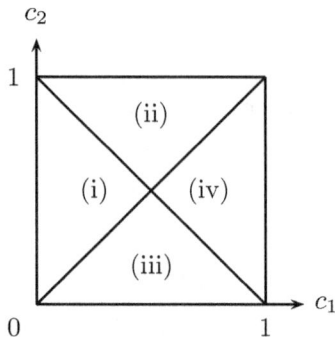

Figure 2. The four regions corresponding to different supports for $t$ in (3); see Table 1.

See also [Pearl 2000] Theorem 9.2.10. The range of $t$, or equivalently the support for $p(1,1)$, is one of four intervals, as shown in Table 1. These cases correspond to

|  | $c_1 \leq 1 - c_2$ |  | $c_1 \geq 1 - c_2$ |  |
|---|---|---|---|---|
| $c_1 \leq c_2$ | (i) | $t \in [0, c_1]$ | (ii) | $t \in [c_1 + c_2 - 1, c_1]$ |
| $c_1 \geq c_2$ | (iii) | $t \in [0, c_2]$ | (iv) | $t \in [c_1 + c_2 - 1, c_2]$ |

Table 1. The support for $t$ in (3) in each of the four cases relating $c_1$ and $c_2$.

the four regions show in Figure 2.

Finally, we note that since for $c_1, c_2 \in [0,1]$, $\max\{0, (c_1 + c_2) - 1\} \leq \min\{c_1, c_2\}$, it follows that $\{\langle c_1, c_2 \rangle \mid \mathcal{P}_{c_1,c_2} \neq \emptyset\} = [0,1]^2$. Thus for every pair of values $\langle c_1, c_2 \rangle$ there exists a joint distribution $p(U_1, U_2)$ for which $p(U_i = 1) = c_i$, $i = 1, 2$.

## 2.2 Two quantities with a specified average

We now consider the set:

$$\mathcal{Q}_{c,\alpha} = \{\langle u, v \rangle \mid \alpha u + (1 - \alpha)v = c, \ u, v \in [0,1]\}$$

where $c, \alpha \in [0,1]$. In words, $\mathcal{Q}_{c,\alpha}$ is the set of pairs of values $\langle u, v \rangle$ in $[0,1]$ which are such that the weighted average $\alpha u + (1 - \alpha)v$ is $c$.

It is simple to see that this describes a line segment in the unit square. Further consideration shows that for any value of $\alpha \in [0, 1]$, the segment will pass through the point $\langle c, c \rangle$ and will be contained within the union of two rectangles:

$$([c, 1] \times [0, c]) \cup ([0, c] \times [1, c]).$$

The slope of the line is negative for $\alpha \in (0, 1)$. For $\alpha \in (0, 1)$ the line segment may

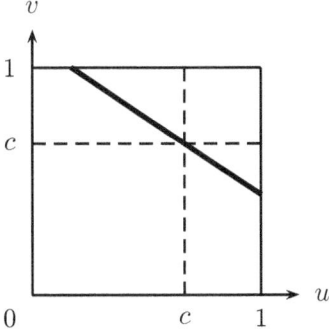

Figure 3. Illustration of $\mathcal{Q}_{c,\alpha}$.

be parametrized as follows:

$$\left\{ \begin{array}{rl} u & = (c-t(1-\alpha))/\alpha, \\ v & = t, \end{array} \quad t \in \left[ \max\left(0, \frac{c-\alpha}{1-\alpha}\right), \min\left(\frac{c}{1-\alpha}, 1\right) \right] \right\}.$$

The left and right endpoints of the line segment are:

$$\langle u, v \rangle = \left\langle \max(0, 1+(c-1)/\alpha),\ \min(c/(1-\alpha), 1) \right\rangle$$

and

$$\langle u, v \rangle = \left\langle \min(c/\alpha, 1),\ \max(0, (c-\alpha)/(1-\alpha)) \right\rangle$$

respectively. See Figure 3.

### 2.3 Three quantities with two averages specified

We now extend the discussion in the previous section to consider the set:

$$\mathcal{Q}_{(c_1,\alpha_1)(c_2,\alpha_2)} = \{ \langle u, v, w \rangle \mid \alpha_1 u + (1-\alpha_1)w = c_1,$$
$$\alpha_2 v + (1-\alpha_2)w = c_2,\ u, v, w \in [0, 1] \}.$$

In words, this consists of the set of triples $\langle u, v, w \rangle \in [0,1]^3$ for which pre-specified averages of $u$ and $w$ (via $\alpha_1$), and $v$ and $w$ (via $\alpha_2$) are equal to $c_1$ and $c_2$ respectively.

If this set is not empty, it is a line segment in $[0,1]^3$ obtained by the intersection of two rectangles:

$$\Big( \{ \langle u, w \rangle \in \mathcal{Q}_{c_1,\alpha_1} \} \times \{ v \in [0,1] \} \Big) \cap \Big( \{ \langle v, w \rangle \in \mathcal{Q}_{c_2,\alpha_2} \} \times \{ u \in [0,1] \} \Big); \quad (4)$$

see Figures 4 and 5. For $\alpha_1, \alpha_2 \in (0, 1)$ we may parametrize the line segment (4) as follows:

$$\left\{ \begin{array}{rl} u & = (c_1 - t(1-\alpha_1))/\alpha_1, \\ v & = (c_2 - t(1-\alpha_2))/\alpha_2, \quad t \in [t_l, t_u] \\ w & = t, \end{array} \right\},$$

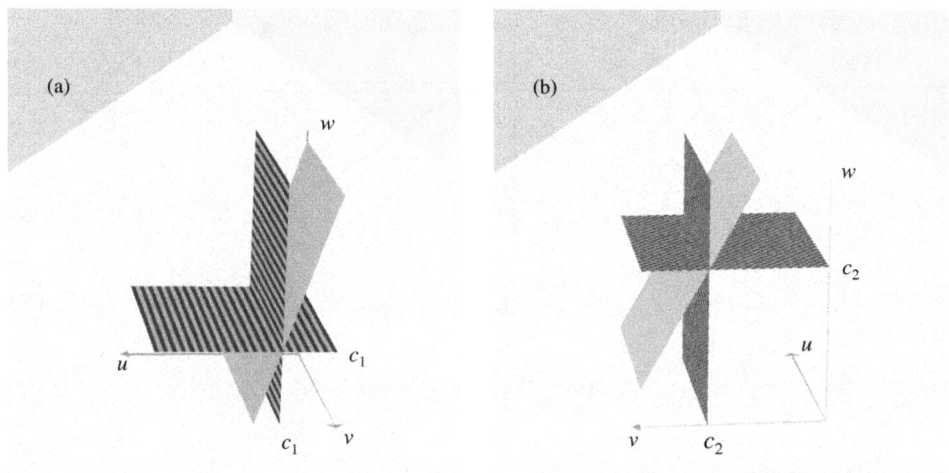

Figure 4. (a) The plane without stripes is $\alpha_1 u + (1 - \alpha_1)w = c_1$. (b) The plane without checks is $\alpha_2 v + (1 - \alpha_2)w = c_2$.

where

$$t_l \equiv \max\left\{0, \frac{c_1 - \alpha_1}{1 - \alpha_1}, \frac{c_2 - \alpha_2}{1 - \alpha_2}\right\}, \qquad t_u \equiv \min\left\{1, \frac{c_1}{1 - \alpha_1}, \frac{c_2}{1 - \alpha_2}\right\}.$$

Thus $\mathcal{Q}_{(c_1,\alpha_1)(c_2,\alpha_2)} \neq \emptyset$ if and only if $t_l \leq t_u$. It follows directly that for fixed $c_1$, $c_2$ the set of pairs $\langle \alpha_1, \alpha_2 \rangle \in [0,1]^2$ for which $\mathcal{Q}_{(c_1,\alpha_1)(c_2,\alpha_2)}$ is not empty may be characterized thus:

$$\mathcal{R}_{c_1,c_2} \equiv \left\{\langle \alpha_1, \alpha_2\rangle \,\middle|\, \mathcal{Q}_{(c_1,\alpha_1)(c_2,\alpha_2)} \neq \emptyset\right\}$$

$$= [0,1]^2 \cap \bigcap_{\substack{i \in \{1,2\} \\ i^* = 3-i}} \{\langle \alpha_1, \alpha_2 \rangle \mid (\alpha_i - c_i)(\alpha_{i^*} - (1 - c_{i^*})) \leq c_i^*(1 - c_i)\}. \qquad (5)$$

In fact, as shown in Figure 6 at most one constraint is active, so simplification is possible: let $k = \arg\max_j c_j$, and $k^* = 3 - k$, then

$$\mathcal{R}_{c_1,c_2} = [0,1]^2 \cap \{\langle \alpha_1, \alpha_2 \rangle \mid (\alpha_k - c_k)(\alpha_{k^*} - (1 - c_{k^*})) \leq c_k^*(1 - c_k)\}.$$

(If $c_1 = c_2$ then $\mathcal{R}_{c_1,c_2} = [0,1]^2$.)

In the two dimensional analysis in §2.2 we observed that for fixed $c$, as $\alpha$ varied, the line segment would always remain inside two rectangles, as shown in Figure 3. In the three dimensional situation, the line segment (4) will stay within three boxes:

(i) If $c_1 < c_2$ then the line segment (4) is within:

$$([0, c_1] \times [0, c_2] \times [c_2, 1]) \cup ([0, c_1] \times [c_2, 1] \times [c_1, c_2]) \cup ([c_1, 1] \times [c_2, 1] \times [0, c_1]).$$

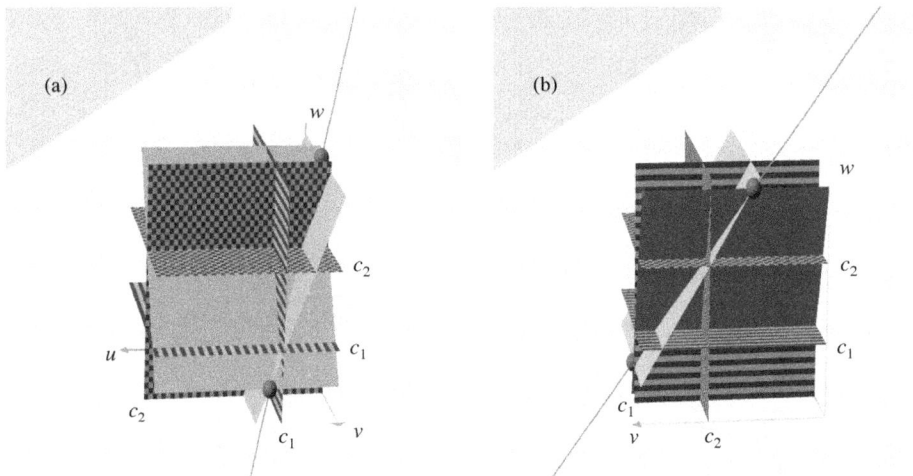

Figure 5. $\mathcal{Q}_{(c_1,\alpha_1)(c_2,\alpha_2)}$ corresponds to the section of the line between the two marked points; (a) view towards $u$-$w$ plane; (b) view from $v$-$w$ plane. (Here $c_1 < c_2$.)

This may be seen as a 'staircase' with a 'corner' consisting of three blocks, descending clockwise from $\langle 0,0,1 \rangle$ to $\langle 1,1,0 \rangle$; see Figure 7(a). The first and second boxes intersect in the line segment joining the points $\langle 0, c_2, c_2 \rangle$ and $\langle c_1, c_2, c_2 \rangle$; the second and third intersect in the line segment joining $\langle c_1, c_2, c_1 \rangle$ and $\langle c_1, 1, c_1 \rangle$.

(ii) If $c_1 > c_2$ then the line segment is within:

$$([0,c_1] \times [0,c_2] \times [c_1,1]) \cup ([c_1,1] \times [0,c_2] \times [c_2,c_1]) \cup ([c_1,1] \times [c_2,1] \times [0,c_2]).$$

This is a 'staircase' of three blocks, descending counter-clockwise from $\langle 0,0,1 \rangle$ to $\langle 1,1,0 \rangle$; see Figure 7(b). The first and second boxes intersect in the line segment joining the points $\langle c_1, 0, c_1 \rangle$ and $\langle c_1, c_2, c_1 \rangle$; the second and third intersect in the line segment joining $\langle c_1, c_2, c_2 \rangle$ and $\langle 1, c_2, c_2 \rangle$.

(iii) If $c_1 = c_2 = c$ then the 'middle' box disappears and we are left with

$$([0,c] \times [0,c] \times [c,1]) \cup ([c,1] \times [c,1] \times [0,c]).$$

In this case the two boxes touch at the point $\langle c, c, c \rangle$.

Note however, that the number of 'boxes' within which the line segment (4) lies may be 1, 2 or 3 (or 0 if $\mathcal{Q}_{(c_1,\alpha_1)(c_2,\alpha_2)} = \emptyset$). This is in contrast to the simpler case considered in §2.2 where the line segment $\mathcal{Q}_{c,\alpha}$ always intersected exactly two rectangles; see Figure 3.

## 3 Characterization of compatible distributions of type

Returning to the Instrumental Variable model introduced in §2, for a given patient the values taken by $Y$ and $X$ are deterministic functions of $Z$, $t_X$ and $t_Y$. Conse-

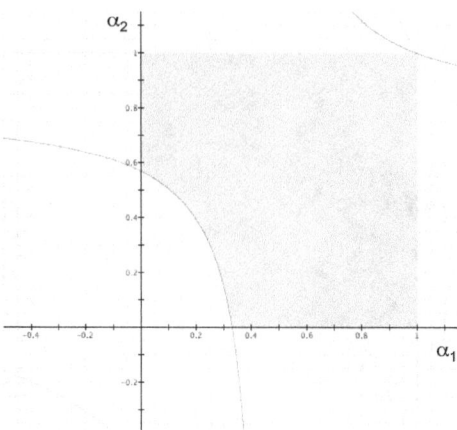

Figure 6. $\mathcal{R}_{c_1,c_2}$ corresponds to the shaded region. The hyperbola of which one arm forms a boundary of this region corresponds to the active constraint; the other hyperbola to the inactive constraint.

quently, under randomization (2), a distribution over $\mathbb{D}$ determines the conditional distributions $p(x, y \mid z)$ for $z \in \{0, 1\}$. However, since distributions on $\mathbb{D}$ form a 15 dimensional simplex, while $p(x, y \mid z)$ is of dimension 6, it is clear that the reverse does not hold; thus many different distributions over $\mathbb{D}$ give rise to the same distributions $p(x, y \mid z)$. In what follows we precisely characterize the set of distributions over $\mathbb{D}$ corresponding to a given distribution $p(x, y \mid z)$.

We will accomplish this in the following steps:

1. We first characterize the set of distributions $\pi_X$ on $\mathbb{D}_X$ compatible with a given distribution $p(x \mid z)$.

2. Next we use the technique used for Step 1 to reduce the problem of characterizing distributions $\pi_{Y\mid X}$ compatible with $p(x, y \mid z)$ to that of characterizing the values of $p(y_x = 1 \mid \mathsf{t}_X)$ compatible with $p(x, y \mid z)$.

3. For a fixed marginal distribution $\pi_X$ on $\mathbb{D}_X$ we then describe the set of values for $p(y_x = 1 \mid x, \mathsf{t}_X)$ compatible with the observed distribution $p(y \mid x, z)$.

4. In general, some distributions $\pi_X$ on $\mathbb{D}_X$ and observed distributions $p(y \mid x, z)$ may be incompatible in that there are no compatible values for $p(y_x = 1 \mid \mathsf{t}_X)$. We use this to find the set of distributions $\pi_X$ on $\mathbb{D}_X$ compatible with $p(y, x \mid z)$ (by restricting the set of distributions found at step 1).

5. Finally we describe the values for $p(y_x = 1 \mid \mathsf{t}_X)$ compatible with the distributions $\pi$ over $\mathbb{D}_X$ found at the previous step.

We now proceed with the analysis.

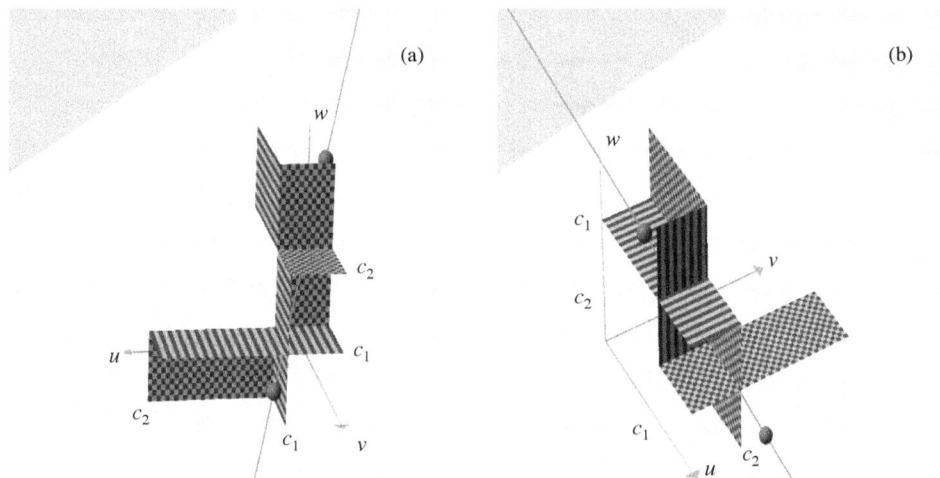

Figure 7. 'Staircases' of three boxes illustrating the possible support for $\mathcal{Q}_{(c_1,\alpha_1)(c_2,\alpha_2)}$; (a) $c_1 < c_2$; (b) $c_2 < c_1$. Sides of the boxes that are formed by (subsets of) faces of the unit cube are not shown. The line segments shown are illustrative; in general they may not intersect all 3 boxes.

### 3.1 Distributions $\pi_X$ on $\mathbb{D}_X$ compatible with $p(x \mid z)$

Under random assignment we have

$$p(x=1 \mid z=0) = p(X_{z=0}=1, X_{z=1}=0) + p(X_{z=0}=1, X_{z=1}=1)$$
$$= p(\text{DE}) + p(\text{AT}),$$

$$p(x=1 \mid z=1) = p(X_{z=0}=0, X_{z=1}=1) + p(X_{z=0}=1, X_{z=1}=1)$$
$$= p(\text{CO}) + p(\text{AT}).$$

Letting $U_{i+1} = X_{z=i}$, $i=0,1$ and $c_{j+1} = p(x=1 \mid z=j)$, $j=0,1$, it follows directly from the analysis in §2.1 that the set of distributions $\pi_X$ on $\mathbb{D}_X$ that are compatible with $p(x \mid z)$ are thus given by

$$\mathcal{P}_{c_1,c_2} = \left\{ \begin{array}{ll} \pi_{\text{AT}} = t, \\ \pi_{\text{DE}} = c_1 - t, \\ \pi_{\text{CO}} = c_2 - t, \\ \pi_{\text{NT}} = 1 - c_1 - c_2 + t, \end{array} \quad t \in \left[\max\{0, (c_1+c_2)-1\}, \min\{c_1, c_2\}\right] \right\}. \qquad (6)$$

### 3.2 Reduction step in characterizing distributions $\pi_{Y|X}$ compatible with $p(x, y \mid z)$

Suppose that we were able to ascertain the set of possible values for the eight quantities:

$$\gamma^i_{\mathbf{t}_X} \equiv p(y_{x=i}=1 \mid \mathbf{t}_X), \text{ for } i \in \{0,1\} \text{ and } \mathbf{t}_X \in \mathbb{D}_X,$$

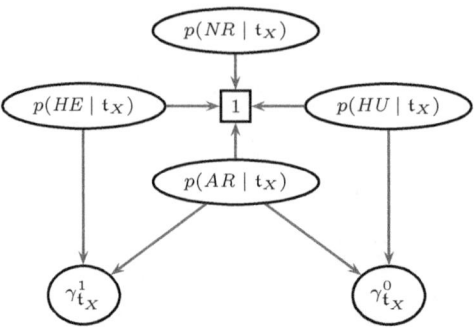

Figure 8. A graph representing the functional dependencies used in the reduction step in §3.2. The rectangular node indicates that the probabilities are required to sum to 1.

that are compatible with $p(x, y \mid z)$. Note that $p(y_{x=i} = 1 \mid t_X)$ is written as $p(y = 1 \mid \text{do}(x=i), t_X)$ using Pearl's do($\cdot$) notation. It is then clear that the set of possible distributions $\pi_{Y\mid X}$ that are compatible with $p(x, y \mid z)$ simply follows from the analysis in §2.1, since

$$\begin{aligned}
\gamma_{t_X}^0 &= p(y_{x=0} = 1 \mid t_X) \\
&= p(HU \mid t_X) + p(AR \mid t_X), \\
\gamma_{t_X}^1 &= p(y_{x=1} = 1 \mid t_X) \\
&= p(HE \mid t_X) + p(AR \mid t_X).
\end{aligned}$$

These relationships are also displayed graphically in Figure 8: in this particular graph all children are simple sums of their parents; the boxed 1 represents the 'sum to 1' constraint.

Thus, by §2.1, for given values of $\gamma_{t_X}^i$ the set of distributions $\pi_{Y\mid X}$ is given by:

$$\left\{\begin{array}{rl}
p(AR \mid t_X) &\in \left[\max\left\{0, (\gamma_{t_X}^0 + \gamma_{t_X}^1) - 1\right\}, \min\left\{\gamma_{t_X}^0, \gamma_{t_X}^1\right\}\right], \\
p(NR \mid t_X) &= 1 - \gamma_{t_X}^0 - \gamma_{t_X}^1 + p(AR \mid t_X), \\
p(HE \mid t_X) &= \gamma_{t_X}^1 - p(AR \mid t_X), \\
p(HU \mid t_X) &= \gamma_{t_X}^0 - p(AR \mid t_X)
\end{array}\right\}. \quad (7)$$

It follows from the discussion at the end of §2.1 that the values of $\gamma_{t_X}^0$ and $\gamma_{t_X}^1$ are not restricted by the requirement that there exists a distribution $p(\cdot \mid t_X)$ on $\mathbb{D}_Y$. Consequently we may proceed in two steps: first we derive the set of values for the eight parameters $\{\gamma_{t_X}^i\}$ and the distribution on $\pi_X$ (jointly) without consideration of the parameters for $\pi_{Y\mid X}$; second we then derive the parameters $\pi_{Y\mid X}$, as described above.

Finally we note that many causal quantities of interest, such as the average causal effect (ACE), and relative risk (RR) of $X$ on $Y$, for a given response type $t_X$, may

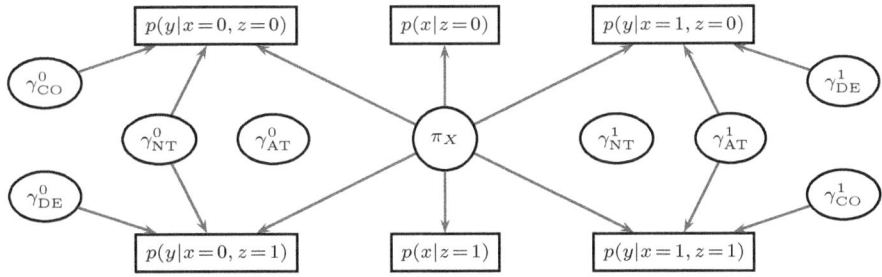

Figure 9. A graph representing the functional dependencies in the analysis of the binary IV model. Rectangular nodes are observed; oval nodes are unknown parameters. See text for further explanation.

be expressed in terms of the $\gamma^i_{t_X}$ parameters:

$$\text{ACE}(t_X) = \gamma^1_{t_X} - \gamma^0_{t_X}, \qquad \text{RR}(t_X) = \gamma^1_{t_X}/\gamma^0_{t_X}.$$

Consequently, for many purposes it may be unnecessary to consider the parameters $\pi_{Y|X}$ at all.

### 3.3 Values for $\{\gamma^i_{t_X}\}$ compatible with $\pi_X$ and $p(y \mid x, z)$

We will call a specification of values for $\pi_X$, *feasible for the observed distribution* if (a) $\pi_X$ lies within the set described in §3.1 of distributions compatible with $p(x \mid z)$ and (b) there exists a set of values for $\gamma^i_{t_X}$ which results in the distribution $p(y \mid x, z)$.

In the next section we give an explicit characterization of the set of feasible distributions $\pi_X$; in this section we characterize the set of values of $\gamma^i_{t_X}$ compatible with a fixed feasible distribution $\pi_X$ and $p(y \mid x, z)$.

PROPOSITION 1. *The following equations relate* $\pi_X$, $\gamma^0_{\text{CO}}$, $\gamma^0_{\text{DE}}$, $\gamma^0_{\text{NT}}$ *to* $p(y \mid x=0, z)$:

$$p(y=1 \mid x=0, z=0) = (\gamma^0_{\text{CO}}\pi_{\text{CO}} + \gamma^0_{\text{NT}}\pi_{\text{NT}})/(\pi_{\text{CO}} + \pi_{\text{NT}}), \qquad (8)$$

$$p(y=1 \mid x=0, z=1) = (\gamma^0_{\text{DE}}\pi_{\text{DE}} + \gamma^0_{\text{NT}}\pi_{\text{NT}})/(\pi_{\text{DE}} + \pi_{\text{NT}}), \qquad (9)$$

*Similarly, the following relate* $\pi_X$, $\gamma^1_{\text{CO}}$, $\gamma^1_{\text{DE}}$, $\gamma^1_{\text{AT}}$ *to* $p(y \mid x=1, z)$:

$$p(y=1 \mid x=1, z=0) = (\gamma^1_{\text{DE}}\pi_{\text{DE}} + \gamma^1_{\text{AT}}\pi_{\text{AT}})/(\pi_{\text{DE}} + \pi_{\text{AT}}), \qquad (10)$$

$$p(y=1 \mid x=1, z=1) = (\gamma^1_{\text{CO}}\pi_{\text{CO}} + \gamma^1_{\text{AT}}\pi_{\text{AT}})/(\pi_{\text{CO}} + \pi_{\text{AT}}). \qquad (11)$$

Equations (8)–(11) are represented in Figure 9. Note that the parameters $\gamma^0_{\text{AT}}$ and $\gamma^1_{\text{NT}}$ are completely unconstrained by the observed distribution since they describe, respectively, the effect of non-exposure ($X = 0$) on Always Takers, and exposure ($X = 1$) on Never Takers, neither of which ever occur. Consequently, the set

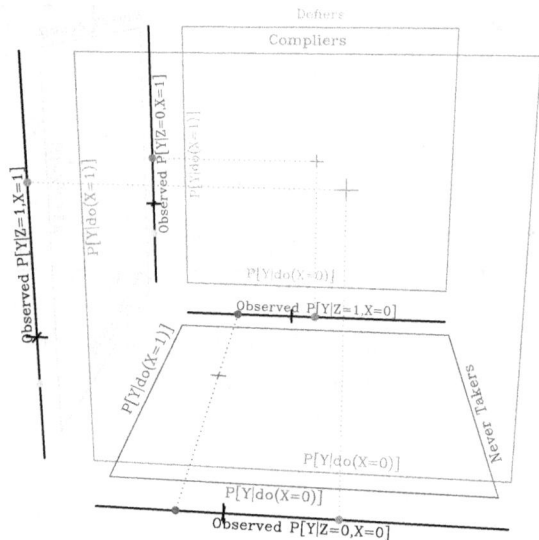

Figure 10. Geometric picture illustrating the relation between the $\gamma^i_{t_x}$ parameters and $p(y \mid x, z)$. See also Figure 9.

of possible values for each of these parameters is always $[0, 1]$. Graphically this corresponds to the disconnection of $\gamma^0_{\text{AT}}$ and $\gamma^1_{\text{NT}}$ from the remainder of the graph.

As shown in Proposition 1 the remaining six parameters may be divided into two groups, $\{\gamma^0_{\text{NT}}, \gamma^0_{\text{DE}}, \gamma^0_{\text{CO}}\}$ and $\{\gamma^1_{\text{AT}}, \gamma^1_{\text{DE}}, \gamma^1_{\text{CO}}\}$, depending on whether they relate to unexposed subjects, or exposed subjects. Furthermore, as the graph indicates, for a fixed feasible value of $\pi_X$, compatible with the observed distribution $p(x, y \mid z)$ (assuming such exists), these two sets are variation independent. Thus, for a fixed feasible value of $\pi_X$ we may analyze each of these sets separately.

A geometric picture of equations (8)–(11) is given in Figure 10: there is one square for each compliance type, with axes corresponding to $\gamma^0_{t_x}$ and $\gamma^1_{t_x}$; the specific value of $\langle \gamma^0_{t_x}, \gamma^1_{t_x} \rangle$ is given by a cross in the square. There are four lines corresponding to the four observed quantities $p(y = 1 \mid x, z)$. Each of these observed quantities, which is denoted by a cross on the respective line, is a weighted average of two $\gamma^i_{t_x}$ parameters, with weights given by $\pi_X$ (the weights are not depicted explicitly).

*Proof of Proposition 1:* We prove (8); the other proofs are similar. Subjects for whom $X = 0$ and $Z = 0$ are either Never Takers or Compliers. Hence

$$
\begin{aligned}
p(y=1 \mid x=0, z=0) &= p(y=1 \mid x=0, z=0, t_X=\text{NT})p(t_X=\text{NT} \mid x=0, z=0) \\
&\quad + p(y=1 \mid x=0, z=0, t_X=\text{CO})p(t_X=\text{CO} \mid x=0, z=0) \\
&= p(y_{x=0}=1 \mid x=0, z=0, t_X=\text{NT})p(t_X=\text{NT} \mid t_X \in \{\text{CO}, \text{NT}\}) \\
&\quad + p(y_{x=0}=1 \mid x=0, z=0, t_X=\text{CO})p(t_X=\text{CO} \mid t_X \in \{\text{CO}, \text{NT}\}) \\
&= p(y_{x=0}=1 \mid z=0, t_X=\text{NT}) \times \pi_{\text{NT}}/(\pi_{\text{NT}} + \pi_{\text{CO}}) \\
&\quad + p(y_{x=0}=1 \mid z=0, t_X=\text{CO}) \times \pi_{\text{CO}}/(\pi_{\text{NT}} + \pi_{\text{CO}}) \\
&= p(y_{x=0}=1 \mid t_X=\text{NT}) \times \pi_{\text{NT}}/(\pi_{\text{NT}} + \pi_{\text{CO}}) \\
&\quad + p(y_{x=0}=1 \mid t_X=\text{CO}) \times \pi_{\text{CO}}/(\pi_{\text{NT}} + \pi_{\text{CO}}) \\
&= (\gamma_{\text{CO}}^0 \pi_{\text{CO}} + \gamma_{\text{NT}}^0 \pi_{\text{NT}})/(\pi_{\text{CO}} + \pi_{\text{NT}}).
\end{aligned}
$$

Here the first equality is by the chain rule of probability; the second follows by consistency; the third follows since Compliers and Never Takers have $X = 0$ when $Z = 0$; the fourth follows by randomization (2). □

**Values for $\gamma_{\text{CO}}^0, \gamma_{\text{DE}}^0, \gamma_{\text{NT}}^0$ compatible with a feasible $\pi_X$**

Since (8) and (9) correspond to three quantities with two averages specified, we may apply the analysis in §2.3, taking $\alpha_1 = \pi_{\text{CO}}/(\pi_{\text{CO}} + \pi_{\text{NT}})$, $\alpha_2 = \pi_{\text{DE}}/(\pi_{\text{DE}} + \pi_{\text{NT}})$, $c_i = p(y = 1 \mid x = 0, z = i-1)$ for $i = 1, 2$, $u = \gamma_{\text{CO}}^0$, $v = \gamma_{\text{DE}}^0$ and $w = \gamma_{\text{NT}}^0$. Under this substitution, the set of possible values for $\langle \gamma_{\text{CO}}^0, \gamma_{\text{DE}}^0, \gamma_{\text{NT}}^0 \rangle$ is then given by $\mathcal{Q}_{(c_1,\alpha_1)(c_2,\alpha_2)}$.

**Values for $\gamma_{\text{CO}}^1, \gamma_{\text{DE}}^1, \gamma_{\text{AT}}^1$ compatible with a feasible $\pi_X$**

Likewise since (10) and (11) contain three quantities with two averages specified we again apply the analysis from §2.3, taking $\alpha_1 = \pi_{\text{CO}}/(\pi_{\text{CO}} + \pi_{\text{AT}})$, $\alpha_2 = \pi_{\text{DE}}/(\pi_{\text{DE}} + \pi_{\text{AT}})$, $c_i = p(y = 1 \mid x = 1, z = 2-i)$ for $i = 1, 2$, $u = \gamma_{\text{CO}}^1$, $v = \gamma_{\text{DE}}^1$ and $w = \gamma_{\text{AT}}^1$. The set of possible values for $\langle \gamma_{\text{CO}}^1, \gamma_{\text{DE}}^1, \gamma_{\text{AT}}^1 \rangle$ is then given by $\mathcal{Q}_{(c_1,\alpha_1)(c_2,\alpha_2)}$.

### 3.4 Values of $\pi_X$ compatible with $p(x, y \mid z)$

In §3.1 we characterized the distributions $\pi_X$ compatible with $p(x \mid z)$ as a one dimensional subspace of the three dimensional simplex, parameterized in terms of $t \equiv \pi_{\text{AT}}$; see (6). We now incorporate the additional constraints on $\pi_X$ that arise from $p(y \mid x, z)$. These occur because some distributions $\pi_X$, though compatible with $p(x \mid z)$, lead to an empty set of values for $\langle \gamma_{\text{CO}}^1, \gamma_{\text{DE}}^1, \gamma_{\text{AT}}^1 \rangle$ or $\langle \gamma_{\text{CO}}^0, \gamma_{\text{DE}}^0, \gamma_{\text{NT}}^0 \rangle$ and thus are infeasible.

**Constraints on $\pi_X$ arising from $p(y \mid x = 0, z)$**

Building on the analysis in §3.3 the set of values for

$$
\begin{aligned}
\langle \alpha_1, \alpha_2 \rangle &= \langle \pi_{\text{CO}}/(\pi_{\text{CO}} + \pi_{\text{NT}}), \pi_{\text{DE}}/(\pi_{\text{DE}} + \pi_{\text{NT}}) \rangle \\
&= \langle \pi_{\text{CO}}/p_{x_0|z_0}, \pi_{\text{DE}}/p_{x_0|z_0} \rangle
\end{aligned} \quad (12)
$$

compatible with $p(y \mid x=0, z)$ (i.e. for which the corresponding set of values for $\langle \gamma_{CO}^0, \gamma_{DE}^0, \gamma_{NT}^0 \rangle$ is non-empty) is given by $\mathcal{R}_{c_1^*, c_2^*}$, where $c_i^* = p(y=1 \mid x=0, z=i-1)$, $i = 1, 2$ (see §2.3). The inequalities defining $\mathcal{R}_{c_1^*, c_2^*}$ may be translated into upper bounds on $t \equiv \pi_{AT}$ in (6), as follows:

$$t \leq \min \left\{ 1 - \sum_{j \in \{0,1\}} p(y=j, x=0 \mid z=j),\ 1 - \sum_{k \in \{0,1\}} p(y=k, x=0 \mid z=1-k) \right\}. \quad (13)$$

*Proof:* The analysis in §3.3 implied that for $\mathcal{R}_{c_1^*, c_2^*} \neq \emptyset$ we require

$$\frac{c_1^* - \alpha_1}{1 - \alpha_1} \leq \frac{c_2^*}{1 - \alpha_2} \quad \text{and} \quad \frac{c_2^* - \alpha_2}{1 - \alpha_2} \leq \frac{c_1^*}{1 - \alpha_1}. \quad (14)$$

Taking the first of these and plugging in the definitions of $c_1^*$, $c_2^*$, $\alpha_1$ and $\alpha_2$ from (12) gives:

$$\frac{p_{y_1|x_0,z_0} - (\pi_{CO}/p_{x_0|z_0})}{1 - (\pi_{CO}/p_{x_0|z_0})} \leq \frac{p_{y_1|x_0,z_1}}{1 - (\pi_{DE}/p_{x_0|z_1})}$$

$(\Leftrightarrow) \quad (p_{y_1|x_0,z_0} - (\pi_{CO}/p_{x_0|z_0}))(1 - (\pi_{DE}/p_{x_0|z_1})) \leq p_{y_1|x_0,z_1}(1 - (\pi_{CO}/p_{x_0|z_0}))$

$(\Leftrightarrow) \quad (p_{y_1,x_0|z_0} - \pi_{CO})(p_{x_0|z_1} - \pi_{DE}) \leq p_{y_1,x_0|z_1}(p_{x_0|z_0} - \pi_{CO}).$

But $p_{x_0|z_1} - \pi_{DE} = p_{x_0|z_0} - \pi_{CO} = \pi_{NT}$, hence these terms may be cancelled to give:

$$(p_{y_1,x_0|z_0} - \pi_{CO}) \leq p_{y_1,x_0|z_1}$$
$(\Leftrightarrow) \quad \pi_{AT} - p_{x_1|z_1} \leq p_{y_1,x_0|z_1} - p_{y_1,x_0|z_0}$
$(\Leftrightarrow) \quad \pi_{AT} \leq 1 - p_{y_0,x_0|z_1} - p_{y_1,x_0|z_0}.$

A similar argument applied to the second constraint in (14) to derive that

$$\pi_{AT} \leq 1 - p_{y_0,x_0|z_0} - p_{y_1,x_0|z_1},$$

as required. $\square$

**Constraints on $\pi_X$ arising from $p(y \mid x=1, z)$**

Similarly using the analysis in §3.3 the set of values for

$$\langle \alpha_1, \alpha_2 \rangle = \langle \pi_{CO}/(\pi_{CO} + \pi_{AT}), \pi_{DE}/(\pi_{DE} + \pi_{AT}) \rangle$$

compatible with $p(y \mid x=1, z)$ (i.e. that the corresponding set of values for $\langle \gamma_{CO}^1, \gamma_{DE}^1, \gamma_{AT}^1 \rangle$ is non-empty) is given by $\mathcal{R}_{c_1^{**}, c_2^{**}}$, where $c_i^{**} = p(y=1 \mid x=1, z=2-i)$, $i = 1, 2$ (see §2.3). Again, we translate the inequalities which define $\mathcal{R}_{c_1^{**}, c_2^{**}}$ into further upper bounds on $t = \pi_{AT}$ in (6):

$$t \leq \min \left\{ \sum_{j \in \{0,1\}} p(y=j, x=1 \mid z=j),\ \sum_{k \in \{0,1\}} p(y=k, x=1 \mid z=1-k) \right\}. \quad (15)$$

The proof that these inequalities are implied, is very similar to the derivation of the upper bounds on $\pi_{AT}$ arising from $p(y \mid x = 0, z)$ considered above.

**The distributions $\pi_X$ compatible with the observed distribution**

It follows that the set of distributions on $\mathbb{D}_X$ that are compatible with the observed distribution, which we denote $\mathcal{P}_X$, may be given thus:

$$\mathcal{P}_X = \left\{ \begin{array}{rcl} \pi_{AT} & \in & [l\pi_{AT}, u\pi_{AT}], \\ \pi_{NT}(\pi_{AT}) & = & 1 - p(x=1 \mid z=0) - p(x=1 \mid z=1) + \pi_{AT}, \\ \pi_{CO}(\pi_{AT}) & = & p(x=1 \mid z=1) - \pi_{AT}, \\ \pi_{DE}(\pi_{AT}) & = & p(x=1 \mid z=0) - \pi_{AT} \end{array} \right\}, \quad (16)$$

where

$$l\pi_{AT} = \max\{0, p(x=1 \mid z=0) + p(x=1 \mid z=1) - 1\};$$

$$u\pi_{AT} = \min \left\{ \begin{array}{ll} p(x=1 \mid z=0), & p(x=1 \mid z=1), \\ 1 - \sum_j p(y=j, x=0 \mid z=j), & 1 - \sum_k p(y=k, x=0 \mid z=1-k), \\ \sum_j p(y=j, x=1 \mid z=j), & \sum_k p(y=k, x=1 \mid z=1-k) \end{array} \right\}.$$

Observe that unlike the upper bound, the lower bound on $\pi_{AT}$ (and $\pi_{NT}$) obtained from $p(x, y \mid z)$ is the same as the lower bound derived from $p(x \mid z)$ alone.

We define $\pi_X(\pi_{AT}) \equiv \langle \pi_{NT}(\pi_{AT}), \pi_{CO}(\pi_{AT}), \pi_{DE}(\pi_{AT}), \pi_{AT} \rangle$, for use below. Note the following:

PROPOSITION 2. *When $\pi_{AT}$ (equivalently $\pi_{NT}$) is minimized then either $\pi_{NT} = 0$ or $\pi_{AT} = 0$.*

*Proof:* This follows because, by the expression for $l\pi_{AT}$, either $l\pi_{AT} = 0$, or $l\pi_{AT} = p(x = 1 \mid z = 0) + p(x = 1 \mid z = 1) - 1$, in which case $l\pi_{NT} = 0$ by (16). □

## 4 Projections

The analysis in §3 provides a complete description of the set of distributions over $\mathbb{D}$ compatible with a given observed distribution. In particular, equation (16) describes the one dimensional set of compatible distributions over $\mathbb{D}_X$; in §3.3 we first gave a description of the one dimensional set of values over $\langle \gamma_{CO}^0, \gamma_{DE}^0, \gamma_{NT}^0 \rangle$ compatible with the observed distribution and a specific feasible distribution $\pi_X$ over $\mathbb{D}_X$; we then described the one dimensional set of values for $\langle \gamma_{CO}^1, \gamma_{DE}^1, \gamma_{AT}^1 \rangle$. Varying $\pi_X$ over the set $\mathcal{P}_X$ of feasible distributions over $\mathbb{D}_X$, describes a set of lines, forming two two-dimensional manifolds which represent the space of possible values for $\langle \gamma_{CO}^0, \gamma_{DE}^0, \gamma_{NT}^0 \rangle$ and likewise for $\langle \gamma_{CO}^1, \gamma_{DE}^1, \gamma_{AT}^1 \rangle$. As noted previously, the parameters $\gamma_{AT}^0$ and $\gamma_{NT}^1$ are unconstrained by the observed data. Finally, if there is interest in distributions over response types, there is a one-dimensional set of such distributions associated with each possible pair of values from $\gamma_{t_x}^0$ and $\gamma_{t_x}^1$.

For the purposes of visualization it is useful to look at projections. There are many such projections that could be considered, here we focus on projections that display the relation between the possible values for $\pi_X$ and $\gamma_{t_x}^x$. See Figure 11.

We make the following definition:

$$\alpha_{t_x}^{ij}(\pi_X) \equiv p(t_x \mid X_{z=i} = j),$$

where $\pi_X = \langle \pi_{\text{NT}}, \pi_{\text{CO}}, \pi_{\text{DE}}, \pi_{\text{AT}} \rangle \in \mathcal{P}_X$, as before. For example, $\alpha_{\text{NT}}^{00}(\pi_X) = \pi_{\text{NT}}/(\pi_{\text{NT}} + \pi_{\text{CO}})$, $\alpha_{\text{NT}}^{10}(\pi_X) = \pi_{\text{NT}}/(\pi_{\text{NT}} + \pi_{\text{DE}})$.

### 4.1 Upper and Lower bounds on $\gamma_{t_x}^x$ as a function of $\pi_X$

We use the following notation to refer to the upper and lower bounds on $\gamma_{\text{NT}}^0$ and $\gamma_{\text{AT}}^1$ that were derived earlier. If $\pi_X$ is such that $\pi_{\text{NT}} > 0$, so $\alpha_{\text{NT}}^{00}, \alpha_{\text{NT}}^{10} > 0$ then we define:

$$l\gamma_{\text{NT}}^0(\pi_X) \equiv \max\left\{0, \frac{p_{y_1|x_0z_0} - \alpha_{\text{CO}}^{00}(\pi_X)}{\alpha_{\text{NT}}^{00}(\pi_X)}, \frac{p_{y_1|x_0z_1} - \alpha_{\text{DE}}^{10}(\pi_X)}{\alpha_{\text{NT}}^{10}(\pi_X)}\right\},$$

$$u\gamma_{\text{NT}}^0(\pi_X) \equiv \min\left\{\frac{p_{y_1|x_0z_0}}{\alpha_{\text{NT}}^{00}(\pi_X)}, \frac{p_{y_1|x_0z_1}}{\alpha_{\text{NT}}^{10}(\pi_X)}, 1\right\},$$

while if $\pi_{\text{NT}} = 0$ then we define $l\gamma_{\text{NT}}^0(\pi_X) \equiv 0$ and $u\gamma_{\text{NT}}^0(\pi_X) \equiv 1$. Similarly, if $\pi_X$ is such that $\pi_{\text{AT}} > 0$ then we define:

$$l\gamma_{\text{AT}}^1(\pi_X) \equiv \max\left\{0, \frac{p_{y_1|x_1z_1} - \alpha_{\text{CO}}^{11}(\pi_X)}{\alpha_{\text{AT}}^{11}(\pi_X)}, \frac{p_{y_1|x_1z_0} - \alpha_{\text{DE}}^{01}(\pi_X)}{\alpha_{\text{AT}}^{01}(\pi_X)}\right\},$$

$$u\gamma_{\text{AT}}^1(\pi_X) \equiv \min\left\{\frac{p_{y_1|x_1z_1}}{\alpha_{\text{AT}}^{11}(\pi_X)}, \frac{p_{y_1|x_1z_0}}{\alpha_{\text{AT}}^{01}(\pi_X)}, 1\right\},$$

while if $\pi_{\text{AT}} = 0$ then let $l\gamma_{\text{AT}}^1(\pi_X) \equiv 0$ and $u\gamma_{\text{AT}}^1(\pi_X) \equiv 1$.

We note that Table 2 summarizes the upper and lower bounds, as a function of $\pi_X \in \mathcal{P}_X$, on each of the eight parameters $\gamma_{t_x}^x$ that were derived earlier in §3.3. These are shown by the thicker lines on each of the plots forming the upper and lower boundaries in Figure 11 ($\gamma_{\text{AT}}^0$ and $\gamma_{\text{NT}}^1$ are not shown in the Figure).

The upper and lower bounds on $\gamma_{\text{NT}}^0$ and $\gamma_{\text{AT}}^1$ are relatively simple:

PROPOSITION 3. $l\gamma_{\text{NT}}^0(\pi_X)$ and $l\gamma_{\text{AT}}^1(\pi_X)$ are non-decreasing in $\pi_{\text{AT}}$ and $\pi_{\text{NT}}$. Likewise $u\gamma_{\text{NT}}^0(\pi_X)$ and $u\gamma_{\text{AT}}^1(\pi_X)$ are non-increasing in $\pi_{\text{AT}}$ and $\pi_{\text{NT}}$.

*Proof:* We first consider $l\gamma_{\text{NT}}^0$. By (16), $\pi_{\text{NT}} = 1 - p(x = 1 \mid z = 0) - p(x = 1 \mid z = 1) + \pi_{\text{AT}}$, hence a function is non-increasing [non-decreasing] in $\pi_{\text{AT}}$ iff it is non-increasing [non-decreasing] in $\pi_{\text{NT}}$. Observe that for $\pi_{\text{NT}} > 0$,

$$(p_{y_1|x_0z_0} - \alpha_{\text{CO}}^{00}(\pi_X))/\alpha_{\text{NT}}^{00}(\pi_X) = (p_{y_1|x_0z_0}(\pi_{\text{NT}} + \pi_{\text{CO}}) - \pi_{\text{CO}})/\pi_{\text{NT}}$$
$$= p_{y_1|x_0z_0} - p_{y_0|x_0z_0}(\pi_{\text{CO}}/\pi_{\text{NT}})$$
$$= p_{y_1|x_0z_0} + p_{y_0|x_0z_0}(1 - (p_{x_0|z_0}/\pi_{\text{NT}}))$$

Binary Instrumental Variable Model

|  | Lower Bound | Upper Bound |
|---|---|---|
| $\gamma_{\text{NT}}^0$ | $l\gamma_{\text{NT}}^0(\pi_X)$ | $u\gamma_{\text{NT}}^0(\pi_X)$ |
| $\gamma_{\text{CO}}^0$ | $(p_{y_1|x_0z_0} - u\gamma_{\text{NT}}^0(\pi_X) \cdot \alpha_{\text{NT}}^{00})/\alpha_{\text{CO}}^{00}$ | $(p_{y_1|x_0z_0} - l\gamma_{\text{NT}}^0(\pi_X) \cdot \alpha_{\text{NT}}^{00})/\alpha_{\text{CO}}^{00}$ |
| $\gamma_{\text{DE}}^0$ | $(p_{y_1|x_0z_1} - u\gamma_{\text{NT}}^0(\pi_X) \cdot \alpha_{\text{NT}}^{10})/\alpha_{\text{DE}}^{10}$ | $(p_{y_1|x_0z_1} - l\gamma_{\text{NT}}^0(\pi_X) \cdot \alpha_{\text{NT}}^{10})/\alpha_{\text{DE}}^{10}$ |
| $\gamma_{\text{AT}}^0$ | 0 | 1 |
| $\gamma_{\text{NT}}^1$ | 0 | 1 |
| $\gamma_{\text{CO}}^1$ | $(p_{y_1|x_1z_1} - u\gamma_{\text{AT}}^1(\pi_X) \cdot \alpha_{\text{AT}}^{11})/\alpha_{\text{CO}}^{11}$ | $(p_{y_1|x_1z_1} - l\gamma_{\text{AT}}^1(\pi_X) \cdot \alpha_{\text{AT}}^{11})/\alpha_{\text{CO}}^{11}$ |
| $\gamma_{\text{DE}}^1$ | $(p_{y_1|x_1z_0} - u\gamma_{\text{AT}}^1(\pi_X) \cdot \alpha_{\text{AT}}^{01})/\alpha_{\text{DE}}^{01}$ | $(p_{y_1|x_1z_0} - l\gamma_{\text{AT}}^1(\pi_X) \cdot \alpha_{\text{AT}}^{01})/\alpha_{\text{DE}}^{01}$ |
| $\gamma_{\text{AT}}^1$ | $l\gamma_{\text{AT}}^1(\pi_X)$ | $u\gamma_{\text{AT}}^1(\pi_X)$ |

Table 2. Upper and Lower bounds on $\gamma_{\mathbf{t}_x}^x$, as a function of $\pi_X \in \mathcal{P}_X$. If for some $\pi_X$ an expression giving a lower bound for a quantity is undefined then the lower bound is 0; conversely if an expression for an upper bound is undefined then the upper bound is 1.

which is non-decreasing in $\pi_{\text{NT}}$. Similarly,

$$(p_{y_1|x_0z_1} - \alpha_{\text{DE}}^{10}(\pi_X))/\alpha_{\text{NT}}^{10}(\pi_X) = p_{y_1|x_0z_1} + p_{y_0|x_0z_1}(1 - (p_{x_0|z_1}/\pi_{\text{NT}})).$$

The conclusion follows since the maximum of a set of non-decreasing functions is non-decreasing.

The other arguments are similar. □

We note that the bounds on $\gamma_{\text{CO}}^x$ and $\gamma_{\text{DE}}^x$ need not be monotonic in $\pi_{\text{AT}}$.

PROPOSITION 4. *Let $\pi_X^{\min}$ be the distribution in $\mathcal{P}_X$ for which $\pi_{\text{AT}}$ and $\pi_{\text{NT}}$ are minimized then either:*

(1) $\pi_{\text{NT}}^{\min} = 0$, *hence* $l\gamma_{\text{NT}}^0(\pi_X^{\min}) = 0$ *and* $u\gamma_{\text{NT}}^0(\pi_X^{\min}) = 1$; *or*

(2) $\pi_{\text{AT}}^{\min} = 0$, *hence* $l\gamma_{\text{AT}}^1(\pi_X^{\min}) = 0$ *and* $u\gamma_{\text{AT}}^1(\pi_X^{\min}) = 1$.

*Proof:* This follows from Proposition 2, and the fact that if $\pi_{\mathbf{t}_x} = 0$ then $\gamma_{\mathbf{t}_x}^i$ is not identified (for any $i$). □

## 4.2 Upper and Lower bounds on $p(\text{AT})$ as a function of $\gamma_{\text{NT}}^0$

The expressions given in Table 2 allow the range of values for each $\gamma_{\mathbf{t}_x}^i$ to be determined as a function of $\pi_X$, giving the upper and lower bounding curves in Figure 11. However it follows directly from (8) and (9) that there is a bijection between the three shapes shown for $\gamma_{\text{CO}}^0$, $\gamma_{\text{DE}}^0$ and $\gamma_{\text{NT}}^0$ (top row of Figure 11).

In this section we describe this bijection by deriving curves corresponding to fixed values of $\gamma_{NT}^0$ that are displayed in the plots for $\gamma_{CO}^0$ and $\gamma_{DE}^0$. Similarly it follows from (10) and (11) that there is a bijection between the three shapes shown for $\gamma_{CO}^1, \gamma_{DE}^1, \gamma_{AT}^1$ (bottom row of Figure 11). Correspondingly we add curves to the plots for $\gamma_{CO}^1$ and $\gamma_{DE}^1$ corresponding to fixed values of $\gamma_{AT}^1$. (The expressions in this section are used solely to add these curves and are not used elsewhere.)

As described earlier, for a given distribution $\pi_X \in \mathcal{P}_X$ the set of values for $\langle \gamma_{CO}^0, \gamma_{DE}^0, \gamma_{NT}^0 \rangle$ forms a one dimensional subspace. For a given $\pi_X$ if $\pi_{CO} > 0$ then $\gamma_{CO}^0$ is a deterministic function of $\gamma_{NT}^0$, likewise for $\gamma_{DE}^0$.

It follows from Proposition 3 that the range of values for $\gamma_{NT}^0$ when $\pi_X = \pi_X^{\min}$ contains the range of possible values for $\gamma_{NT}^0$ for any other $\pi_X \in \mathcal{P}_X$. The same holds for $\gamma_{AT}^1$. Thus for any given possible value of $\gamma_{NT}^0$, the minimum compatible value of $\pi_{AT} = l\pi_{AT} \equiv \max\{0, p_{x_1|z_0} + p_{x_1|z_1} - 1\}$. This is reflected in the plots in Figure 11 for $\gamma_{NT}^0$ and $\gamma_{AT}^1$ in that the left hand endpoints of the thinner lines (lying between the upper and lower bounds) all lie on the same vertical line for which $\pi_{AT}$ is minimized.

In contrast the upper bounds on $\pi_{AT}$ vary as a function of $\gamma_{NT}^0$ (also $\gamma_{AT}^1$). The upper bound for $\pi_{AT}$ as a function of $\gamma_{NT}^0$ occurs when one of the thinner horizontal lines in the plot for $\gamma_{NT}^0$ in Figure 11 intersects either $u\gamma_{NT}^0(\pi_X)$, $l\gamma_{NT}^0(\pi_X)$, or the vertical line given by the global upper bound, $u\pi_{AT}$, on $\pi_{AT}$:

$$u\pi_{AT}(\gamma_{NT}^0) \equiv \max\{\pi_{AT} \mid \gamma_{NT}^0 \in [l\gamma_{NT}^0(\pi_X), u\gamma_{NT}^0(\pi_X)]\}$$
$$= \min\left\{ p_{x_1|z_1} - p_{x_0|z_0}\left(1 - \frac{p_{y_1|x_0z_0}}{\gamma_{NT}^0}\right), p_{x_1|z_0} - p_{x_0|z_1}\left(1 - \frac{p_{y_1|x_0z_1}}{\gamma_{NT}^0}\right), \right.$$
$$\left. p_{x_1|z_1} - p_{x_0|z_0}\left(1 - \frac{p_{y_0|x_0z_0}}{1-\gamma_{NT}^0}\right), p_{x_1|z_0} - p_{x_0|z_1}\left(1 - \frac{p_{y_0|x_0z_1}}{1-\gamma_{NT}^0}\right), u\pi_{AT} \right\};$$

similarly we have

$$u\pi_{AT}(\gamma_{AT}^1) \equiv \max\{\pi_{AT} \mid \gamma_{AT}^1 \in [l\gamma_{AT}^1(\pi_X), u\gamma_{AT}^1(\pi_X)]\}$$
$$= \min\left\{ u\pi_{AT}, \frac{p_{x_1|z_1}p_{y_1|x_1z_1}}{\gamma_{AT}^1}, \frac{p_{x_1|z_0}p_{y_1|x_1z_0}}{\gamma_{AT}^1}, \frac{p_{x_1|z_1}p_{y_0|x_1z_1}}{1-\gamma_{AT}^1}, \frac{p_{x_1|z_0}p_{y_0|x_1z_0}}{1-\gamma_{AT}^1} \right\}.$$

The curves added to the unexposed plots for Compliers and Defiers in Figure 11 are as follows:

$$\gamma_{CO}^0(\pi_X, \gamma_{NT}^0) \equiv (p_{y_1|x_0z_0} - \gamma_{NT}^0 \cdot \alpha_{NT}^{00})/\alpha_{CO}^{00},$$
$$c\gamma_{CO}^0(\pi_{AT}, \gamma_{NT}^0) \equiv \{\langle \pi_{AT}, \gamma_{CO}^0(\pi_X(\pi_{AT}), \gamma_{NT}^0)\rangle\}; \qquad (17)$$

$$\gamma_{DE}^0(\pi_X, \gamma_{NT}^0) \equiv (p_{y_1|x_0z_1} - \gamma_{NT}^0 \cdot \alpha_{NT}^{10})/\alpha_{DE}^{10},$$
$$c\gamma_{DE}^0(\pi_{AT}, \gamma_{NT}^0) \equiv \{\langle \pi_{AT}, \gamma_{DE}^0(\pi_X(\pi_{AT}), \gamma_{NT}^0)\rangle\}; \qquad (18)$$

for $\gamma_{NT}^0 \in [l\gamma_{NT}^0(\pi_X^{\min}), u\gamma_{NT}^0(\pi_X^{\min})]$; $\pi_{AT} \in [l\pi_{AT}, u\pi_{AT}(\gamma_{NT}^0)]$. The curves added

Table 3. Flu Vaccine Data from [McDonald, Hiu, and Tierney 1992].

| Z | X | Y | count |
|---|---|---|---|
| 0 | 0 | 0 | 99 |
| 0 | 0 | 1 | 1027 |
| 0 | 1 | 0 | 30 |
| 0 | 1 | 1 | 233 |
| 1 | 0 | 0 | 84 |
| 1 | 0 | 1 | 935 |
| 1 | 1 | 0 | 31 |
| 1 | 1 | 1 | 422 |
|   |   |   | 2,861 |

to the exposed plots for Compliers and Defiers in Figure 11 are given by:

$$\begin{aligned}\gamma^1_{\text{CO}}(\pi_X, \gamma^1_{\text{AT}}) &\equiv (p_{y_1|x_1 z_1} - \gamma^1_{\text{AT}} \cdot \alpha^{11}_{\text{AT}})/\alpha^{11}_{\text{CO}}, \\ c\gamma^1_{\text{DE}}(\pi_{\text{AT}}, \gamma^1_{\text{AT}}) &\equiv \{\langle \pi_{\text{AT}}, \gamma^1_{\text{CO}}(\pi_X(\pi_{\text{AT}}), \gamma^1_{\text{AT}})\rangle\};\end{aligned} \quad (19)$$

$$\begin{aligned}\gamma^1_{\text{DE}}(\pi_X, \gamma^1_{\text{AT}}) &\equiv (p_{y_1|x_1 z_0} - \gamma^1_{\text{AT}} \cdot \alpha^{01}_{\text{AT}})/\alpha^{01}_{\text{DE}}, \\ c\gamma^1_{\text{DE}}(\pi_{\text{AT}}, \gamma^1_{\text{AT}}) &\equiv \{\langle \pi_{\text{AT}}, \gamma^1_{\text{DE}}(\pi_X(\pi_{\text{AT}}), \gamma^1_{\text{AT}})\rangle\};\end{aligned} \quad (20)$$

for $\gamma^1_{\text{AT}} \in [l\gamma^1_{\text{AT}}(\pi_X^{\min}), u\gamma^1_{\text{AT}}(\pi_X^{\min})]$; $\pi_{\text{AT}} \in [l\pi_{\text{AT}}, u\pi_{\text{AT}}(\gamma^1_{\text{AT}})]$.

### 4.3 Example: Flu Data

To illustrate some of the constructions described we consider the influenza vaccine dataset [McDonald, Hiu, and Tierney 1992] previously analyzed by [Hirano, Imbens, Rubin, and Zhou 2000]; see Table 3. Here the instrument $Z$ was whether a patient's physician was sent a card asking him to remind patients to obtain flu shots, or not; $X$ is whether or not the patient did in fact get a flu shot. Finally $Y = 1$ indicates that a patient was *not* hospitalized. Unlike the analysis of [Hirano, Imbens, Rubin, and Zhou 2000] we ignore baseline covariates, and restrict attention to displaying the set of parameters of the IV model that are compatible with the empirical distribution.

The set of values for $\pi_X$ vs. $\langle \gamma^0_{\text{CO}}, \gamma^0_{\text{DE}}, \gamma^0_{\text{NT}} \rangle$ (upper row), and $\pi_X$ vs. $\langle \gamma^1_{\text{CO}}, \gamma^1_{\text{DE}}, \gamma^1_{\text{AT}} \rangle$ corresponding to the empirical distribution for $p(x, y \mid z)$ are shown in Figure 11. The empirical distribution is not consistent with there being no Defiers (though the scales in Figure 11 show 0 as one endpoint for the proportion $\pi_{\text{DE}}$ this is merely a consequence of the significant digits displayed; in fact the true lower bound on this proportion is 0.0005).

We emphasize that this analysis merely derives the logical consequences of the empirical distribution under the IV model and ignores sampling variability.

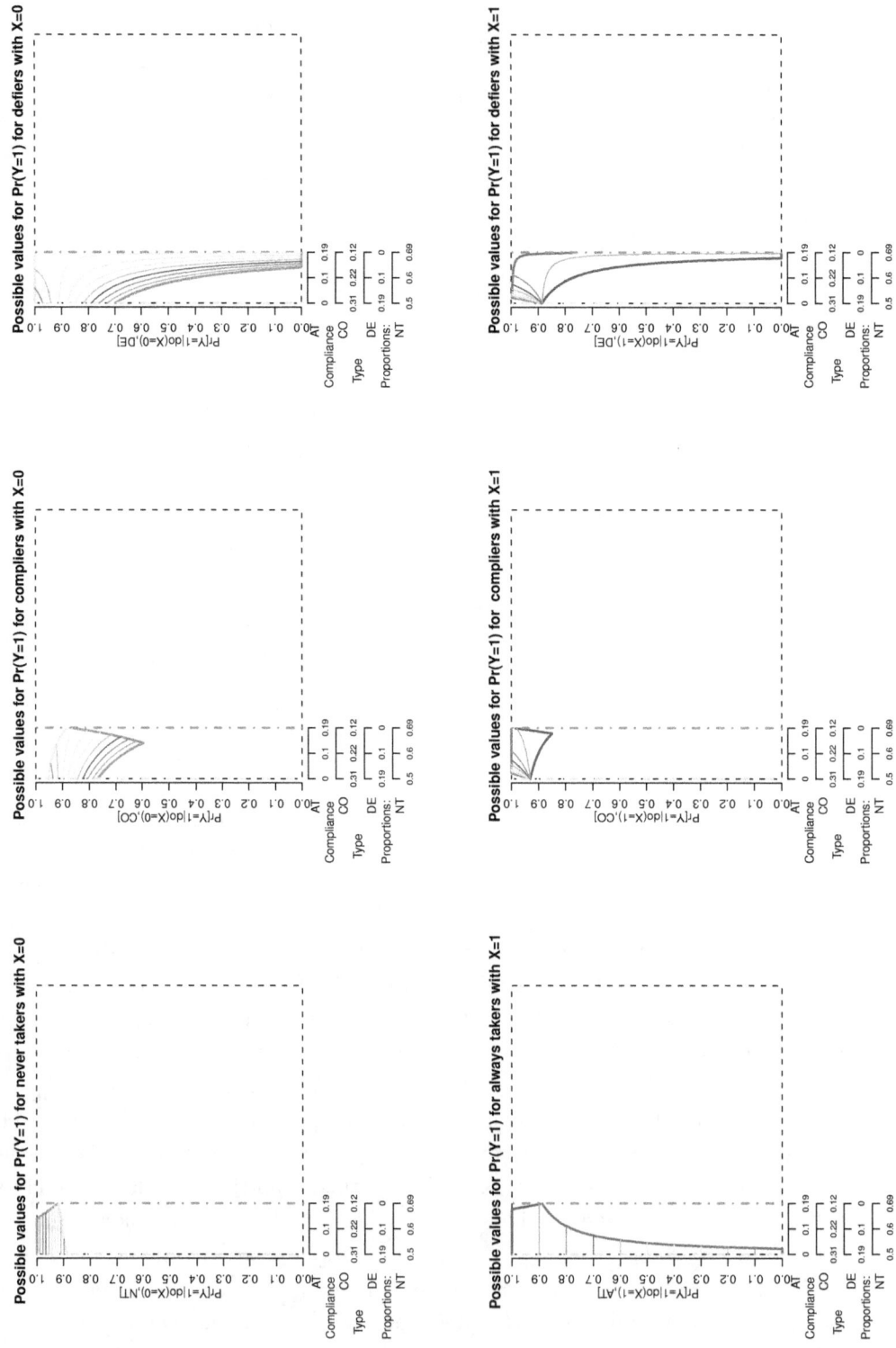

Figure 11. Depiction of the set of values for $\pi_X$ vs. $\langle \gamma_{CO}^0, \gamma_{DE}^0, \gamma_{NT}^0 \rangle$ (upper row), and $\pi_X$ vs. $\langle \gamma_{CO}^1, \gamma_{DE}^1, \gamma_{AT}^1 \rangle$ for the flu data.

## 5 Bounding Average Causal Effects

We may use the results above to obtain bounds on average causal effects, for different complier strata:

$$\text{ACE}_{t_x}(\pi_X, \gamma^0_{t_x}, \gamma^1_{t_x}) \equiv \gamma^1_{t_x}(\pi_X) - \gamma^0_{t_x}(\pi_X),$$

$$l\text{ACE}_{t_x}(\pi_X) \equiv \min_{\gamma^0_{t_x}, \gamma^1_{t_x}} \text{ACE}_{t_x}(\pi_X, \gamma^0_{t_x}, \gamma^1_{t_x}),$$

$$u\text{ACE}_{t_x}(\pi_X) \equiv \max_{\gamma^0_{t_x}, \gamma^1_{t_x}} \text{ACE}_{t_x}(\pi_X, \gamma^0_{t_x}, \gamma^1_{t_x}),$$

as a function of a feasible distribution $\pi_X$; see Table 5. As shown in the table, the values of $\gamma^0_{\text{NT}}$ and $\gamma^1_{\text{AT}}$ which maximize (minimize) $\text{ACE}_{\text{CO}}$ and $\text{ACE}_{\text{DE}}$ are those which minimize (maximize) $\text{ACE}_{\text{NT}}$ and $\text{ACE}_{\text{AT}}$; this is an immediate consequence of the negative coefficients for $\gamma^0_{\text{NT}}$ and $\gamma^1_{\text{AT}}$ in the bounds for $\gamma^x_{\text{CO}}$ and $\gamma^x_{\text{DE}}$ in Table 2.

ACE bounds for the four compliance types are shown for the flu data in Figure 12. The ACE bounds for Compliers indicate that, under the observed distribution, the possibility of a zero ACE for Compliers is consistent with all feasible distributions over compliance types, except those for which the proportion of Defiers in the population is small.

Following [Pearl 2000; Robins 1989; Manski 1990; Robins and Rotnitzky 2004] we also consider the average causal effect on the entire population:

$$\text{ACE}_{\text{global}}(\pi_X, \{\gamma^x_{t_x}\}) \equiv \sum_{t_x \in \mathbb{D}_X} (\gamma^1_{t_x}(\pi_X) - \gamma^0_{t_x}(\pi_X))\pi_{t_x};$$

upper and lower bounds taken over $\{\gamma^x_{t_x}\}$ are defined similarly. The bounds given for $\text{ACE}_{t_x}$ in Table 5 are an immediate consequence of equations (8)–(11) which relate $p(y \mid x, z)$ to $\pi_X$ and $\{\gamma^x_{t_x}\}$. Before deriving the ACE bounds we need the following observation:

LEMMA 5. *For a given feasible $\pi_X$ and $p(y, x \mid z)$,*

$$\text{ACE}_{\text{global}}(\pi_X, \{\gamma^x_{t_x}\})$$
$$= p_{y_1, x_1 \mid z_1} - p_{y_1, x_0 \mid z_0} + \pi_{\text{DE}}(\gamma^1_{\text{DE}} - \gamma^0_{\text{DE}}) + \pi_{\text{NT}}\gamma^1_{\text{NT}} - \pi_{\text{AT}}\gamma^0_{\text{AT}} \qquad (21)$$
$$= p_{y_1, x_1 \mid z_0} - p_{y_1, x_0 \mid z_1} + \pi_{\text{CO}}(\gamma^1_{\text{CO}} - \gamma^0_{\text{CO}}) + \pi_{\text{NT}}\gamma^1_{\text{NT}} - \pi_{\text{AT}}\gamma^0_{\text{AT}}. \qquad (22)$$

*Proof:* (21) follows from the definition of $\text{ACE}_{\text{global}}$ and the observation that $p_{y_1, x_1 \mid z_1} = \pi_{\text{CO}}\gamma^1_{\text{CO}} + \pi_{\text{AT}}\gamma^1_{\text{AT}}$ and $p_{y_1, x_0 \mid z_0} = \pi_{\text{CO}}\gamma^0_{\text{CO}} + \pi_{\text{NT}}\gamma^0_{\text{NT}}$. The proof of (22) is similar. □

PROPOSITION 6. *For a given feasible $\pi_X$ and $p(y, x \mid z)$, the compatible distribution which minimizes [maximizes] $\text{ACE}_{\text{global}}$ has*

| Group | ACE Lower Bound | ACE Upper Bound |
|---|---|---|
| NT | $0 - u\gamma_{\text{NT}}^0(\pi_X)$ | $1 - l\gamma_{\text{NT}}^0(\pi_X)$ |
| CO | $l\gamma_{\text{CO}}^1(\pi_X) - u\gamma_{\text{CO}}^0(\pi_X)$<br>$= \gamma_{\text{CO}}^1(\pi_X, u\gamma_{\text{AT}}^1(\pi_X))$<br>$\quad - \gamma_{\text{CO}}^0(\pi_X, l\gamma_{\text{NT}}^0(\pi_X))$ | $u\gamma_{\text{CO}}^1(\pi_X) - l\gamma_{\text{CO}}^0(\pi_X)$<br>$= \gamma_{\text{CO}}^1(\pi_X, l\gamma_{\text{AT}}^1(\pi_X))$<br>$\quad - \gamma_{\text{CO}}^0(\pi_X, u\gamma_{\text{NT}}^0(\pi_X))$ |
| DE | $l\gamma_{\text{DE}}^1(\pi_X) - u\gamma_{\text{DE}}^0(\pi_X)$<br>$= \gamma_{\text{DE}}^1(\pi_X, u\gamma_{\text{AT}}^1(\pi_X))$<br>$\quad - \gamma_{\text{DE}}^0(\pi_X, l\gamma_{\text{NT}}^0(\pi_X))$ | $u\gamma_{\text{DE}}^1(\pi_X) - l\gamma_{\text{DE}}^0(\pi_X)$<br>$= \gamma_{\text{DE}}^1(\pi_X, l\gamma_{\text{AT}}^1(\pi_X))$<br>$\quad - \gamma_{\text{DE}}^0(\pi_X, u\gamma_{\text{NT}}^0(\pi_X))$ |
| AT | $l\gamma_{\text{AT}}^1(\pi_X) - 1$ | $u\gamma_{\text{AT}}^1(\pi_X) - 0$ |
| global | $p_{y_1,x_1|z_1} - p_{y_1,x_0|z_0}$<br>$\quad + \pi_{\text{DE}} \cdot l\text{ACE}_{\text{DE}}(\pi_X) - \pi_{\text{AT}}$<br>$= p_{y_1,x_1|z_0} - p_{y_1,x_0|z_1}$<br>$\quad + \pi_{\text{CO}} \cdot l\text{ACE}_{\text{CO}}(\pi_X) - \pi_{\text{AT}}$ | $p_{y_1,x_1|z_1} - p_{y_1,x_0|z_0}$<br>$\quad + \pi_{\text{DE}} \cdot u\text{ACE}_{\text{DE}}(\pi_X) + \pi_{\text{NT}}$<br>$= p_{y_1,x_1|z_0} - p_{y_1,x_0|z_1}$<br>$\quad + \pi_{\text{CO}} \cdot u\text{ACE}_{\text{CO}}(\pi_X) + \pi_{\text{NT}}$ |

Table 4. Upper and Lower bounds on average causal effects for different groups, as a function of a feasible $\pi_X$. Here $\pi_{\text{NT}}^c \equiv 1 - \pi_{\text{NT}}$

$$\langle \gamma_{\text{NT}}^0, \gamma_{\text{AT}}^1 \rangle = \langle l\gamma_{\text{NT}}^0, u\gamma_{\text{AT}}^1 \rangle \quad [\langle u\gamma_{\text{NT}}^0, l\gamma_{\text{AT}}^1 \rangle]$$
$$\langle \gamma_{\text{NT}}^1, \gamma_{\text{AT}}^0 \rangle = \langle 0, 1 \rangle \quad [\langle 1, 0 \rangle]$$

thus also minimizes [maximizes] $\text{ACE}_{\text{CO}}$ and $\text{ACE}_{\text{DE}}$, and conversely maximizes [minimizes] $\text{ACE}_{\text{AT}}$ and $\text{ACE}_{\text{NT}}$.

*Proof:* The claims follow from equations (21) and (22), together with the fact that $\gamma_{\text{AT}}^0$ and $\gamma_{\text{NT}}^1$ are unconstrained, so $\text{ACE}_{\text{global}}$ is minimized by taking $\gamma_{\text{AT}}^0 = 1$ and $\gamma_{\text{NT}}^1 = 0$, and maximized by taking $\gamma_{\text{AT}}^0 = 0$ and $\gamma_{\text{NT}}^1 = 1$. □

It is of interest here that although the definition of $\text{ACE}_{\text{global}}$ treats the four compliance types symmetrically, the compatible distribution which minimizes [maximizes] this quantity (for a given $\pi_X$) does not: it always corresponds to the scenario in which the treatment has the smallest [greatest] effect on Compliers and Defiers.

The bounds on the global ACE for the flu vaccine data of [Hirano, Imbens, Rubin, and Zhou 2000] are shown are shown in Figure 13.

Finally we note that it would be simple to develop similar bounds for other measures such as the Causal Relative Risk and Causal Odds Ratio.

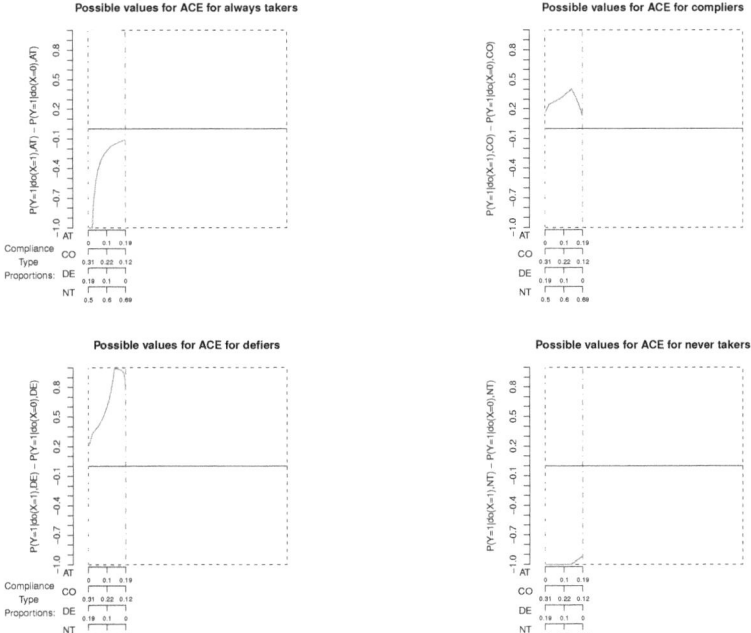

Figure 12. Depiction of the set of values for $\pi_X$ vs. $\mathrm{ACE}_{\mathbf{t}_X}(\pi_X)$ for $\mathbf{t}_X \in \mathbb{D}_X$ for the flu data.

## 6 Instrumental inequalities

The expressions involved in the upper bound on $\pi_{\mathrm{AT}}$ in (16) appear similar to those which occur in Pearl's instrumental inequalities. Here we show that the requirement that $\mathcal{P}_X \neq \emptyset$, or equivalently, $l\pi_{\mathrm{AT}} \leq u\pi_{\mathrm{AT}}$ is in fact equivalent to the instrumental inequality. This also provides an interpretation as to what may be inferred from the violation of a specific inequality.

THEOREM 7. *The following conditions place equivalent restrictions on $p(x \mid z)$ and $p(y \mid x=0, z)$:*

(a1) $\max\{0, \, p(x=1 \mid z=0) + p(x=1 \mid z=1) - 1\} \leq$
$\qquad \min\left\{1 - \sum_j p(y=j, x=0 \mid z=j), \, 1 - \sum_k p(y=k, x=0 \mid z=1-k)\right\}$;

(a2) $\max\left\{\sum_j p(y=j, x=0 \mid z=j), \, \sum_k p(y=k, x=0 \mid z=1-k)\right\} \leq 1$.

*Similarly, the following place equivalent restrictions on $p(x \mid z)$ and $p(y \mid x=1, z)$:*

(b1) $\max\{0, p(x=1 \mid z=0) + p(x=1 \mid z=1) - 1\} \leq$
$\qquad \min\left\{\sum_j p(y=j, x=1 \mid z=j), \, \sum_k p(y=k, x=1 \mid z=1-k)\right\}$;

(b2) $\max\left\{\sum_j p(y=j, x=1 \mid z=j), \, \sum_k p(y=k, x=1 \mid z=1-k)\right\} \leq 1$.

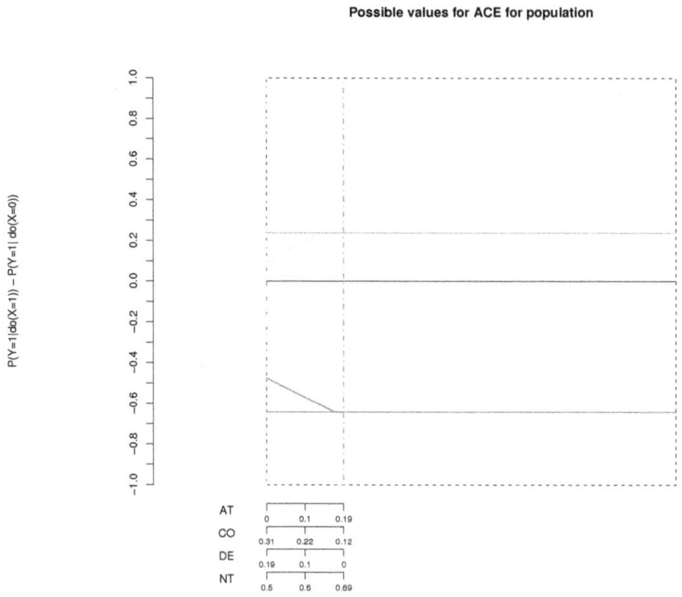

**Possible values for ACE for population**

Figure 13. Depiction of the set of values for $\pi_X$ vs. the global ACE for the flu data. The horizontal lines represent the overall bounds on the global ACE due to Pearl.

Thus the instrumental inequality (a2) corresponds to the requirement that the upper bounds on $p(\text{AT})$ resulting from $p(x \mid z)$ and $p(y=1 \mid x=0, z)$ be greater than the lower bound on $p(\text{AT})$ (derived solely from $p(x \mid z)$). Similarly for (b2) and the upper bounds on $p(\text{AT})$ resulting from $p(y=1 \mid x=1, z)$.

*Proof:* [(a1) $\Leftrightarrow$ (a2)] We first note that:

$$1 - \sum_j p(y=j, x=0 \mid z=j) \geq \left(\sum_j p(x=1 \mid z=j)\right) - 1$$
$$\Leftrightarrow \sum_j (1 - p(y=j, x=0 \mid z=j)) \geq \sum_j p(x=1 \mid z=j)$$
$$\Leftrightarrow \sum_j (p(y=1-j, x=0 \mid z=j) + p(x=1 \mid z=j)) \geq \sum_j p(x=1 \mid z=j)$$
$$\Leftrightarrow \sum_j p(y=j, x=0 \mid z=j) \geq 0.$$

which always holds. By a symmetric argument we can show that it always holds that:

$$1 - \sum_j p(y=j, x=0 \mid z=1-j) \geq \left(\sum_j p(x=1 \mid z=j)\right) - 1.$$

Thus if (a1) does not hold then $\max\{0, p(x=1 \mid z=0) + p(x=1 \mid z=1) - 1\} = 0$. It is then simple to see that (a1) does not hold iff (a2) does not hold.

[(b1) $\Leftrightarrow$ (b2)] It is clear that neither of the sums on the RHS of (b1) are negative, hence if (b1) does not hold then $\max\{0, p(x=1 \mid z=0) + p(x=1 \mid z=1) - 1\} =$

$\left( \sum_j p(x\!=\!1 \mid z\!=\!j) \right) - 1$. Now

$$\sum_j p(y\!=\!j, x\!=\!1 \mid z\!=\!j) < \left( \sum_j p(x\!=\!1 \mid z\!=\!j) \right) - 1$$
$$\Leftrightarrow \quad 1 < \sum_j p(y\!=\!j, x\!=\!1 \mid z\!=\!1-j).$$

Likewise

$$\sum_j p(y\!=\!j, x\!=\!1 \mid z\!=\!1-j) < \left( \sum_j p(x\!=\!1 \mid z\!=\!j) \right) - 1$$
$$\Leftrightarrow \quad 1 < \sum_j p(y\!=\!j, x\!=\!1 \mid z\!=\!j).$$

Thus (b1) fails if and only if (b2) fails. □

This equivalence should not be seen as surprising since [Bonet 2001] states that the instrument inequalities (a2) and (b2) are sufficient for a distribution to be compatible with the binary IV model. This is not the case if, for example, $X$ takes more than 2 states.

## 6.1 Which alternatives does a test of the instrument inequalities have power against?

[Pearl 2000] proposed testing the instrument inequalities (a2) and (b2) as a means of testing the IV model; [Ramsahai 2008] develops tests and analyzes their properties. It is then natural to ask what should be inferred from the failure of a specific instrumental inequality. It is, of course, always possible that randomization has failed. If randomization is not in doubt, then the exclusion restriction (1) must have failed in some way. The next result implies that tests of the inequalities (a2) and (b2) have power, respectively, against failures of the exclusion restriction for Never Takers (with $X = 0$) and Always Takers (with $X = 1$):

THEOREM 8. *The conditions* (RX), (RY$_{X=0}$) *and* (E$_{X=0}$) *described below imply* (a2); *similarly* (RX), (RY$_{X=1}$) *and* (E$_{X=1}$) *imply* (b2).

(RX)  $Z \perp\!\!\!\perp t_X$ equivalently $Z \perp\!\!\!\perp X_{z=0}, X_{z=1}$ :

(RY$_{X=0}$)  $Z \perp\!\!\!\perp Y_{x=0,z=0} \mid t_X = \text{NT}; \quad Z \perp\!\!\!\perp Y_{x=0,z=1} \mid t_X = \text{NT};$

(RY$_{X=1}$)  $Z \perp\!\!\!\perp Y_{x=1,z=1} \mid t_X = \text{AT}; \quad Z \perp\!\!\!\perp Y_{x=1,z=1} \mid t_X = \text{AT};$

(E$_{X=0}$)  $p(Y_{x=0,z=0} = Y_{x=0,z=1} \mid t_X = \text{NT}) = 1;$

(E$_{X=1}$)  $p(Y_{x=1,z=0} = Y_{x=1,z=1} \mid t_X = \text{AT}) = 1.$

Conditions (RX) and (RY$_{X=x}$) correspond to the assumption of randomization with respect to compliance type and response type. For the purposes of technical clarity we have stated condition (RY$_{X=x}$) in the weakest form possible. However, we know of no subject matter knowledge which would lead one to believe that (RX) and (RY$_{X=x}$) held, without also implying the stronger assumption (2). In contrast, the exclusion restrictions (E$_{X=x}$) are significantly weaker than (1), e.g. one could conceive of situations where assignment had an effect on the outcome for Always

Takers, but not for Compliers. It should be noted that tests of the instrument inequalities have no power to detect failures of the exclusion restriction for Compliers or defier.

We first prove the following Lemma, which also provides another characterization of the instrument inequalities:

LEMMA 9. *Suppose* (RX) *holds and* $Y \perp\!\!\!\perp Z \mid t_X = \text{NT}$ *then* (a2) *holds. Similarly, if* (RX) *holds and* $Y \perp\!\!\!\perp Z \mid t_X = \text{AT}$ *then* (b2) *holds.*

Note that the conditions in the antecedent make no assumption regarding the existence of counterfactuals for $Y$.

*Proof:* We prove the result for Never Takers; the other proof is similar. By hypothesis we have:

$$p(Y=1 \mid Z=0, t_X = \text{NT}) = p(Y=1 \mid Z=1, t_X = \text{NT}) \equiv \gamma_{\text{NT}}^0. \qquad (23)$$

In addition,

$$\begin{aligned}
p(Y=1 \mid Z=0, X=0) \\
= \; & p(Y=1 \mid Z=0, X=0, X_{z=0}=0) \\
= \; & p(Y=1 \mid Z=0, X_{z=0}=0) \\
= \; & p(Y=1 \mid Z=0, t_X = \text{CO})\, p(t_X = \text{CO} \mid Z=0, X_{z=0}=0) \\
& + p(Y=1 \mid Z=0, t_X = \text{NT})\, p(t_X = \text{NT} \mid Z=0, X_{z=0}=0) \\
= \; & p(Y=1 \mid Z=0, t_X = \text{CO})\, p(t_X = \text{CO} \mid X_{z=0}=0) \\
& + \gamma_{\text{NT}}^0\, p(t_X = \text{NT} \mid X_{z=0}=0).
\end{aligned} \qquad (24)$$

The first three equalities here follow from consistency, the definition of the compliance types and the law of total probability. The final equality uses (RX). Similarly, it may be shown that

$$\begin{aligned}
p(Y=1 \mid Z=1, X=0) \\
= \; & p(Y=1 \mid Z=1, t_X = \text{DE})\, p(t_X = \text{DE} \mid X_{z=1}=0) \\
& + \gamma_{\text{NT}}^0\, p(t_X = \text{NT} \mid X_{z=1}=0).
\end{aligned} \qquad (25)$$

Equations (24) and (25) specify two averages of three quantities, thus taking $u = p(Y=1 \mid Z=0, t_X = \text{CO})$, $v = p(Y=1 \mid Z=1, t_X = \text{DE})$ and $w = \gamma_{\text{NT}}^0$, we may apply the analysis of §2.3. This then leads to the upper bound on $\pi_{\text{AT}}$ given by equation (15). (Note that the lower bounds on $\pi_{\text{AT}}$ are derived from $p(x \mid z)$ and hence are unaffected by dropping the exclusion restriction.) The requirement that there exist some feasible distribution $\pi_X$ then implies equation (a2) which is shown in Theorem 7 to be equivalent to (b2) as required. □

Binary Instrumental Variable Model

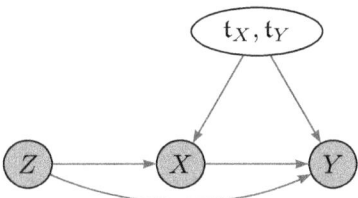

Figure 14. Graphical representation of the model given by the randomization assumption (2) alone. It is no longer assumed that $Z$ does not have a direct effect on $Y$.

*Proof of Theorem 8:* We establish that (RX), (RY$_{X=0}$), (E$_{X=0}$) $\Rightarrow$ (a2). The proof of the other implication is similar. By Lemma 9 it is sufficient to establish that $Y \perp\!\!\!\perp Z \mid t_X = \text{NT}$.

$$
\begin{aligned}
p(Y = 1 \mid Z = 0, t_X = \text{NT}) & \\
= \; & p(Y = 1 \mid Z = 0, X = 0, t_X = \text{NT}) & \text{definition of NT;} \\
= \; & p(Y_{x=0,z=0} = 1 \mid Z = 0, X = 0, t_X = \text{NT}) & \text{consistency;} \\
= \; & p(Y_{x=0,z=0} = 1 \mid Z = 0, t_X = \text{NT}) & \text{definition of NT;} \\
= \; & p(Y_{x=0,z=0} = 1 \mid t_X = \text{NT}) & \text{by (RY}_{X=0}\text{);} \\
= \; & p(Y_{x=0,z=1} = 1 \mid t_X = \text{NT}) & \text{by (E}_{X=0}\text{);} \\
= \; & p(Y_{x=0,z=1} = 1 \mid Z = 1, t_X = \text{NT}) & \text{by (RY}_{X=0}\text{);} \\
= \; & p(Y = 1 \mid Z = 1, t_X = \text{NT}) & \text{consistency, NT.}
\end{aligned}
$$

□

A similar result is given in [Cai, Kuroki, Pearl, and Tian 2008], who consider the *Average Controlled Direct Effect*, given by:

$$\text{ACDE}(x) \equiv p(Y_{x,z=1}=1) - p(Y_{x,z=0}=1),$$

under the model given solely by the equation (2), which corresponds to the graph in Figure 14. Cai et al. prove that under this model the following bounds obtain:

$$\text{ACDE}(x) \geq p(y=0, x \mid z=0) + p(y=1, x \mid z=1) - 1, \tag{26}$$
$$\text{ACDE}(x) \leq 1 - p(y=0, x \mid z=1) - p(y=1, x \mid z=0). \tag{27}$$

It is simple to see that $\text{ACDE}(x)$ will be bounded away from 0 for some $x$ iff one of the instrumental inequalities is violated. This is as we would expect: the IV model of Figure 1 is a sub-model of Figure 14, but if $\text{ACDE}(x)$ is bounded away from 0 then the $Z \to Y$ edge is present, and hence the exclusion restriction (1) is incompatible with the observed distribution.

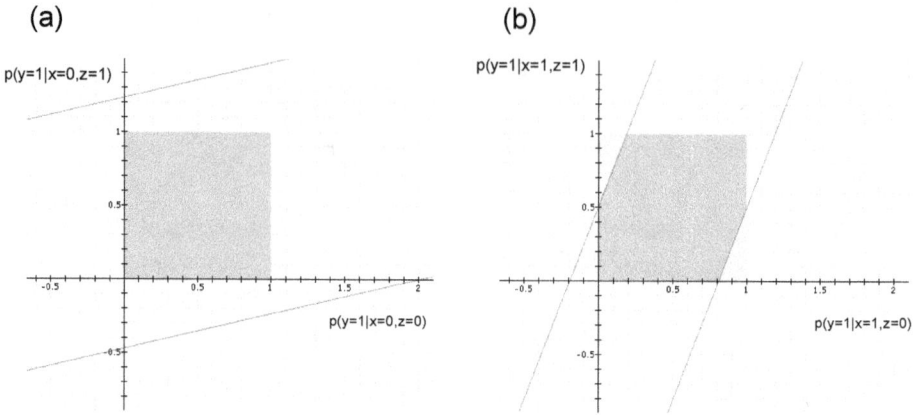

Figure 15. Illustration of the possible values for $p(y \mid x, z)$ compatible with the instrument inequalities, for a given distribution $p(x|z)$. The darker shaded region satisfies the inequalities: (a) $X = 0$, inequalities (a2); (b) $X = 1$, inequalities (b2). In this example $p(x = 1 \mid z = 0) = 0.84$, $p(x = 1 \mid z = 1) = 0.32$. Since $0.84/(1 - 0.32) > 1$, (a2) is trivially satisfied; see proof of Theorem 10.

## 6.2 How many instrument inequalities may be violated by a single distribution?

THEOREM 10. *For any distribution $p(x, y \mid z)$, at most one of the four instrument inequalities:*

(a2.1) $\sum_j p(y=j, x=0 \mid z=j) \leq 1$;    (a2.2) $\sum_j p(y=j, x=0 \mid z=1-j) \leq 1$;

(b2.1) $\sum_j p(y=j, x=1 \mid z=j) \leq 1$;    (b2.2) $\sum_j p(y=j, x=1 \mid z=1-j) \leq 1$;

*is violated.*

*Proof:* We first show that at most one of (a2.1) and (a2.2) may be violated. Letting $\theta_{ij} = p(y = 1 \mid x = j, z = i)$ we may express these inequalities as:

$$\theta_{10} \cdot p_{x_0|z_1} - \theta_{00} \cdot p_{x_0|z_0} \leq p_{x_1|z_0}, \quad \text{(a2.1)}$$
$$\theta_{10} \cdot p_{x_0|z_1} - \theta_{00} \cdot p_{x_0|z_0} \geq -p_{x_1|z_1}, \quad \text{(a2.2)}$$

giving two half-planes in $(\theta_{00}, \theta_{10})$-space (see Figure 15(a)). Since the lines defining the half-planes are parallel, it is sufficient to show that the half-planes always intersect, and hence that the regions in which (a2.1) and (a2.2) are violated are disjoint. However, this is immediate since the (non-empty) set of points for which $\theta_{10} \cdot p_{x_0|z_1} - \theta_{00} \cdot p_{x_0|z_0} = 0$ always satisfy both inequalities.

The proof that at most one of (b2.1) and (b2.2) may be violated is symmetric.

We now show that the inequalities (a2.1) and (a2.2) place non-trivial restrictions on $(\theta_{00}, \theta_{10})$ iff (b2.1) and (b2.2) place trivial restrictions on $(\theta_{01}, \theta_{11})$. The line corresponding to (a2.1) passes through $(\theta_{00}, \theta_{10}) = (-p_{x_1|z_0}/p_{x_0|z_0}, 0)$ and

$(0, p_{x_1|z_0}/p_{x_0|z_1})$; since the slope of the line is non-negative, it has non-empty intersection with $[0,1]^2$ iff $p_{x_1|z_0}/p_{x_0|z_1} \leq 1$. Thus there are values of $(\theta_{01}, \theta_{11}) \in [0,1]^2$ which fail to satisfy (a2.1) iff $p_{x_1|z_0}/p_{x_0|z_1} < 1$. By a similar argument it may be shown that (a2.2) is non-trivial iff $p_{x_1|z_1}/p_{x_0|z_0} < 1$, which is equivalent to $p_{x_1|z_0}/p_{x_0|z_1} < 1$.

The proof is completed by showing that (b2.1) and (b2.2) are non-trivial if and only if $p_{x_1|z_0}/p_{x_0|z_1} > 1$. □

COROLLARY 11. *Every distribution $p(x,y \mid z)$ is consistent with randomization (RX) and (2), and at least one of the exclusion restrictions $E_{X=0}$ or $E_{X=1}$.*

**Flu Data Revisited**

For the data in Table 3, all of the instrument inequalities hold. Consequently there is no evidence of a direct effect of $Z$ on $Y$. (Again we emphasize that unlike [Hirano, Imbens, Rubin, and Zhou 2000], we are not using any information on baseline covariates in the analysis.) Finally we note that, since all of the instrumental inequalities hold, maximum likelihood estimates for the distribution $p(x,y \mid z)$ under the IV model are given by the empirical distribution. However, if one of the IV inequalities were to be violated then the MLE would not be equal to the empirical distribution, since the latter would not be a law within the IV model. In such a circumstance a fitting procedure would be required; see [Ramsahai 2008, Ch. 5].

## 7 Conclusion

We have built upon and extended the work of Pearl, displaying how the range of possible distributions over types compatible with a given observed distribution may be characterized and displayed geometrically. Pearl's bounds on the global ACE are sometimes objected to on the grounds that they are too extreme, since for example, the upper bound presupposes a 100% success rate among Never Takers if they were somehow to receive treatment, likewise a 100% failure rate among Always Takers were they not to receive treatment. Our analysis provides a framework for performing a sensitivity analysis. Lastly, our analysis relates the IV inequalities to the bounds on direct effects.

**Acknowledgements**

This research was supported by the U.S. National Science Foundation (CRI 0855230) and U.S. National Institutes of Health (R01 AI032475) and Jesus College, Oxford where Thomas Richardson was a Visiting Senior Research Fellow in 2008. The authors used Avitzur's *Graphing Calculator* software (www.pacifict.com) to construct two and three dimensional plots. We thank McDonald, Hiu and Tierney for giving us permission to use their flu vaccine data.

# References

Balke, A. and J. Pearl (1997). Bounds on treatment effects from studies with imperfect compliance. *Journal of the American Statistical Association 92*, 1171–1176.

Bonet, B. (2001). Instrumentality tests revisited. In *Proceedings of the 17$^{th}$ Conference on Uncertainty in Artificial Intelligence*, pp. 48–55.

Cai, Z., M. Kuroki, J. Pearl, and J. Tian (2008). Bounds on direct effects in the presence of confounded intermediate variables. *Biometrics 64*, 695–701.

Chickering, D. and J. Pearl (1996). A clinician's tool for analyzing non-compliance. In *AAAI-96 Proceedings*, pp. 1269–1276.

Erosheva, E. A. (2005). Comparing latent structures of the Grade of Membership, Rasch, and latent class models. *Psychometrika 70*, 619–628.

Fienberg, S. E. and J. P. Gilbert (1970). The geometry of a two by two contingency table. *Journal of the American Statistical Association 65*, 694–701.

Hirano, K., G. W. Imbens, D. B. Rubin, and X.-H. Zhou (2000). Assessing the effect of an influenza vaccine in an encouragement design. *Biostatistics 1*(1), 69–88.

Manski, C. (1990). Non-parametric bounds on treatment effects. *American Economic Review 80*, 351–374.

McDonald, C., S. Hiu, and W. Tierney (1992). Effects of computer reminders for influenza vaccination on morbidity during influenza epidemics. *MD Computing 9*, 304–312.

Pearl, J. (2000). *Causality*. Cambridge, UK: Cambridge University Press.

Ramsahai, R. (2008). *Causal Inference with Instruments and Other Supplementary Variables*. Ph. D. thesis, University of Oxford, Oxford, UK.

Robins, J. (1989). The analysis of randomized and non-randomized AIDS treatment trials using a new approach to causal inference in longitudinal studies. In L. Sechrest, H. Freeman, and A. Mulley (Eds.), *Health Service Research Methodology: A focus on AIDS*. Washington, D.C.: U.S. Public Health Service.

Robins, J. and A. Rotnitzky (2004). Estimation of treatment effects in randomised trials with non-compliance and a dichotomous outcome using structural mean models. *Biometrika 91*(4), 763–783.

# 26
# Pearl Causality and the Value of Control
ROSS SHACHTER AND DAVID HECKERMAN

## 1 Introduction

We welcome this opportunity to acknowledge the significance of Judea Pearl's contributions to uncertain reasoning and in particular to his work on causality. In the decision analysis community causality had long been "taboo" even though it provides a natural framework to communicate with decision makers and experts [Shachter and Heckerman 1986]. Ironically, while many of the concepts and methods of causal reasoning are foundational to decision analysis, scholars went to great lengths to avoid causal terminology in their work. Judea Pearl's work is helping to break this barrier, allowing the exploration of some fundamental principles. We were inspired by his work to understand exactly what assumptions are being made in his causal models, and we would like to think that our subsequent insights have contributed to his and others' work as well.

In this paper, we revisit our previous work on how a decision analytic perspective helps to clarify some of Pearl's notions, such as those of the *do* operator and *atomic intervention*. In addition, we show how influence diagrams [Howard and Matheson 1984] provide a general graphical representation for cause. Decision analysis can be viewed simply as determining what interventions we want to make in the world to improve the prospects for us and those we care about, an inherently causal concept. As we shall discuss, causal models are naturally represented within the framework of decision analysis, although the causal aspects of issues about counterfactuals and causal mechanisms that arise in computing the value of clairvoyance [Howard 1990], were first presented by Heckerman and Shachter [1994, 1995]. We show how this perspective helps clarify decision-analytic measures of sensitivity, such as the value of control and the value of revelation [Matheson 1990; Matheson and Matheson 2005].

## 2 Decision-Theoretic Foundations

In this section we introduce the relevant concepts from [Heckerman and Shachter 1995], the framework for this paper, along with some extensions to those concepts.

Our approach rests on a simple but powerful primitive concept of *unresponsiveness*. An uncertain variable is unresponsive to a set of decisions if its value is unaffected by our choice for the decisions. It is unresponsive to those decisions in worlds limited by other variables if the decisions cannot affect the uncertain variable without also changing one of the other variables.

We can formalize this by introducing concepts based on Savage [1954]. We consider three different kinds of distinctions, which he called *acts, consequences*, and *possible states of the world*. We have complete control over the acts but no control over the uncertain state of the world. We might have some level of control over consequences, which are logically determined, after we act, by the state of the world. Therefore, a consequence can be represented as a deterministic function of acts and the state of the world, inheriting uncertainty from the state of the world while affected, more or less, by our choice of action.

In practice, it is convenient to represent acts and consequences with variables in our model. We call a variable describing a set of mutually exclusive and collectively exhaustive acts a *decision*, and we denote the set of decisions by $D$. We call a variable describing a consequence *uncertain*, and we denote the set of uncertain variables by $U$. At times we will distinguish between the uncertain variables that serve as our objectives or *value variables*, $V$, and the other uncertain variables which we call *chance variables*, $C = U \setminus V$. Finally, in this section we will use the variables $S$ to represent the possible states of the world. As a convention we will refer to single variables with lower-case ($x$ or $d$), sets of variables with upper-case ($D$ or $V$), and particular instances of variables with bold ($\mathbf{x}$ or $\mathbf{D}$). In this notation, the set of uncertain variables $X$ takes value $X[\mathbf{S}, \mathbf{D}]$ deterministically when $\mathbf{D}$ is chosen and $\mathbf{S}$ is the state of the world.

DEFINITION 1 (Unresponsiveness). Given a decision problem described by uncertain variables $U$, decision variables $D$, and state of the world $S$, and variable sets $X \subseteq U$ and $Y \subseteq D \cup U$, $X$ is said to be *unresponsive to* $D$, denoted $X \not\leftarrow D$, if we believe that

$$\forall \mathbf{S} \in S, \mathbf{D_1} \in D, \mathbf{D_2} \in D : X[\mathbf{S}, \mathbf{D_1}] = X[\mathbf{S}, \mathbf{D_2}]$$

and, if not, $X$ is said to be *responsive to* $D$.

Furthermore, $X$ is said to be *unresponsive to* $D$ *in worlds limited by* $Y$, denoted $X \not\leftarrow_Y D$, if we believe that

$$\forall \mathbf{S} \in S, \mathbf{D_1} \in D, \mathbf{D_2} \in D : Y[\mathbf{S}, \mathbf{D_1}] = Y[\mathbf{S}, \mathbf{D_2}] \implies X[\mathbf{S}, \mathbf{D_1}] = X[\mathbf{S}, \mathbf{D_2}]$$

and, if not, $X$ is said to be *responsive to* $D$ *in worlds limited by* $Y$.

The distinctions of unresponsiveness and limited unresponsiveness seem natural for decision makers to consider. Unresponsiveness is related to independence, in that any uncertain variables $X$ that are unresponsive to decisions $D$ are independent of $D$. Although it is not necessarily the case that $X$ independent of $D$ is unresponsive to $D$, that implication is often assumed [Spirtes, Glymour, and Scheines 1993]. In contrast, there is no such general correspondence between limited unresponsiveness and conditional independence.

To illustrate these concepts graphically, we introduce influence diagrams [Howard and Matheson 1984]. An *influence diagram* is an acyclic directed graph $G$ with

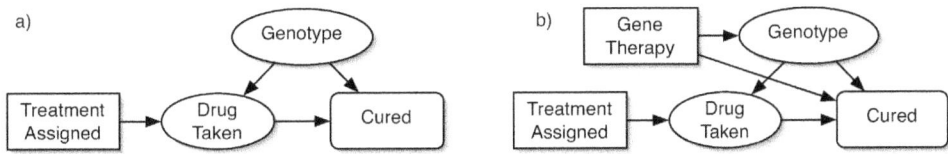

Figure 1. The treatment assignment only cures the patient if it affects whether the drug is taken, but genotype does not have a causal effect unless it is responsive to decisions.

nodes corresponding to the variables, rectangles for decisions, ovals for chance variables, and rounded rectangles for value variables. Arcs into chance and value nodes, are *conditional*. For each uncertain variable $x$ there is a conditional probability distribution for $x$ given its parents, $Pa(x)$. If the distribution is a deterministic function, we represent that in the graph by a double oval or double rounded rectangle. Arcs into decisions are *informational*, representing that the parent variables will be observed before the decision is made. Although there are significant issues involving informational arcs, we will focus primarily on models in which there are no informational arcs and all of the decisions could be made in any order, before any of the uncertain variables are observed.

We allow multiple value nodes, all with no children, assuming that their values will be summed. We assume that the criterion for making decisions is either the total value or an increasing exponential utility function of the total. This simplifies the valuation of a proposed change to a decision problem because the most a decision maker should be willing to pay for the change is the difference in the values of the diagrams with and without the proposed change.

Although we have defined unresponsiveness without regard to a graphical representation, there is an intuitive graphical interpretation (with some technical exceptions described in Heckerman and Shachter [1985]). The uncertain descendants of decisions are usually responsive to them, and the other uncertain variables are usually unresponsive. Also, $X$ is usually unresponsive to $D$ in worlds limited by $Y$ if all of the directed paths from $D$ to $X$ include nodes in $Y$. When these rules of thumb are all satisfied, we say that an influence diagram is causal.

DEFINITION 2 (Causal Influence Diagram). An influence diagram with graph $G$ and decision nodes $D$, chance nodes $C$, and value nodes $V$, is said to be *causal* if we believe that uncertain variables $X \subseteq C \cup V$ are unresponsive to decisions $D$, $X \not\leftarrow D$, whenever there is no directed path from $D$ to $X$, and $X$ is unresponsive to decisions $D$ in worlds limited by $Y$, $X \not\leftarrow_Y D$, whenever every directed path from $D$ to $X$ includes a node from $Y$.

Consider the influence diagram shown in Figure 1a which we believe is causal. In this case, we believe that *Drug Taken* and *Cured* are responsive to *Treatment Assigned* while *Genotype* is unresponsive to *Treatment Assigned*. We also believe that *Cured* is unresponsive to *Treatment Assigned* in worlds limited by *Drug Taken*. Note that *Treatnent Assigned* is *not* independent of *Genotype* or *Cured* given *Drug Taken*.

The concept of limited unresponsiveness allows us to define how one variable can cause another in a way that is natural for decision makers to understand.

DEFINITION 3 (Cause with Respect to Decisions). Given a decision problem described by uncertain variables $U$ and decision variables $D$, and a variable $x \in U$, the set of variables $Y \subseteq D \cup U \setminus \{x\}$ is said to be a *cause for $x$ with respect to $D$* if $Y$ is a minimal set of variables such that $x \not\leftarrow_Y D$.

Defining cause with respect to a particular set of decisions adds clarity. Consider again the causal influence diagram shown in Figure 1a. With respect to the decision *Treatment Assigned*, the cause of *Cured* is either {*Treament Assigned*} or {*Drug Taken*}, while the cause of *Genotype* is {}. Because we believe that *Genotype* is unresponsive to *Treatment Assigned* it has no cause with respect to $D$. On the other hand, we believe that *Cured* is responsive to *Treatment Assigned* but not in worlds limited by *Drug Taken*, so {*Drug Taken*} is a cause of *Cured* with respect to $D$.

Consider now the causal influence diagram shown in Figure 1b, in which we have added the decision *Gene Therapy*. Because *Genotype* is now responsive to $D$, the cause of *Genotype* is {*Gene Therapy*} with respect to $D$. If the gene therapy has some side effect on whether the patient is cured, then {*Gene Therapy*, *Drug Taken*} but not {*Genotype*, *Drug Taken*} would be a cause of *Cured* with respect to the decisions, because *Cured* is unresponsive to $D$ in worlds limited by the former but not the latter.

The concept of limited unresponsiveness also allows us to formally define direct and atomic interventions. A set of decision $I$ is a direct intervention on a set of uncertain variables $X$ if the effects of $I$ on all other uncertain variables are mediated through their effects on $X$.

DEFINITION 4 (Direct Intervention). Given a decision problem described by uncertain variables $U$ and decision variables $D$, a set of decisions $I \subseteq D$ is said to be a *direct intervention on $X \subseteq U$ with respect to $D$* if (1) $x \hookleftarrow I$ for all $x \in X$, and (2) $y \not\leftarrow_X I$ for all $y \in U$.

In a causal influence diagram every node in $I$ has children only in $X$ and there is a directed path from $I$ to every node in $X$. In the causal influence diagram shown in Figure 1b, *Treatment Assigned* is a direct intervention on *Drug Taken*, and the set of decisions is a direct intervention on all three uncertain variables. Note that whether a decision is a direct intervention depends on the underlying causal mechanism. If the gene therapy had no side effect then *Gene Therapy* would be a direct intervention on *Genotype*, but regardless whether there is a side effect, *Gene Therapy* is a direct intervention on {*Genotype*, *Cured*}.

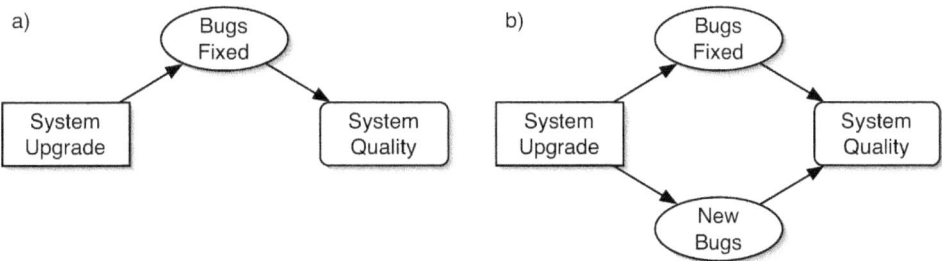

Figure 2. We believe that a system upgrade will affect system quality by fixing bugs unless new bugs are introduced in the process.

DEFINITION 5 (Atomic Intervention). Given a decision problem described by uncertain variables $U$ and decision variables $D$, a decision $do(x) \in D$ is said to be a *atomic intervention on $x \in U$ with respect to $D$* if (1) $do(x)$ is a direct invention on $x$ with respect to $D$, and (2) $do(x)$ has precisely the instances (a) **idle**, which corresponds to no intervention, and (b) $\mathbf{do(x)}$ for every instance $\mathbf{x}$ of $x$, where $x = \mathbf{x}$ whenever $do(x) = \mathbf{do(x)}$.

This is precisely the atomic intervention described without definition in Pearl [1993]. The assumptions underlying it are quite strong. The causal influence diagram shown in Figure 2a assumes that we can upgrade our system and improve the quality by fixing the bugs, but the diagram shown in (b) illustrates the all too familiar situation when new bugs are introduced in the process, compromising system quality. In that case, *System Upgrade* is not a direct intervention on *Bugs Fixed* and {*Bugs Fixed*} is not a cause of *System Quality* with respect to $D$. Although the system upgrade was intended to be an atomic intervention, it can have unintended and undesirable consequences.

We can now represent the relationship between an uncertain variable $x$ and other variables $Y$, such as its parents in a causal influence diagram. We consider the uncertain function $x(Y)$ as a variable, and now $x$ is a deterministic function of $Y$ and $x(Y)$. In fact, if $Y$ is a cause of $x$ with respect to $D$ then $x(Y)$ must be unresponsive to $D$.

DEFINITION 6 (Mapping Variable). Given a decision problem described by uncertain variables $U$ and decision variables $D$, $x \in U$ and variables $Y$ such that for every $y \in Y \cap U$ there exists an atomic intervention $do(y) \in D$, the *mapping variable* $x(Y)$ is the chance variable that represents all possible mappings from $Y$ to $x$.

Finally, we have developed the machinery to characterize a *Pearl causal model* and structural equations [Pearl 1993]. Given uncertain variables $U$, suppose the decisions $D$ comprise an atomic intervention $do(x)$ on every $x \in U$. Given a graph

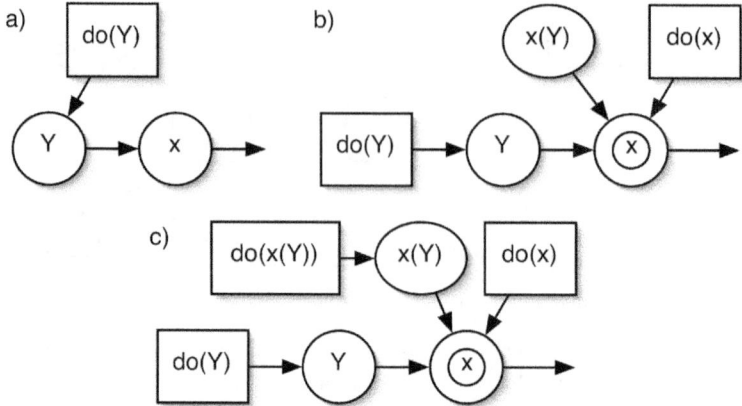

Figure 3. The partial influence diagram for $x$ in a causal model, shown in (a) with parents $Y$, becomes the diagram shown in (b) explicitly representing the structural equation for $x$, and, when $Y$ is nonempty, the diagram shown in (c) with an explicit atomic intervention on the mapping variable.

$G$ with nodes $U$, such that $Pa(x) \cup \{do(x)\}$ is a cause for $x$ with respect to $D$. Then

$$x = f_x(Pa(x), do(x), x(Pa(x)))$$

for all $x \in U$ where $f_x$ is a deterministic function such that $x = \mathbf{x}$ if $do(x) = \mathbf{do(x)}$.

We can extend this to allow manipulation of a mapping variable for $x \in U$ with parents to obtain a *Pearl causal model with an atomic intervention for mapping variable* $x(Pa(x))$. The decisions $D$ now also include a atomic intervention $do(x(Pa(x)))$. As a result, $Pa(x) \cup \{do(x), do(x(Pa(x)))\}$ is now a cause for $x$ with respect to $D$ and $x(Pa(x)) = \mathbf{x}(Pa(x))$ when $do(x(Pa(x))) = \mathbf{do}(\mathbf{x}(Pa(x)))$.

The causal model is represented by the partial influence diagrams shown in Figure 3 with $Y = Pa(x) \subseteq C$ in the graph $G$. We assume in (a) that there are atomic interventions $do(y)$ on each $y \in Y$ represented as $do(Y)$. The diagram shown in (b) explicitly represents the structural equation for $x$ as a deterministic function of $Y$, an atomic intervention, $do(x)$, and the mapping variable, $x(Y)$. The influence diagram is causal, showing that $Y \cup \{do(x)\}$ is a cause for $x$ with respect to $D$. We can extend the model by adding an atomic intervention for the mapping variable, $do(x(Y))$. If $Y$ is empty then nothing needs to be added, as $do(x)$ is the same atomic intervention as $do(x())$, but otherwise we obtain the diagram shown in (c). Now $Y \cup \{do(x), do(x(Y))\}$ is a cause for $x$ with respect to $D$.

An influence diagram is said to be in *canonical form* if each uncertain variable responsive to a decision is a descendant of that decision and represented as a deterministic node. Each decision, including atomic interventions, is explicit. Each uncertain variable that is responsive to $D$ is a deterministic function of its parents, including any decisions that are direct interventions on it, and a mapping variable. As an example, the influence diagram shown in Figure 3b is in canonical form.

In the next section we apply these concepts to define and contrast different measures for the value to a decision maker of manipulating (or observing) an uncertain variable.

## 3 Value of Control

When assisting a decision maker developing a model, sensitivity analysis measures help the decision maker to validate the model. One popular measure is the *value of clairvoyance*, the most a decision maker should be willing to pay to observe a set of uncertain variables before making particular decisions [Howard 1967]. Our focus of attention is another measure, the value of control (or wizardry), the most a decision maker should be willing to pay a hypothetical wizard to optimally control the distribution of an uncertain variable [Matheson 1990], [Matheson and Matheson 2005]. We consider and contrast the value of control with two other measures, the value of do, and the value of revelation, and we develop the conditions under which the different measures are equal.

In formalizing the value of control, it is natural to consider the value of an atomic intervention on uncertain variable $x$, in particular $\mathbf{do}(\mathbf{x}^*)$, that would set it to $\mathbf{x}^*$ the instance yielding the most valuable decision situation, rather than to **idle**. We call the most the decision maker should be willing to pay for such an intervention the *value of do* and compute it as the difference in the values of the diagrams.

DEFINITION 7 (Value of Do). Given a decision problem including an atomic intervention on uncertain variable $x \in U$, the *value of do for* $x$, denoted by $VoD(\mathbf{x}^*)$, is the most one should be willing to pay for an atomic intervention on uncertain variable $x$ to the best possible deterministic instance, $\mathbf{do}(\mathbf{x}^*)$, instead of to **idle**.

Our goal in general is to value the optimal manipulation of the *conditional* distribution of a target uncertain variable $x$ in a causal influence diagram, $P\{x|Y\}$, and the most we should be willing to pay for such an intervention is the *value of control*. The simplest case is when $\{do(x)\}$ is a cause of $x$ with respect to $D$, $Y = \{\}$, so the optimal distribution is equivalent to an atomic intervention on $x$ to $\mathbf{x}^*$, and control and *do* are the same intervention. Otherwise, the *do* operation effectively severs the arcs from $Y$ to $x$ and replaces the previous causal mechanism with the new atomic one. By contrast, the control operation is an atomic intervention on the mapping variable $x(Y)$ to its optimal value $\mathbf{do}(\mathbf{x}^*(Y))$ rather than to **idle**.

DEFINITION 8 (Value of Control). Given a decision problem including variables $Y$, a mapping variable $x(Y)$ for uncertain variable $x \in U$, and atomic interventions $do(x)$ and $do(x(Y))$ such that $Y \cup \{do(x), do(x(Y))\}$ is a cause of $x$ with respect to $D$, the *value of control for* $x$, denoted by $VoC(\mathbf{x}^*(Y))$, is the most one should be willing to pay for an atomic intervention on the mapping variable for uncertain variable $x$ to the best possible deterministic function of $Y$, $\mathbf{do}(\mathbf{x}^*(Y))$, instead of to **idle**.

If $Y = \{\}$, then $do(x)$ is the same atomic intervention as $do(x(Y))$, and the values of do and control for $x$ are equal, $VoD(\mathbf{x}^*) = VoC(\mathbf{x}^*())$.

In many cases, while it is tempting to assume atomic interventions, they can be cumbersome or implausible. In an attempt to avoid such issues, Ronald A. Howard has suggested an alternative passive measure, the *value of revelation*: how much better off the decision maker should be by observing that the uncertain variable in question obtained its most desirable value. This is only well-defined for variables unresponsive to $D$, except for those atomic interventions that are set to **idle**, because otherwise the observation would be made before decisions it might be responsive to. Under our assumptions this can be computed as the difference in value between two situations, but it is hard to describe it as a willingness to pay for this difference as it is more passive than intentional. (The value of revelation is in fact an intermediate term in the computation of the value of clairvoyance.)

DEFINITION 9 (Value of Revelation). Given a decision problem including uncertain variable $x \in U$ and a (possibly empty) set of atomic interventions, $A$, that is a cause for $x$ with respect to $D$, the *value of revelation for* uncertain variable $x \in U$, denoted by $VoR(\mathbf{x}^*)$, is the increase in the value of the situation with $d = $ **idle** for all $d \in A$, if one observed that uncertain variable $x = \mathbf{x}^*$, the best possible deterministic instance, instead of not observing $x$.

To illustrate these three measures we, consider a partial causal influence diagram including $x$ and its parents, $Y$, which we assume for this example are uncertain and nonempty, as shown in Figure 4a. There are atomic interventions $do(x)$ on $x$, $do(x(Y))$ on mapping variable $x(Y)$, and $do(y)$ on each $y \in Y$ represented as $do(Y)$. The variable $x$ is a deterministic function of $Y$, $do(x)$ and $x(Y)$. In this model, $Y \cup \{do(x), do(x(Y))\}$ is a cause of $x$ with respect to $D$. The dashed line from $x$ to values $V$ suggests that there might be some directed path from $x$ to $V$. If not, $V$ would be unresponsive to $do(x)$ and $do(x(Y))$ and the values of do and control would be zero.

To obtain the reference diagram for our proposed changes, we set all of the atomic interventions to **idle** as shown in Figure 4b1. We can compute the value of this diagram by eliminating the idle decisions and absorbing the mapping variable into $x$, yielding the simpler diagram shown in (b2). To compute the value of do for x, we can compute the value of the diagram with $\mathbf{do}(\mathbf{x}^*)$ by setting the other atomic interventions to **idle**, as shown in (c1). But since that is making the optimal choice for $x$ with no interventions on $Y$ or $x(Y)$, we can now think of $x$ as a decision variable as indicated in the diagram shown in (c2). We shall use this shorthand in many of the examples that we consider. To compute the value of control for x, we can compute the value of the diagram with $\mathbf{do}(\mathbf{x}^*(Y))$ by setting the other atomic interventions to **idle**, as shown in (d1). But since that is making the optimal choice for $x(Y)$ with none of the other interventions, we can compute its value with

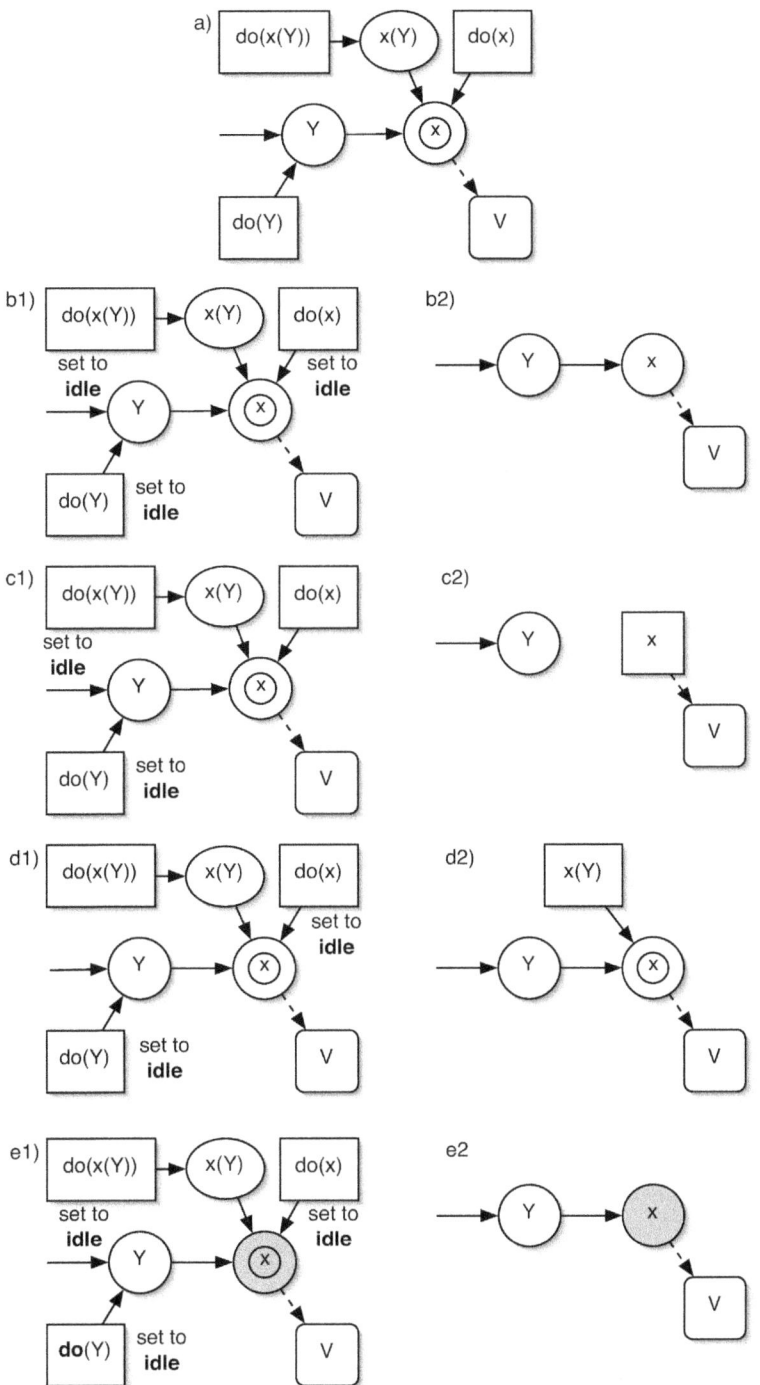

Figure 4. Partial causal influence diagrams to compute the values of do, control, and revelation for $x$ when $Y$ is nonempty.

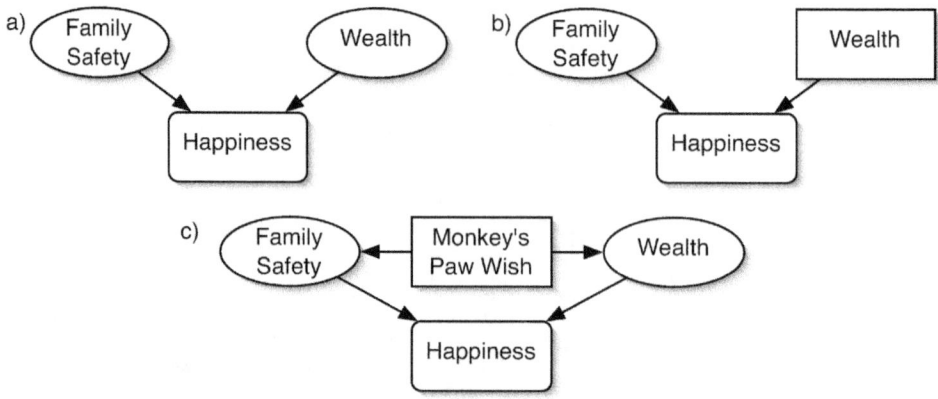

Figure 5. Unless the intervention is direct there can be disastrous side effects.

the simpler influence diagram shown in (d2), again using our shorthand. Finally, to compute the value of revelation for x, we can compute the value of the diagram with $x = x^*$ and all of the atomic interventions **idle**, as shown in (e1). The observation is well-defined because all of the interventions are **idle**, but that also means that we can compute its value with the simpler influence diagram shown in (e2).

Each of the three measures requires evaluation of two influence diagrams to determine its value, the reference diagram with all of the atomic interventions set to **idle** and a revised one, a diagram with either an atomic intervention or an observation. The values of these diagrams can be computed using simpler influence diagrams, with either one new decision, an atomic one made with no observations, or a new observation made before any decision, and the simpler diagram for the reference value has neither new decisions nor observations. These simpler diagrams are well-defined even if there are other decisions elsewhere and some observations prior to some of the other decisions [Shachter 1986]. Note that care must be taken in computing the value of control because there can be an exponential number of instances for the mapping variable.

The assumption of a direct intervention is crucial. Matheson and Matheson [2005]( refer to it as "pure" and to an atomic intervention as "perfect".) There is a classic horror story of a man granted three wishes on a monkey's paw [Jacobs 1902]. He chooses to be wealthy and his wish is granted, tragically, through the death of his son. This corresponds to the causal influence diagrams shown in Figure 5. The value of his situation with no intervention is represented by the diagram in (a). The atomic intervention on *Wealth* he intends would yield the same value as a diagram in which *Wealth* is a decision as in (b), but the value with his intervention actually equals the value of the diagram shown in (c). The wish decision he actually made was not the direct intervention on *Wealth* he desired. The lesson is clear: in

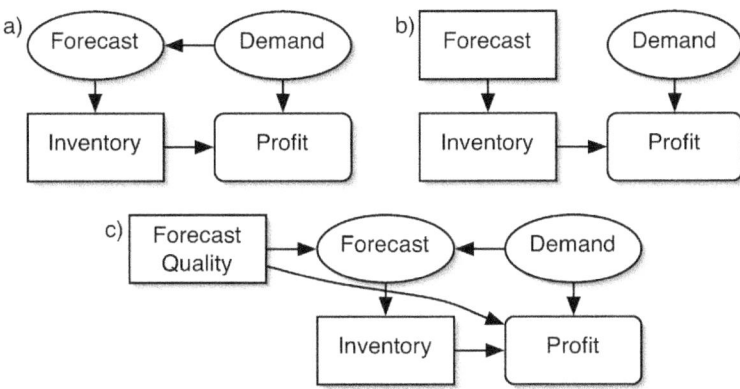

Figure 6. When we intervene on a forecast, we want to improve its quality, rather than to obtain a single most desirable instance.

manipulating our situation, we must beware of the unintended consequences.

Suppose the uncertain variable is being used to provide information, such as a forecast. Consider the causal influence diagram shown in Figure 6. This situation corresponds to one in which inventory decisions must be made before demand is observed, but a forecast relevant to demand will be observed before choosing inventory as shown in (a). Alas, an atomic intervention setting the forecast to our most desirable value ("highest demand") as in (b) does not improve profit since it tells us nothing about the real demand. What we would like to manipulate is the quality of the forecast, having it represent the best possible signal about demand as in (c). In this case, the value of do for *Forecast* is zero, but the value of control for *Forecast* should be positive. In fact, if there are as many instances for *Forecast* as there are for *Demand*, the highest quality forecast possible is clairvoyance on the demand, and the value of control would be equal to the value of clairvoyance. In the diagram *Forecast Quality* might not be an atomic intervention, both because there might only be a choice among imperfect information sources, and because there might be different costs associated with those different information sources.

Consider the causal influence diagrams shown in Figure 7, in which we believe that *Product Quality* is unresponsive to direct interventions (not shown) on *Sales* or *Profit*. We would like to understand how much we would improve our profit by manipulating our product quality. The diagram shown in (a) treats quality and sales as uncertain with its atomic interventions set to **idle**, and its value is the reference for any changes. The diagram shown in (b) has the same value as an atomic intervention on *Product Quality* to its optimal instance, and because that intervention is the cause of *Product Quality* with respect to $D$, the difference in values of this diagram relative to the one in (a) is both the value of do and the value of control for *Product Quality*. Alternatively, in (c) if we observed that *Product*

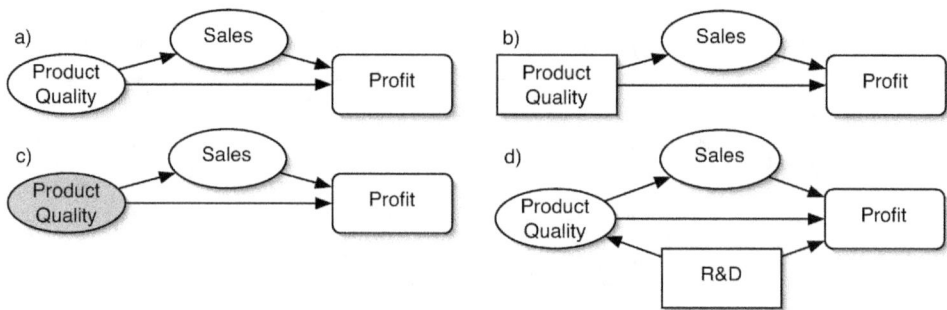

Figure 7. In this causal influence diagram the values of do, control, and revelation for *Product Quality* are equal.

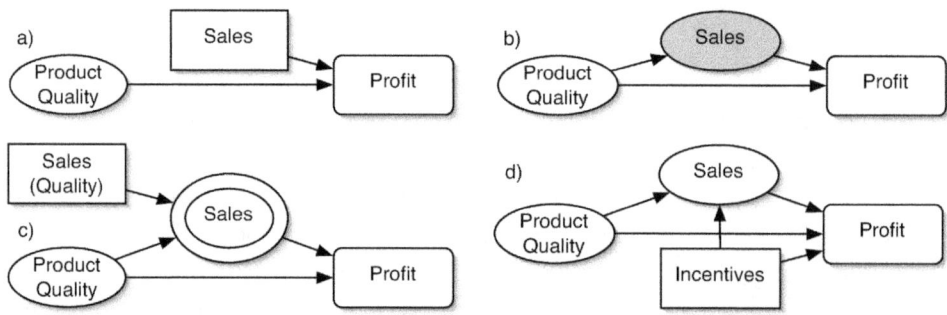

Figure 8. The values of do, control, and revelation for *Sales* might not be equal.

*Quality* takes the best possible value, this diagram has the same value as the one in (b). As a result, the value of revelation is equal to the other two values. Finally, in (d) we could contemplate a research and development effort that might lead to higher product quality. Because the diagram in (d) is causal, {*Product Quality*} is a cause of *Sales* with respect to $D$.

Now consider the causal influence diagrams shown in Figure 8, in which we are manipulating sales rather than product quality to improve our profit. We obtain the diagram shown in (a) by assuming that *Product Quality* is unresponsive to an atomic intervention on *Sales*. In (b) we could observe that *Sales* takes that same value, but this observation updates our belief about the *Product Quality*, and the value of this diagram might not be equal to the value of the diagram in (a). We obtain the diagram shown in (c) by an atomic intervention on the mapping variable for *Sales*, not determining sales but rather how it depends on quality (assuming that there is an atomic intervention on *Product Quality*). In this situation the values of do, control, and revelation could all be different! Finally, in (d) we consider offering incentives to boost sales, recognizing that it might affect our profits both directly

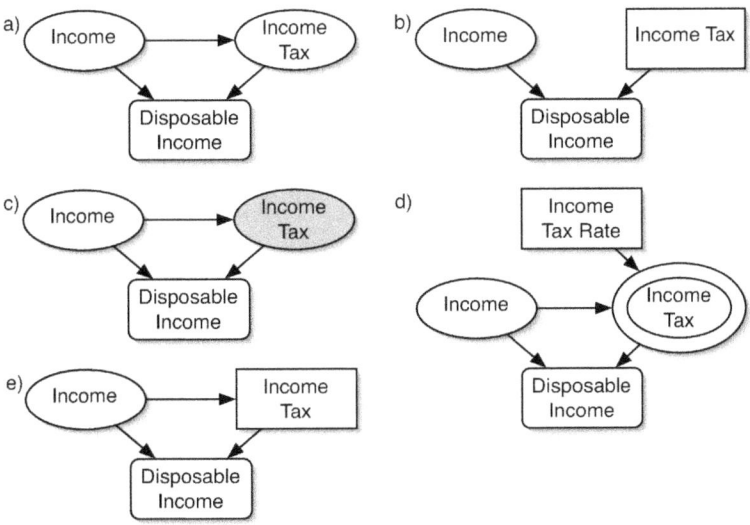

Figure 9. The values of do, control, and revelation are different for *Income Tax*.

and indirectly.

There can be a significant difference between passive observation of uncertain variable $x$ and intervention on $x$. Consider the causal influence diagrams shown in Figure 9 representing disposable income after taxes. We believe that *Income* is unresponsive to a direct intervention on *Income Tax*, but *Income Tax* might be responsive to a direct intervention on *Income*. However, the value of do, the difference between the values of the diagrams in (b) and (a), is quite different from the value of revelation based on (c) and (a). Being able to choose not to pay any tax is quite different from learning that you will pay no tax, since it is more likely in the latter case that you have lost your job. Alternatively, we can consider setting the income tax rate as shown in (d), which would lead to the value of control. In this case, we can simplify the calculation in (d) that searches all possible mapping variable instances, to a simpler decision shown in (e), recognizing that in this case there is no interaction among the components of the mapping variable, and therefore we can independently search for the best possible instance for *Income Tax* for each possible instance of *Income*.

The correspondence between passive observation and intervention has been studied, primarily to identify causal effects from observational data [Robins 1986], [Pearl 1993] and [Spirtes, Glymour, and Scheines 1993]. In our framework, a set of variables $Y$ is said to satisfy the *back door condition* for $x$ if $Y$ is unresponsive to $do(x)$ while $do(x)$ is d-separated from $V$ by $\{x\} \cup Y$. When $Y$ satisfies the back door condition, there is a correspondence among the values of do, control and revelation,

in that
$$P\{V|\mathbf{Y},\mathbf{x}^*\} = P\{V|\mathbf{Y},\mathbf{do}(\mathbf{x}^*)\} = P\{V|\mathbf{Y},\mathbf{do}(\mathbf{x}^*(Y))\}.$$

However, in valuing the decision situation we do not get to observe $Y$ and thus $P\{V|\mathbf{x}^*\}$ might not be equal to $P\{V|\mathbf{do}(\mathbf{x}^*)\}$. Consider the diagrams shown in Figure 9. Because *Income* satisfies the back door criterion relative to *Income Tax*, the values of do, control and revelation on *Income Tax* would all be the same if we observed *Income*. But we do not know what our *Income* will be and the values of do, control, and revelation can all be different.

Nonetheless, if we make a stronger assumption, that $Y$ is d-separated from $V$ by $x$, the three measures will be equal. The atomic intervention on $x$ or its mapping variable only affects the value $V$ through the descendants of $x$ in a causal model, and all other variables are unresponsive to the intervention in worlds limited by $x$. However, the atomic interventions might not be independent of $V$ given $x$ unless $Y$ is d-separated from $V$ by $x$. Otherwise, observing $x$ or an atomic intervention on the mapping variable for $x$ can lead to a different value for the diagram than an atomic intervention on $x$.

We establish this result in two steps for both general situations and for Pearl causal models. By assuming that $do(x)$ is independent of $V$ given $x$, we first show that the values of do and revelation are equal. If we then assume that $Y$ is d-separated from $V$ by $x$, we show that the values of do and control are equal. The conditions under which these two different comparisons can be made are not identical either. To be able to compute the value of revelation for $x$ we must set to **idle** all interventions that $x$ is responsive to, while to compute the value of control for $x$ we need to be ensure that we have an atomic intervention on a mapping variable for $x$.

THEOREM 10 (Equal Values of Do and Revelation). *Given a decision problem including uncertain variable $x \in U$, if there is a set of atomic interventions $A$, including $do(x)$, that is a cause of $x$ with respect to $D$, and $do(x)$ is independent of $V$ given $x$, then the values of do and revelation for $x$ are equal, $VoD(\mathbf{x}^*) = VoR(\mathbf{x}^*)$.*

*If $\{do(x)\}$ is a cause of $x$ with respect to $D$, then they are also equal to the value of control for $x$, $VoC(\mathbf{x}^*()) = VoD(\mathbf{x}^*) = VoR(\mathbf{x}^*)$.*

**Proof.** Consider the probability of $V$ after the intervention $\mathbf{do}(\mathbf{x}^*)$ with all other interventions in $A$ set to **idle**. Because $x$ is determined by $\mathbf{do}(\mathbf{x}^*)$, and $do(x)$ is independent of $V$ given $x$,

$$P\{V|\mathbf{do}(\mathbf{x}^*)\} = P\{V|\mathbf{x}^*,\mathbf{do}(\mathbf{x}^*)\} = P\{V|\mathbf{x}^*\} = P\{V|\mathbf{x}^*,do(x) = \mathbf{idle}\}.$$

If $\{do(x)\}$ is is a cause of $x$ with respect to $D$ then the values of do and control for $x$ are equal. □

COROLLARY 11. *Given a decision problem described by a Pearl causal model including uncertain variable $x \in U$, if $Pa(x)$ is d-separated from $V$ by $x$, then the* values of do and revelation for $x$ are equal, $VoD(\mathbf{x}^*) = VoR(\mathbf{x}^*)$. If $x$ has no parents, then the the values of do, control, and revelation for $x$ are equal,

$$VoD(\mathbf{x}^*) = VoC(\mathbf{x}^*()) = VoR(\mathbf{x}^*).$$

THEOREM 12 (Equal Values of Do and Control). *Given a decision problem described by an influence diagram including uncertain variable $x \in U$, and nonempty set of variables $Y$. If there are atomic interventions $do(x)$ for $x$, $do(y)$ for every $y \in Y \cap U$, and $do(x(Y))$ for the mapping variable $x(Y)$, $Y \cup \{do(x), do(x(Y))\}$ is a cause of $x$ with respect to $D$, and $Y$ is d-separated from $V$ by $x$, then the values of do and control are equal,*

$$VoD(\mathbf{x}^*) = VoC(\mathbf{x}^*(Y)).$$

**Proof.** We know that $Y \cup \{do(x), do(x(Y))\}$ is independent of $V$ given $x$, because otherwise $Y$ would not be d-separated from $V$ by $x$. Because $do(x)$ is an atomic intervention on $x$ and $do(x)$ is independent of $V$ given $x$, as in Theorem 10, $P\{V|\mathbf{do}(\mathbf{x}^*)\} = P\{V|\mathbf{x}^*, \mathbf{do}(\mathbf{x}^*)\} = P\{V|\mathbf{x}^*\}$. Now consider the probability of $V$ after the intervention $\mathbf{do}(\mathbf{x}^*(Y))$. Because $x = \mathbf{x}^*(\mathbf{Y})$ is determined by $\mathbf{do}(\mathbf{x}^*(Y))$ and $\mathbf{Y}$, and $Y \cup \{do(x(Y))\}$ is independent of $V$ given $x$,

$$\begin{aligned}P\{V|\mathbf{do}(\mathbf{x}^*(Y)), \mathbf{Y}\} &= P\{V|x = \mathbf{x}^*(\mathbf{Y}), \mathbf{do}(\mathbf{x}^*(Y)), \mathbf{Y}\} \\ &= P\{V|x = \mathbf{x}^*(\mathbf{Y})\},\end{aligned}$$

The optimal choice of $x(Y)$ does not depend on $Y$, $\mathbf{x}^*(Y) = \mathbf{x}^*$, yielding

$$P\{V|\mathbf{do}(\mathbf{x}^*(Y)), \mathbf{Y}\} = P\{V|\mathbf{x}^*\}.$$

As a result,

$$\begin{aligned}P\{V|\mathbf{do}(\mathbf{x}^*(Y))\} &= \sum_Y P\{V, \mathbf{Y}|\mathbf{do}(\mathbf{x}^*(Y))\} \\ &= \sum_Y P\{V|\mathbf{do}(\mathbf{x}^*(Y)), \mathbf{Y}\} P\{\mathbf{Y}|\mathbf{do}(\mathbf{x}^*(Y))\} \\ &= \sum_Y P\{V|\mathbf{x}^*\} P\{\mathbf{Y}|\mathbf{do}(\mathbf{x}^*(Y))\} \\ &= P\{V|\mathbf{x}^*\} \sum_Y P\{\mathbf{Y}|\mathbf{do}(\mathbf{x}^*(Y))\} \\ &= P\{V|\mathbf{x}^*\}\end{aligned}$$

□

COROLLARY 13. *Given an uncertain variable $x \in U$ with parents in a decision problem described by a Pearl causal model with an atomic intervention for mapping*

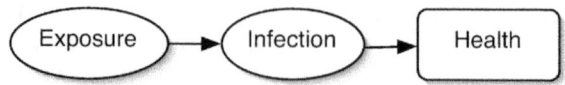

Figure 10. The values of do, control, and revelation are equal for each uncertain variable.

variable $x(Pa(x))$, if $Pa(x)$ is d-separated from $V$ by $x$, then the values of do, control, and revelation for $x$ are equal, $VoD(\mathbf{x}^*) = VoC(\mathbf{x}^*(Pa(x))) = VoR(\mathbf{x}^*)$.

Consider the causal influence diagrams shown in Figure 10, concerning a communicable disease, for which we believe that *Exposure* is unresponsive to any direct intervention on *Infection*, and both of them are unresponsive to any direct intervention on *Health*, but all of the uncertain variables might be responsive to a direction intervention on *Exposure*. Because *Exposure* has no parents, the values of do, control, and revelation for it will be equal. Furthermore, in this case, even though *Infection* has a parent, the values of do, control, and revelation for it will be also equal, because *Exposure* is independent of *Health* given *Infection*. Likewise, there will be equal values of do, control, and revelation for *Health*.

## 4 Conclusions

We have sharpened the distinctions underlying the value of control and related value of revelation and value of do, and shown that they are equivalent when the target variable $x$ in a causal influence diagram either has no parents, or its parents, $Pa(x)$ are d-separated from the value $V$ by $x$.

The general problem, which have only touched upon, permits multiple decisions and information sets at those other decisions. In that case, there is a question of how to recognize when $Pa(x)$ in d-separated from $V$ by $x$. We can address this in general by either constructing the normal form diagram [Bhattacharjya and Shachter 2007] or by building a policy diagram, iteratively substituting deterministic policies for decisions starting with the latest decision [Shachter 1999]. These approaches exploit the causal structure and the separable value function represented in the influence diagram.

## References

Bhattacharjya, D. and R. Shachter (2007). Evaluating influence diagrams with decision circuits. In R. Parr and L. van der Gaag (Eds.), *Proceedings of the Twenty-Third Conference on Uncertainty in Artificial Intelligence*, pp. 9–16. Oregon: AUAI Press.

Heckerman, D. and R. Shachter (1995). Decision-theoretic foundations for causal reasoning. *Journal of Artificial Intelligence Research 3*, 405–430.

Heckerman, D. E. and R. D. Shachter (1994). A decision-based view of causality.

In R. Lopez de Mantaras and D. Poole (Eds.), *Uncertainty in Artificial Intelligence: Proceedings of the Tenth Conference*, pp. 302–310. San Mateo, CA: Morgan Kaufmann.

Howard, R. (1967). Value of information lotteries. *IEEE Transa. Systems Sci. Cybernetics SSC-3*(1), 54–60.

Howard, R. A. (1990). From influence to relevance to knowledge. In R. M. Oliver and J. Q. Smith (Eds.), *Influence Diagrams, Belief Nets, and Decision Analysis*, pp. 3–23. Chichester: Wiley.

Howard, R. A. and J. E. Matheson (1984). Influence diagrams. In R. A. Howard and J. E. Matheson (Eds.), *The Principles and Applications of Decision Analysis*, Volume II. Menlo Park, CA: Strategic Decisions Group.

Jacobs, W. W. (1902, September). The monkey's paw. *Harper's Monthly 105*, 634–639.

Matheson, D. and J. Matheson (2005). Describing and valuing interventions that observe or control decision situations. *Decision Analysis 2*(3), 165–181.

Matheson, J. E. (1990). Using influence diagrams to value information and control. In R. M. Oliver and J. Q. Smith (Eds.), *Influence Diagrams, Belief Nets, and Decision Analysis*, pp. 25–48. Chichester: Wiley.

Pearl, J. (1993). Comment: Graphical models, causality, and intervention. *Statistical Science 8*, 266–269.

Robins, J. (1986). A new approach to causal inference in mortality studies with sustained exposure results. *Mathematical Modeling 7*, 1393–1512.

Savage, L. (1954). *The Foundations o Statistics*. New York: Wiley.

Shachter, R. D. (1986). Evaluating influence diagrams. *Operations Research 34*(November-December), 871–882.

Shachter, R. D. (1999). Efficient value of information computation. In *Uncertainty in Artificial Intelligence: Proceedings of the Fifteenth Conference*, pp. 594–601. San Francisco, CA: Morgan Kaufmann.

Shachter, R. D. and D. E. Heckerman (1986). A backwards view for assessment. In *Workshop on Uncertainty in Artificial Intelligence*, University of Pennsylvania, Philadelphia, pp. 237–242.

Spirtes, P., C. Glymour, and R. Scheines (1993). *Causation, Prediction, and Search*. New York: Springer-Verlag.

# 27
# Cause for Celebration, Cause for Concern

YOAV SHOHAM

It is truly a pleasure to contribute to this collection, celebrating Judea Pearl's scientific contributions. My focus, as well as that of several other contributors, is on his work in the area of causation and causal reasoning. Any student of these topics who ignores Judea's evolving contributions, culminating in the seminal [Pearl 2009], does so at his or her peril. In addition to the objective content of these contributions, Judea's unique energy and personality have led to his having unparalleled impact on the subject, in a diverse set of disciplines far transcending AI, his home turf. This body of work is truly a cause for celebration, and accounts for the first half of the title of this piece.

The second half of the title refers to a concern I have about the literature in AI regarding causation. As an early contributor to this literature I wade back into this area gingerly, aware of many of the complexities involved and difficulties encountered by earlier attempts to capture the notion formally. I am also aware of the fact that many developments have taken place in the past decade, indeed many associated with Judea himself, and only some of which I am familiar with. Still, it seems to me that the concern merits attention. The concern is not specific to Judea's work, and certainly applies to my own work in the area. It has to do with the yardsticks by which we judge this or that theory of causal representation or reasoning.

A number of years ago, the conference on Uncertainty in AI (UAI) held a panel on causation, chaired by Judea, in which I participated. In my remarks I listed a few requirements for a theory of causation in AI. One of the other panelists, whom I greatly respect, responded that he couldn't care less about such requirements; if the theory was useful that was good enough for him. In hindsight that was a discussion worth developing further then, and I believe it still is now.

Let us look at a specific publication, [Halpern and Pearl 2001]. This selection is arbitrary and I might as well have selected any number of other publications to illustrate my point, but it is useful to examine a concrete example. In this paper Halpern and Pearl present an account of "actual cause" (as opposed to "generic cause"; "the lighting last night caused the fire" versus "lightnings cause fire"). This account is also the basis for Chapter 10 of [Pearl 2009]. Without going into their specific (and worthwhile) account, let me focus on how they argue in its favor. In the third paragraph they say

> While it is hard to argue that our definition (or any other definition, for

that matter) is the "right definition", we show that it deals well with the difficulties that have plagued other approaches in the past, especially those exemplified by the rather extensive compendium of [Hall 2004][1].

The reference is to a paper by a philosopher, and indeed of the thirteen references in the paper to work other than by the authors themselves, eight are to work by philosophers.

This orientation towards philosophy is evident throughout the paper, in particular in their relying strongly on particularly instructive examples that serve as test cases. This is an established philosophical tradition. The "morning star – evening star" example [Kripke 1980] catalyzed discussion of cross-world identity in first-order modal logic (you may have different beliefs regarding the star seen in the morning from those regarding the star seen in the evening, even though, unbeknownst to you, they are in fact the same star – Venus). Similarly, the example of believing that you will win the lottery and coincidentally later actually winning it served to disqualify the definition of knowledge as true belief, and a similar example argues against defining knowledge as *justified* true belief [Gettier 1963].

Such "intuition pumps" clearly guide the theory in [Halpern and Pearl 2001], as evidenced by the reference to [Hall 2004] mentioned earlier, and the fact that over four out of the paper's ten pages are devoted to examples. These examples can be highly instructive, but the question is what role they play. In philosophy they tend to serve as necessary but insufficient conditions for a theory. They are necessary in the sense that each of them is considered sufficient grounds for disqualifying a theory (namely, a theory which does not treat the example in an intuitively satisfactory manner). And they are insufficient since new examples can always be conjured up, subjecting the theory to ever-increasing demands.

This is understandable from the standpoint of philosophy, to the extent that it attempts to capture a complex, natural notion (be it knowledge or causation) it its full glory. But is this also the goal for such theories in AI? If not, what is the role of these test cases?

If taken seriously, the necessary-but-insufficient interpretation of the examples presents an impossible challenge to formal theory; a theoretician would never win in this game, in which new requirements may surface at any moment. Indeed, most of the philosophical literature is much less formal than the literature in AI, in particular [Halpern and Pearl 2001]. So where does this leave us?

This is not the first time computer scientists have faced this dilemma. Consider knowledge, for example. The S5 logic of knowledge [Fagin, Halpern, Moses, and Vardi 1994] captures well certain aspects of knowledge in idealized form, but the terms "certain" and "idealized" are important here. The logic has nothing to say about belief (as opposed to knowledge), nor about the dynamic aspects of knowledge (how it changes over time). Furthermore, even with regard to the static aspects of

---

[1]They actually refer to an earlier, unpublished version of Hall's paper from 1998.

knowledge, it is not hard to come up with everyday counterexamples to each of its axioms.

And yet, the logic proves useful to reason about certain aspects of distributed systems, and the mismatch between the properties of the modal operator $K$ and the everyday word "know" does not get in the way, within these confines. All this changes as one switches the context. For example, if one wishes to consider cryptographic protocols, the K axiom ($Kp \wedge K(p \supset q) \supset Kq$, valid in any normal modal logic, and here representing logical omniscience) is blatantly inappropriate. Similarly, when one considers knowledge and belief together, axiom 5 of the logic ($\neg Kp \supset K \neg Kp$, representing negative introspection ability) seems impossible to reconcile with any reasonable notion of belief, and hence one is forced to retreat back from the S5 system to something weaker.

The upshot of all this is the following criterion for a formal theory of natural concepts: One should be explicit about the intended use of the theory, and within the scope of this intended use one should require that everyday intuition about the natural concepts be a useful guide in thinking about their formal counterparts.

A concrete interpretation of the above principle is what in [Shoham 2009] I called the *artifactual* perspective.[2] Artifactual theories attempt to shed light on the operation of a specific artifact, and use the natural notion almost as a mere visual aid. In such theories there is a precise interpretation of the natural notion, which presents a precise requirement for the formal theory. One example is indeed the use of "knowledge" to reason about protocols governing distributed systems. Another, discussed in [Shoham 2009], is the use of "intention" to reason about a database serving an AI planner.

Is there a way to instantiate the general criterion above, or more specifically the artifactual perspective, in the context of causation? I don't know the answer, but it seems to me worthy of investigation. If the answer is "yes" then we will be in a position to devise provably correct theories, and the various illustrative examples will be relegated to the secondary role of showing greater or lesser match with the everyday concept.

**Acknowledgments:** This work was supported by NSF grant IIS-0205633-001.

# References

Fagin, R., J. Y. Halpern, Y. Moses, and M. Y. Vardi (1994). *Reasoning about Knowledge*. MIT Press.

Gettier, E. L. (1963). Is justified true belief knowledge? *Analysis 23*, 121–123.

Hall, N. (2004). Two concepts of causation. In J. Collins, N. Hall, and L. A. Paul (Eds.), *Causation and Counterfactuals*. MIT Press.

---

[2] The discussion there is done in the context of formal models of intention, but the considerations apply here just as well.

Halpern, J. Y. and J. Pearl (2001). Causes and explanations: A structural-model approach. part I: Causes. In *Proceedings of the 17th Annual Conference on Uncertainty in Artificial Intelligence (UAI-01)*, San Francisco, CA, pp. 194–202. Morgan Kaufmann.

Kripke, S. A. (1980). *Naming and necessity* (Revised and enlarged ed.). Blackwell, Oxford.

Pearl, J. (2009). *Causality.* Cambridge University Press. Second edition.

Shoham, Y. (2009). Logics of intention and the database perspective. *Journal of Philosophical Logic 38*(6), 633–648.

# 28

# Automated Search for Causal Relations – Theory and Practice

PETER SPIRTES, CLARK GLYMOUR, RICHARD SCHEINES, AND ROBERT TILLMAN

## 1      Introduction

The rapid spread of interest in the last two decades in principled methods of search or estimation of causal relations has been driven in part by technological developments, especially the changing nature of modern data collection and storage techniques, and the increases in the speed and storage capacities of computers. Statistics books from 30 years ago often presented examples with fewer than 10 variables, in domains where some background knowledge was plausible. In contrast, in new domains, such as climate research where satellite data now provide daily quantities of data unthinkable a few decades ago, fMRI brain imaging, and microarray measurements of gene expression, the number of variables can range into the tens of thousands, and there is often limited background knowledge to reduce the space of alternative causal hypotheses. In such domains, non-automated causal discovery techniques appear to be hopeless, while the availability of faster computers with larger memories and disc space allow for the practical implementation of computationally intensive automated search algorithms over large search spaces. Contemporary science is not your grandfather's science, or Karl Popper's.

Causal inference without experimental controls has long seemed as if it must somehow be capable of being cast as a kind of statistical inference involving estimators with some kind of convergence and accuracy properties under some kind of assumptions. Until recently, the statistical literature said *not*. While parameter estimation and experimental design for the effective use of data developed throughout the $20^{th}$ century, as recently as 20 years ago the methodology of causal inference without experimental controls remained relatively primitive. Besides a cessation of hostilities from the majority of the statistical and philosophical communities (which has still only partially happened), several things were needed for theories of causal estimation to appear and to flower: well defined mathematical objects to represent causal relations; well defined connections between aspects of these objects and sample data; and a way to compute those connections. A sequence of studies beginning with Dempster's work on the factorization of probability distributions [Dempster 1972] and culminating with Kiiveri and Speed's [Kiiveri & Speed 1982] study of linear structural equation models, provided the first, in the form of directed acyclic graphs, and the second, in the form of the "local" Markov condition. Pearl and his students [Pearl 1988], and independently, Stefan

Lauritzen and his collaborators [Lauritzen, Dawid, Larsen, & Leimer 1990], provided the third, in the form of the "global" Markov condition, or d-separation in Pearl's formulation, and the assumption of its converse, which came to be known as "stability" or "faithfulness." Further fundamental conceptual and computational tools were needed, many of them provided by Pearl and his associates; for example, the characterization and representation of Markov equivalence classes and the idea of "inducing paths," essential to understanding the properties of models with unrecorded variables. Initially, most of these authors, including Pearl, did not connect directed graphical models with a causal interpretation (in the sense of representing outcomes of interventions). This connection between graphs and interventions was drawn from an earlier tradition in econometrics [Strotz & Wold 1960], and in our work [Spirtes, Glymour, & Scheines 1993]. With this connection, and the pieces Speed, Lauritzen, Pearl and others had established, a principled theory of causal estimation could, and did, begin around 1990, and Pearl and his students have made important contributions to it. Pearl has become the foremost advocate in the universe for reconceiving the relations between causality and statistics. Once begun for special cases, the understanding of search methods for causal relations has expanded to a variety of scientific and statistical settings, and in many scientific enterprises—neuroimaging for example—causal representations and search are treated as almost routine.

The theory of interventions also provided a coherent normative theory of inference using causal premises. That effort can also be traced back to Strotz and Wold [Strotz & Wold 1960], then to our own work [Spirtes, Glymour, & Scheines 1993] on prediction from classes of causal graphs, and then to the full development of a non-parametric theory of prediction for graphical models by Pearl and his collaborators [Shpitser & Pearl 2008]. Pearl brilliantly turned philosopher and developed the theory of interventions into a general account of counterfactual reasoning. Although we will not discuss it further, we think there remain interesting open problems about prediction algorithms for various parametric classes of graphical causal models.

The following paper surveys a broad range of causal estimation problems and algorithms, concentrating especially on those that can be illustrated with empirical examples that we and our students and collaborators have analyzed. This has naturally led to a concentration on the algorithms and tools that we have developed. The kinds of causal estimation problems and algorithms discussed are broadly representative of the most important developments in methods for estimating causal structure since 1990, but it is not a comprehensive survey. There have been so many improvements to the basic algorithms that we describe here there is not room to discuss them all. A good resource for a description of further research in this area is the Proceedings of the Conferences on Uncertainty in Artificial Intelligence, at http://uai.sis.pitt.edu.

The dimensions of the problems, as we have long understood them, are these:
1. Finding computationally and statistically feasible methods for discovering causal information for large numbers of variables, provably correct under standard sampling assumptions, assuming no confounding by unrecorded variables.

2. The same when the "no confounding" assumption is abandoned.
3. Finding methods for obtaining causal information when there is systematic sample selection bias — when values of some of the variables of interest are associated with sample membership.
4. Finding methods for establishing the existence of unobserved causes and estimating *their* causal relations with one another.
5. Finding methods for discovering causal relations in data produced by feedback systems.
6. Finding methods for discovering causal relations in time series data.
7. Finding methods for discovering causal relations in linear and in non-linear non-Gaussian systems with continuous variables.
8. Finding methods for discovering causal relations using distributed, multiple data sets.
9. Finding methods for merging the above with experimental design.

## 2   Assumptions

We assume the reader's familiarity with the standard notions used in discussions of graphical causal model search: conditional independence, Markov properties, d-separation, Markov equivalence, patterns, distribution equivalence, causal sufficiency, etc. The appendix gives a brief review of the essential definitions, assumptions and theorems required for known proofs of correctness of the algorithms we will discuss.

## 3   Model Search Assuming Causal Sufficiency

The assumption of causal sufficiency (roughly no unrecorded confounders) is often unrealistic, but it is useful in explicating search because the concepts and methods used in search algorithms that make more realistic assumptions are more complex versions of ideas that are used in searches that assume causal sufficiency.

### 3.1   The PC Algorithm

The PC algorithm is a constraint-based search that attempts to find the pattern that most closely entails all and only the conditional independence constraints judged to hold in the population. The SGS algorithm [Spirtes & Glymour 1991] and the IC algorithm [Verma & Pearl 1990] were early versions of this algorithm that were statistically and computationally feasible only on data sets with few variables because they required conditioning on all possible subsets of variables.) The PC algorithm solved both difficulties in typical cases.

The PC algorithm has an adjacency phase in which the adjacencies are determined, and an orientation phase in which as many edges as possible are oriented. The adjacency phase is stated below, and illustrated in Figure 1. Let **Adjacencies**$(G,A)$ be the set of vertices adjacent to $A$ in undirected graph $G$. (In the algorithm, the graph $G$ is continually updated, so **Adjacencies**$(G,A)$ may change as the algorithm progresses.)

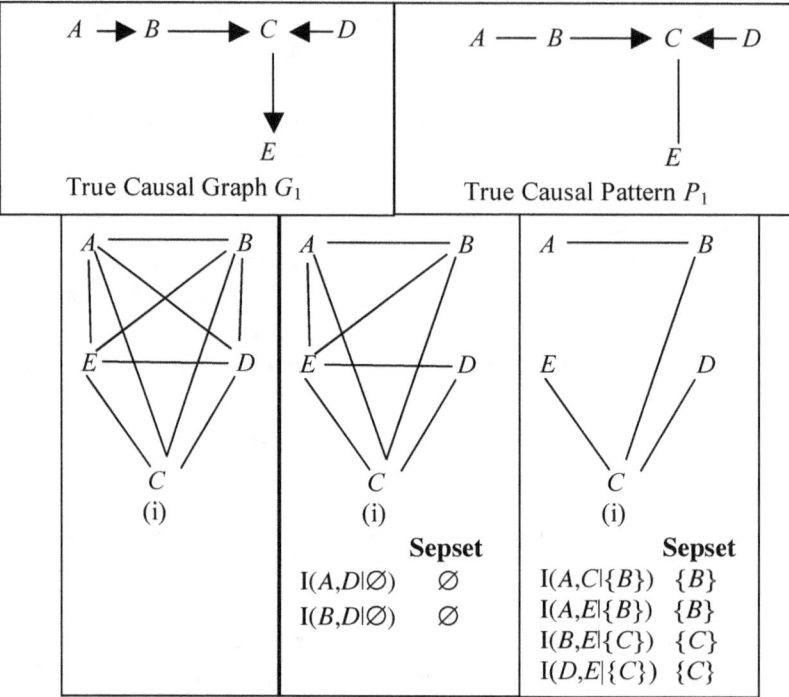

Figure 1: Constraint based search, where correct pattern is $P_1$

**Adjacency Phase of PC Algorithm:**
Form an undirected graph $G$ in which every pair of vertices in **V** is adjacent.
$n := 0$.
repeat
    repeat
        Select an ordered pair of variables $X$ and $Y$ that are adjacent in $G$ such that **Adjacencies**$(G,X)\backslash\{Y\}$ has cardinality greater than or equal to $n$, and a subset **S** of **Adjacencies**$(G,X)\backslash\{Y\}$ of cardinality $n$, and if $X$ and $Y$ are independent conditional on **S** delete edge $X \text{---} Y$ from $C$ and record **S** in **Sepset**$(X,Y)$ and **Sepset**$(Y,X)$;
    until all ordered pairs of adjacent variables $X$ and $Y$ such that **Adjacencies**$(G,X)\backslash\{Y\}$ has cardinality greater than or equal to $n$ and all subsets **S** of **Adjacencies**$(G,X)\backslash\{Y\}$ of cardinality $n$ have been tested for conditional independence;
    $n := n + 1$;
until for each ordered pair of adjacent vertices $X$, $Y$, **Adjacencies**$(G,X)\backslash\{Y\}$ is of cardinality less than $n$.

After the adjacency phase of the algorithm, the orientation phase of the algorithm is performed. The orientation phase of the algorithm is illustrated in Figure 2.

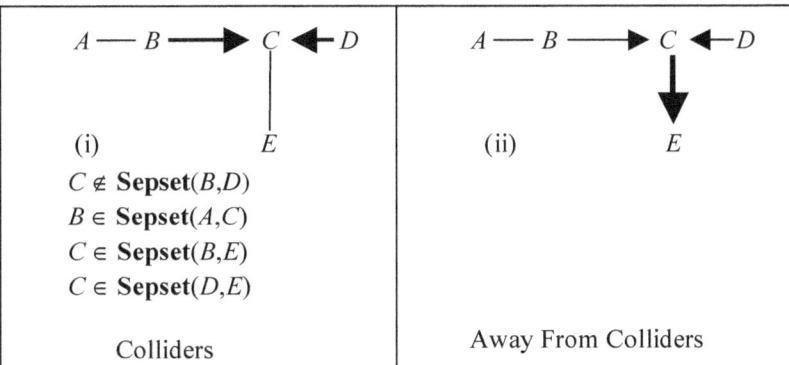

**Figure 2: Orientation phase of PC algorithm, assuming true pattern is $P_1$**

The orientation phase of the PC algorithm is stated more formally below. The last two orientation rules (Away from Cycles, and Double Triangle) are not used in the example, but are sound because if the edges were oriented in ways that violated the rules, there would be a directed cycle in the pattern, which would imply a directed cycle in the graph (which in this section is assumed to be impossible). The orientation rules are complete [Meek 1995], i.e. every edge that has the same orientation in every member of a DAG conditional independence equivalence class is oriented by these rules.

**Orientation Phase of PC Algorithm**
For each triple of vertices $X$, $Y$, $Z$ such that the pair $X$, $Y$ and the pair $Y$, $Z$ are each adjacent in graph $G$ but the pair $X$, $Z$ are not adjacent in $G$, orient $X-Y-Z$ as $X \to Y \leftarrow Z$ if and only if $Y$ is not in **Sepset**$(X,Z)$.
repeat
    Away from colliders: If $A \to B - C$, and $A$ and $C$ are not adjacent, then orient as $B \to C$.
    Away from cycles: If $A \to B \to C$ and $A - C$, then orient as $A \to C$.
    Double Triangle: If $A \to B \leftarrow C$, $A$ and $C$ are not adjacent, $A - D - C$, and there is an edge $B - D$, orient $B - D$ as $D \to B$.
until no more edges can be oriented.

The tests of conditional independence can be performed in the usual way. Conditional independence among discrete variables can be tested using the $G^2$ statistic; conditional independence among multivariate Gaussian variables can be tested using Fisher's Z-transformation of the partial correlations [Spirtes, Glymour, & Scheines 2001]. Section 3.4 describes more general tests of conditional independence. Such tests require specifying a significance level for the test, which is a user-specified parameter of the algorithm. Because the PC algorithm performs a sequence of tests without adjustment, the significance level does not represent any (easily calculable) statistical feature of the output, but should only be understood as a parameter used to guide the search.

Assuming that the causal relations can be represented by a directed acyclic graph, the Causal Markov Assumption, the Causal Faithfulness Assumption, and consistent tests of conditional independence, in the large sample (i.i.d.) limit for a causally sufficient set of variables, the PC algorithm outputs a pattern that represents the true causal graph.

The PC algorithm has been shown to apply to very high dimensional data sets (under a stronger version of the Causal Faithfulness Assumption), both for finding causal structure [Kalisch & Buhlmann 2007] and for classification [Aliferis, Tsamardinos, & Statnikov 2003]. A version of the algorithm controlling the false discovery rate is available [Junning & Wang 2009].

### 3.1.1 Example - Foreign Investment

This example illustrates how the PC algorithm can find plausible alternatives to a model built from domain knowledge. Timberlake and Williams used regression to claim foreign investment in third-world countries promotes dictatorship [Timberlake & Williams 1984]. They measured political exclusion (*PO*) (i.e., dictatorship), foreign investment penetration in 1973 (*FI*), energy development in 1975 (*EN*), and civil liberties (*CV*) for 72 countries. *CV* was measured on an ordered scale from 1 to 7, with lower values indicating greater civil liberties.

Their inference is unwarranted. Their model (with the relations between the regressors omitted) and the pattern obtained from the PC algorithm using a 0.12 significance level to test for vanishing partial correlations) are shown in Figure 3.[1] We typically run the algorithms at a variety of different significance levels, and compare the results to see if any of the features of the output are constant.

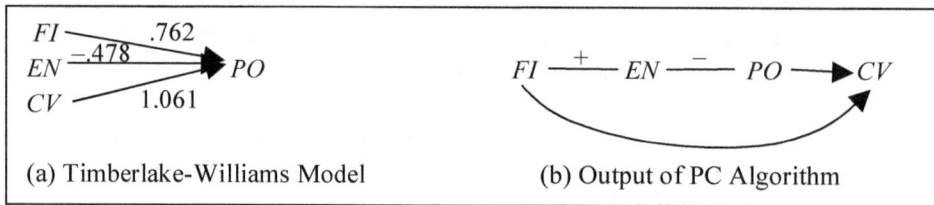

(a) Timberlake-Williams Model        (b) Output of PC Algorithm

**Figure 3: Two Models of Foreign Investment**

The PC Algorithm will not orient the *FI – EN* and *EN – PO* edges, and assumes that the edges are not due to an unmeasured common cause. Maximum likelihood estimates of any linear, Gaussian parameterization of any DAG represented by the pattern output by the PC algorithm requires that the influence of *FI* on *PO* (if any) be negative, and the models easily pass a likelihood ratio test. If any of these SEMs is correct, Timberlake and William's regression model appears to be a case in which an effect of the outcome variable is taken as a regressor.

Given the small sample size, and the uncertainty about the distributional assumptions, we do not present the alternative models suggested by the PC algorithm as particularly well supported by the evidence. However, we do think that they are at least

---

[1] Searches at lower significance levels remove the adjacency between *FI* and *EN*.

as well supported as the regression model, and hence serve to cast doubt upon conclusions drawn from that model.

### 3.1.2  Example - Spartina Biomass

This example illustrates a case where the PC algorithm output received some experimental confirmation. A textbook on regression [Rawlings 1988] skillfully illustrates regression principles and techniques for a biological study from a dissertation [Linthurst 1979] in which it is reasonable to think there is a causal process at work relating the variables. The question at issue is plainly causal: among a set of 14 variables, which have the most influence on an outcome variable, the biomass of Spartina grass? Since the example is the principle application given for an entire textbook on regression, the reader who reaches the $13^{th}$ chapter may be surprised to find that the methods yield almost no useful information about that question.

According to Rawlings, Linthurst obtained five samples of Spartina grass and soil from each of nine sites on the Cape Fear Estuary of North Carolina. Besides the mass of Spartina (*BIO*), fourteen variables were measured for each sample:

- Free Sulfide ($H_2S$)
- Salinity (*SAL*)
- Redox potentials at pH 7 ($EH_7$)
- Soil pH in water (*PH*)
- Buffer acidity at pH 6.6 (*BUF*)
- Phosphorus concentration (*P*)
- Potassium concentration (*K*)
- Calcium concentration (*CA*)
- Magnesium concentration (*MG*)
- Sodium concentration (*NA*)
- Manganese concentration (*MN*)
- Zinc concentration (*ZN*)
- Copper concentration (*CU*)
- Ammonium concentration ($NH_4$)

The aim of the data analysis was to determine for a later experimental study which of these variables most influenced the biomass of Spartina in the wild. Greenhouse experiments would then try to estimate causal dependencies out in the wild. In the best case one might hope that the statistical analyses of the observational study would correctly select variables that influence the growth of Spartina in the greenhouse. In the worst case, one supposes, the observational study would find the wrong causal structure, or would find variables that influence growth in the wild (e.g., by inhibiting or promoting growth of a competing species) but have no influence in the greenhouse.

Using the SAS statistical package, Rawlings analyzed the variable set with a multiple regression and then with two stepwise regression procedures from the SAS package. A search through all possible subsets of regressors was not carried out, presumably because the candidate set of regressors is too large. The results were as follows:

(i) a multiple regression of *BIO* on all other variables gives only *K* and *CU* significant regression coefficients;

(ii) two stepwise regression procedures[2] both yield a model with *PH*, *MG*, *CA* and *CU* as the only regressors, and multiple regression on these variables alone gives them all significant coefficients;

(iii) simple regressions one variable at a time give significant coefficients to *PH*, *BUF*, *CA*, *ZN* and $NH_4$.

What is one to think? Rawling's reports that "None of the results was satisfying to the biologist; the inconsistencies of the results were confusing and variables expected to be biologically important were not showing significant effects." (p. 361).

This analysis is supplemented by a ridge regression, which increases the stability of the estimates of coefficients, but the results for the point at issue--identifying the important variables--are much the same as with least squares. Rawlings also provides a principal components factor analysis and various geometrical plots of the components. These calculations provide no information about which of the measured variables influence Spartina growth.

Noting that *PH*, for example, is highly correlated with *BUF*, and using *BUF* instead of *PH* along with *MG*, *CA* and *CU* would also result in significant coefficients, Rawlings effectively gives up on this use of the procedures his book is about:

> Ordinary least squares regression tends either to indicate that none of the variables in a correlated complex is important when all variables are in the model, or to arbitrarily choose one of the variables to represent the complex when an automated variable selection technique is used. A truly important variable may appear unimportant because its contribution is being usurped by variables with which it is correlated. Conversely, unimportant variables may appear important because of their associations with the real causal factors. It is particularly dangerous in the presence of collinearity to use the regression results to impart a "relative importance," whether in a causal sense or not, to the independent variables. (p. 362)

Rawling's conclusion is correct in spirit, but misleading and even wrong in detail. If we apply the PC algorithm to the Linthurst data then there is one robust conclusion: the only variable that may *directly* influence biomass in this population[3] is *PH*; *PH* is distinguished from all other variables by the fact that the correlation of every other variable (except *MG*) with *BIO* vanishes or vanishes when *PH* is conditioned on.[4] The relation is not symmetric; the correlation of *PH* and *BIO*, for example, does not vanish when *BUF* is controlled. The algorithm finds *PH* to be the only variable adjacent to *BIO*

---

[2] The "maximum R-square" and "stepwise" options in PROC REG in the SAS program.

[3] Although the definition of the population in this case is unclear, and must in any case be drawn quite narrowly.

[4] More exactly, at .05, with the exception of *MG* the partial correlation of every regressor with *BIO* vanishes when some set containing *PH* is controlled for; the correlation of *MG* with *BIO* vanishes when *CA* is controlled for.

no matter whether we use a significance level of .05 to test for vanishing partial correlations, or a level of 0.1, or a level of 0.2. In all of these cases, the PC algorithm (and the FCI algorithm, which allows for the possibility of latent variables in section 4.2 ) yields the result that *PH* and only *PH* can be directly connected with *BIO*. If the system is linear normal and the Causal Markov Assumption obtains, then in this population any influence of the other regressors on *BIO* would be blocked if *PH* were held constant. Of course, over a larger range of values of the variables there is little reason to think that *BIO* depends linearly on the regressors, or that factors that have no influence in producing variation within this sample would continue to have no influence.

Although the analysis cannot conclusively rule out possibility that *PH* and *BIO* are confounded by one or more unmeasured common causes, in this case the principles of the theory and the data argue against it. If *PH* and *BIO* have a common unmeasured cause $T$, say, and any other variable, $Z_i$, among the 13 others either causes *PH* or has a common unmeasured cause with *PH* (Figure 4, in which we do not show connections among the **Z** variables), then $Z_i$ and *BIO* should be correlated conditional on *PH*, which is statistically not the case.

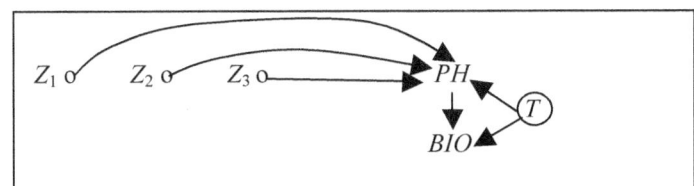

**Figure 4 : *PH* and *BIO* Confounding?**

The program and theory lead us to expect that if *PH* is forced to have values like those in the sample--which are almost all either below *PH* 5 or above *PH* 7-- then manipulations of other variables within the ranges evidenced in the sample will have no effect on the growth of Spartina. The inference is a little risky, since growing plants in a greenhouse under controlled conditions may not be a direct manipulation of the variables relevant to growth in the wild. If, for example, in the wild variations in *PH* affect Spartina growth chiefly through their influence on the growth of competing species not present in the greenhouse, a greenhouse experiment will not be a direct manipulation of *PH* for the system.

The fourth chapter of Linthurst's thesis partly confirms the PC algorithm's analysis. In the experiment Linthurst describes, samples of Spartina were collected from a salt marsh creek bank (presumably at a different site than those used in the observational study). Using a 3 x 4 x 2 (*PH* x *SAL* x *AERATION*) randomized complete block design with four blocks, after transplantation to a greenhouse the plants were given a common nutrient solution with varying values *PH* and *SAL* and *AERATION*. The *AERATION* variable turned out not to matter in this experiment. Acidity values were *PH* 4, 6 and 8. *SAL* for the nutrient solutions was adjusted to 15, 25, 35 and 45 ‰.

Linthurst found that growth varied with *SAL* at *PH* 6 but not at the other *PH* values, 4 and 8, while growth varied with *PH* at all values of *SAL* (p. 104). Each variable was correlated with plant mineral levels. Linthurst considered a variety of mechanisms by which extreme *PH* values might control plant growth:

> At pH 4 and 8, salinity had little effect on the performance of the species. The pH appeared to be more dominant in determining the growth response. However, there appears to be no evidence for any causal effects of high or low tissue concentrations on plant performance unless the effects of pH and salinity are also accounted for. (p.108)
> The overall effect of pH at the two extremes is suggestive of damage to the root, thereby modifying its membrane permeability and subsequently its capacity for selective uptake. (p. 109).

A comparison of the observational and experimental data suggests that the PC Algorithm result was essentially correct and can be extrapolated through the variation in the populations sampled in the two procedures, but cannot be extrapolated through *PH* values that approach neutrality. The result of the PC search was that in the non-experimental sample, observed variations in aerial biomass were perhaps caused by variations in *PH*, but were not caused (at least not directly, relative to *PH*) by variations in other variables. In the observational data Rawlings reports (p. 358) almost all *SAL* measurements are around 30--the extremes are 24 and 38. Compared to the experimental study rather restricted variation was observed in the wild sample. The observed values of *PH* in the wild, however, are clustered at the two extremes; only four observations are within half a *PH* unit of 6, and no observations at all occurred at *PH* values between 5.6 and 7.1. For the observed values of *PH* and *SAL*, the experimental results appear to be in very good agreement with our results from the observational study: small variations in *SAL* have no effect on Spartina growth if the *PH* value is extreme.

### 3.1.3 College Plans

Sewell and Shah [Sewell & Shah 1968] studied five variables from a sample of 10,318 Wisconsin high school seniors.[5] The variables and their values are:

- *SEX* — male = 0, female = 1
- *IQ* = Intelligence Quotient, — lowest = 0, highest = 3
- *CP* = college plans — yes = 0, no = 1
- *PE* = parental encouragement — low = 0, high = 1
- *SES* = socioeconomic status — lowest = 0, highest = 3

The question of interest is what the causes of college plans are. This data set is of interest because it has been used by a variety of different search algorithms that make different assumption. The different results illustrate the role that the different assumptions make in the output and are discussed in subsequent sections.

---

[5] Examples of the analysis of the Sewell and Shah data using Bayesian networks are given in Spirtes et al. (2001), and Heckerman (1998).

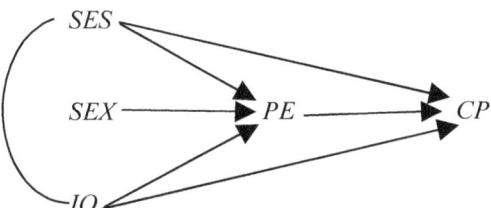

**Figure 5: Model of Causes of College Plans**

The pattern produced as the output of the PC algorithm is shown in Figure 5. The model predicts that *SEX* affects *CP* only indirectly via *PE*.

It is possible to predict the effects of some manipulations from the pattern, but not others. For example, because the pattern is compatible both with *SES* → *IQ* and with *SES* ← *IQ*, it is not possible to determine if *SES* is a cause or an effect of *IQ*, and hence it is not possible to predict the effect of manipulating *SES* on *IQ* from the pattern. On the other hand, it can be shown that all of the models in the conditional independence equivalence class represented by the pattern entail the same predictions about the quantitative effects of manipulating *PE* on *CP*. When *PE* is manipulated, in the manipulated distribution: $P(CP=0|PE=0) = .095$; $P(CP=1|PE=0) = .905$; $P(CP=0|PE=1) = .484$; $P(CP=1|PE=1) = .516$ [Spirtes, Scheines, Glymour, & Meek 2004].

## 3.2  Greedy Equivalence Search Algorithm

Algorithms that maximize a score have certain advantages over constraint-based algorithms such as PC. When the data are not Gaussian, but the system is linear, extensive unpublished simulations find that at least one such algorithm, the Greedy Equivalence Search (GES) algorithm [Meek 1997] outperforms PC. GES can be used with a number of different scores for patterns, including posterior probabilities (for some parametric families and under some priors), and the Bayesian Information Criterion (BIC), which is an approximation of a class of posterior distributions in the large sample limit. The BIC score [Schwarz 1978] is: $-2 \ln(ML) + k \ln(n)$, where ML is the likelihood of the data at the maximum likelihood estimate of the parameters, $k$ is the dimension of the model and $n$ is the sample size. For uniform priors on models and smooth priors on the parameters, the posterior probability conditional on the data is a monotonic function of BIC in the large sample limit. In the forward stage of the search, starting with an initial (possibly empty) pattern, at each stage GES selects the pattern that is the one-edge addition compatible with the current pattern and has the highest score. The forward stage continues until no further additions improve the score. Then a reverse procedure is followed that removes edges according to the same criterion, until no improvement is found. The computational and convergence advantages of the algorithm depend on the fact that it searches over Markov equivalence classes of DAGs rather than individual DAGs, and that only one forward stage and one backward stage are required for an asymptotically correct search. In the large sample limit, GES identifies the Markov equivalence class of the true graph if the assumptions above are met [Chickering 2002].

GES has proved especially valuable in searches for latent structure (GESMIMBuild) and in searches with multiple data sets (IMaGES). Examples are discussed in sections 4.4 and 5.3 .

### 3.3 LiNGAM

Standard implementations of the constraint-based and score-based algorithms above usually assume that continuous variables have multivariate Gaussian distributions. This assumption is inappropriate in many contexts such as EEG analysis where variables are known to deviate from Gaussianity.

The LiNGAM (Linear Non-Gaussian Acyclic Model) algorithm [Shimizu, Hoyer, Hyvärinen, & Kerminen 2006] is appropriate specifically for cases where each variable in a set of measured variables can be written as a linear function of other measured variables plus an independent noise component, where at most one of the measured variables' noise components may be Gaussian. For example, consider the system with the causal graph shown in Figure 6 and assume $X$, $Y$, and $Z$ are determined as follows, where $a$, $b$, and $c$ are real-valued coefficients and $\varepsilon_x$, $\varepsilon_y$, and $\varepsilon_z$ are independent noise components of which at least two are non-Gaussian.

(1) $X = \varepsilon_x$
(2) $Y = aX + \varepsilon_y$
(3) $Z = bX + cY + \varepsilon_z$

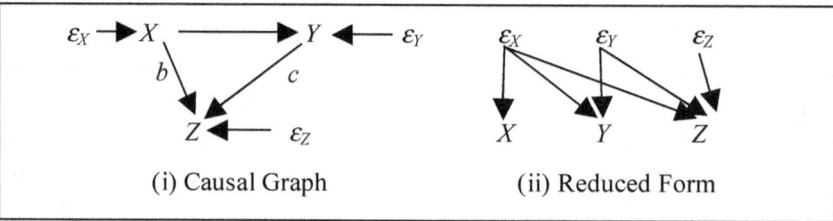

(i) Causal Graph    (ii) Reduced Form

**Figure 6: Causal Graph and Reduced Form**

The equations can be rewritten in what economists called reduced form, also shown in Figure 6:

(4)     $X = \varepsilon_X$
(5)     $Y = a\varepsilon_X + \varepsilon_Y$
(6)     $Z = b\varepsilon_X + ac\varepsilon_X + c\varepsilon_Y + \varepsilon_Z$

The standard Independent Components Analysis (ICA) procedure [Hyvärinen & Oja, 2000] can be used to recover a matrix containing the real-valued coefficients $a$, $b$, and $c$ from an i.i.d. sample of data generated from the above system of equations. The LiNGAM algorithm finds the correct matching of coefficients in this ICA matrix to variables and prunes away any insignificant coefficients using statistical criteria.

The procedure yields correct values even if the coefficients were to perfectly cancel, and hence the variables such as $X$, $Z$ above were to be uncorrelated. Since coefficients are determined for each variable, we can always reconstruct the true unique DAG, instead of its Markov equivalence class. The procedure converges (at least) pointwise to the true DAG and coefficients assuming: (1) there are no unmeasured common causes; (2) the dependencies among measured variables are linear; (3) none of the relations among measured variables are deterministic; (4) i.i.d. sampling; (5) the Markov Condition; (6) at most one error or disturbance term is Gaussian. We do not know its complexity properties.

The LiNGAM procedure can be generalized to estimate causal relations among observables when there are latent common causes [Hoyer, Shimizu, & Kerminen 2006], although the result is not in general a unique DAG, and LiNGAM has been combined [Shimizu, Hoyer, & Hyvarinen 2009] with Silva's clustering procedure (section 4.4 ) for locating latent variables to estimate a unique DAG among latent variables, and also with search for cyclic graphs [Lacerda, Spirtes, Ramsey, & Hoyer 2008], and combined with the PC and GES algorithms when more than one disturbance term is Gaussian [Hoyer et al. 2008].

### 3.4 The kPC Algorithm

The kPC algorithm [Tillman, Gretton, & Spirtes, 2009] relaxes distributional assumptions further, allowing not only non-Gaussian noise with continuous variables, but also nonlinear dependencies. In many cases, kPC will return a unique DAG (even when there is more than one DAG in the Markov equivalence class. However, unlike LiNGAM there is no requirement that a certain number of variables be non-Gaussian.

kPC consists of two stages. In the first stage of kPC, the standard PC algorithm is applied to the data using efficient implementations of the Hilbert-Schmidt Independence Criteria [Gretton, Fukumizu, Teo, Song, Scholkopf, & Smola, 2008], a nonparametric independence test and an extension of this test to the conditional cases based on the dependence measure given in [Fukumizu, Gretton, Sun, & Scholkopf, 2008]. This produces a pattern. Additional orientations are then possible if the true causal model, or a submodel (after removing some variables) of the true causal model is an *additive noise model* [Hoyer, Janzing, Mooij, Peters, & Scholkopf, 2009] that is *noninvertible*.

A set of variables is an additive noise model if (i) the function form of each variable can be expressed as a (possible nonlinear) smooth function of its parents in the true causal model plus an additive (Gaussian or non-Gaussian) noise component and (ii) the additive noise components are mutually independent. An additive noise model is noninvertible if we cannot reverse any edges in the model and still obtain smooth functional forms for each variable and mutually independent additive noise components that fit the data.

For example, consider the two variable case where $X \rightarrow Y$ is the true DAG and we have the following function forms and additive noise components for $X$ and $Y$:

$X = \varepsilon_x$, $Y = \sin(\pi X) + \varepsilon_x$, $\varepsilon_x \sim Uniform(-1,1)$, $\varepsilon_y \sim Uniform(-1,1)$

If we fit a nonparametric regression model for Y regressed on X, the forward model, Figure 7a, and for X regressed on Y, the backward model, Figure 7b, we observe $I(\hat{\varepsilon}_y, X)$ and $\neg I(\hat{\varepsilon}_x, Y)$ since this additive noise model is noninvertible.

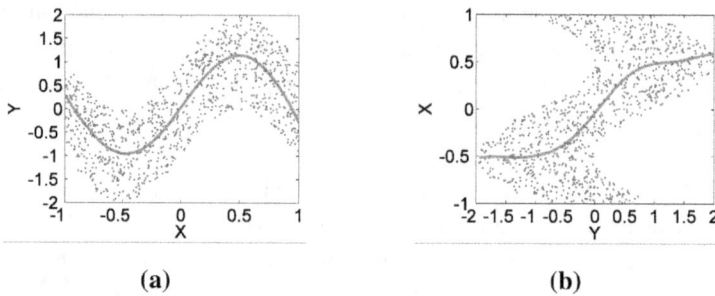

**Figure 7:** Nonparametric regressions of (a) Y on X, and (b) X on Y with the data overlayed for nonlinear non-Gaussian case

Thus in this case, we can conclude that $X \rightarrow Y$ is the true DAG from the data since the additive noise model fits in only one direction, i.e. it is noninvertible. However, consider the following linear Gaussian case:

$$X = \varepsilon_x, \ Y = 2.4 \cdot X + \varepsilon_y, \ \varepsilon_x \sim N(0,1), \ \varepsilon_y \sim N(0,1)$$

After fitting nonparametric regression models for both directions, Figure 8, we find $I(\hat{\varepsilon}_y, X)$ and $I(\hat{\varepsilon}_x, Y)$ so we cannot determine whether $X \rightarrow Y$ or $Y \rightarrow X$ is the correct DAG.

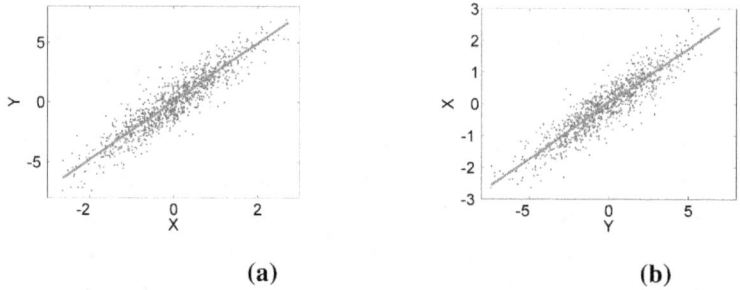

**Figure 8:** Nonparametric regressions of (a) Y on X, and (b) X on Y with the data overlayed for linear Gaussian case

[Zhang and Hyvarinen, 2009] show that only a few special cases, other than the linear Gaussian case, exist where the additive noise model is invertible.

The second stage of kPC consists of searches for submodels that are consistent with the pattern learned in the first stage of kPC that may be noninvertible additive noise models. If such models are discovered, then further orientations of edges can be made resulting in an equivalence class of possible DAGs that is a proper subset of the Markov equivalence class. In many cases, only a few variables need be nonlinear or non-Gaussian to obtain a unique DAG using kPC.

kPC requires the following additional assumption:

**Weak Additivity Assumption:** If the relationship between $X$ and **Parents**$(G,X)$ in the true DAG $G$ cannot be expressed as a noninvertible additive noise model, there does not exist a $Y$ in **Parents**$(G,X)$ and alternative DAG $G'$ such that $Y$ and **Parents**$(G',Y)$ can be expressed as a noninvertible additive noise model where $X$ is included in **Parents**$(G',Y)$.

This assumption does rule out invertible additive noise models or many cases where noise may not be additive, only the hypothetical case where we can fit an additive noise model to the data, but only in the incorrect direction. Weak additivity can be considered an extension of the simplicity intuitions underlying the causal faithfulness assumption, i.e. a complicated true model will not generate data resembling a different simpler model. Faithfulness can fail, but under a broad range of distributions, violations are Lebesgue measure zero [Spirtes, Glymour, & Scheines 2000]. Whether a similar justification can be given for the weak additivity assumption is an open question.

kPC is both correct and complete, i.e. it converges to the correct DAG or smallest possible equivalence class of DAGs in the limit under weak additivity and the assumptions of the PC algorithm.

### 3.4.1  Example - Auto MPG

Figure 9 shows the structures learned for the Auto MPG dataset, which records *MPG* fuel consumption of 398 automobiles in 1983 with 8 characteristics from the UCI database (Asuncion & Newman, 2007). The nominal variables *Year* and *Origin* were excluded.

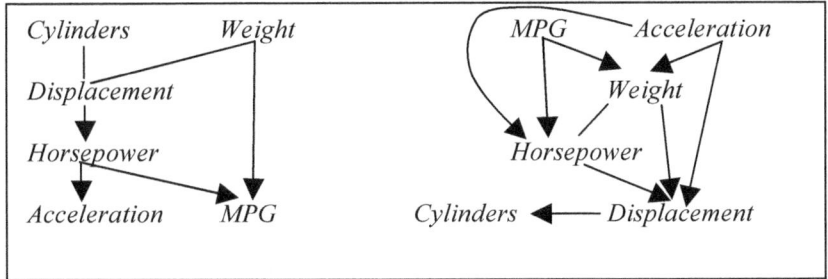

**Figure 9: Automobile Models**

The PC result indicates *MPG* causes *Weight* and *Horsepower*, and *Acceleration* causes *Weight*, *Horsepower*, and *Displacement*, which are clearly false. kPC finds the more plausible chain *Displacement* → *Horsepower* → *Acceleration* and finds *Horsepower* and *Weight* cause *MPG*.

### 3.4.2  Example - Forest Fires

The Forest Fires dataset contains 517 recordings of meteorological for forest fires observed in northeast Portugal and the total area burned (*Area*) [Asuncion & Newman 2007]. We again exclude nominal variables *Month* and *Year*. Figure 10 shows the structures learned by PC and kPC for this dataset. kPC finds every variable other than *Area* is a cause of *Area*, which is sensible since each of these variables were included in the dataset by domain experts as predictors which influence the total area burned by forest fires.

The PC structure, however, indicates that Area is not associated with any of the variables, which are all assumed to be predictors by experts.

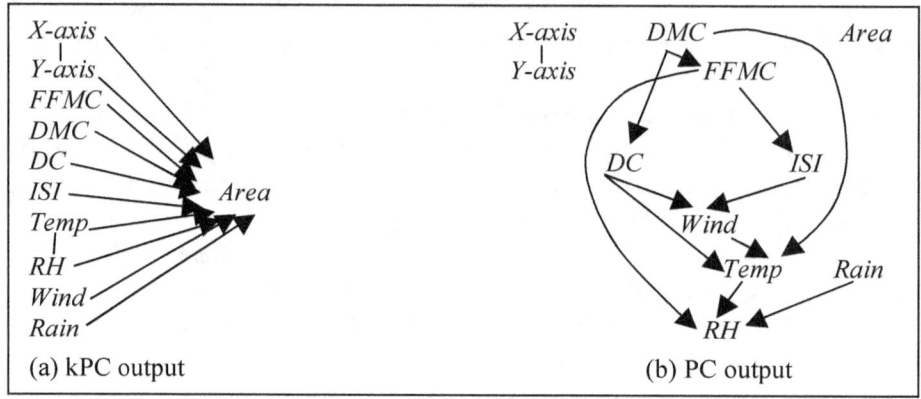

Figure 10: kPC and PC Forest Fires

## 4 Search For Latent Variable Models

The assumption that the observed variables are causally sufficient is usually unwarranted. In this section, we describe searches that do not make this assumption.

### 4.1 Distribution and Conditional Independence Equivalence

Let **O** be the set of observed variables, which may not be causally sufficient. If $G_1$ is a DAG over $\mathbf{V}_1$, $G_2$ is a DAG over $\mathbf{V}_2$, $\mathbf{O} \subseteq \mathbf{V}_1$, and $\mathbf{O} \subseteq \mathbf{V}_2$, $G_1$ and $G_2$ are **O**-conditional independence equivalent, if they both entail the same set of conditional independence relations among the variables in **O** (i.e. they have the same d-separation relations among the variables in **O**). $<G_1,\Theta_1>$ and $<G_2,\Theta_2>$ are **O-distribution equivalent** with respect to the parametric families $\Theta_1$ and $\Theta_2$ if and only if they represent the same set of marginal distributions over **O**.

It is possible that two directed graphs are conditional independence equivalent, or even distributionally equivalent (relative to given parametric families) but are not **O**-distributionally equivalent (relative to the same parametric families), as long as at least one of them contains a latent variable. Although there are algebraic techniques that determine when two Bayesian networks with latent variables are **O**-distributionally equivalent for some parametric families, or find features common to an **O**-distributional equivalence class, known algorithms to do so are not computationally feasible [Geiger & Meek 1999] for models with more than a few variables. In addition, if an unlimited number of latent variables are allowed, the number of DAGs that are **O**-distributionally equivalent may be infinite. Hence, instead of searching for **O**-distribution equivalence classes of models, we will describe how to search for **O**-conditional independence classes of models. This is not as informative as the computationally infeasible strategy of searching for **O**-distribution equivalence classes, but is nevertheless correct.

It is often far from intuitive what constitutes a complete set of graphs **O**-conditional independence equivalent to a given graph although algorithms for deciding this now exist [Ali, Richardson, & Spirtes 2009].

## 4.2 The Fast Causal Inference Algorithm

The PC algorithm gives an asymptotically correct representation of the conditional independence equivalence class of a DAG without latent variables by outputting a pattern that represents all of the features that the DAGs in the equivalence class have in common. The same basic strategy can be used without assuming causal sufficiency, but the rules for detecting adjacencies and orientations are much more complicated, so we will not describe them in detail. The FCI algorithm[6] outputs an asymptotically correct representation of the **O**-conditional independence equivalence class of the true causal DAG (assuming the Causal Markov and Causal Faithfulness Principles), in the form of a graphical structure called a partial ancestral graph (PAG) that represents some of the features that the DAGs in the equivalence class have in common. The FCI algorithm takes as input a sample, distributional assumptions, optional background knowledge (e.g. time order), and a significance level, and outputs a partial ancestral graph. Because the algorithm uses only tests of conditional independence among sets of observed variables, it avoids the computational problems involved in calculating posterior probabilities or scores for latent variable models.

Just as the pattern can be used to predict the effects of some manipulations, a partial ancestral graph can also be used to predict the effects of some manipulations. Instead of calculating the effects of manipulations for which every member of the **O**-distribution equivalence class agree, we can calculate the effects only of those manipulations for which every member of the **O**-conditional independence equivalence agree. This will typically predict the effects of fewer manipulations than could be predicted given the **O**-distributional equivalence class (because a larger set of graphs have to make the same prediction), but the predictions made will still be correct.

Even though the set $S$ of DAGs in an **O**-conditional independence equivalence class is infinite, it is still possible to extract the features that the members of $S$ have in common. For example, every member of the conditional independence class over **O** that contains the DAG in Figure 11 has a directed path from *PE* to *CP* and no latent common cause of *PE* and *CP*. This is informative because even though the data do not help choose between members of the equivalence class, insofar as the data are evidence for the disjunction of the members in the equivalence class, they are evidence that *PE* is a cause of *CP*.

---

[6]The FCI algorithm is similar to Pearl's IC* algorithm [Pearl 2000] in many respects, and uses concepts bases on IC*; however IC* is computationally and statistically feasible only for a few variables.

A partial ancestral graph is analogous to a pattern, and represents the features common to an **O**-conditional independence class. Figure 11 shows an example of a DAG and the corresponding partial ancestral graph over **O** = {*IQ, SES, PE, CP, SEX*}. Two variables *A* and *B* are adjacent in a partial ancestral graph that represents an **O**-conditional independence class, when *A* and *B* are not entailed to be independent (i.e. they are d-connected) conditional on any subset of the variables in **O**\{*A,B*} for each DAG in the **O**-conditional independence class. The "–" endpoint of the *PE* → *CP* edge means that *PE* is an ancestor of *CP* in every DAG in the **O**-conditional independence class. The ">" endpoint of the *PE* → *CP* edges means that *CP* is not an ancestor of *PE* in any member of the **O**-conditional independence class. The "o" endpoint of the *SES* o–o *IQ* edge makes no claim about whether *SES* is an ancestor of *IQ* or not.

Applying the FCI algorithm to the Sewell and Shah data yields the PAG in Figure 11. The output predicts that when *PE* is manipulated, the following conditional probabilities hold: $P(CP=0|PE=0) = .063$; $P(CP=1|PE=0) = .937$; $P(CP=0|PE=1) = .572$; $P(CP=1PE=1) = .428$. These estimates are close to the estimates given by the output of the PC algorithm, although unlike the PC algorithm the output of the FCI algorithm posits the existence of latent variables. A bootstrap test of the output run at significance level 0.001 yielded the same results on 8 out of 10 samples. In the other two samples, the algorithm could not calculate the effect of the manipulation.

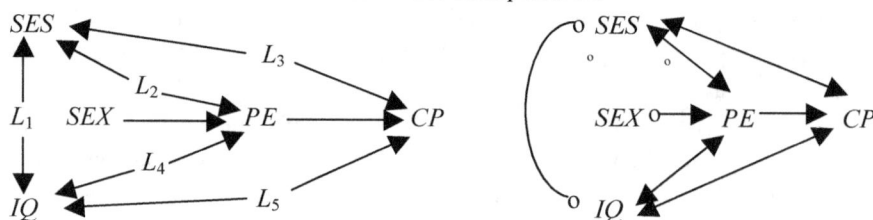

**Figure 11: DAG and Partial Ancestral Graph**

### 4.2.1 Online Course

Data from an online course provides an example where there was some experimental confirmation of the FCI causal model. Carnegie Mellon University offers a full semester online course that serves as a tutor on the subject of causal reasoning.[7] The course contains a number of different modules that contain both text and interactive online exercises that illustrate various concepts. Each module ends with a quiz that students must take. The interactive exercises are purely voluntary and play no role in calculating the student's final grade. It is possible to print the text from the online modules, but a student who studies from the printed text cannot use the online interactive exercises. The following variables were measured for each student:
- Pre-test (%)
- Print-outs (% modules printed)
- Quiz Scores (avg. %)

---

[7]See http://oli.web.cmu.edu/openlearning/forstudents/freecourses/csr

- Voluntary Exercises (% completed)
- Final Exam (%)
- 9 other variables

Using data from 2002, and some background knowledge about causal order, the output of the FCI algorithm was the PAG shown in Figure 12a. That model predicts that interventions that stops students from printing out the text and encourages students to use the online interactive exercises should raise the final grade in the class.

In 2003, students were advised that completing the voluntary exercises seemed to be important in helping grades, but that printing out the modules seemed to prevent completing the voluntary exercises. They were advised that, if they printed out the text they should make extra effort to go online and complete the interactive online exercises. Data on the same variables was gathered in 2003, and the output of the FCI algorithm is shown Figure 12b. The interventions to discourage printing and encourage the use of the online interactive exercises were largely successful, and the PAG output by the FCI algorithm from the 2003 data is exactly the PAG one would expect after intervening on the PAG output by the FCI algorithm from the 2002 data.

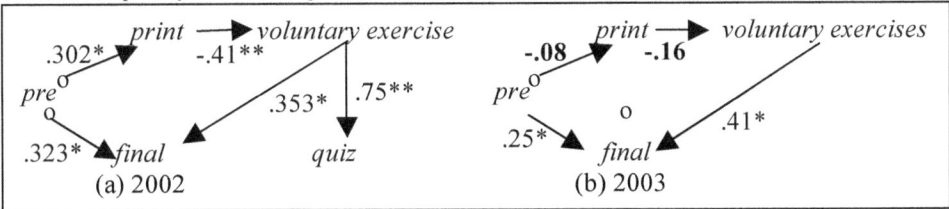

**Figure 12: Online Course Printing**

## 4.3 Errors in Variables: Combining Constraint Based Search and Bayesian Reasoning

In some cases the parameters of the output of the FCI algorithm are not identifiable or it is important to find not a particular latent variable model, but an equivalence class of latent variable models. In some of those cases the FCI algorithm can be combined with Bayesian methods.

### 4.3.1 Example - Lead and IQ

The next example shows how the FCI algorithm can be used to find a PAG, which can then be used as a starting point for a search for a latent variable DAG model and Bayesian estimation of parameters. It also illustrates how such a procedure produces different results than simply applying regression or using regression to generate more sophisticated models, such as errors-in-variables models.

By measuring the concentration of lead in a child's baby teeth, Herbert Needleman was the first epidemiologist to even approximate a reliable measure of cumulative lead exposure. His work helped convince the United States to eliminate lead from gasoline and most paint [Needleman 1979]. In their 1985 article in *Science* [Needleman, Geiger, & Frank 1985], Needleman, Geiger and Frank gave results for a multivariate linear regression of children's IQ on lead exposure. Having started their analysis with almost 40 covariates, they were faced with a variable selection problem to which they applied backwards-stepwise variable selection, arriving at a final regression model involving lead and five of the original 40 covariates. The covariates were measures of genetic contributions to the child's IQ (the parent's IQ), the amount of environmental stimulation in the child's early environment (the mother's education), physical factors that might compromise the child's cognitive endowment (the number of previous live births), and the parent's age at the birth of the child, which might be a proxy for many factors. The measured variables they used are as follows:

- *ciq* - child's verbal IQ score         *piq* - parent's IQ scores
- *lead* - measured concentration in baby teeth  *mab* - mother's age at child's birth
- *med* - mother's level of education in years   *fab* - father's age at child's birth
- *nlb* - number of live births previous to the sampled child

The standardized regression solution[8] is as follows, with t-ratios in parentheses. Except for *fab*, which is significant at 0.1, all coefficients are significant at 0.05, and $R^2$ = .271.

$$\hat{ciq} = -.143\ lead + .219\ med + .247\ piq + .237\ mab - .204\ fab - .159\ nlb$$

(2.32)     (3.08)     (3.87)     (1.97)     (1.79)     (2.30)

This analysis prompted criticism from Steve Klepper and Mark Kamlet, economists at Carnegie Mellon [Klepper, 1988/Klepper, Kamlet, & Frank 1993]. Klepper and Kamlet correctly argued that Needleman's statistical model (a linear regression) neglected to account for measurement error in the regressors. That is, Needleman's measured regressors were in fact imperfect proxies for the actual but latent causes of variations in IQ, and in these circumstances a regression analysis gives a biased estimate of the desired causal coefficients and their standard errors. Klepper and Kamlet constructed an errors-in-variables model to take into account the measurement error. See Figure 13, where the latent variables are in boxes, and the relations between the regressors are unconstrained.

Unfortunately, an errors-in-variables model that explicitly accounts for Needleman's measurement error is "underidentified," and thus cannot be estimated by classical techniques without making additional assumptions. Klepper, however, worked out an ingenious technique to bound the estimates, provided one could reasonably bound the amount of measurement error contaminating certain measured regressors [Klepper, 1988; Klepper et al. 1993]. The required measurement error bounds vary with each problem, however, and those required in order to bound the effect of actual lead exposure below 0 in Needleman's model seemed wholly unreasonable. Klepper concluded that the statistical evidence for Needleman's hypothesis was indeed weak. A Bayesian analysis, based on Gibbs sampling techniques, found that several posteriors corresponding to different priors lead to similar results. Although the size of the Bayesian point estimate

---

[8] The covariance data for this reanalysis was originally obtained from Needleman by Steve Klepper, who generously forwarded it. In this, and all subsequent analyses described, the correlation matrix was used.

for lead's influence on *IQ* moved up and down slightly, its sign and significance (the 95% central region in the posterior over the *lead-iq* connection always included zero) were robust.

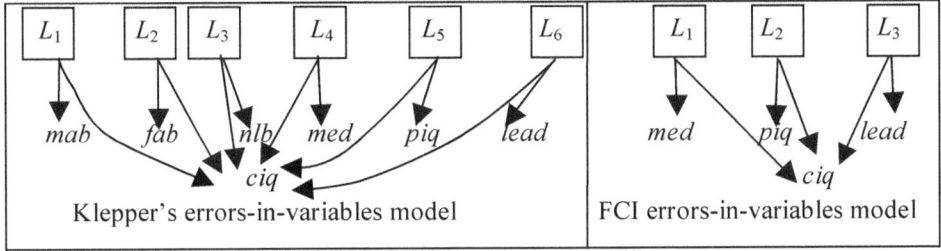

Figure 13: Errors-in-Variables Models

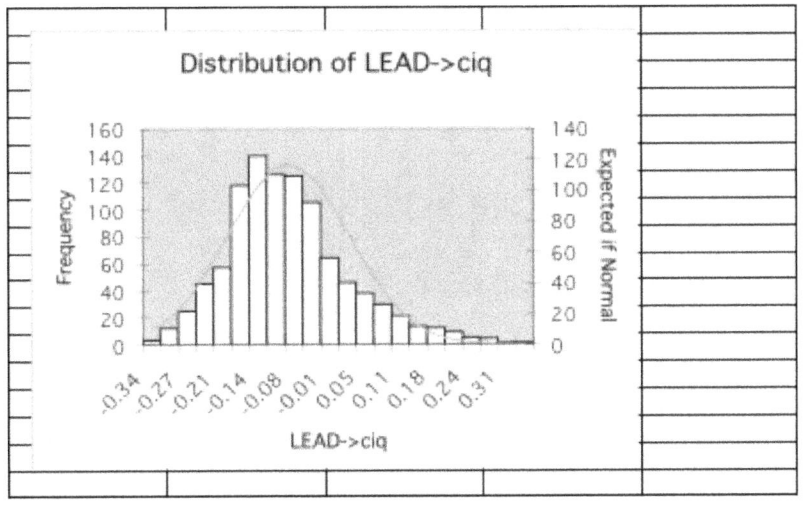

Figure 14: Posterior for Klepper's Model

A reanalysis using the FCI algorithm produced different results [Scheines 2000]. Scheines first used the FCI algorithm to generate a PAG, which was subsequently used as the basis for constructing an errors-in-variables model. The FCI algorithm produced a PAG that indicated that *mab*, *fab*, and *nlb* are *not* adjacent to *ciq*, contrary to Needleman's regression.[9] If we construct an errors-in-variables model compatible with the PAG produced by the FCI algorithm, the model does not contain *mab*, *fab*, or *nlb*. See Figure 13. (We emphasize that there are other models compatible with the PAG, which are not errors-in-variables models; the selection of an error-in-variables model from the

---

[9] The fact that *mab* had a significant regression coefficient indicates that *mab* and *ciq* are correlated conditional on the other variables; the FCI algorithm concluded that *mab* is not a cause of *ciq* because *mab* and *ciq* are unconditionally uncorrelated.

set of models represented by the PAG is an assumption.) In fact the variables that the FCI algorithm eliminated were precisely those, which required unreasonable measurement error assumptions in Klepper's analysis. With the remaining regressors, Scheines specified an errors-in-variables model to parameterize the effect of actual lead exposure on children's IQ. This model is still underidentified but under several priors, nearly all the mass in the posterior was over negative values for the effect of actual lead exposure (now a latent variable) on measured IQ. In addition, applying Klepper's bounds analysis to this model indicated that the effect of actual lead exposure on *ciq* was bounded below zero given reasonable assumptions about the degree of measurement error.

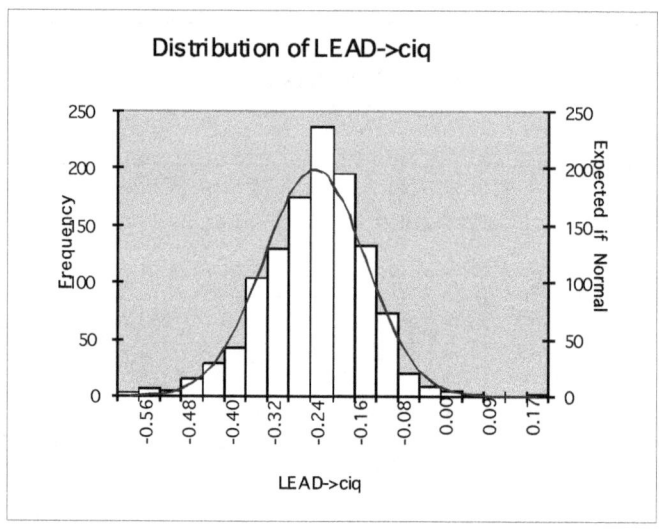

**Figure 15: Posterior for FCI model**

## 4.4 BuildPureClusters and MIMBuild

Searches using conditional independence constraints are correct, but completely uninformative for some common kinds of data sets. Consider the model $S$ in Figure 16. The data comes from a survey of test anxiety indicators administered to 335 grade 12 male students in British Columbia [Gierl & Todd 1996]. The survey contains 20 measures of symptoms of anxiety under test conditions. Each question is about a symptom of anxiety. For example, question 8 is about how often one feels "jittery when taking tests". The answer is observed on a four-point approximately Likert scale (almost never, sometimes, often, or almost always). As in many such analyses, we will assume that the variables are approximately Gaussian.

Each $X$ variable represents an answer to a question on the survey. For reasons to be explained later, not all of the questions on the test have been included in the model. There are three unobserved common causes in the model: *Emotionality*, *Care about achieving* (which will henceforth be referred to as *Care*) and *Self-defeating*. The test questions are of little interest in themselves; of more interest is what information they reveal about some unobserved psychological traits. If $S$ is correct, there are no conditional independence relations among the $X$ variables alone - the only entailed conditional independencies require conditioning on an unobserved common cause. Hence the FCI algorithm would return a completely unoriented PAG in which every pair of variables in **X** is adjacent. Such a PAG makes no predictions at all about the effects of manipulations of the observed variables.

Furthermore, in this case, the effects of manipulating the observed variables (answers to test questions) are of no interest - the interesting questions are about the effects of manipulating the unobserved variables and the qualitative causal relationships between them.

Although PAGs can reveal the existence of latent common causes (as by the double-headed arrows in Figure 11 for example), before one could make a prediction about the effect of manipulating an unobserved variable(s), one would have to identify what the variable (or variables) is, which is never possible from a PAG.

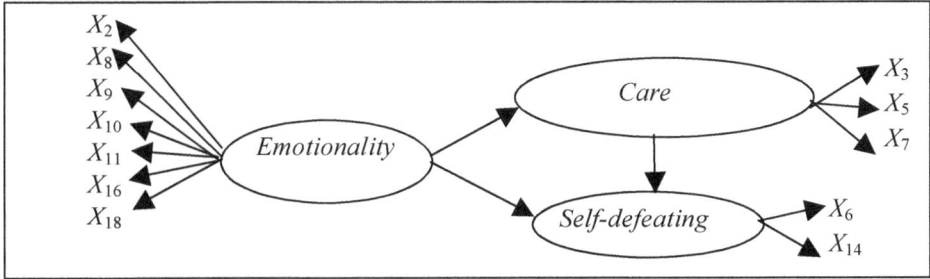

**Figure 16: SEM S**

Models such as $S$ are *multiple indicator models*, and can be divided into two parts: the measurement model, which contains the edges between the unobserved variables and the observed variables (e.g. *Emotionality* $\rightarrow X_2$), and the structural model, which contains the edges between the unobserved variables (e.g. *Emotionality* $\rightarrow$ *Care*).

The **X** variables in $S$ ($\{X_2, X_3, X_5, X_6, X_7, X_8, X_9, X_{10}, X_{11}, X_{14}, X_{16}, X_{18}\}$) were chosen with the idea that they indirectly measure some psychological trait that cannot be directly observed. Ideally, the **X** variables can be broken into clusters, where each variable in the cluster is caused by one unobserved cause common to the members of the cluster, and a unique error term uncorrelated with the other error terms, and nothing else. From the values of the variables in the cluster, it is then possible to make inferences about the value of the unobserved common cause. Such a measurement model is called *pure*. In psychometrics, pure measurement models satisfy the property of local independence: each measured variable is independent of all other variables, conditional on the unobserved variable it measures. In Figure 16, the measurement model of $S$ is pure.

If the measurement model is impure (i.e. there are multiple common causes of a pair of variables in **X**, or some of the **X** variables cause each other) then drawing inferences about the values of the common causes is much more difficult. Consider the set **X'** = **X** $\cup$ $\{X_{15}\}$. If $X_{15}$ indirectly measured (was a direct effect of) the unobserved variable *Care*, but $X_{10}$ directly caused $X_{15}$, then the measurement model over the expanded set of variables would not be pure. If a measurement model for a set **X'** of variables is not pure, it is nevertheless possible that some subset of **X'** (such as **X**) has a pure measurement model. If the only reason that the measurement model is impure is that $X_{10}$ causes $X_{15}$ then $\mathbf{X} = \mathbf{X'}\backslash\{X_{15}\}$ does have a pure measurement model, because all the "impurities" have been removed. $S$ does not contain all of the questions on the survey precisely because various tests described below indicated that they some of them needed to be excluded in order to have a pure measurement model.

The task of searching for a multiple indicator model can then be broken into two parts: first finding clusters of variables so that the measurement model is pure; second, use the pure measurement model to make inferences about the structural model.

Factor analysis is often used to determine the number of unmeasured common causes in a multiple indicator model, but there are important theoretical and practical problems in using factor analysis in this way. Factor analysis constructs models with unobserved common causes (factors) of the observed **X** variables. However, factor analysis models typically connect each unobserved common cause (factor) to each **X** variable, so the measurement model is not pure. A major difficulty with giving a causal interpretation to factor analytic models is that the observed distribution does not determine the covariance matrix among the unobserved factors. Hence, a number of different factor analytic models are compatible with the same observed data [Harman 1976]. In order to reduce the underdetermination of the factor analysis model by the data, it is often assumed that the unobserved factors are independent of each other; however, this is clearly not an appropriate assumption for unobserved factors that are supposed to represent actual causes that may causally interact with each other. In addition, simulation studies indicate that factor analysis is not a reliable tool for estimating the correct number of unobserved common causes [Glymour 1998].

On this data set, factor analysis indicates that there are 2 unobserved direct common causes, rather than 3 unobserved direct common causes [Bartholomew, Steele, Moustaki, & Galbraith 2002]. If a pure measurement model is constructed from the factor analytic model by associating each observed $X$ variable only with the factor that it is most strongly associated with, the resulting model fails a statistical test (has a p-value of zero) [Silva, Scheines, Glymour, & Spirtes 2006]. A search for pure measurement models that depends upon testing vanishing tetrad constraints is an alternative to factor analysis. Conceptually, the task of building a pure measurement model from the observed variables can be broken into 3 separate tasks:

1. Select a subset of the observed variables that form a pure measurement model.
2. Determine the number of clusters (i.e. the number of unobserved common causes) that the observed variables measure.
3. Cluster the observed variables into the proper groups (so each group has exactly one unobserved direct common cause.)

It is possible to construct pure measurement models using vanishing tetrad constraints as a guide [Silva et al. 2006]. A *vanishing tetrad constraint* holds among $X$, $Y$,

$Z$, $W$ when $\text{cov}(X,Y) \cdot \text{cov}(Z,W) - \text{cov}(X,Z) \cdot \text{cov}(Y,W) = 0$. A pure measurement model entails that each $X_i$ variables is independent of every other $X_j$ variable conditional on its unobserved parent, e.g. $S$ entails $X_2$ is independent of $X_j$ conditional on *Emotionality*. These conditional independence relations cannot be directly tested, because *Emotionality* is not observed. However, together with the other conditional independence relations involving unobserved variables entailed by $S$, they imply vanishing tetrad constraints on the observed variables that reveal information about the measurement model that does not depend upon the structural model among the unobserved common causes. The basic idea extends back to Spearman's attempts to use vanishing tetrad constraints to show that there was a single unobserved factor of intelligence that explained a variety of observed competencies [Spearman 1904].

Because $X_2$ and $X_8$ have one unobserved direct common cause (*Emotionality*), and $X_3$ and $X_5$ have a different unobserved direct common cause (*Care*), $S$ entails $\text{cov}_S(X_2, X_3) \cdot \text{cov}_S(X_5, X_8) = \text{cov}_S(X_2, X_5) \cdot \text{cov}_S(X_3, X_8) \neq \text{cov}_S(X_2, X_8) \cdot \text{cov}_S(X_3, X_5)$ for all values of the model's free parameters (here $\text{cov}_S$ is the covariance matrix entailed by $S$).[10] On the other hand, because $X_2$, $X_8$, $X_9$, and $X_{10}$ all have one unobserved common cause (*Emotionality*) as a direct common cause, the following vanishing tetrad constraints are entailed by $S$: $\text{cov}_S(X_2, X_8) \cdot \text{cov}_S(X_9, X_{10}) = \text{cov}_S(X_2, X_9) \cdot \text{cov}_S(X_8, X_{10}) = \text{cov}_S(X_2, X_{10}) \cdot \text{cov}_S(X_8, X_9)$ [Spirtes et al. 2001]. The BuildPureClusters algorithm uses the vanishing tetrad constraints as a guide to the construction of pure measurement models, and in the large sample limit reliably succeeds if there is a pure measurement model among a large enough subset of the observed variables [Silva et al. 2006].

In this example, BuildPureClusters automatically constructed the measurement model corresponding to the measurement model of $S$. The clustering on statistical grounds makes substantive sense, as indicated by the fact that it is similar to a prior theory-based clustering based on background knowledge about the content of the questions; however BuildPureClusters removes some questions, and splits one of the clusters of questions constructed from domain knowledge into two clusters.

Once a pure measurement model has been constructed, there are several algorithms for finding the structural model. One way is to estimate the covariances among the unobserved common causes, and then input the estimated covariances to the FCI algorithm. The output is then a PAG among the unobserved common causes. Alternative searches for the structural model include the MIMBuild and GESMIMBuild algorithms, which output patterns [Silva et al. 2006].

In this particular analysis, the MIMBuild algorithm, which also employs vanishing tetrad constraints, was used to construct a variety of output patterns corresponding to different values of the search parameters. The best pattern returned contains an undirected edge between every pair of unobserved common causes. ($S$ is an example that is compatible with the pattern, but any other orientation of the edges among the three

---

[10] The inequality is based on an extension of the Causal Faithfulness Assumption that states that vanishing tetrad constraints that are not entailed for all values of the free parameters by the true causal graph are assumed not to hold.

unobserved common causes that does not create a cycle is also compatible with the pattern.) The resulting model (or set of models) passes a statistical test with a p-value of 0.47.

### 4.4.1 Example - Religion and Depression

Data relating religion and depression provides an example that shows how an automated causal search produces a model that is compatible with background knowledge, but fits much better than a model that was built from theories about the domain.

Bongjae Lee from the University of Pittsburgh organized a study to investigate religious/spiritual coping and stress in graduate students [Silva & Scheines 2004]. In December of 2003, 127 Masters in Social Works students answered a questionnaire intended to measure three main factors:

- *Stress*, measured with 21 items, each using a 7-point scale (from "not all stressful" to "extremely stressful") according to situations such as: "fulfilling responsibilities both at home and at school"; "meeting with faculty"; "writing papers"; "paying monthly expenses"; "fear of failing"; "arranging childcare";
- *Depression*, measured with 20 items, each using a 4-point scale (from "rarely or none" to "most or all the time") according to indicators as: "my appetite was poor"; "I felt fearful"; "I enjoyed life" "I felt that people disliked me"; "my sleep was restless";
- *Spiritual coping*, measured with 20 items, each using a 4-point scale (from "not at all" to "a great deal") according to indicators such as: "I think about how my life is part of a larger spiritual force"; "I look to God (high power) for strength in crises"; "I wonder whether God (high power) really exists"; "I pray to get my mind off of my problems";

The goal of the original study was to use graphical models to quantify how *Spiritual coping* moderates the association of *Stress* and *Depression*, and hypothesized that *Spiritual coping* reduces the association of *Stress* and *Depression*. The theoretical model (Figure 17) fails a chi-square test: p = 0. The measurement model produced by BuildPureClusters is shown in Figure 18. Note that the variables selected automatically are proper subsets of Lee's substantive clustering. The full model automatically produced with GESMIMBuild with the prior knowledge that *Stress* is not an effect of other latent variables is given in Figure 19. This model passes a chi square test, p = 0.28, even though the algorithm itself does not try to directly maximize the fit. Note that it supports the hypothesis that *Depression* causes *Spiritual Coping* rather than the other way around. Although this conclusion is not conclusive, the example does illustrate how the algorithm can find a theoretically plausible alternative model that fits the data well.

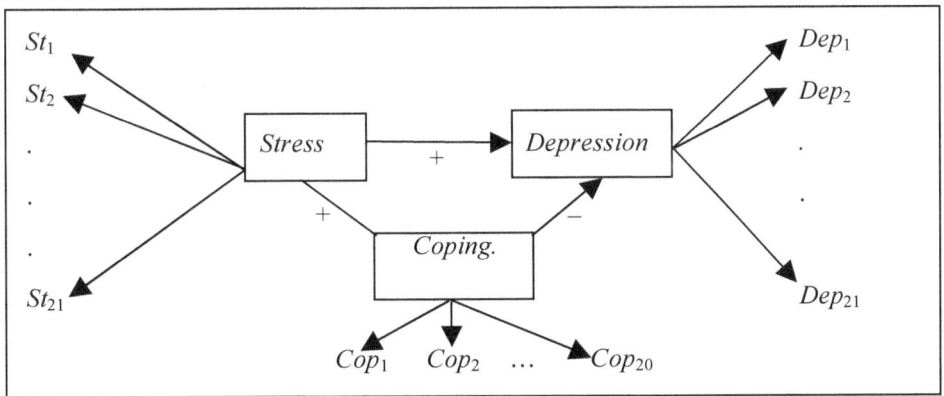

**Figure 17: Model from Theory**

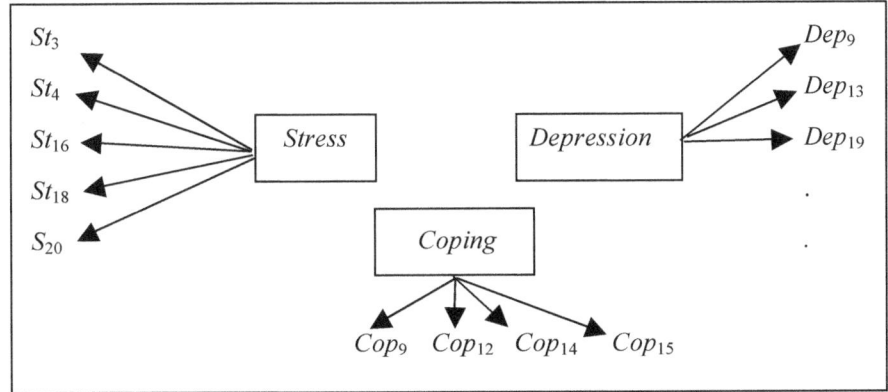

**Figure 18: Output of BuildPureClusters**

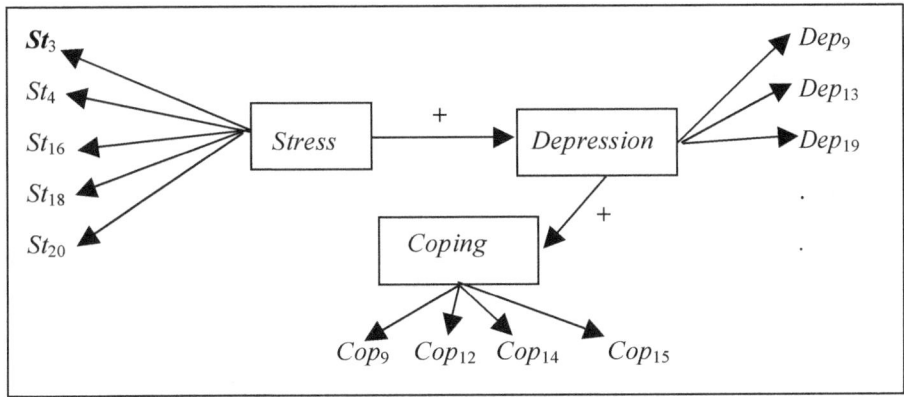

**Figure 19: Output of GESMIMBuild**

## 5 Time Series and Feedback

The models described so far are for "equilibrium." That is, they assume that an intervention fixes the values of a variable or variables, and that the causal process results in stable values of effect variables, so that time can be ignored. When time cannot be ignored, representation, interventions and search are all more complicated.

Time series models with a causal interpretation are naturally represented by directed acyclic graphs in at least three different forms: A graph whose variables are indexed by time, a "unit" graph giving a substructure that is repeated in the time indexed graph, and a finite graph that may be cyclic. Models of the first kind have been described as "Dynamical Causal Models" but the description does not address the difficulties of search. Pursuing a strategy of the PC or FCI kind, for example, requires a method of correctly estimating conditional independence relations.

### 5.1 Time series models

Chu and Glymour [2008] describe conditional independence tests for additive models, and use these tests in a slight modification of the PC and FCI algorithms. The series data is examined by standard methods to determine the requisite number of lags. The data are then replicated a number of times equal to the lags, delaying the first replicant by one time step, the second by two time steps, and so on, and conditional independence tests applied to the resulting sets of data. They illustrate the algorithm with climate data.

Climate teleconnections are associations of geospatially remote climate phenomena produced by atmospheric and oceanic processes. The most famous, and first established teleconnection, is the association of El Nino/Southern Oscillation (*ENSO*) with the failure of monsoons in India. A variety of associations have been documented among sea surface temperatures (*SST*), atmospheric pressure at sea level (*SLP*), land surface temperatures (*LST*) and precipitation over land areas. Since the 1970s data from a sequence of satellites have provided monthly (and now daily) measurements of such variables, at resolutions as small as 1 square kilometer. Measurements in particular spatial regions have been clustered into time-indexed indices for the regions, usually by principal components analysis, but also by other methods. Climate research has established that some of these phenomena are exogenous drivers of others, and has sought physical mechanisms for the teleconnections.

Chu and Glymour (2008) consider data from the following 6 ocean climate indices, recorded monthly from 1958 to 1999, each forming a time series of 504 time steps:
- *QBO* (*Quasi Biennial Oscillation*): Regular variation of zonal stratospheric winds above the equator
- *SOI* (*Southern Oscillation*): Sea Level Pressure (*SLP*) anomalies between Darwin and Tahiti
- *WP* (*Western Pacific*): Low frequency temporal function of the 'zonal dipole' *SLP* spatial pattern over the North Pacific.
- *PDO* (*Pacific Decadal Oscillation*): Leading principal component of monthly Sea Surface Temperature (*SST*) anomalies in the North Pacific Ocean, poleward of 20° N

- *AO (Arctic Oscillation)*: First principal component of SLP poleward of 20° N
- *NAO (North Atlantic Oscillation)* Normalized SLP differences between Ponta Delgada, Azores and Stykkisholmur, Iceland

Some connections among these variables are reasonably established, but are not assumed in the analysis that follows. In particular, *SO* and *NAO* are thought to be exogenous drivers.

After testing for stationarity, the PC algorithm yields the structure for the climate data shown in Figure 20. The double-headed arrows indicate the hypothesis of common unmeasured causes.[11] So far as the exogenous drivers are concerned, the algorithm output is in accord with expert opinion.

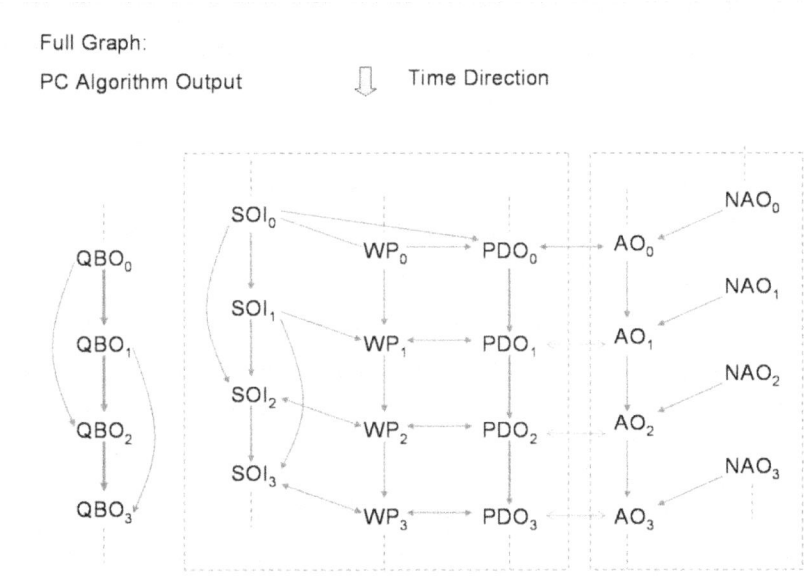

**Figure 20: Climate Time Series**

Monthly time series of temperatures and pressures at the sea surface present a case in which one might think that the causal processes take place more rapidly that the sampling rate. If so, then the causal structure in between time samples, the "contemporaneous" causal structure, should look much like a unit of the time series causal structure. When we sample at intervals of time as in economic, climate, and other time series, can we discover what goes on in the intervals between samples? Swanson and Granger suggested that an autoregression be used to remove the effects on each variable of variables at previous times, and a search could then be applied to the residual correlations [Swanson & Granger 1997]. The search they suggested was to assume a chain and to test

---

[11] When under the usual assumptions, the PC algorithm produces double headed arrows, they reliably indicate common unobserved causes as will FCI. But unlike FCI, PC is not complete in this respect.

it by methods described in [Glymour, Scheines, Spirtes, & Kelly 1987], some of the work whose aims and methods Cartwright previously sought to demonstrate is impossible [Cartwright 1994]. But a chain model of contemporaneous causes is far too special a case. Hoover & Demiralp, and later, Moneta & Spirtes, proposed applying PC to the residuals [Hoover & Demiralp 2003; Moneta & Spirtes 2006]. (Moneta also worked out the statistical corrections to the correlations required by the fact that they are obtained as residuals from regressions.) When that is done for model above, the result is the unit structure of the time series: $QBO \quad SOI \to WP \leftrightarrow PDO \leftrightarrow AO \leftarrow NA$.

## 5.2 Cyclic Graphs

Since the 1950s, the engineering literature has developed methods for analyzing the statistical properties of linear systems described by cyclic graphs. The literature on search is more recent. Spirtes showed that linear systems with independent noises satisfy a simple generalization of d-separation, and the idea of faithfulness is well-defined for such systems [Spirtes 1995]; Pearl & Dechter extended these results to discrete variable systems [Pearl & Dechter 1996]. Richardson proved some of the essential properties of such graphs, and developed a pointwise consistent PC style algorithm for search [Richardson 1996]. More recently, an extension of the LiNGAM algorithm for linear, cyclic, non-Gaussian models has been developed [Lacerda et al. 2008].

## 5.3 Distributed Multiple Data Sets: ION and IMaGES

Data mining has focused on learning from a single database, but inferences from multiple databases are often needed in social science, multiple subject time series in physiological and psychological experiments, and to exploit archived data in many subjects. Such data sets typically pose missing variable problems: some of what is measured in one study or for one subject, may not be measured in another. In many cases such multiple data sets cannot, for physical, sociological or statistical reasons, be merged into a single data set with missing variables. There are two strategies for this kind of problem: learn a structure or set of structures separately for each data set and then find the set of structures consistent with the several "marginal" structures, or learn a single set of structures by evaluating steps in a search procedure using all of the data sets. The first strategy could be carried out using PC, kPC GES, FCI, LiNGAM or other procedure on each data set, and then using an algorithm that returns a description of the set of all graphs, or mixed graphs, consistent with the results from each database [Tillman, Danks, & Glymour 2008]. Tillman, Danks and Glymour have used such a procedure in combination with GES and FCI. The result in some (surprising) cases is a unique partial ancestral graph, and in other cases a large set of alternatives collectively carrying little information. The second strategy has been implemented in the IMaGES algorithm [Ramsey et al. 2009]. The algorithm uses GES, but at each step in the evaluation of a candidate edge addition or removal, the candidate is scored separately by BIC on each data set and the average of the BIC scores is used by the algorithm in edge addition or deletion choices. The IMaGES strategy is more limited—no consistency proof is available when the samples are from mixed distributions, and a proof of convergence of averages of BIC scores to a function of posteriors is only available when the sample sizes of several data sets are equal. Nonetheless, IMaGES has been applied to fMRI data from multiple subjects with remarkably good results. For example, an fMRI study of responses to visually presented rhyming and non-rhyming words and non-words should produce a left hemisphere cascade leading to right hemisphere effects, which is exactly what IMaGES finds, using only the prior knowledge that the input variable is not an effect of other variables.

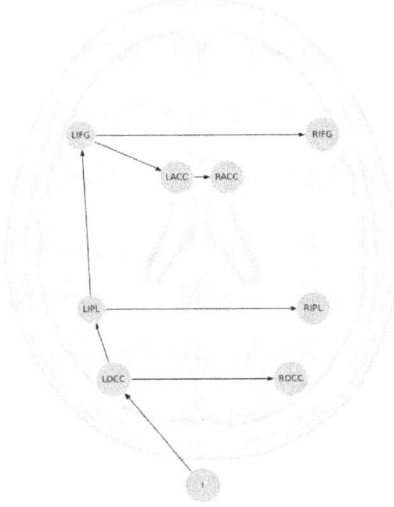

**Figure 21: IMaGES Output for fMRI Data**

## 6   Conclusion

The discovery of d-separation, and the development of several related notions, has made possible principled search for causal relations from observational and quasi-experimental data in a host of disciplines. New insights, algorithms and applications have appeared almost every year since 1990, and they continue. We are seeing a revolution in understanding of what is and is not possible to learn from data, but the insights and methods have seeped into statistics and applied science only slowly. We hope that pace will quicken.

## 7   Appendix

A *directed graph* (e.g. $G_1$ of Figure 22) consists of a set of vertices and a set of directed edges, where each edge is an ordered pair of vertices. In $G_1$, the vertices are $\{A,B,C,D,E\}$, and the edges are $\{B \rightarrow A, B \rightarrow C, D \rightarrow C, C \rightarrow E\}$. In $G_1$, $B$ is a *parent* of $A$, $A$ is a *child* of $B$, and $A$ and $B$ are *adjacent* because there is an edge $A \rightarrow B$. A *path* in a directed graph is a sequence of adjacent edges (i.e. edges that share a single common endpoint). A *directed path* in a directed graph is a sequence of adjacent edges all pointing in the same direction. For example, in $G_1$, $B \rightarrow C \rightarrow E$ is a directed path from $B$ to $E$. In contrast, $B \rightarrow C \leftarrow D$ is a path, but not a directed path in $G_1$ because the two edges do not point in the same direction; in addition, $C$ is a *collider* on the path because both edges on the path are directed into C. A triple of vertices <B,C,D> is a *collider* if there are edges B → C ← D in $G_1$; <B,C,D> is an *unshielded collider* if in addition there is no edge between B and D. E is a *descendant* of B (and B is an *ancestor* of E) because there is a directed path from B to E; in addition, by convention, each vertex is a descendant (and ancestor) of itself. A directed graph is *acyclic* when there is no directed path from any vertex to itself: in that case the graph is a directed acyclic graph, or DAG for short.

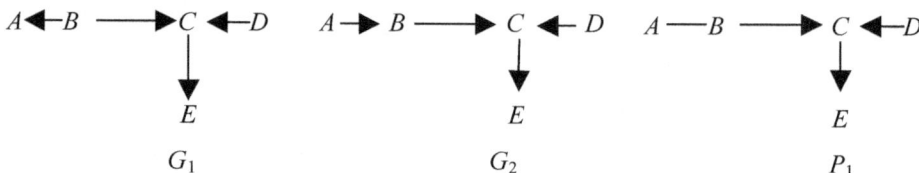

**Figure 22**: $G_1$, $G_2$, and $P_1$ (pattern for $G_1$ and $G_2$)

A probability distribution P(**V**) satisfies the *local Markov condition* for a DAG $G_1$ with vertices **V** when each variable is independent of its non-parental non-descendants conditional on its parents. A *Bayesian network* is an ordered pair of a directed acyclic graph G and a set of probability distributions that satisfy the local Markov condition for G.

The graphical relationship among sets of variables in a DAG G called "d-separation" determines which conditional independence relations are entailed by satisfying the local directed Markov property). Following [Pearl 1988], in a DAG G, for disjoint variable sets **X**, **Y**, and **Z**, **X** and **Y** are *d-separated* conditional on **Z** in G if and only if there exists no path U between an X ∈ **X** and a Y ∈ **Y** such that (i) every collider on U has a descendent in **Z** and (ii) no other vertex on U is in **Z**. An important theorem in [Pearl 1988] is that a DAG G entails that **X** is independent of **Y** conditional on **Z** if and only if **X** is d-separated from **Y** conditional on **Z** in G.

A Bayesian network restricted to a parametric family <G,**Q**> where G is a DAG and **Q** is some parameterization of the DAG, e.g. multivariate Gaussian, has two distinct interpretations. First, it has a probabilistic interpretation as a distribution over the variables in G, for distributions that satisfy the local Markov condition for G. Under this interpretation, it is a useful tool for calculating conditional probabilities.

Second, it has a causal interpretation, and can be used to calculate the effects of manipulations. Intuitively, a manipulation of a variable is an exogenous action that *forces* a value (or a distribution over values) upon a variable in the system, e.g. as in a randomized experiment - if no exogenous action is taken on variable X, X is said to have undergone a null manipulation. An example of a manipulation is a randomized experiment, in which a distribution for some variables (e.g. ½ of the subjects take a given drug, and ½ of the subjects do not take the drug) is imposed from outside. The kinds of manipulations that we will consider are ideal in the sense that a manipulation of X directly affects only X.

$X$ is a *direct cause* of $Y$ relative to a set of variables **V** if there is a pair of manipulations (including possibly null manipulations, and including hypothetical manipulations in the many circumstances where no actual manipulations are feasible) of the values of the variables in $\mathbf{V}\backslash\{Y\}$ that differ only in the value assigned to $X$, but that have different distributions for $Y$. This is in accord with the idea that the gold standard for determining causation is randomized experiments. (This is not a reduction of causality to non-causal concepts, because manipulation is itself a causal concept that we have taken as primitive.) Under the causal interpretation of DAGs, there is an edge $X \rightarrow Y$ when $X$ is a direct cause of $Y$ relative to the set of variables in the DAG. A set of variables **V** is *causally sufficient* if every direct cause (relative to **V**) of any pair of variables in **V**, is also in **V**. We will assume that causally interpreted DAGs are causally sufficient, although we will not generally all of the variables in a causally interpreted DAG are measured.

In automated causal search, the goal is to discover as much as possible about the true causal graph for a population from a sample from the joint probability distribution over the population, together with background knowledge (e.g. parametric assumptions, time order, etc.) This requires having some assumptions that link (samples from) probability distributions on the one hand, and causal graphs on the other hand. Extensive discussions of the following assumptions that we will make, including arguments for making the assumptions as well as limitations of the assumptions can be found in *Causation, Prediction, & Search* [Spirtes et al. 2001].

## 7.1 Causal Markov Assumption

The Causal Markov Assumption is a generalization of two commonly made assumptions: the immediate past screens off the present from the more distant past; and if $X$ does not cause $Y$ and $Y$ does not cause $X$, then $X$ and $Y$ are independent conditional on their common causes. It presupposes that while the random variables of a unit in the population may causally interact, the units themselves are not causally interacting with each other.

**Causal Markov Assumption:** Let $G$ be a causal graph with causally sufficient vertex set **V** and let $P$ be a probability distribution over the vertices in **V** generated by the causal structure represented by $G$. $G$ and $P$ satisfy the Causal Markov Assumption if and only if for every $W$ in **V**, $W$ is independent of its non-parental non-descendants conditional on its parents in $G$.

In graphical terms, the Causal Markov Assumption states that in the population distribution over a causally sufficient set of variables, each variable is independent of its non-descendants and non-parents, conditional on its parents in the true causal graph.

While the Causal Markov Assumption allows for some causal conclusions from sample data, it only supports inferences that some causal connections exist - it does not support inferences that some causal connections do not exist. The following assumption does support the latter kind of inference.

## 7.2 Causal Faithfulness Assumption

Often the set of distributions that satisfy the local Markov condition for $G$ is restricted to some parametric family $\Theta$ (e.g. Gaussian). In those cases, the set of distributions belonging to the Bayesian network will be denoted as $f(<G,\Theta>)$, and $f(<G,\theta>)$ will denote a member of $f(<G,\Theta>)$ for the particular value $\theta \in \Theta$ (and $f(<G,\theta>)$ is **represented** by $<G,\Theta>$). Let $I_f(\mathbf{X},\mathbf{Y}|\mathbf{Z})$ denote that $\mathbf{X}$ is independent of $\mathbf{Y}$ conditional on $\mathbf{Z}$ in a distribution $f$.

If a DAG $G$ does not entail that $I_{f(G,\theta)}(\mathbf{X},\mathbf{Y}|\mathbf{Z})$ for *all* $\theta \in \Theta$, nevertheless there may be *some* parameter values $\theta$ such that $I_{f(G,\theta)}(\mathbf{X},\mathbf{Y}|\mathbf{Z})$. In that case say that $f(<G,\theta>)$ is **unfaithful** to $G$. In Pearl's terminology the distribution is *unstable* [Pearl 1988]. This would happen for example if taking birth control pills increased the probability of blood clots directly, but decreased the probability of pregnancy which in turn increased the probability of blood clots, and the two causal paths exactly cancelled each other. We will assume that such unfaithful distributions do not happen - that is there may be such canceling causal paths, but the causal paths do not exactly cancel each other.

**Causal Faithfulness Assumption:** For a true causal graph $G$ over a causally sufficient set of variables $\mathbf{V}$, and probability distribution $P(\mathbf{V})$ generated by the causal structure represented by $G$, if $G$ does not entail that $\mathbf{X}$ is independent of $\mathbf{Y}$ conditional on $\mathbf{Z}$ then $\mathbf{X}$ is not independent of $\mathbf{Y}$ conditional on $\mathbf{Z}$ in $P(\mathbf{V})$.

## 7.3 Conditional Independence Equivalence

Let $\mathbf{I}(<G,\Theta>)$ be the set of all conditional independence relations entailed by satisfying the local Markov condition. For any distribution that satisfies the local directed Markov property for $G$, all of the conditional independence relations in $\mathbf{I}(<G,\Theta>)$ hold. Since these independence relations don't depend upon the particular parameterization but only on the graphical structure and the local directed Markov property, they will henceforth be denoted by $\mathbf{I}(G)$.

$G_1$ and $G_2$ are *conditional independence equivalent* if and only if $\mathbf{I}(G_1) = \mathbf{I}(G_2)$. This occurs if and only if $G_1$ and $G_2$ have the same d-separation relations. A set of graphs that are all conditional independence equivalent to each other is a *conditional independence equivalence class*. If the graphs are all restricted to be DAGs, then they form a *DAG conditional independence equivalence class*. Two DAGs are conditional independence equivalent if and only if they have the same d-separation relations.

**Theorem 1 (Pearl, 1988):** Two directed acyclic graphs are conditional independence equivalent if and only if they contain the same vertices, the same adjacencies, and the same unshielded colliders.

For example, Theorem 1 entails that the set consisting of $G_1$ and $G_2$ in Figure 22 is a DAG conditional independence equivalence class. The fact that $G_1$ and $G_2$ are conditional independence equivalent, but are different causal models, indicates that in general any algorithm that relies only on conditional independence relations to discover the causal graph cannot (without stronger assumptions or more background knowledge) reliably output a single DAG. A reliable algorithm could at best output the DAG conditional independence equivalence class, e.g. $\{G_1, G_2\}$.

Fortunately, Theorem 1 is also the basis of a simple representation called a pattern [Verma & Pearl 1990] of a DAG conditional independence equivalence class. Patterns can be used to determine which predicted effects of a manipulation are the same in every member of a DAG conditional independence equivalence class and which are not.

The adjacency phase of the PC algorithm is based on the following two theorems, where **Parents**$(G,A)$ is the set of parents of $A$ in $G$.

**Theorem 2**: If $A$ and $B$ are d-separated conditional on any subset **Z** in DAG $G$, then $A$ and $B$ are not adjacent in $G$.

**Theorem 3**: $A$ and $B$ are not adjacent in DAG $G$ if and only if $A$ and $B$ are d-separated conditional on **Parents**$(G,A)$ or **Parents**$(G,B)$ in $G$.

The justification of the orientation phase of the PC algorithm is based on Theorem 4.

**Theorem 4**: If in a DAG $G$, $A$ and $B$ are adjacent, $B$ and $C$ are adjacent, but $A$ and $C$ are not adjacent, either $B$ is in every subset of variables **Z** such that $A$ and $C$ are d-separated conditional on **Z**, in which case $<A,B,C>$ is not a collider, or $B$ is in no subset of variables **Z** such $A$ and $C$ are d-separated conditional on Z, in which case $<A,B,C>$ is a collider.

A **pattern** (also known as a PDAG) $P$ represents a DAG conditional independence equivalence class **X** if and only if:

1. $P$ contains the same adjacencies as each of the DAGs in **X**;
2. each edge in $P$ is oriented as $X \rightarrow Z$ if and only if the edge is oriented as $X \rightarrow Z$ in every DAG in **X**, and as $X - Z$ otherwise.

There are simple algorithms for generating patterns from a DAG [Meek, 1995; Andersson, Madigan, & Perlman 1997; Chickering 1995]. The pattern $P_1$ for the DAG conditional independence equivalence class containing $G_1$ is shown in Figure 22. It contains the same adjacencies as $G_1$, and the edges are the same except that the edge between $A$ and $B$ is undirected in the pattern, because it is oriented as $A \leftarrow B$ in $G_1$, and oriented as $A \rightarrow B$ in $G_2$.

## 7.4 Distributional Equivalence

For multi-variate Gaussian distributions and for multinomial distributions, every distribution that satisfies the set of conditional independence relations in $\mathbf{I}(<G,\Theta>)$ is also a member of $f(<G,\Theta>)$. However, for other families of distributions, it is possible that there are distributions that satisfy the conditional independence relations in $\mathbf{I}(<G,\Theta_\alpha>)$, but are not in $f(<G,\Theta_\alpha>)$, i.e. the parameterization imposes constraints that are not conditional independence constraints [Lauritzen et al. 1990; Pearl 2000; Spirtes et al. 2001].

It can be shown that when restricted to multivariate Gaussian distributions, $G_1$ and $G_2$ in Figure 22 represent exactly the same set of probability distributions, i.e. $f(<G_1,\Theta_1>)$

$= f(<G_2,\Theta_2>)$. In that case say that $<G_1,\Theta_1>$ and $<G_2,\Theta_2>$ are *distributionally equivalent* (relative to the parametric family). Whether two models are distributionally equivalent depends not only on the graphs in the models, but also on the parameterization families of the models. A set of models that are all distributionally equivalent to each other is a *distributional equivalence class*. If the graphs are all restricted to be DAGs, then they form a *DAG distributional equivalence class*.

In contrast to conditional independence equivalence, distribution equivalence depends upon the parameterization families as well as the graphs. Conditional independence equivalence of $G_1$ and $G_2$ is a necessary, but not always sufficient condition for the distributional equivalence of $<G_1,\Theta_A>$ and $<G_2,\Theta_B>$.

Algorithms that rely on constraints beyond conditional independence may be able to output subsets of conditional independence equivalence classes, although without further background knowledge or stronger assumptions they could at best reliably output a DAG distribution equivalence class. In general, it would be preferable to take advantage of the non conditional independence constraints to output a subset of the conditional independence equivalence class, rather than simply outputting the conditional independence equivalence class. For some parametric families it is known how to take advantage of the non conditional independence constraints (sections 3.4 and 4.4 ); however in other parametric families, either there are no non conditional independence constraints, or it is not known how to take advantage of the non conditional independence constraints.

**Acknowledgements**: Clark Glymour and Robert Tillman thanks the James S. McDonnell Foundation for support of their research.

## References

Ali, A. R., Richardson, T. S., & Spirtes, P. (2009). Markov Equivalence for Ancestral Graphs. *Annals of Statistics*, *37(5B)*, 2808-2837.

Aliferis, C. F., Tsamardinos, I., & Statnikov, A. (2003). HITON: A Novel Markov Blanket Algorithm for Optimal Variable Selection. *Proceedings of the 2003 American Medical Informatics Association Annual Symposium*, Washington, DC, 21-25.

Andersson, S. A., Madigan, D., & Perlman, M. D. (1997). A characterization of Markov equivalence classes for acyclic digraphs. *Ann Stat*, *25*(2), 505-541.

Asuncion, A. & Newman, D. J. (2007). UCI Machine Learning Repository.

Bartholomew, D. J., Steele, F., Moustaki, I., & Galbraith, J. I. (2002). *The Analysis and Interpretation of Multivariate Data for Social Scientists (Texts in Statistical Science Series)*. Chapman & Hall/CRC.

Cartwright, N. (1994). *Nature's Capacities and Their Measurements (Clarendon Paperbacks)*. Oxford University Press, USA.

Chickering, D. M. (2002). Optimal Structure Identification with Greedy Search. *Journal of Machine Learning Research*, *3*, 507-554.

Chickering, D. M. (1995). A transformational characterization of equivalent Bayesian network structures. *Proceedings of the Eleventh Conference on Uncertainty in Artificial Intelligence*, 87-98.

Chu, T., & Glymour, C. (2008). Search for Additive Nonlinear Time Series Causal Models. *Journal of Machine Learning Research*, 9(May):967-991.

Dempster, A. (1972). Covariance selection. *Biometrics, 28*, 157-175.

Fukumizu, K., Gretton, A., Sun, X., & Scholkopf, B. (2008). Kernel Measures of Conditional Dependence. *Advances in Neural Information Processing Systems 20*.

Geiger, D. & Meek, C. (1999). Quantifier Elimination for Statistical Problems. *Proceedings of the 15th Conference on Uncertainty in Artificial Intelligence*, Stockholm, Sweden, 226-233.

Gierl, M. J. & Todd, R. W. (1996). A Confirmatory Factor Analysis of the Test Anxiety Inventory Using Canadian High School Students. *Educational and Psychological Measurement, 56*(2), 315-324.

Glymour, C. (1998). What Went Wrong? Reflections on Science by Observation and the Bell Curve. *Philosophy of Science, 65*(1), 1-32.

Glymour, C., Scheines, R., Spirtes, P., & Kelly, K. (1987). *Discovering Causal Structure: Artificial Intelligence, Philosophy of Science, and Statistical Modeling.* Academic Press.

Gretton, A., Fukumizu, K., Teo, C. H., Song, L., Scholkopf, B., & Smola, A. J. (2008) A kernel statistical test of independence, In *Advances in Neural Information Processing Systems 20*, 585-592.

Harman, H. H. (1976). *Modern Factor Analysis.* University Of Chicago Press.

Hoover, K. & Demiralp, S. (2003). Searching for the Causal Structure of a Vector Autoregression. *Oxford Bulletin of Economics and Statistics 65 (Supplement), 65*, 745-767.

Hoyer, P. O., Janzing, D., Mooij, J. M., Peters, J., & Scholkopf, B. (2009). Nonlinear causal discovery with additive noise models. *Advances in Neural Information Processing Systems 21*, 689-696.

Hoyer, P. O., Shimizu, S., & Kerminen, A. (2006). Estimation of linear, non-gaussian causal models in the presence of confounding latent variables. *Third European Workshop on Probabilistic Graphical Models*, 155-162.

Hoyer, P. O., Hyvärinen, A., Scheines, R., Spirtes, P., Ramsey, J., Lacerda, G. &Shimizu, S. (2008). Causal discovery of linear acyclic models with arbitrary distributions. *Proceedings of the Twentyfourth Annual Conference on Uncertainty in Artificial Intelligence*, 282-289.

Hyvärinen, A., & Oja, E. (2000). Independent component analysis: Algorithms and applications. *Neural Networks*, 13(4-5): 411 - 430.

Junning, L. & Wang, Z. (2009). Controlling the False Discovery Rate of the Association/Causality Structure Learned with the PC Algorithm. *Journal of Machine Learning Research*, 475 - 514.

Kalisch, M. & Buhlmann, P. (2007). Estimating high dimensional directed acyclic graphs with the PC algorithm. *Journal of Machine Learning Research, 8*, 613-636.

Kiiveri, H. & Speed, T. (1982). Structural analysis of multivariate data: A review. In S. Leinhardt (Ed.), *Sociological Methodology 1982*. San Francisco: Jossey-Bass.

Klepper, S. (1988). Regressor Diagnostics for the Classical Errors-in-Variables Model. *J Econometrics, 37*(2), 225-250.

Klepper, S., Kamlet, M., & Frank, R. (1993). Regressor Diagnostics for the Errors-in-Variables Model - An Application to the Health Effects of Pollution. *J Environ Econ Manag, 24*(3), 190-211.

Lacerda, G., Spirtes, P., Ramsey, J., & Hoyer, P. O. (2008). Discovering Cyclic Causal Models by Independent Component Analysis. *Proceedings of the 24th Conference on Uncertainty In Artificial Intelligence*, 366-374.

Lauritzen, S. L., Dawid, A. P., Larsen, B. N., & Leimer, H. G. (1990). Independence properties of directed Markov fields. *Networks, 20*, 491-505.

Linthurst, R. A. (1979). Aeration, nitrogen, pH and salinity as factors affecting Spartina Alterniflora growth and dieback, Ph.D. dissertation, North Carolina State University.

Meek, C. (1995). Causal inference and causal explanation with background knowledge. *Proceedings of the Eleventh Conference on Uncertainty in Artificial Intelligence*, 403-411.

Meek, C. (1997). A Bayesian approach to learning Bayesian networks with local structure. *Proceedings of the Thirteenth Conference on Uncertainty in Artificial Intelligence*, 80-89.

Moneta, A. & Spirtes, P. (2006). Graphical Models for the Identification of Causal Structures in Multivariate Time Series Model. Paper presented at the *2006 Joint Conference on Information Sciences*.

Needleman, H. L. (1979). Deficits in psychologic and classroom performance of children with elevated dentine lead levels. N *Engl J Med, 300(13)*, 689-695.

Needleman, H. L., Geiger, S. K., & Frank, R. (1985). Lead and IQ scores: a reanalysis. *Science, 227(4688)*(4688), 701-2, 704.

Pearl, J. & Dechter, R. (1996). Identifying independencies in causal graphs with feedback. *Proceedings of the Twelfth Conference on Uncertainty in Artificial Intelligence*, Portland, OR, 420-426.

Pearl, J. (1988). *Probabilistic Reasoning in Intelligent Systems: Networks of Plausible Inference*. Morgan Kaufmann.

Pearl, J. (2000). *Causality: Models, Reasoning, and Inference*. Cambridge University Press.

Ramsey, J.D., Hanson, S.J., Hanson, C., Halchenko, Y.O., Poldrack, R.A., & Glymour, C. (2010). Six problems for causal inference from fMRI. *NeuroImage*, 49, 1545–1558.

Rawlings, J. (1988). *Applied Regression Analysis*. Belmont, CA: Wadsworth.

Richardson, T. S. (1996). A discovery algorithm for directed cyclic graphs. *Proceedings of the Twelfth Conference on Uncertainty in Artificial Intelligence*, Portland, OR., 454-462.

Scheines, R. (2000). Estimating Latent Causal Influences: TETRAD III Variable Selection and Bayesian Parameter Estimation: the effect of Lead on IQ. In P. Hayes (Ed.), *Handbook of Data Mining*. Oxford University Press.

Schwarz, G. E. (1978). Estimating the dimension of a model. *Annals of Statistics*, 6(2), 461-464.

Sewell, W. H. & Shah, V. P. (1968). Social Class, Parental Encouragement, and Educational Aspirations. *Am J Sociol*, 73(5), 559-572.

Shimizu, S., Hoyer, P. O., & Hyvärinen, A. (2009). Estimation of linear non-Gaussian acyclic models for latent factors. *Neurocomputing*, 72(7-9), 2024-2027.

Shimizu, S., Hoyer, P. O., Hyvärinen, A., & Kerminen, A. (2006). A Linear Non-Gaussian Acyclic Model for Causal Discovery. *Journal of Machine Learning Research*, 7, 2003-2030.

Shpitser, I. & Pearl, J. (2008). Complete Identification Methods for the Causal Hierarchy. *Journal of Machine Learning Research*, 9, 1941-1979.

Silva, R. & Scheines, R. (2004). Generalized Measurement Models. *reports-archive.adm.cs.cmu.edu*.

Silva, R., Scheines, R., Glymour, C., & Spirtes, P. (2006). Learning the structure of linear latent variable models. *J Mach Learn Res*, 7, 191-246.

Spearman, C. (1904). General Intelligence objectively determined and measured. *American Journal of Psychology*, 15, 201-293.

Spirtes, P. (1995). Directed cyclic graphical representations of feedback models. *Proceedings of the Eleventh Conference on Uncertainty in Artificial Intelligence*, Montreal, Canada, 491-499.

Spirtes, P. & Glymour, C. (1991). An Algorithm for Fast Recovery of Sparse Causal Graphs. *Social Science Computer Review*, 9(1), 67-72.

Spirtes, P., Glymour, C., & Scheines, R. (1993). *Causation, Prediction, and Search*. Spring-Verlag Lectures in Statistics.

Spirtes, P., Glymour, C., & Scheines, R. (2001). *Causation, Prediction, and Search, Second Edition (Adaptive Computation and Machine Learning)*. The MIT Press.

Spirtes, P., Scheines, R., Glymour, C., & Meek, C. (2004). Causal Inference. In D. Kaplan (Ed.), *SAGE Handbook of Quantitative Methodology*. (pp. 447-477). SAGE Publications.

Strotz, R. H. & Wold, H. O. A. (1960). Recursive VS Nonrecursive Systems- An Attempt At Synthesis. *Econometrica*, 28(2), 417-427.

Sun, X. (2008). *Causal inference from statistical data*. MPI for Biological Cybernetics.

Swanson, N. R. & Granger, C. W. J. (1997). Impulse Response Function Based on a Causal Approach to Residual Orthogonalization in Vector Autoregressions. *Journal of the American Statistical Association*, 92(437), 357-367.

Tillman, R. E., Danks, D., & Glymour, C. (2009). Integrating Locally Learned Causal Structures with Overlapping Variables. In *Advances in Neural Information Processing Systems 21*, 1665-1672.

Tillman, R. E., Gretton, A., & Spirtes, P. (2009). Nonlinear directed acyclic structure learning with weakly additive noise models. In *Advances in Neural Information Processing Systems 22*.

Timberlake, M. & Williams, K. R. (1984). Dependence, Political Exclusion And Government Repression - Some Cross-National Evidence. *Am Sociol Rev, 49*(1), 141-146.

Verma, T. S. & Pearl, J. (1990). Equivalence and Synthesis of Causal Models. In *Proceedings of the 6th Conference on Uncertainty in Artificial Intelligence*, 220-227.

Zhang, K. & Hyvarinen, A. (2009). On the Identifiability of the Post-Nonlinear Causal Model. *Proceedings of the 26th International Conference on Uncertainty in Artificial Intelligence*, 647-655.

# 29

# The Structural Model and the Ranking Theoretic Approach to Causation: A Comparison

WOLFGANG SPOHN

## 1 Introduction

Large parts of Judea Pearl's very rich work lie outside philosophy; moreover, basically being a computer scientist, his natural interest was in computational efficiency, which, as such, is not a philosophical virtue. Still, the philosophical impact of Judea Pearl's work is tremendous and often immediate; for the philosopher of science and the formal epistemologist few writings are as relevant as his. Fully deservedly, this fact is reflected in some philosophical contributions to this Festschrift; I am glad I can contribute as well.

For decades, Judea Pearl and I were pondering some of the same topics. We both realized the importance of the Bayesian net structure and elaborated on it; his emphasis on the graphical part was crucial, though. We both saw the huge potential of this structure for causal theorizing, in particular for probabilistic causation. We both felt the need for underpinning the probabilistic account by a theory of deterministic causation; this is, after all, the primary notion. And we both came up with relevant proposals. Judea Pearl approached these topics from the Artificial Intelligence side, I from the philosophy side. Given our different proveniences, overlap and congruity are surprisingly large.

Nevertheless, it slowly dawned upon me that the glaring similarities are deceptive, and that we fill the same structure with quite different contents. It is odd how much divergence can hide underneath so much similarity. I have identified no less than fifteen different, though interrelated points of divergence, and, to be clear, I am referring here only to our accounts of deterministic causation, the structural model approach so richly developed by Judea Pearl and my (certainly idiosyncratic) ranking-theoretic approach. In this brief paper I just want to list the points of divergence in a more or less descriptive mood, without much argument. Still, the paper may serve as a succinct reference list of the many crucial points that are at issue when dealing with causation and may thus help future discussion.

At bottom, my comparison refers, on the one hand, to the momentous book of Pearl (2000), the origins of which reach back to the other momentous book of Pearl (1988) and many important papers in the 80's and 90's,s and, on the other hand, to the chapters 14 and 15 of Spohn (forthcoming) on causation, the origins of which reach back to Spohn (1978, sections 3.2 - 3, and 1983) and a bunch of subsequent papers. For ease of access, though, I shall substantially refer to Halpern, Pearl (2005) and Spohn (2006) where the relevant accounts are presented in a more compact way. Let me start with reproducing the basic explications in section 2 and then proceed to my list of points of comparison in section 3. Section 4 concludes with a brief moral.

## 2 The Accounts to be Compared

For all those taking philosophical talk of events not too seriously (the vast majority among causal theorists) the starting point is a *frame*, a (non-empty, finite) set **U** of *variables*; $X, Y, Z, W$, etc. denote members of **U**, $\vec{X}, \vec{Y}, \vec{Z}, \vec{W}$, etc. subsets of **U**. Each variable $X \in U$ has a *range* $\Omega_X$ of values and is a function from some *possibility space* $\Omega$ into its range $\Omega_X$. For simplicity, we may assume that $\Omega$ is the Cartesian product of all $\Omega_X$ and $X$ the projection from $\Omega$ to $\Omega_X$. For $x \in \Omega_X$ and $A \subseteq \Omega_X$, $\{X = x\} = \{\omega \in \Omega \mid X(\omega) = x\}$ and $\{X \in A\} = \{\omega \mid X(\omega) \in A)$ are *propositions* (or events), and all those propositions generate a propositional algebra **A** over $\Omega$. For $\vec{X} = \{X_1, ..., X_n\}$ and $\vec{x} = \langle x_1, ..., x_n \rangle$ $\{\vec{X} = \vec{x}\}$ is short for $\{X_1 = x_1$ and ... and $X_n = x_n\}$. How a variable is to be precisely understood may be exemplified in the usual ways; however, we shall see that it is one of the issues still to be discussed.

The causal theorist may or may not presuppose a temporal order among variables; I shall. So, let $\prec$ be a linear order on the frame **U** representing temporal precedence. Linearity excludes simultaneous variables. The issue of simultaneous causation is pressing, but not one dividing us; therefore I put it to one side. Let, e.g., $\{\prec Y\}$ denote $\{Z \in U \mid Z \prec Y\}$, that is, the set of variables preceding $Y$. So much for the algebraic groundwork.

A further ingredient is needed in order to explain causal relations. In the structural-model approach it is a set of structural equations, in the ranking-theoretic approach it is a ranking function.

A set **F** of *structural equations* is just a set of functions $F_Y$ that specifies for each variable $Y$ in some subset $\vec{V}$ of **U** how $Y$ (essentially) functionally depends on some subset $\vec{X}$ of **U**; thus $F_Y$ maps $\Omega_{\vec{X}}$ into $\Omega_Y$. $\vec{V}$ is the set of *endogenous* variables, $\vec{U} = \mathbf{U} - \vec{V}$ the set of *exogenous* variables. The only condition on **F** is that no $Y$ in $\vec{V}$ indirectly functionally depends on itself via the equations in **F**. Thus, **F** induces a DAG on **U** such that, if $F_Y$ maps $\Omega_{\vec{X}}$ into $\Omega_Y$, $\vec{X}$ is the set of parents of $Y$. (In their appendix A.4 Halpern, Pearl (2005) generalize their account by dropping the assumption of the acyclicity of the structural equations.) The idea is that **F** provides a set of laws that govern the variables in **U**, though, again, the precise interpretation of **F** will have to be discussed below. $\langle \mathbf{U}, \mathbf{F} \rangle$ is then called a *structural model* (*SM*). Note that a SM does not fix the values of any variables. However, once we fix the values $\vec{u}$ of all the exogenous variables in $\vec{U}$, the equations in **F** determine the values $\vec{v}$ of all the endogenous variables in $\vec{V}$. Let us call $\langle \mathbf{U}, \mathbf{F}, \vec{u} \rangle$ a *contextualized structural model* (*CSM*). Thus, each CSM determines a specific world or course of events $\omega = \langle \vec{u}, \vec{v} \rangle$ in $\Omega$. Accordingly, each proposition $A$ in **A** is *true* or *false* in a CSM $\langle \mathbf{U}, \mathbf{F}, \vec{u} \rangle$, depending on whether or not $\omega \in A$ for the $\omega$ thereby determined.

## The Structural Model and the Ranking Theoretic Approach to Causation

For the structural model approach, causation is essentially related to intervention. Therefore we must first explain the latter notion. An *intervention* always intervenes on a CSM $\langle \mathbf{U}, \mathbf{F}, \vec{u} \rangle$, more specifically, on a certain set $\vec{X} \subseteq \vec{V}$ of endogenous variables, thereby setting the values of $\vec{X}$ to some fixed values $\vec{x}$; that intervention or setting is denoted by $\vec{X} \leftarrow \vec{x}$. What such an intervention $\vec{X} \leftarrow \vec{x}$ does is to turn the CSM $\langle \mathbf{U}, \mathbf{F}, \vec{u} \rangle$ into another CSM. The variables in $\vec{X}$ are turned into exogenous variables; i.e., the set $\mathbf{F}$ of structural equations is reduced to the set $\mathbf{F}^{\vec{X}}$, as I denote it, that consists of all the equations in $\mathbf{F}$ for the variables in $\vec{V} - \vec{X}$. Correspondingly, the context $\vec{u}$ of the original CSM is enriched by the chosen setting $\vec{x}$ for the new exogenous variables in $\vec{X}$. In short, the intervention $\vec{X} \leftarrow \vec{x}$ changes the CSM $\langle \mathbf{U}, \mathbf{F}, \vec{u} \rangle$ into the CSM $\langle \mathbf{U}, \mathbf{F}^{\vec{X}}, \langle \vec{u}, \vec{x} \rangle \rangle$. Again, it will be an issue what this precisely means.

Now, we can proceed to Pearl's explication of actual causation; this is definition 3.1 of Halpern, Pearl (2005, p. 853) slightly adapted to the notation introduced so far (see also Halpern, Hitchcock (2010, Section 3)). Not every detail will be relevant to my further discussion below; I reproduce it here only for reasons of accuracy:

SM DEFINITION: $\{\vec{X} = \vec{x}\}$ is an *actual cause* of $\{Y = y\}$ in the CSM $\langle \mathbf{U}, \mathbf{F}, \vec{u} \rangle$ iff the following three conditions hold:

(1) $\{\vec{X} = \vec{x}\}$ and $\{Y = y\}$ are true in $\langle \mathbf{U}, \mathbf{F}, \vec{u} \rangle$.

(2) There exists a partition $\langle Z, W \rangle$ of $\vec{V}$ with $\vec{X} \subseteq \vec{Z}$ and some setting $\langle \vec{x}', \vec{w}' \rangle$ of the variables in $\vec{X}$ and $\vec{W}$ such that if $\{Z = \vec{z}\}$ is true in $\langle \mathbf{U}, \mathbf{F}, \vec{u} \rangle$, then both of the following conditions hold:

  (a) $\{Y = y\}$ is false in the intervention $\langle X, W \rangle \leftarrow \langle \vec{x}', \vec{w}' \rangle$ on $\langle \mathbf{U}, \mathbf{F}, \vec{u} \rangle$, i.e., in $\langle \mathbf{U}, \mathbf{F}^{\vec{X},\vec{W}}, \langle \vec{u}, \vec{x}', \vec{w}' \rangle \rangle$. In other words, changing $\langle X, W \rangle$ from $\langle \vec{x}, \vec{w} \rangle$ to $\langle \vec{x}', \vec{w}' \rangle$ changes $\{Y = y\}$ from true to false.

  (b) $\{Y = y\}$ is true in $\langle \mathbf{U}, \mathbf{F}^{\vec{X},\vec{W}',\vec{Z}'}, \langle \vec{u}, \vec{x}, \vec{w}', \vec{z}' \rangle \rangle$ for all subsets $\vec{W}'$ of $\vec{W}$ and all subsets $\vec{Z}'$ of $\vec{Z}$, where $\vec{z}'$ is the subsequence of $\vec{z}$ pertaining to $\vec{Z}'$.

(3) $\vec{X}$ is minimal; i.e., no subset of $\vec{X}$ satisfies conditions (1) and (2).

This is not as complicated as it may look. Condition (1) says that the cause and the effect actually occur in the relevant CSM $\langle \mathbf{U}, \mathbf{F}, \vec{u} \rangle$ and, indeed, had to occur given the structural equations in $\mathbf{F}$ and the context $\vec{u}$. Condition (2a) says that if the cause variables in $\vec{X}$ had been set differently, the effect $\{Y = y\}$ would not have occurred. It is indeed more liberal in allowing that also the variables in $\vec{W}$ outside $\vec{X}$ are set to different values, the reason being that the effect of $\vec{X}$ on $Y$ may be hidden, as it were, by the actual values of $\vec{W}$, and uncovered only by setting $\vec{W}$ to different values. However, this alone would be too liberal; perhaps the failure of the effect $\{Y = y\}$ to occur is due only to the change of $\vec{W}$ rather than that of $\vec{X}$. Condition (2b) counteracts this permissiveness, and ensures that basically the change in $\vec{X}$ alone brings about the change of *Y*. Condition (3), finally, is to guarantee that the cause $\{\vec{X} = \vec{x}\}$ does not contain irrelevant parts; for the change described in (2a) all the variables in $\vec{X}$ are required. Note that $\vec{X}$ is a set of variables so that $\{\vec{X} = \vec{x}\}$ should be called a *total cause* of $\{Y = y\}$; its parts $\{X_i = x_i\}$ for $X_i \in \vec{X}$ may then be called *contributory causes*.

The details of the SM definition are mainly motivated by an adequate treatment of various troubling examples much discussed in the literature. It would take us too far to go into all of them. I should also mention that the SM definition is only preliminary in Halpern, Pearl (2005); but again, the details of their more refined definition presented on p. 870 will not be relevant for the present discussion.

The basics of the ranking-theoretic account may be explained in an equally brief way: A *negative ranking function* $\kappa$ for $\Omega$ is just a function $\kappa$ from $\Omega$ into $\mathbf{N} \cup \{\infty\}$ such that $\kappa(\omega) = 0$ for at least one $\omega \in \Omega$. It is extended to propositions in $\mathbf{A}$ by defining $\kappa(A) = \min\{\kappa(\omega) \mid \omega \in A\}$ and $\kappa(\emptyset) = \infty$; and it is extended to conditional ranks by defining $\kappa(B \mid A) = \kappa(A \cap B) - \kappa(A)$ for $\kappa(A) \neq \infty$. Negative ranks express degrees of disbelief: $\kappa(A) > 0$ says that $A$ is disbelieved, so that $\kappa(\overline{A}) > 0$ expresses that $A$ is believed in $\kappa$; however, we may well have $\kappa(A) = \kappa(\overline{A}) = 0$. It is useful to have both belief and disbelief represented in one function. Hence, we define the *two-sided rank* $\tau(A) = \kappa(\overline{A}) - \kappa(A)$, so that $A$ is believed, disbelieved, or neither according to whether $\tau(A) > 0$, $< 0$, or $= 0$. Again, we have conditional two-sided ranks: $\tau(B \mid A) = \kappa(\overline{B} \mid A) - \kappa(B \mid A)$. The *positive relevance* of a proposition $A$ to a proposition $B$ is then defined by $\tau(B \mid A) > \tau(B \mid \overline{A})$, i.e., by the fact that $B$ is more firmly believed or less firmly disbelieved given $A$ than given $\overline{A}$; we might also say in this case that *A confirms* or *is a reason for B*. Similarly for negative relevance and irrelevance (= independence).

Like a set of structural equations, a ranking function $\kappa$ induces a DAG on the frame $\mathbf{U}$ conforming with the given temporal order $\prec$. The procedure is the same as with probabilities: we simply define the set of parents of a variable $Y$ as the unique minimal set $\vec{X} \subseteq \{\prec Y\}$ such that $Y$ is independent of $\{\prec Y\} - \vec{X}$ given $\vec{X}$ relative to $\kappa$, i.e., such that $Y$ is independent of all the other preceding variables given $\vec{X}$. If $\vec{X}$ is empty, $Y$ is exogenous; if $\vec{X} \neq \emptyset$, $Y$ is endogenous. The reading that $Y$ directly causally depends on its parents will be justified later on.

Now, for me, being a cause is just being a special kind of conditional reason, i.e., being a reason given the past. In order to express this, for a subset $\vec{X}$ of $\mathbf{U}$ and a course of events $\omega \in \Omega$ let $^{\omega}[\vec{X}]$ denote the proposition that the variables in $\vec{X}$ behave as they do in $\omega$. (So far, we could denote such a proposition by $\{\vec{X} = \vec{x}\}$, if $\vec{X}(\omega) = \vec{x}$, but we shall see in a moment that this notation is now impractical.) Then the basic definition of the ranking-theoretic account is this:

RT DEFINITION 1: For $A \subseteq W_X$ and $B \subseteq W_Y$ $\{X \in A\}$ is a *direct cause* of $\{Y \in B\}$ in $\omega \in \Omega$ relative to the ranking function $\kappa$ (or the associated $\tau$) iff
(a) $X \prec Y$,
(b) $X(\omega) \in A$ and $Y(\omega) \in B$, i.e., $\{X \in A\}$ and $\{Y \in B\}$ are facts in $\omega$,
(c) $\tau(\{Y \in B\} \mid \{X \in A\} \cap {}^{\omega}[\{\prec Y\} - \{X\}]) > \tau(\{Y \in B\} \mid \{X \in \overline{A}\} \cap {}^{\omega}[\{\prec Y\} - \{X\}])$; i.e., $\{X \in A\}$ is a reason for $\{Y \in B\}$ given the rest of the past of $Y$ as it is in $\omega$.

It is obvious that the SM and the RT definition deal more or less with the same explicandum; both are after actual causes, where actuality is represented either by the context $\vec{u}$ of a CSM $\langle \mathbf{U}, \mathbf{F}, \vec{u} \rangle$ in the SM definition or by the course of events $\omega$ in the RT definition. A noticeable difference is that in the RT definition the cause $\{X \in A\}$ refers only to a single variable $X$. Thus, the RT definition grasps what has been called a contributory cause, a total cause of $\{Y \in B\}$ then being something like the conjunction of its contributory causes. As mentioned, the SM definition proceeds the other way around.

Of course, the major differences lie in the explicantia; this will be discussed in the next section. A further noticeable difference in the definienda is that the RT definition 1 explains only direct causation; indeed, if $\{X \in A\}$ would be an indirect cause of $\{Y \in B\}$, we could not expect $\{X \in A\}$ to be positively relevant to $\{Y \in B\}$ conditional on the rest of the past of $Y$ in $\omega$, since that condition would not keep open the causal path from $X$ to $Y$, but fix it to its actual state in $\omega$. Hence, the RT definition 1 is restricted accordingly. As the required extension, I propose the following

RT DEFINITION 2: $\{X \in A\}$ is a (*direct or indirect*) *cause* of $\{Y \in B\}$ in $\omega \in \Omega$ relative to $\kappa$ (or $\tau$) iff there are $Z_i \in \mathbf{U}$ and $C_i \in \Omega_{Z_i}$ ($i = 1, ..., n \geq 2$) such that $X = Z_1$, $A = C_1$, $Y = Z_n$, $B = C_n$, and $\{Z_i \in C_i\}$ is a direct cause of $\{Z_{i+1} \in C_{i+1}\}$ in $\omega$ relative to $\kappa$ for all $i = 1, ..., n - 1$.

In other words, causation in $\omega$ is just the transitive closure of direct causation in $\omega$.

We may complete the ranking-theoretic account by explicating causal dependence between variables:

RT DEFINITION 3: $Y \in \mathbf{U}$ (*directly*) *causally depends* on $X \in \mathbf{U}$ relative $\kappa$ iff there are $A \subseteq W_X$, $B \subseteq W_Y$, and $\omega \in \Omega$ such that $\{X \in A\}$ is a (direct) cause of $\{Y \in B\}$ in $\omega$ relative to $\kappa$.

One consequence of RT definition 3 is that the set of parents of $Y$ in the DAG generated by $\kappa$ and $\prec$ consists precisely of all the variables on which $Y$ directly causally depends.

So much for the two accounts to be compared. There are all the differences that meet the eye. As we shall see, there are even more. Still, let me conclude this section by pointing out that there are also less differences than meet the eye. I have already mentioned that both accounts make use of the DAG structure of causal graphs. And when we supplement the probabilistic versions of the two accounts, they further converge. In the structural-model approach we would then replace the context $\vec{u}$ of a CSM $\langle \mathbf{U}, \mathbf{F}, \vec{u} \rangle$ by a probability distribution over the exogenous variables rendering them independent and extending via the structural equations to a distribution for the whole of $\mathbf{U}$, thus forming a *pseudo-indeterministic system*, as Spirtes et al. (1993, pp. 38f.) call it, and hence a Bayesian net in which the probabilities agree with the causal graph. In the ranking-theoretic approach, we would replace the ranking function by a probability measure for $\mathbf{U}$ (or over $\mathbf{A}$) that, together with the temporal order of the variables, would again induce a DAG or a causal graph so as to form a Bayesian net. In this way, the basic ingredient of both accounts would become the same: a probability measure; the remaining differences appear to be of a merely technical nature.

Indeed, as I see the recent history of the theory of causation, this large agreement initially dominated the picture of probabilistic causation. However, the need for underpinning the probabilistic by a deterministic account was obvious; after all, the longer history of the notion was an almost entirely deterministic one up to the recent counterfactual accounts following Lewis (1973). And so the surprising ramification sketched above came about, both branches of which well agree with their probabilistic origins. The ramification is revealing since it makes explicit dividing lines that were hard to discern within the probabilistic harmony. Indeed, the points of divergence between the structural-model and the ranking-theoretic approach to be discussed in the next section apply to their probabilistic sisters as well, a claim that is quite suggestive, though I shall not elaborate on it.

## 3  Fifteen Points of Comparison

All in all, I shall come up with fifteen clearly distinguishable, though multiply connected points of comparison. The theorist of causation must take a stance towards all of them, and even more; my list is pertinent to the present comparison and certainly not exhaustive. Let us go through the list point for point:

(1) The most obvious instances provoking comparison and divergence are provided by *examples*, about preemption and prevention, overdetermination and switches, etc. The literature abounds in cases challenging all theories of causation and examples designed for discriminating among them, a huge bulk still awaiting systematic classification (though I attempted one in my (1983, ch. 3) as far as possible at that time). A theory of causation must do well with these examples in order to be acceptable. No theory, though, will reach a perfect score, all the more as many examples are contested by themselves, and do not provide a clear-cut criterion of adequacy. And what a 'good score' would be cannot be but vague. Therefore, I shall not even open this unending field of comparison regarding the two theories at hand.

(2) The main reason why examples provide only a soft criterion is that it is ultimately left to *intuition* to judge whether an example has been adequately treated. There are strong intuitions and weak ones. They often agree and often diverge. And they are often hard to compromise. Indeed, intuitions play an indispensable and important role in assessing theories of causation; they seem to provide the ultimate unquestionable grounds for that assessment.

Still, I have become cautious about the role of intuitions. Quite often I felt that the intuitions authors claim to have are guided by their theory; their intuitions seem to be what their theory suggests they should be. Indeed, the more I dig into theories of causation and develop my own, the harder it is for me to introspectively discern whether or not I share certain intuitions independently of any theorizing. So, again, the appeal to intuitions must be handled with care, and I shall not engage into a comparison of the relevant theories on an intuitive level.

(3) Another large field of comparison is the *proximity to* and the *applicability in scientific practice*. No doubt, the SM account fares much better in this respect than the RT approach. Structural modeling is something many scientists really do, whereas ranking theory is unknown in the sciences and it may be hard to say why it should be known outside epistemology. The point applies to other accounts as well. The regularity theory of causation seems close to the sciences, since they seem to state laws and regularities, whereas counterfactual analyses seem remote, since counterfactual claims are not an official part of scientific theories, even though, unofficially, counterfactual talk is ubiquitous. And probabilistic theories maintain their scientific appearance by ecumenically hiding disputes about the interpretation of probability.

Again, the importance of this criterion is undeniable; the causal theorist is well advised to appreciate the great expertise of the sciences, in general and specifically concerning causation. Still, I tend to downplay this criterion, not only in order to keep the RT account as a running candidate. The point is rather that the issue of causation is of a kind for which the sciences are not so well prepared. The counterfactual analysis is a case in point. If it should be basically correct, then the counterfactual idiom can no longer be treated as a second-rate vernacular (to use Quine's term), as the sciences do, but must be squarely faced in a systematic way, as, e.g., Pearl (2000, ch. 7) does, but qua philosopher, not qua scientist. Probabilities are a similar case. Mathematicians and statisticians by far know best how to deal with them. However, when it comes to say what probabilities mean, they are not in a privileged position.

The point of these three remarks is to claim primacy for theoretical issues about causation as such. External considerations are relevant and helpful, but they cannot release us from the task of taking some stance or other towards these theoretical issues. So, let us turn to them.

(4) Both, the SM and the RT account, are based on a *frame* providing a *framework of variables* and appertaining *facts*. I am not sure, however, whether we interpret it in the same way. A (random) variable is a function from some state space into some range of values, usually the reals; this is mathematical standard. That a variable takes a certain value is a proposition, and if the value is the true one (in some model), the proposition is a fact (in that model); so much is clear. However, the notion of a variable is ambiguous, and it is so since its statistic origins. A variable may vary over a given population as its state space and take on a certain value for each item in the population. E.g., size varies among Germans and takes (presently) the value 6' 0" for me. This is what I call a *generic variable*. Or a variable may vary over a set of possibilities as its state space and take values accordingly. For example, *my* (present) size is a variable in this sense and actually takes the value 6' 0", though it takes other values in other possibilities; I might (presently) have a different size. I call this a *singular variable* representing the possibility range of a

given single case. For each German (and time), size is such a singular variable. The generic variable of size, then, is formed by the actual values of all these singular variables.

The above RT account exclusively speaks about singular variables and their realizations; generic variables simply are out of the picture. By contrast, the ambiguity seems to afflict the SM account. I am sure everybody is fully clear about the ambiguity, but this clarity seems insufficiently reflected in the terminology. For instance, the equations of a SM represent laws or ceteris paribus laws or invariances in Woodward's (2003) terms or statistical laws, if supplemented by statistical 'error' terms, and thus state relations between generic variables. It is contextualization by which the model gets applied to a given single case; then, the variables should rather be taken as singular ones; their taking certain values then are specific facts. There is, however, no terminological distinction of the two interpretations; somehow, the notion of a variable seems to be intended to play both roles. In probabilistic extensions we find the same ambiguity, since probabilities may be interpreted as statistical distributions over populations or as realization propensities of the single case.

(5) I would not belabor the point if it did not extend to the causal relations we try to capture. We have causation among facts, as analyzed in the SM definition and the RT definitions 1 - 2; they are bound to apply to the single case. And we have causal relations among variables, i.e., causal dependence (though often and in my view confusingly the term "cause" is used here as well), and we find here the same ambiguity. Causal dependence between generic variables is a matter of causal laws or of *general causation*. However, there is also causal dependence between singular variables, something rarely made explicit, and it is a matter of *singular causation* applying to the single case just as much as causation between facts. Since its inception the discussion of probabilistic causality was caught in this ambiguity between singular and general causation; and I am wondering whether we can still observe the aftermath of that situation.

In any case, structural equations are intended to capture causal order, and the order among generic variables thus given pertains to general causation. Derivatively these equations may be interpreted as stating causal dependencies also between singular variables. In the SM account, though, singular causation is explicitly treated only as pertaining to facts. By contrast, the RT definition 3 explicates only causal dependence between singular variables. The RT account is so far silent about general causation and can grasp it only by generalizing over the causal relations in the single case. These remarks are not just pedantry; I think it is important to observe these differences for an adequate comparison of the accounts.

(6) I see these differences related to the issue of the role of *time* in an analysis of causation. The point is simply that generic variables as such are not temporally ordered, since their arguments, the items to which they apply, may have varying temporal positions; usually, statistical data do not come temporally ordered. By contrast, singular variables are temporally ordered, since their variable realizability across possibilities is tied to a fixed time. As a consequence, the SM definition makes no explicit reference to time,

whereas the RT definitions make free use of that reference. While I think that this point has indeed disposed Judea Pearl and me to our diverging perspectives on the relation between time and causation, it must be granted that the issue takes on much larger dimensions that open enough room for indecisive defenses of both perspectives.

Many points are involved: (i) Issues of analytic adequacy: while Pearl (2000, pp. 249ff.) argues that reference to time does not sufficiently further the analytic project and proposes ingenious alternatives (sections 2.3 - 4 + 8 - 9), I am much more optimistic about the analytic prospects of referring to time (see my 1990, section 3, and forthcoming, section 14.4). (ii) Issues of analytic policy (see also point 10 below): Is it legitimate to refer to time in an analysis of causation? I was never convinced by the objections. Or should the two notions be analytically decoupled? Or should the analytic order be even reversed by constructing a causal theory of time? Pearl (2000, section 2.8) shows sympathies for the latter project, although he suggests an evolutionary explanation, rather than Reichenbach's (1956) physical explanation for relating temporal direction with causal directionality. (iii) The issue of causal asymmetry: Is the explanation of causal asymmetry by temporal asymmetry illegitimate? Or incomplete? Or too uninformative, as far as it goes? If any of these, what is the alternative?

(7) Causation always is causation within given *circumstances*. What do the accounts say what the circumstances are? The RT definition 1 explicitly takes the entire past of the effect except the cause as the circumstances of a direct causal relationship, something apparently much too large and hence inadequate, but free of conceptual circularity, as I have continuously emphasized. In contrast, Pearl (2000, pp. 250ff.) endorses the circular explanation of Cartwright (1979) that those circumstances consist of the other causes of the effect and hence, in the case of direct causation, of the realizations of the other parents of the effect variable in the causal graph. Pearl thus accepts also Cartwright's conclusion that the reference to the obtaining circumstances does not help explicating causation; he thinks that this reference at best provides a kind of consistency test. I argue that the explicatory project is not doomed thereby, since Cartwright's circular explanation may be derived from my apparently inadequate definition (cf. Spohn 1990, section 4). As for the circumstances of indirect causation, the RT definition 2 is entirely silent, since it relies on transitivity; however, in Spohn (1990, Theorems 14 and 16) I explored how much I can say about them. In contrast, the SM definition contains an implicit account of the circumstances that applies to indirect causal relationships as well; it is hidden in the partition $\langle Z, W \rangle$ of the set $\vec{V}$ of endogenous variables. However, it still accepts Cartwright's circular explanation, since it presupposes the causal graph generated by the structural equations. So, this is a further respect in which our accounts are diametrically opposed.

(8) The preceding point contains two further issues. One concerns the distinction of *direct and indirect causation*. The SM approach explicates causation without attending to this distinction. Of course, it could account for it, but it does not acquire a basic importance. By contrast, the distinction receives analytic significance within the RT approach

that first defines direct causation and then, only on that basis, indirect causation. The reason is that, in this way, the RT approach hopes to reach a non-circular explication of causation, whereas the SM approach has given up on this hope (see also point 10 below) and thus sees no analytic rewards in this distinction.

(9) The other issue already alluded to in (7) is the issue of transitivity. This is a most vexed topic, and the community seems unable to find a stable attitude. Transitivity had to be given up, it seemed, within probabilistic causation (cf. Suppes 1970, p. 58), while it was derivable from a regularity account and was still defended by Lewis (1973) for deterministic causation. In the meantime the situation has reversed; transitivity has become more respectable within the probabilistic camp; e.g., Spirtes et al. (1993, p. 44) simply assume it in their definition of "indirect cause". By contrast, more and more tend to reject it for deterministic causation (cf., e.g., McDermott 1995 and Hitchcock 2001).

This uncertainty is also reflected in the present comparison. Pearl (2000, p. 237) rejects transitivity of causal dependence among variables, but, as the argument shows, only in the sense of what Woodward (2003, p. 51) calls "total cause". Still, Woodward (2003, p. 59), in his concluding explication **M**, accepts the transitivity of causal dependence among variables in the sense of "contributory cause", and I have not found any indication in Pearl (2000) or Halpern, Pearl (2005) that they would reject Woodward's account of contributory causation. However, all of them deny the transitivity of actual causation between facts.

I see it just the other way around. The RT definition 2 stipulates the transitivity of causation (with arguments, though; cf. Spohn 1990, p. 138, and forthcoming, section 14.12), whereas the RT definition 3 entails the transitivity of causal dependence among variables in the contributory sense only under (mild) additional assumptions. Another diametrical opposition.

(10) A much grander issue is looming behind the previous points, the issue of *analytic policy*. The RT approach starts defining direct causation between singular facts, proceeds to indirect causation and then to causal dependence between singular variables, and finally only hopes to thereby grasp general causation as well. It thus claims to give a non-circular explication or a reductive analysis of causation. The SM approach proceeds in the opposite direction. It presupposes an account of general causation that is contained in the structural equations, transfers this to causal dependence between singular variables (I mentioned in points 4 and 5 that this step is not fully explicit), and finally arrives at actual causation between facts. The claim is thereby to give an illuminating analysis of causation, but not a reductive one.

Now, one may have an argument about conceptual order: which causal notions to explicate on the basis of which? I admit I am bewildered by the SM order. The deeper issue, though, or perhaps the deepest, is the feasibility of reductive analysis. Nobody doubts that it would be most welcome to have one; therefore the history of the topic is full of attempts at such an analysis. Perhaps, though, they are motivated by wishful thinking. How to decide? One way of assessing the issue is by inspecting the proposals. The proponents are certainly confident of their analyses, but their inspection revealed so many problems that doubts preponderate. However, this does not prove their failure. Also, one may advance principled arguments such as Cartwright's (1979) that one cannot avoid being entangled in conceptual circles. For such reasons, the majority, it seems, has acquiesced in non-reductive analysis; cf., e.g., Woodward (2003, pp. 104ff.) for an apology of non-reductivity or Glymour (2004) for a eulogy of the, as he calls it, Euclidean as opposed to the Socratic ideal.

Another way of assessing the issue is more philosophical. Are there any more basic features of reality to which causation may reduce? One may well say no, and thereby justify the rejection of reductive analysis. Or one may say yes. Laws may be such a more basic feature; this, however, threatens to result either in an inadequate regularity theory of causation or in an inability to say what laws are beyond regularities. Objective probabilities may be such a feature – if we only knew what they are. What else is there on offer? On the other hand, it is not so easy to simply accept causation as a basic phenomenon; after all, the point has deeply worried philosophers for centuries after Hume.

In any case, all these issue are involved in settling for a certain analytic policy. It will become clearer in the subsequent points why I nevertheless maintain the possibility of reductive analysis.

(11) The most conspicuous difference of the SM and the RT approach is a direct consequence of their different policies. The SM account bases its analysis on *structural models* or *equations*, whereas the RT account explicates causation in terms of *ranking functions*. These are entirely different things!

Prima facie, structural equations are easier to grasp. Despite its non-reductive procedure the SM approach incurs the obligation, though, to somehow explain how the structural equations can establish causal order among generic variables. They can do this, because Pearl (2000, pp. 157ff.) explicitly gives them an interventionistic interpretation that, in turn, is basically a counterfactual one, as is entirely clear to Pearl; most interventions are only counterfactual. Woodward (2003) repeatedly emphasizes the point that the interventionistic account clarifies the counterfactual approach by forcing a specific interpretation of the multiply ambiguous counterfactual idiom. Still, despite Woodward's (2003, pp. 121f.) claim to use counterfactuals only when they are clearly true of false, and despite Pearl's (2000, section 7.1) attempt to account for counterfactuals within structural models, the issue how counterfactuals acquire truth conditions remains a mystery in my view.

By contrast, it is quite bewildering to base an analysis of causation on ranking functions that are avowedly to be understood only as doxastic states, i.e., in a purely epistemological way. One of my reasons for doing so is that the closer inspection envisaged in (10) comes out, on the whole, more satisfactorily than for other accounts, that is, the overall score in dealing with examples is better. The other reason why I find ranking functions not so implausible a starting point lies in my profoundly Humean strategy in dealing with causation. There is no more basic feature of reality to which causation might reduce. The issue rather is how modal facts come into the world – where modal facts pertain to lawhood, causation, counterfactuals, probabilities, etc. We do not find 'musts' and 'cans' in the world as we find apples and pears; this was Hume's crucial challenge. And his answer was what is now called Hume's projectivism (cf. Blackburn 1993, in particular the essays in part I). Ranking functions are well suited for laying out this projectivist answer in detail. This fundamental difference between the SM and the RT approach further unfolds in the final four points.

(12) A basic idea in our notion of causation between facts is, very roughly, that the cause does something for its effect, contributes to it, makes it possible or necessary or more likely, in short: that the cause is somehow positively relevant to its effect. One fact could also be negatively relevant to another, in which case the second obtains despite the first. As for causal dependence between variables, it is only required that the one is relevant for the other. What are the notions of *relevance* and *positive relevance* provided by the SM and the RT approach?

Ranking theory has a rich notion of positive and negative relevance, analogous and equivalent in formal behavior to the probabilistic notions. Its relevance notion is much richer and, I find, more adequate to the needs of causal theorizing than those provided by the key terms of other approaches to deterministic causation: laws, counterfactuals, interventions, structural equations, or whatever. This fact grounds my optimism that the RT approach is, on the whole, better able to cope with all the examples and problem cases.

I just said that the relevance notion provided by the SM approach is poorer. What is it? Clause (2b) of the SM definition says, in a way, that the effect $\{Y = y\}$ had to occur given the cause $\{\vec{X} = \vec{x}\}$ occurs, and clause (2a) says that the effect might not have occurred if the cause does not occur and, indeed, would not have occurred if the cause variable(s) $\vec{X}$ would have been realized in a suitable alternative way. In traditional terms, we could say that the cause is a necessary and sufficient condition of the effect provided the circumstances – where the subtleties of the SM approach lie in the proviso; that's the SM positive relevance notion. So, roughly, in SM terms, the only 'action' a cause can do is making its effect necessary, whereas ranking theory allows many more 'actions'. This is what I mean by the SM approach being poorer. For instance, it is not clear how a fact could be negatively relevant to another fact in the SM approach, or how one fact could be positively and another negatively relevant to a third one. And so forth.

(13) Let's take a closer look at what "action" could mean in the previous paragraph. In the RT approach it means comparing ranks conditional on the cause $\{X \in A\}$ and on its negation $\{X \in \overline{A}\}$; the rank raising showing up in that comparison is what the cause 'does'. In the SM approach we do not conditionalize on the cause $\{\vec{X} = \vec{x}\}$ and some alternative $\{\vec{X} = \vec{x}'\}$; rather, in clauses (2a-b) of the SM definition we look at the consequences of the interventions $\vec{X} \leftarrow \vec{x}$ and $\vec{X} \leftarrow \vec{x}'$, i.e., by replacing the structural equation(s) for $\vec{X}$ by the stipulation $\vec{X} = \vec{x}$ or, respectively, $= \vec{x}'$. The received view by now is that *intervention* is quite different from *conditionalization* (cf., e.g., Goldszmidt, Pearl 1992, and Meek, Glymour 1994), the suggestion being that interven-

tion is what causal theorizing requires, and that all approaches relying on conditionalization such as the RT approach therefore are misguided (cf. also Pearl 2000, section 3.2).

The difference looks compelling: intervention is a real activity, whereas conditionalization is only a mental, suppositional activity. But once we grant that intervention is mostly counterfactual (i.e., also suppositional), the difference shrinks. Indeed, I tend to say that there never is a real intervention in a given single case; after a real intervention we deal with a different single case than before. Hence, I think the difference the received view assumes is spurious; rather, interventions may be construed in terms of conditionalization:

Of course, the intervention $\vec{X} \leftarrow \vec{x}$ differs from conditioning on $\{\vec{X} = \vec{x}\}$; in this, the received view is correct. However, the RT and other conditioning approaches do not simply conditionalize on the cause, but on much more. What the intervention $X_1 \leftarrow x_1$ on the single variable $X_1$ does is change the value of $X_1$ to $x_1$ while at the same time keeping fixed the values of all temporally preceding variables as they are in the given context, or, if only a causal graph and not temporal order is available, either of all ancestors of $X_1$ or of all non-descendants of $X_1$ (which comes to the same thing in structural models, and also in probabilistic terms given the common cause principle). Thus, the intervention is equivalent to conditioning on $\{X_1 = x_1\}$ *and* on the fixed values of those other variables.

Similarly for a double intervention $\langle X_1, X_2 \rangle \leftarrow \langle x_1, x_2 \rangle$. For assessing the behavior of the variables temporally between $X_1$ and $X_2$ (or being descendants of $X_1$, but not of $X_2$) under the double intervention, we have to look at the same conditionalization as in the single intervention $X_1 \leftarrow x_1$, whereas for the variables later than $X_2$ (or descending from both $X_1$ and $X_2$) we have to condition on $\{X_1 = x_1\}$, $\{X_2 = x_2\}$, the past of $X_1$ as it is in the given context, and on those intermediate variables taking the values as they are after the intervention $X_1 \leftarrow x_1$. And so forth for multiple interventions (that are so crucial for the SM approach).

Given this translation, this kind of difference between the SM and the RT approach vanishes, I think. Consider, e.g., the definition of direct causal dependence of Woodward (2003, p. 55): $Y$ directly causally depends on $X$ iff an intervention on $X$ can make a difference to $Y$, provided the values of all other variables in the given frame **U** are somehow fixed by intervention. Translate this as proposed, and you arrive at the conditionalization I use in the above RT definitions to characterize direct causation.

(14) The preceding argument has a gap that emerges when we attend to another topic that I find crucial, but nowhere thoroughly discussed: the *frame-relativity* of causation. Everybody agrees that the distinction between direct and indirect causation is frame-relative; of course, a direct causal relationship relative to a coarse-grained frame may turn indirect under refinements. What about causation itself, though? One may try some moderate antirealism, e.g., general thoughts to the effect that science only produces models of reality and never truly represents reality as it really is; then causation would be model-relative, too.

However, this is not what I have in mind. The point is quite specific: The RT definition 1 refers, in a way I had explained in point 7, to the obtaining circumstances, however only insofar as they are represented in the given frame **U**. This entails a genuine frame-relativity of causation as such; $\{X = x\}$ may be a (direct) cause of $\{Y = y\}$ within one frame, but not within another or more refined frame. As Halpern, Hitchock (2010, Section 4.1) argue, this phenomenon may also show up within the SM approach.

I do not think that this agrees with Pearl's intention in pursuing the SM account; an actual cause should not cease to be an actual cause simply by refining the frame. Perhaps, the intention was to arrive at a frame-independent notion of causation by assuming a frame-independent notion of intervention. My translation of the intervention $X_1 \leftarrow x_1$ into conditionalization referred to the past (or the ancestors or the non-descendants) of $X_1$ as far as they are represented in the given frame **U**, and thus reproduced only a frame-relative notion of intervention. However, the intention presumably is to refer to the entire past of $X_1$ absolutely, not leaving any hole for the supposition of $\{X_1 = x_1\}$ to backtrack. If so, there is another sharp difference between the SM and the RT approach with repercussions on the previous point.

Of course, I admit that our intuitive notion of causation is not frame-relative; we aim at an absolute notion. However, this aim bars us from having a reductive analysis of causation, since the analysis would have to refer then to the rest of the world, as it were, to many things outside the frame that are thus prevented from entering the analysis. In fact, any rigorous causal theorizing is thereby frustrated in my view. For, how can you theoretically deal with all those don't-know-what's? For this reason I always preferred to work with a fixed frame, to pretend that this frame is all there is, and then to say everything about causation that can be said within this frame. This procedure at least allows a reductive analysis of a frame-relative notion.

How, then, can we get rid of the frame-relativity? I propose, by ever more fine-graining and extending the frame, studying the frame-relative causal relations within all these well-defined frames, and finding out what remains stable across all these refinements; we may hope, then, that these stable features are preserved even in the maximally refined, universal frame (cf. Spohn forthcoming, section 14.9; for Halpern, Hitchcock (2010, Section 4.1) this stability is also crucial). I would not know how else to deal with the challenge posed by frame-relativity, and I suspect that considerable problems in causal theorizing result from not explicitly facing this challenge.

(15) The various points may be summarized in the final opposition: whether causation is to be *subjectivistically* or *objectivistically* conceived. Common sense, Judea Pearl, and many others are on the objectivistic side: "I now take causal relationships to be the fundamental building blocks both of physical reality and of human understanding of that reality" (Pearl 2000, pp. xiiif.). And insofar as structural equations are objective, the SM approach shares this objectivism. By contrast, frame-relativity is an element of subject-relativity; frames are chosen by us. And the use of only epistemically interpretable ranking functions involves a much deeper subjectivization of the topic of causation. (The issue of relevance, point 12, is related, by the way, since in my view only epistemic relevance is rich enough a concept.)

The motive of the subjectivistic RT approach was, I said, Hume's challenge. And the gain, I claimed, is the feasibility of a reductive analysis. Any objectivistic approach has to tell how else to cope with that challenge and how to make peace with non-reductivity. Still, we cannot simply acquiesce in subjectivism, since it flies in the face of everyone keeping some sense of reality. The general philosophical strategy to escape pure subjectivism has been aptly described by Blackburn (1993, part I) as Humean projectivism leading to so-called quasi-realism that is indistinguishable from 'real' realism.

This general strategy may be precisely explicated in the case of causation: I had indicated in the previous point how I propose to get rid of frame-relativity. And in Spohn (forthcoming, ch. 15) I develop an objectification theory for ranking functions, according to which some ranking functions, the objectifiable ones, may be said, to truly (or falsely) represent causal relations. No doubt, this objectification theory is disputable, but it shows that the subjectivistic starting point need not preclude us from objectivistic aims. Maybe, though, these aims are more convincingly served by approaching them in a more direct and realistic way, as the SM account does.

## 4 Conclusion

On none of the fifteen differences above could I seriously start discussion; obviously nothing below book length would do. Indeed, discussing these points was not my aim at all, let alone treating anyone conclusively (though, of course, I could not hide where my sympathies are). My first intention was simply to display the differences, not all of which are clearly seen in the literature; already the sheer number is surprising. And I expressed my second intention between point 3 and point 4: namely to show that there are many internal theoretical issues in the theory of causation. On all of them one must take and argue a stance, a most demanding requirement. My hunch is that those theoretical considerations will eventually override issues of exemplification and application. All the more important it is to take some stance; no less will do for reaching a considered judgment. Judea Pearl has paradigmatically shown how to do this. His brilliant theoretical developments have not closed, but tremendously advanced our understanding of all these issues pertaining to causation.

**Acknowledgment:** I am indebted to Joe Halpern for providing most useful comments and correcting my English.

## References

Blackburn, S. (1993). *Essays in Quasi-Realism*, Oxford: Oxford University Press.

Cartwright, N. (1979). Causal laws and effective strategies. *Noûs 13*, 419-437.

Glymour, C. (2004). Critical notice on: James Woodward, Making Things Happen, *British Journal for the Philosophy of Science 55*, 779-790.

Goldszmidt, M., and J. Pearl (1992). Rank-based systems: A simple approach to belief revision, belief update, and reasoning about evidence and actions. In B. Nebel,

C. Rich, and W. Swartout (Eds.), *Proceedings of the Third International Conference on Knowledge Representation and Reasoning*, San Mateo, CA: Morgan Kaufmann, pp. 661-672.

Halpern, J. Y., and C. Hitchcock (2010). Actual causation and the art of modeling. This volume, chapter 22.

Halpern, J. Y., and J. Pearl (2005). Causes and explanations: A structural-model approach. Part I: Causes. *British Journal for the Philosophy of Science 56*, 843-887.

Hitchcock, C. (2001). The intransitivity of causation revealed in equations and graphs. *Journal of Philosophy 98*, 273-299.

Lewis, D. (1973). Causation. *Journal of Philosophy 70*, 556-567.

McDermott, M. (1995). Redundant causation. *British Journal for the Philosophy of Science 46*, 523-544.

Meek, C., and C. Glymour (1994). Conditioning and intervening. *British Journal for the Philosophy of Science 45*, 1001-1021.

Reichenbach, H. (1956). *The Direction of Time*. Los Angeles: The University of California Press.

Pearl, J. (1988). *Probabilistic Reasoning in Intelligent Systems: Networks of Plausible Inference*. San Mateo, CA: Morgan Kaufmann.

Pearl, J. (2000). *Causality. Models, Reasoning, and Inference*. Cambridge: Cambridge University Press.

Spirtes, P., C. Glymour, and R. Scheines (1993). *Causation, Prediction, and Search*. Berlin: Springer, 2nd ed. 2000.

Spohn, W. (1978). *Grundlagen der Entscheidungstheorie*, Kronberg/Ts.: Scriptor. Out of print, pdf-version at: http://www.uni-konstanz.de/FuF/Philo/Philosophie/philosophie/files/ge.buch.gesamt.pdf.

Spohn, W. (1983). *Eine Theorie der Kausalität*. Unpublished Habilitationsschrift, University of München, pdf-version at: http://www.uni-konstanz.de/FuF/Philo/Philosophie/philosophie/files/habilitation.pdf.

Spohn, W. (1990). Direct and indirect causes. *Topoi 9*, 125-145.

Spohn, W. (2006). Causation: An alternative. *British Journal for the Philosophy of Science 57*, 93-119.

Spohn, W. (forthcoming). *Ranking Theory. A Tool for Epistemology*.

Suppes, P. (1970). *A Probabilistic Theory of Causality*. Amsterdam: North-Holland.

Woodward, J. (2003). *Making Things Happen. A Theory of Causal Explanation*. Oxford: Oxford University Press.

# 30
# On Identifying Causal Effects
JIN TIAN AND ILYA SHPITSER

## 1 Introduction

This paper deals with the problem of inferring cause-effect relationships from a combination of data and theoretical assumptions. This problem arises in diverse fields such as artificial intelligence, statistics, cognitive science, economics, and the health and social sciences. For example, investigators in the health sciences are often interested in the effects of treatments on diseases; policymakers are concerned with the effects of policy decisions; AI research is concerned with effects of actions in order to design intelligent agents that can make effective plans under uncertainty; and so on.

To estimate causal effects, scientists normally perform randomized experiments where a sample of units drawn from the population of interest is subjected to the specified manipulation directly. In many cases, however, such a direct approach is not possible due to expense or ethical considerations. Instead, investigators have to rely on observational studies to infer effects. A fundamental question in causal analysis is to determine when effects can be inferred from statistical information, encoded as a joint probability distribution, obtained under normal, intervention-free behavior. A key point here is that it is not possible to make causal conclusions from purely probabilistic premises – it is necessary to make causal assumptions. This is because without any assumptions it is possible to construct multiple "causal stories" which can disagree wildly on what effect a given intervention can have, but agree precisely on all observables. For instance, smoking may be highly correlated with lung cancer either because it causes lung cancer, or because people who are genetically predisposed to smoke may also have a gene responsible for a higher cancer incidence rate. In the latter case there will be no effect of smoking on cancer.

In this paper, we assume that the causal assumptions will be represented by directed acyclic causal graphs [Pearl, 2000; Spirtes *et al.*, 2001] in which arrows represent the potential existence of direct causal relationships between the corresponding variables and some variables are presumed to be unobserved. Our task will be to decide whether the qualitative causal assumptions represented in any given graph are sufficient for assessing the strength of causal effects from nonexperimental data.

This problem of identifying causal effects has received considerable attention in the statistics, epidemiology, and causal inference communities [Robins, 1986;

Robins, 1987; Pearl, 1993; Robins, 1997; Kuroki and Miyakawa, 1999; Glymour and Cooper, 1999; Pearl, 2000; Spirtes *et al.*, 2001]. In particular Judea Pearl and his colleagues have made major contributions in solving the problem. In his seminal paper Pearl (1995) established a *calculus of interventions* known as *do-calculus* – three inference rules by which probabilistic sentences involving interventions and observations can be transformed into other such sentences, thus providing a syntactic method of deriving claims about interventions. Later, *do*-calculus was shown to be complete for identifying causal effects, that is, every causal effects that can be identified can be derived using the three *do*-calculus rules [Shpitser and Pearl, 2006a; Huang and Valtorta, 2006b]. Pearl (1995) also established the popular "back-door" and "front-door" criteria – sufficient graphical conditions for ensuring identification of causal effects. Using *do*-calculus as a guide, Pearl and his collaborators developed a number of sufficient graphical criteria: a criterion for identifying causal effects between singletons that combines and expands the front-door and back-door criteria [Galles and Pearl, 1995], a condition for evaluating the effects of plans in the presence of unmeasured variables, each plan consisting of several concurrent or sequential actions [Pearl and Robins, 1995]. More recently, an approach based on c-component factorization has been developed in [Tian and Pearl, 2002a; Tian and Pearl, 2003] and complete algorithms for identifying causal effects have been established [Tian and Pearl, 2003; Shpitser and Pearl, 2006b; Huang and Valtorta, 2006a]. Finally, a general algorithm for identifying arbitrary counterfactuals has been developed in [Shpitser and Pearl, 2007], while the special case of effects of treatment on the treated has been considered in [Shpitser and Pearl, 2009].

In this paper, we summarize the state of the art in identification of causal effects. The rest of the paper is organized as follows. Section 2 introduces causal models and gives formal definition for the identifiability problem. Section 3 presents Pearl's *do*-calculus and a number of easy to use graphical criteria. Section 4 presents the results on identifying (unconditional) causal effects. Section 5 shows how to identify conditional causal effects. Section 6 considers identification of counterfactual quantities which arise when we consider effects of relative interventions. Section 7 concludes the paper.

## 2  Notation, Definitions, and Problem Formulation

In this section we review the graphical causal models framework and introduce the problem of identifying causal effects.

### 2.1  Causal Bayesian Networks and Interventions

The use of graphical models for encoding distributional and causal assumptions is now fairly standard [Heckerman and Shachter, 1995; Lauritzen, 2000; Pearl, 2000; Spirtes *et al.*, 2001]. A *causal Bayesian network* consists of a DAG $G$ over a set $V = \{V_1, \ldots, V_n\}$ of variables, called a *causal diagram*. The interpretation of such a graph has two components, probabilistic and causal. The probabilistic interpretation views $G$ as representing conditional independence assertions: Each variable is independent of all its non-descendants given its direct parents in the graph.[1] These assertions imply that the joint probability function $P(v) = P(v_1, \ldots, v_n)$ factorizes according to the product [Pearl, 1988]

$$P(v) = \prod_i P(v_i|pa_i), \tag{1}$$

where $pa_i$ are (values of) the parents of variable $V_i$ in the graph. Here we use uppercase letters to represent variables or sets of variables, and use corresponding lowercase letters to represent their values (instantiations).

The set of conditional independences implied by the causal Bayesian network can be obtained from the causal diagram $G$ according to the d-separation criterion [Pearl, 1988].

DEFINITION 1 (d-separation). *A path [2] $p$ is said to be blocked by a set of nodes $Z$ if and only if*

1. *$p$ contains a chain $V_i \rightarrow V_j \rightarrow V_k$ or a fork $V_i \leftarrow V_j \rightarrow V_k$ such that the node $V_j$ is in $Z$, or*

2. *$p$ contains an inverted fork $V_i \rightarrow V_j \leftarrow V_k$ such that $V_j$ is not in $Z$ and no descendant of $V_j$ is in $Z$.*

A path not blocked by $Z$ is called *d-connecting* or *active*. A set $Z$ is said to *d-separate* $X$ from $Y$, denoted by $(X \perp\!\!\!\perp Y | Z)_G$, if and only if $Z$ blocks every path from a node in $X$ to a node in $Y$.

We have that if $Z$ d-separates $X$ from $Y$ in the causal diagram $G$, then $X$ is conditionally independent of $Y$ given $Z$ in the distribution $P(v)$ given in Eq. (1).

The causal interpretation views the arrows in $G$ as representing causal influences between the corresponding variables. In this interpretation, the factorization of (1) still holds, but the factors are further assumed to represent autonomous data-generation processes, that is, each parents-child relationship characterized by a conditional probability $P(v_i|pa_i)$ represents a stochastic process by which the values of $V_i$ are assigned in response to the values $pa_i$ (previously chosen for $V_i$'s parents), and the stochastic variation of this assignment is assumed independent of the variations in all other assignments in the model. Moreover, each assignment process remains invariant to possible changes in the assignment processes that govern other variables in the system. This modularity assumption enables us to infer the effects of interventions, such as policy decisions and actions, whenever interventions are described as specific modifications of some factors in the product of (1). The simplest such intervention, called *atomic*, involves fixing a set $T$ of variables to some

---

[1] We use family relationships such as "parents," "children," and "ancestors" to describe the obvious graphical relationships.

[2] A path is a sequence of consecutive edges (of any directionality).

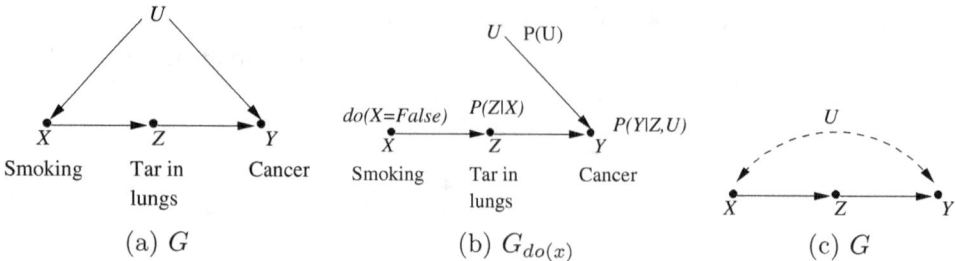

Figure 1. A causal diagram illustrating the effect of smoking on lung cancer

constants $T = t$ denoted by $do(T = t)$ or $do(t)$, which yields the post-intervention distribution[3]

$$P_t(v) = \begin{cases} \prod_{\{i | V_i \notin T\}} P(v_i | pa_i) & v \text{ consistent with } t. \\ 0 & v \text{ inconsistent with } t. \end{cases} \quad (2)$$

Eq. (2) represents a truncated factorization of (1), with factors corresponding to the manipulated variables removed. This truncation follows immediately from (1) since, assuming modularity, the post-intervention probabilities $P(v_i|pa_i)$ corresponding to variables in $T$ are either 1 or 0, while those corresponding to unmanipulated variables remain unaltered. If $T$ stands for a set of treatment variables and $Y$ for an outcome variable in $V \setminus T$, then Eq. (2) permits us to calculate the probability $P_t(y)$ that event $Y = y$ would occur if treatment condition $T = t$ were enforced uniformly over the population. This quantity, often called the "causal effect" of $T$ on $Y$, is what we normally assess in a controlled experiment with $T$ randomized, in which the distribution of $Y$ is estimated for each level $t$ of $T$.

As an example, consider the model shown in Figure 1(a) from [Pearl, 2000] that concerns the relation between smoking ($X$) and lung cancer ($Y$), mediated by the amount of tar ($Z$) deposited in a person's lungs. The model makes qualitative causal assumptions that the amount of tar deposited in the lungs depends on the level of smoking (and external factors) and that the production of lung cancer depends on the amount of tar in the lungs but smoking has no effect on lung cancer except as mediated through tar deposits. There might be (unobserved) factors (say some unknown carcinogenic genotype) that affect both smoking and lung cancer, but the genotype nevertheless has no effect on the amount of tar in the lungs except indirectly (through smoking). Quantitatively, the model induces the joint distribution factorized as

$$P(u, x, z, y) = P(u)P(x|u)P(z|x)P(y|z, u). \quad (3)$$

---

[3][Pearl, 1995; Pearl, 2000] used the notation $P(v|set(t))$, $P(v|do(t))$, or $P(v|\hat{t})$ for the post-intervention distribution, while [Lauritzen, 2000] used $P(v||t)$.

Assume that we could perform an ideal intervention on variable $X$ by banning smoking[4], then the effect of this action is given by

$$P_{X=False}(u,z,y) = P(u)P(z|X=False)P(y|z,u), \quad (4)$$

which is represented by the model in Figure 1(b).

## 2.2 The Identifiability Problem

We see that, whenever all variables in $V$ are observed, given the causal diagram $G$, all causal effects can be computed from the observed distribution $P(v)$ as given by Eq. (2). However, if some variables are not measured, or two or more variables in $V$ are affected by unobserved confounders, then the question of *identifiability* arises. The presence of such confounders would not permit the decomposition of the observed distribution $P(v)$ in (1). For example, in the model shown in Figure 1(a), assume that the variable $U$ (unknown genotype) is unobserved and we have collected a large amount of data summarized in the form of (an estimated) join distribution $P$ over the observed variables $(X, Y, Z)$. We wish to assess the causal effect $P_x(y)$ of smoking on lung cancer.

Let $V$ and $U$ stand for the sets of observed and unobserved variables, respectively. If each $U$ variable is a root node with exactly two observed children, then the corresponding model is called a *semi-Markovian* model. In this paper, we will present results on semi-Markovian models as they allow for simpler treatment. However the results are general as it has been shown that causal effects in a model with arbitrary sets of unobserved variables can be identified by first projecting the model into a semi-Markovian model [Tian and Pearl, 2002b; Huang and Valtorta, 2006a].

In a semi-Markovian model, the observed probability distribution $P(v)$ becomes a mixture of products:

$$P(v) = \sum_u \prod_i P(v_i|pa_i, u^i) P(u) \quad (5)$$

where $Pa_i$ and $U^i$ stand for the sets of the observed and unobserved parents of $V_i$ respectively, and the summation ranges over all the $U$ variables. The post-intervention distribution, likewise, will be given as a mixture of truncated products

$$P_t(v) = \begin{cases} \sum_u \prod_{\{i|V_i \notin T\}} P(v_i|pa_i, u^i)P(u) & v \text{ consistent with } t. \\ 0 & v \text{ inconsistent with } t. \end{cases} \quad (6)$$

And, the question of identifiability arises, i.e., whether it is possible to express some causal effect $P_t(s)$ as a function of the observed distribution $P(v)$, independent of the unknown quantities, $P(u)$ and $P(v_i|pa_i, u^i)$.

---
[4]Whether or not any actual action is an ideal manipulation of a variable (or is feasible at all) is not part of the theory - it is input to the theory.

It is convenient to represent a semi-Markovian model with a graph $G$ that does not show the elements of $U$ explicitly but, instead, represents the confounding effects of $U$ variables using (dashed) bidirected edges. A bidirected edge between nodes $V_i$ and $V_j$ represents the presence of unobserved confounders that may influence both $V_i$ and $V_j$. For example the model in Figure 1(a) will be represented by the graph in Figure 1(c).

In general we may be interested in identifying conditional causal effects $P_t(s|c)$, the causal effects of $T$ on $S$ conditioned on another set $C$ of variables. This problem is important for evaluating *conditional plans* and stochastic plans [Pearl and Robins, 1995], where action $T$ is taken to respond in a specified way to a set $C$ of other variables – say, through a functional relationship $t = g(c)$. The effects of such actions may be evaluated through identifying conditional causal effects in the form of $P_t(s|c)$ [Pearl, 2000, chapter 4].

DEFINITION 2 (Causal-Effect Identifiability). The causal effect of a set of variables $T$ on a disjoint set of variables $S$ conditioned on another set $C$ is said to be identifiable in a causal diagram $G$ if the quantity $P_t(s|c)$ can be computed uniquely from any positive probability $P(v)$ of the observed variables—that is, if $P_t^{M_1}(s|c) = P_t^{M_2}(s|c)$ for every pair of models $M_1$ and $M_2$ with $P^{M_1}(v) = P^{M_2}(v) > 0$.

## 3 *Do*-calculus and Graphical Criteria

In general the identifiability of causal effects can be decided using Pearl's *do*-calculus – a set of inference rules by which probabilistic sentences involving interventions and observations can be transformed into other such sentences. A finite sequence of syntactic transformations, each applying one of the inference rules, may reduce expressions of the type $P_t(s)$ to subscript-free expressions involving observed quantities.

Let $X$, $Y$, and $Z$ be arbitrary disjoint sets of nodes in $G$. We denote by $G_{\overline{X}}$ the graph obtained by deleting from $G$ all arrows pointing to nodes in $X$. We denote by $G_{\underline{X}}$ the graph obtained by deleting from $G$ all arrows emerging from nodes in $X$. Similarly, $G_{\overline{X}\underline{Z}}$ will represent the deletion of both incoming and outgoing arrows.

THEOREM 3 (Rules of *do*-Calculus). *[Pearl, 1995] For any disjoint sets of variables $X, Y, Z$, and $W$ we have the following rules.*

**Rule 1** (*Insertion/deletion of observations*) :

$$P_x(y|z,w) = P_x(y|w) \quad \text{if} \quad (Y \perp\!\!\!\perp Z | X, W)_{G_{\overline{X}}}. \tag{7}$$

**Rule 2** (*Action/observation exchange*) :

$$P_{x,z}(y|w) = P_x(y|z,w) \quad \text{if} \quad (Y \perp\!\!\!\perp Z | X, W)_{G_{\overline{X}\underline{Z}}}. \tag{8}$$

**Rule 3** (*Insertion/deletion of actions*) :

$$P_{x,z}(y|w) = P_x(y|w) \quad \text{if} \quad (Y \perp\!\!\!\perp Z | X, W)_{G_{\overline{X},\overline{Z(W)}}}, \tag{9}$$

where $Z(W)$ is the set of $Z$-nodes that are not ancestors of any $W$-node in $G_{\overline{X}}$.

A key result about do-calculus is that any interventional distribution that is identifiable can be expressed in terms of the observational distribution by means of applying a sequence of do-calculus rules.

THEOREM 4. *[Shpitser and Pearl, 2006a] Do-calculus is complete for identifying causal effects of the form* $P_x(y|z)$.

In practice, do-calculus may be difficult to apply manually in complex causal diagrams, since, as stated, the rules give little guidance for chaining them together into a valid derivation.

Fortunately, a number of graphical criteria have been developed for quickly judging the identifiability by looking at the causal diagram $G$, of which the most influential are Pearl's back-door and front-door criteria. A path from $X$ to $Y$ is called *back-door* (relative to $X$) if it starts with an arrow pointing at $X$.

DEFINITION 5 (Back-Door). A set of variables $Z$ satisfies the back-door criterion relative to an ordered pair of variables $(X_i, X_j)$ in a DAG $G$ if:

(i) no node in $Z$ is a descendant of $X_i$; and

(ii) $Z$ blocks every back-door path from $X_i$ to $X_j$.

Similarly, if $X$ and $Y$ are two disjoint sets of nodes in $G$, then $Z$ is said to satisfy the back-door criterion relative to $(X, Y)$ if it satisfies the criterion relative to any pair $(X_i, X_j)$ such that $X_i \in X$ and $X_j \in Y$.

THEOREM 6 (Back-Door Criterion). *[Pearl, 1995] If a set of variables $Z$ satisfies the back-door criterion relative to $(X, Y)$, then the causal effect of $X$ on $Y$ is identifiable and is given by the formula*

$$P_x(y) = \sum_z P(y|x,z)P(z). \qquad (10)$$

For example, in Figure 1(c) $X$ satisfies the back-door criterion relative to $(Z, Y)$ and we have

$$P_z(y) = \sum_x P(y|x,z)P(x). \qquad (11)$$

DEFINITION 7 (Front-Door). A set of variables $Z$ is said to satisfy the front-door criterion relative to an ordered pair of variables $(X, Y)$ if:

(i) $Z$ intercepts all directed paths from $X$ to $Y$;

(ii) all back-door paths from $X$ to $Z$ are blocked (by empty set); and

(iii) all back-door paths from $Z$ to $Y$ are blocked by $X$.

**THEOREM 8** (Front-Door Criterion). *[Pearl, 1995] If $Z$ satisfies the front-door criterion relative to an ordered pair of variables $(X, Y)$, then the causal effect of $X$ on $Y$ is identifiable and is given by the formula*

$$P_x(y) = \sum_z P(z|x) \sum_{x'} P(y|x', z) P(x'). \tag{12}$$

For example, in Figure 1(c) $Z$ satisfies the front-door criterion relative to $(X, Y)$ and the causal effect $P_x(y)$ is given by Eq. (12).

There is a simple yet powerful graphical criterion for identifying the causal effects of a singleton. For any set $S$, let $An(S)$ denote the union of $S$ and the set of ancestors of the variables in $S$. For any set $C$, let $G_C$ denote the subgraph of $G$ composed only of variables in $C$. Let a path composed entirely of bidirected edges be called a *bidirected path*.

**THEOREM 9.** *[Tian and Pearl, 2002a] The causal effect $P_x(s)$ of a variable $X$ on a set of variables $S$ is identifiable if there is no bidirected path connecting $X$ to any of its children in $G_{An(S)}$.*

In fact, for $X$ and $S$ being singletons, this criterion covers both back-door and front-door criteria, and also the criterion in [Galles and Pearl, 1995].

These criteria are simple to use but are not necessary for identification. In the next sections we present complete systematic procedures for identification.

## 4  Identification of Causal Effects

In this section, we present a systematic procedure for identifying causal effects using so-called c-component decomposition.

### 4.1  C-component decomposition

The set of variables $V$ in $G$ can be partitioned into disjoint groups by assigning two variables to the same group if and only if they are connected by a bidirected path. Assuming that $V$ is thus partitioned into $k$ groups $S_1, \ldots, S_k$, each set $S_j$ is called a *c-component* of $V$ in $G$ or a c-component of $G$. For example, the graph in Figure 1(c) consists of two c-components $\{X, Y\}$ and $\{Z\}$.

For any set $C \subseteq V$, define the quantity $Q[C](v)$ to denote the post-intervention distribution of $C$ under an intervention to all other variables:[5]

$$Q[C](v) = P_{v \setminus c}(c) = \sum_u \prod_{\{i | V_i \in C\}} P(v_i | pa_i, u^i) P(u). \tag{13}$$

In particular, we have $Q[V](v) = P(v)$. If there is no bidirected edges connected with a variable $V_i$, then $U^i = \emptyset$ and $Q[\{V_i\}](v) = P(v_i|pa_i)$. For convenience, we will often write $Q[C](v)$ as $Q[C]$.

The importance of the c-component steps from the following lemma.

---

[5] Set $Q[\emptyset](v) = 1$ since $\sum_u P(u) = 1$.

LEMMA 10 (C-component Decomposition). *[Tian and Pearl, 2002a]* Assuming that $V$ is partitioned into c-components $S_1, \ldots, S_k$, we have

(i) $P(v) = \prod_i Q[S_i]$.

(ii) Each $Q[S_i]$ is computable from $P(v)$. Let a topological order over $V$ be $V_1 < \ldots < V_n$, and let $V^{(i)} = \{V_1, \ldots, V_i\}$, $i = 1, \ldots, n$, and $V^{(0)} = \emptyset$. Then each $Q[S_j]$, $j = 1, \ldots, k$, is given by

$$Q[S_j] = \prod_{\{i | V_i \in S_j\}} P(v_i | v^{(i-1)}) \tag{14}$$

The lemma says that for each c-component $S_i$ the causal effect $Q[S_i] = P_{v \setminus s_i}(s_i)$ is identifiable. For example, in Figure 1(c), we have $P_{x,y}(z) = Q[\{Z\}] = P(z|x)$ and $P_z(x,y) = Q[\{X,Y\}] = P(y|x,z)P(x)$.

Lemma 10 can be generalized to the subgraphs of $G$ as given in the following lemma.

LEMMA 11 (Generalized C-component Decomposition). *[Tian and Pearl, 2003]* Let $H \subseteq V$, and assume that $H$ is partitioned into c-components $H_1, \ldots, H_l$ in the subgraph $G_H$. Then we have

(i) $Q[H]$ decomposes as

$$Q[H] = \prod_i Q[H_i]. \tag{15}$$

(ii) Each $Q[H_i]$ is computable from $Q[H]$. Let $k$ be the number of variables in $H$, and let a topological order of the variables in $H$ be $V_{m_1} < \cdots < V_{m_k}$ in $G_H$. Let $H^{(i)} = \{V_{m_1}, \ldots, V_{m_i}\}$ be the set of variables in $H$ ordered before $V_{m_i}$ (including $V_{m_i}$), $i = 1, \ldots, k$, and $H^{(0)} = \emptyset$. Then each $Q[H_j]$, $j = 1, \ldots, l$, is given by

$$Q[H_j] = \prod_{\{i | V_{m_i} \in H_j\}} \frac{Q[H^{(i)}]}{Q[H^{(i-1)}]}, \tag{16}$$

where each $Q[H^{(i)}]$, $i = 1, \ldots, k$, is given by

$$Q[H^{(i)}] = \sum_{h \setminus h^{(i)}} Q[H]. \tag{17}$$

Lemma 11 says that if the causal effect $Q[H] = P_{v \setminus h}(h)$ is identifiable, then for each c-component $H_i$ of the subgraph $G_H$, the causal effect $Q[H_i] = P_{v \setminus h_i}(h_i)$ is identifiable.

Next, we show how to use the c-component decomposition to identify causal effects.

## 4.2 Computing causal effects

First we present a facility lemma. For $W \subseteq C \subseteq V$, the following lemma gives a condition under which $Q[W]$ can be computed from $Q[C]$ by summing over $C \setminus W$, like ordinary marginalization in probability theory.

LEMMA 12. *[Tian and Pearl, 2003]* Let $W \subseteq C \subseteq V$, and $W' = C \setminus W$. If $W$ contains its own ancestors in the subgraph $G_C$ $(An(W)_{G_C} = W)$, then

$$\sum_{w'} Q[C] = Q[W]. \tag{18}$$

Note that we always have $\sum_c Q[C] = 1$.

Next, we show how to use Lemmas 10–12 to identify the causal effect $P_t(s)$ where $S$ and $T$ are arbitrary (disjoint) subsets of $V$. We have

$$P_t(s) = \sum_{(v \setminus t) \setminus s} P_t(v \setminus t) = \sum_{(v \setminus t) \setminus s} Q[V \setminus T]. \tag{19}$$

Let $D = An(S)_{G_{V \setminus T}}$. Then by Lemma 12, variables in $(V \setminus T) \setminus D$ can be summed out:

$$P_t(s) = \sum_{d \setminus s} \sum_{(v \setminus t) \setminus d} Q[V \setminus T] = \sum_{d \setminus s} Q[D]. \tag{20}$$

Assume that the subgraph $G_D$ is partitioned into c-components $D_1, \ldots, D_l$. Then by Lemma 11, $Q[D]$ can be decomposed into products of $Q[D_i]$'s, and Eq. (20) can be rewritten as

$$P_t(s) = \sum_{d \setminus s} \prod_i Q[D_i]. \tag{21}$$

We obtain that $P_t(s)$ is identifiable if all $Q[D_i]$'s are identifiable.

Let $G$ be partitioned into c-components $S_1, \ldots, S_k$. Then any $D_i$ is a subset of certain $S_j$ since if the variables in $D_i$ are connected by a bidirected path in a subgraph of $G$ then they must be connected by a bidirected path in $G$. Assuming $D_i \subseteq S_j$, $Q[D_i]$ is identifiable if it is computable from $Q[S_j]$. In general, for $C \subseteq T \subseteq V$, whether $Q[C]$ is computable from $Q[T]$ can be determined recursively by repeated applications of Lemmas 12 and 11, as given in the recursive algorithm shown in Figure 2. At each step of the algorithm, we either find an expression for $Q[C]$, find $Q[C]$ unidentifiable, or reduce the problem to a simpler one.

In summary, an algorithm for computing $P_t(s)$ is given in Figure 3, and the algorithm has been shown to be complete, that is, if the algorithm outputs FAIL, then $P_t(s)$ is not identifiable.

THEOREM 13. *[Shpitser and Pearl, 2006b; Huang and Valtorta, 2006a] The algorithm* **ID** *in Figure 3 is complete.*

## 5 Identification of Conditional Causal Effects

An important refinement to the problem of identifying causal effects $P_x(y)$ is concerned with identifying *conditional causal effects*, in other words causal effects in a particular subpopulation where variables $Z$ are known to attain values $z$. These

**Algorithm Identify**$(C, T, Q)$
INPUT: $C \subseteq T \subseteq V$, $Q = Q[T]$. $G_T$ and $G_C$ are both composed of one single c-component.
OUTPUT: Expression for $Q[C]$ in terms of $Q$ or FAIL.

Let $A = An(C)_{G_T}$.

- IF $A = C$, output $Q[C] = \sum_{t \backslash c} Q$.

- IF $A = T$, output FAIL.

- IF $C \subset A \subset T$

    1. Assume that in $G_A$, $C$ is contained in a c-component $T'$.
    2. Compute $Q[T']$ from $Q[A] = \sum_{t \backslash a} Q$ by Lemma 11.
    3. Output Identify$(C, T', Q[T'])$.

Figure 2. An algorithm for determining if $Q[C]$ is computable from $Q[T]$.

**Algorithm ID**$(s, t)$
INPUT: two disjoint sets $S, T \subset V$.
OUTPUT: the expression for $P_t(s)$ or FAIL.
Phase-1:

1. Find the c-components of $G$: $S_1, \ldots, S_k$. Compute each $Q[S_i]$ by Lemma 10.

2. Let $D = An(S)_{G_{V \backslash T}}$ and the c-components of $G_D$ be $D_i$, $i = 1, \ldots, l$.

Phase-2:
For each set $D_i$ such that $D_i \subseteq S_j$:
  Compute $Q[D_i]$ from $Q[S_j]$ by calling **Identify**$(D_i, S_j, Q[S_j])$ in Figure 2. If the function returns FAIL, then stop and output FAIL.

Phase-3: Output $P_t(s) = \sum_{d \backslash s} \prod_i Q[D_i]$.

Figure 3. A complete algorithm for computing $P_t(s)$.

conditional causal effects are written as $P_x(y|z)$, and defined just as regular conditional distributions as

$$P_x(y|z) = \frac{P_x(y,z)}{P_x(z)}$$

Complete closed form algorithms for identifying effects of this type have been developed. One approach [Tian, 2004] generalizes the algorithm for identifying *unconditional* causal effects $P_x(y)$ found in Section 4. There is, however, an easier approach which works.

The idea is to reduce the expression $P_x(y|z)$, which we don't know how to handle to something like $P_{x'}(y')$, which we do know how to handle via the algorithm already presented. This reduction would have to find a way to get rid of variables $Z$ in the conditional effect expression.

Ridding ourselves of some variables in $Z$ can be accomplished via rule 2 of do-calculus. Recall that applying rule 2 to an expression allows us to replace conditioning on some variable set $W \subseteq Z$ by fixing $W$ instead. Rule 2 states that this is possible in the expression $P_x(y|z)$ whenever $W$ contains no back-door paths to $Y$ conditioned on the remaining variables in $Z$ and $X$ (that is $X \cup Z \setminus W$), in the graph where all incoming arrows to $X$ have been cut.

It's not difficult to show the following uniqueness lemma.

LEMMA 14. *[Shpitser and Pearl, 2006a] For every conditional effect $P_x(y|z)$ there exists a unique maximal $W \subseteq Z$ such that $P_x(y|z)$ is equal to $P_{x,w}(y|z \setminus w)$ according to rule 2 of do-calculus.*

Lemma 14 states that we only need to apply rule 2 once to rid ourselves of as many conditioned variables as possible in the effect of interest. However, even after this is done, we may be left with some variables in $Z \setminus W$ past the conditioning bar in our effect expression. If we insist on using unconditional effect identification, we may try to identify the joint distribution $P_{x,w}(y, z \setminus w)$ to obtain an expression $\alpha$, and obtain the conditional distribution $P_{x,w}(y|z \setminus w)$ by taking $\frac{\alpha}{\sum_y \alpha}$. But what if $P_{x,w}(y, z \setminus w)$ is not identifiable? Are there cases where $P_{x,w}(y, z \setminus w)$ is not identifiable, but $P_{x,w}(y|z \setminus w)$ is? Fortunately, it turns out the answer is no.

LEMMA 15. *[Shpitser and Pearl, 2006a] Let $P_x(y|z)$ be a conditional effect of interest, and $W \subseteq Z$ the unique maximal set such that $P_x(y|z)$ is equal to $P_{x,w}(y|z \setminus w)$. Then $P_x(y|z)$ is identifiable if and only if $P_{x,w}(y, z \setminus w)$ is identifiable.*

Lemma 15 gives us a simple algorithm for identifying arbitrary conditional effects by first reducing the problem into one of identifying an unconditional effect – and then invoking the complete algorithm **ID** in Figure 3. This simple algorithm is actually complete since the statement in Lemma 15 is if and only if. The algorithm itself is shown in Fig. 4. The algorithm as shown picks elements of $W$ one at a time, although the set it picks as it iterates will equal the maximal set $W$ due to the following lemma.

Algorithm **IDC**(y, x, z)
INPUT: disjoint sets $X, Y, Z \subset V$.
OUTPUT: Expression for $P_x(y|z)$ in terms of P or FAIL.

1. if $(\exists W \in Z)(Y \perp\!\!\!\perp W | X, Z \setminus \{W\})_{G_{\overline{x},\underline{w}}}$,
   return **IDC**$(y, x \cup \{w\}, z \setminus \{w\})$.

2. else let $P' = $ **ID**$(y \cup z, x)$.
   return $P' / \sum_y P'$.

Figure 4. A complete identification algorithm for conditional effects.

LEMMA 16. *Let $P_x(y|z)$ be a conditional effect of interest in a causal model inducing $G$, and $W \subseteq Z$ the unique maximal set such that $P_x(y|z)$ is equal to $P_{x,w}(y|z \setminus w)$. Then $W = \{W' | P_x(y|z) = P_{x,w'}(y | z \setminus \{w'\})\}$.*

Completeness of the algorithm easily follows from the results we presented.

THEOREM 17. *[Shpitser and Pearl, 2006a] The algorithm **IDC** is complete.*

We note that the procedures **ID** and **IDC** served as a means to prove the completeness of do-calculus (Theorem 4). The proof [Shpitser and Pearl, 2006b] proceeds by reducing the steps in these procedures to sequences of do-calculus derivations.

## 6 Relative Interventions and the Effect of Treatment on the Treated

Interventions considered in the previous sections are what we term "absolute," since the values $x$ to which variables are set by $do(x)$ bear no relationship to whatever natural values were assumed by variables $X$ prior to an intervention. Such absolute interventions correspond to clamping a wire in a circuit to ground, or performing a randomized clinical trial for a drug which does not naturally occur in the body.

By contrast, many interventions are *relative*, in other words, the precise level $x$ to which the variable $X$ is set depends on the values $X$ naturally attains. A typical relative intervention is the addition of insulin to the bloodstream. Since insulin is naturally synthesized by the human body, the effect of such an intervention depends on the initial, pre-intervention concentration of insulin in the blood, even if a constant amount is added for every patient. The insulin intervention can be denoted by $do(i + X)$, where $i$ is the amount of insulin added, and $X$ denotes the random variable representing pre-intervention insulin concentration in the blood. More generally, a relative intervention on a variable $X$ takes the form of $do(f(X))$ for some function $f$.

How are we to make sense of a relative intervention $do(f(X))$ on $X$ applied to a given population where the values of $X$ are not known? Can relative interventions be reduced to absolute interventions? It appears that in general the answer is "no." Consider: if we knew that $X$ attained the value $x$ for a given unit, then the effect of an intervention in question on the outcome variable $Y$ is really $P(y|do(f(x)),x)$. This expression is almost like the (absolute) conditional causal effect of $do(f(x))$ on $y$, except the evidence that is being conditioned on is on the same variable that is being intervened. Since $x$ and $f(x)$ are not in general the same, it appears that this expression contains a kind of value conflict. Are these kinds of probabilities always 0? Are they even well defined?

In fact, expressions of this sort are a special case of a more general notion of a *counterfactual distribution*, which can be derived from *functional causal models* [Pearl, 2000, Chapter 7]. Such models consist of two sets of variables, the observable set $V$ representing the domain of interest, and the unobservable set $U$ representing the background to the model that we are ignorant of. Associated with each observable variable $V_i$ in $V$ is a function $f_i$ which determines the value of $V_i$ in terms of values of other variables in $V \cup U$. Finally, there is a joint probability distribution $P(u)$ over the unobservable variables, signifying our ignorance of the background conditions of the model.

The causal relationships in functional causal models are represented, naturally, by the functions $f_i$; each function causally determines the corresponding $V_i$ in terms of its inputs. Causal relationships entailed by a given model have an intuitive visual representation using a causal diagram. Causal diagrams contain two kinds of edges. Directed edges are drawn from a variable $X$ to a variable $V_i$ if $X$ appears as an input of $f_i$. Directed edges from the same unobservable $U_i$ to two observables $V_j, V_k$ can be replaced by a bidirected edge between $V_j$ to $V_k$. We will consider semi-Markovian models which induce acyclic graphs where $P(u) = \prod_i P(u_i)$, and each $U_i$ has at most two observable children. A graph obtained in this way from a model is said to be induced by said model.

Unlike causal Bayesian networks introduced in Section 2, functional causal models represent fundamentally deterministic causal relationships which only appear stochastic due to our ignorance of background variables. This inherent determinism allows us to define counterfactual distributions which span multiple worlds under different interventions regimes. Formally, a joint counterfactual distribution is a distribution over events of the form $Y_x$ where $Y$ is a post-intervention random variable in a causal model (the intervention in question being $do(x)$). A single joint distribution can contain multiple such events, with different, possibly conflicting interventions.

Such joint distributions are defined as follows:

$$P(Y^1_{x^1} = y^1, ..., Y^k_{x^k} = y^k) = \sum_{\{u | Y^1_{x^1}(u) = y^1 \wedge ... \wedge Y^k_{x^k}(u) = y^k\}} P(u), \qquad (22)$$

where $U$ is the set of unobserved variables in the model. In other words, a joint counterfactual probability is obtained by adding up the probabilities of every setting of unobserved variables in the model that results in the observed values of each counterfactual event $Y_x$ in the expression. The query with the conflict we considered above can then be expressed as a conditional distribution derived from such a joint, specifically $P(Y_{f(x)} = y | X = x) = \frac{P(Y_{f(x)}=y, X=x)}{P(X=x)}$. Queries of this form are well known in the epidemiology literature as the effect of treatment on the treated (ETT) [Heckman, 1992; Robins et al., 2006].

In fact, relative interventions aren't quite the same as ETT since we don't actually know the original levels of $X$. To obtain effects of relative interventions, we simply average over possible values of $X$, weighted by the prior distribution $P(x)$ of $X$. In other words, the relative causal effect $P(y|do(f(X)))$ is equal to $\sum_x P(Y_{f(x)} = y | X = x) P(X = x)$.

Since relative interventions reduce to ETT, and because ETT questions are of independent interest, identification of ETT is an important problem. If interventions are performed over multiple variables, it turns out that identifying ETT questions is almost as intricate as general counterfactual identification [Shpitser and Pearl, 2009; Shpitser and Pearl, 2007]. However, in the case of a singleton intervention, there is a formulation which bypasses most of the complexity of counterfactual identification. This formulation is the subject of this section.

We want to approach identification of ETT in the same way we approached identification of causal effects in the previous sections, namely by providing a graphical representation of conditional independences in joint distributions of interest, and then expressing the identification algorithm in terms of this graphical representation. In the case of causal effects, we were given as input the causal diagram representing the original, pre-intervention world, and we were asking questions about the post-intervention world where arrows pointing to intervened variables were cut. In the case of counterfactuals we are interested in joint distributions that span multiple worlds each with its own intervention. We want to construct a graph for these distributions.

The intuition is that each interventional world is represented by a copy of the original causal diagram, with the appropriate incoming arrows cut to represent the changes in the causal structure due to the intervention. All worlds are assumed to share history up to the moment of divergence due to differing interventions. This is represented by all worlds sharing unobserved variables $U$. In the special case of two interventional worlds the resulting graph is known as the *twin network graph* [Balke and Pearl, 1994b; Balke and Pearl, 1994a].

In the general case, a refinement of the resulting graph (to account for the possibility of duplicate random variables) is known as the *counterfactual graph* [Shpitser and Pearl, 2007]. The counterfactual graph represents conditional independences in the corresponding counterfactual distribution via the d-separation criterion just as the causal diagram represents conditional independences in the observed distribution of the original world. The graph in Figure 5(b) is a counterfactual graph for the query $P(Y_x = y | X = x')$ obtained from the original causal diagram shown in

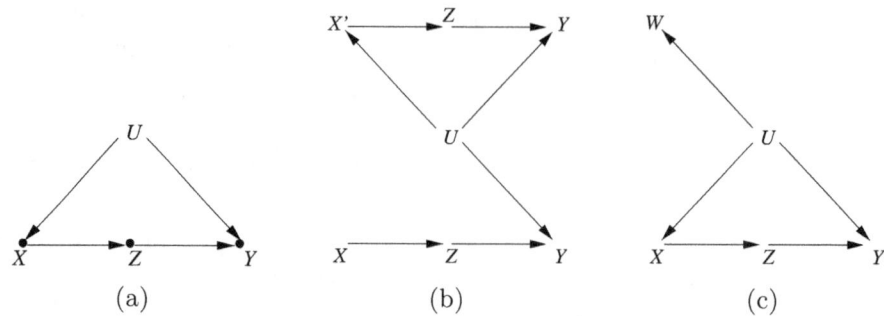

Figure 5. (a) A causal diagram $G$. (b) The counterfactual graph for $P(Y_x = y|x')$ in $G$. (c) The graph $G'$ from Theorem 18.

Figure 5(a).

There exists a rather complicated general algorithm for identifying arbitrary counterfactual distributions from either interventional or observational data [Shpitser and Pearl, 2007; Shpitser and Pearl, 2008], based on ideas from the causal effect identification algorithms given in the previous sections, only applied to the counterfactual graph, rather than the causal diagram. It turns out that while identifying ETT of a single variable $X$ can be represented as an identification problem of ordinary causal effects, ETT of multiple variables is significantly more complex [Shpitser and Pearl, 2009]. In this paper, we will concentrate on single variable ETT with multiple outcome variables $Y$.

What makes single variable ETT $P(Y_x = y|X = x')$ particularly simple is the form of its counterfactual graph. For the case of all ETTs, this graph will have variables from two worlds – the "natural" world where $X$ is observed to have taken the value $x'$ and the interventional world, where $X$ is fixed to assume the value $x$. There are two key points that simplify matters. The first is that no descendant of $X$ (including variables in $Y$) is of interest in the "natural" world, since we are only interested in the outcome $Y$ in the interventional world. The second is that all non-descendants of $X$ behave the same in both worlds (since interventions do not affect non-descendants). Thus, when constructing the counterfactual graph we don't need to make copies of non-descendants of $X$, and we can ignore descendants of $X$ in the "natural" world. But this means the only variable in the "natural" world we will construct is a copy of $X$ itself.

What this implies is that a problem of identifying the ETT $P(Y_x = y|X = x')$ can be rephrased as a problem of identifying a certain conditional causal effect.

THEOREM 18. *[Shpitser and Pearl, 2009] For a singleton variable $X$, and a set $Y$, $P(Y_x = y|X = x')$ is identifiable in $G$ if and only if $P_x(y|w)$ is identifiable in $G'$, where $G'$ is obtained from $G$ by adding a new node $W$ with the same set of parents (both observed and unobserved) as $X$, and no children. Moreover, the estimand for*

$P(Y_x = y | X = x')$ is equal to that of $P_x(y|w)$ with all occurrences of $w$ replaced by $x'$.

We illustrate the application of Theorem 18 by considering the graph $G$ in Fig. 5(a). The query $P(Y_x = y | X = x')$ is identifiable by considering $P_x(y|w)$ in the graph $G'$ shown in Fig. 5(c), while the counterfactual graph for $P(Y_x = y|x')$ is shown in Fig. 5(b). Identifying $P_x(y|w)$ in $G'$ using the algorithm **IDC** in the previous section leads to $\sum_z P(z|x) \sum_x P(y|z,w,x) P(w,x) / P(w)$. Replacing $w$ by $x'$ yields the expression $\sum_z P(z|x) \sum_{x''} P(y|z,x',x'') P(x',x'') / P(x')$.

Ordinarily, we know that $P(y|z,x',x'')$ is undefined if $x'$ is not equal to $x''$. However, in our case, we know that observing $X = x'$ in the natural world implies $X = x'$ in any other interventional world which shares ancestors of $X$ with the natural world. This implies the expression $\sum_{x''} P(y|z,x',x'') P(x',x'') / P(x')$ is equivalent to $P(y|z,x')$, thus our query $P(Y_x = y | X = x')$ is equal to $\sum_z P(y|z,x') P(z|x)$.

It is possible to use Theorem 18 to derive analogues of the back-door and front-door criteria for ETT.

COROLLARY 19 (Back-door Criterion for ETT). *If a set $Z$ satisfies the back-door criterion relative to $(X, Y)$, where $X$ is a singleton variable, then $P(Y_x = y | X = x')$ is identifiable and equal to $\sum_z P(y|z,x) P(z|x')$.*

The intuition for the back-door criterion for ETT is that $Z$, by assumption, screens $X$ and $Y$ from observed values of $X$ in other counterfactual worlds. Thus, the first term in the back-door expression does not change. The second term changes in an obvious way since $Z$ depends on observing $X = x'$.

COROLLARY 20 (Front-door Criterion for ETT). *If a set $Z$ satisfies the front-door criterion relative to $(X, Y)$, where $X, Y$ are singleton variables, then $P(Y_x = y | X = x')$ is identifiable and equal to $\sum_z P(y|z,x') P(z|x)$.*

**Proof.** We will be using a number of graphs in this proof. $G$ is the original graph. $G^w$ is the graph obtained from $G$ by adding a copy of $X$ called $W$ with the same parents (including unobserved parents) as $X$ and no children. $G'$ is a graph representing independences in $P(X, Y, Z)$. It is obtained from $G$ by removing all nodes other than $X, Y, Z$, by adding a directed arrow between any remaining $A$ and $B$ in $X, Y, Z$ if there is a d-connected path containing only nodes not in $X, Y, Z$ which starts with a directed arrow pointing away from $A$ and ends with any arrow pointing to $B$. Similarly, a bidirected arrow is added between any $A$ and $B$ in $X, Y, Z$ if there is a d-connected path containing only nodes not in $X, Y, Z$ which starts with any arrow pointing to $A$ and ends with any arrow pointing to $B$. (This graph is known as a latent projection [Pearl, 2000]). The graphs $G'^w, G'^w_{\overline{x}}$ are defined similarly as above.

We want to identify $P_x(y,z,w)$ in $G'^w$. First, we want to show that no node in $Z$ shares a c-component with $W$ or any node in $Y$ in $G'^w_{\overline{x}}$. This can only happen if a node in $Z$ and $W$ or a node in $Y$ share a bidirected arc in $G'^w_{\overline{x}}$. But this means that either there is a back-door d-connected path from $Z$ to $Y$ in $G_{\overline{x}}$, or there is a back-door d-connected path from $X$ to $Z$ in $G$. Both of these claims are contradicted by our assumption that $Z$ satisfies the front-door criterion for $(X, Y)$.

This implies $P_x(y, z, w) = P_{z,x}(y, w) P_{x,w}(z)$ in $G^w$.

By construction of $G^w$ and the front-door criterion, $P_{x,w}(z) = P_x(z) = P(z|x)$. Furthermore, since no nodes in $Z$ and $Y$ share a c-component in $G'^w$, no node in $Z$ has a bidirected path to $Y$ in $G'^w$. This implies, by Lemma 1 in [Shpitser et al., 2009], that $P_z(y, w, x) = P(y|z, w, x) P(w, x)$.

Since $Z$ intercepts all directed paths from $X$ to $Y$ (by the front-door criterion), $P_{z,x}(y, w) = P_z(y, w) = \sum_x P(y|z, w, x) P(w, x)$.

We conclude that $P_x(y, w)$ is equal to $\sum_z P(z|x) \sum_x P(y|z, w, x) P(w, x)$. Since $P_x(w) = P(w)$ in $G'^w$, $P_x(y|w) = \sum_z P(z|x) \sum_x P(y|z, w, x) P(x|w)$.

Finally, recall that $W$ is just a copy of $X$, and $X$ is observed to attain value $x'$ in the "natural" world. This implies that our expression simplifies to $\sum_z P(z|x) P(y|z, x')$, which proves our result. □

If neither the back-door nor the front-door criteria hold, we must invoke general causal effect identification algorithms from the previous sections. However, in the case of ETT of a single variable, there is a simple complete graphical criterion which works.

THEOREM 21. *[Shpitser and Pearl, 2009] For a singleton variable $X$, and a set $Y$, $P(Y_x = y | X = x')$ is identifiable in $G$ if and only if there is no bidirected path from $X$ to a child of $X$ in $G_{an(y)}$. Moreover, if there is no such bidirected path, the estimand for $P(Y_x = y | X = x')$ is obtained by multiplying the estimand for $\sum_{an(y) \setminus (y \cup \{x\})} P_x(an(y) \setminus x)$ (which exists by Theorem 9) by $\frac{Q[S^x]'}{P(x') \sum_x Q[S^x]}$, where $S^x$ is the c-component in $G$ containing $X$, and $Q[S^x]'$ is obtained from the expression for $Q[S^x]$ by replacing all occurrences of $x$ with $x'$.*

## 7 Conclusion

In this paper we described the state of the art in identification of causal effects and related quantities in the framework of graphical causal models. We have shown how this framework, developed over the period of two decades by Judea Pearl and his collaborators, and presented in Pearl's seminal work [Pearl, 2000], can sharpen causal intuition into mathematical precision for a variety of causal problems faced by scientists.

**Acknowledgments:** Jin Tian was partly supported by NSF grant IIS-0347846. Ilya Shpitser was partly supported by AFOSR grant #F49620-01-1-0055, NSF grant #IIS-0535223, MURI grant #N00014-00-1-0617, and NIH grant #R37AI032475.

## References

A. Balke and J. Pearl. Counterfactual probabilities: Computational methods, bounds, and applications. In R. Lopez de Mantaras and D. Poole, editors,

*Uncertainty in Artificial Intelligence 10*, pages 46–54. Morgan Kaufmann, San Mateo, CA, 1994.

A. Balke and J. Pearl. Probabilistic evaluation of counterfactual queries. In *Proceedings of the Twelfth National Conference on Artificial Intelligence*, volume I, pages 230–237. MIT Press, Menlo Park, CA, 1994.

D. Galles and J. Pearl. Testing identifiability of causal effects. In P. Besnard and S. Hanks, editors, *Uncertainty in Artificial Intelligence 11*, pages 185–195. Morgan Kaufmann, San Francisco, 1995.

C. Glymour and G. Cooper, editors. *Computation, Causation, and Discovery*. MIT Press, Cambridge, MA, 1999.

D. Heckerman and R. Shachter. Decision-theoretic foundations for causal reasoning. *Journal of Artificial Intelligence Research*, 3:405–430, 1995.

J.J. Heckman. Randomization and social policy evaluation. In C. Manski and I. Garfinkle, editors, *Evaluations: Welfare and Training Programs*, pages 201–230. Harvard University Press, 1992.

Y. Huang and M. Valtorta. Identifiability in causal bayesian networks: A sound and complete algorithm. In *Proceedings of the Twenty-First National Conference on Artificial Intelligence*, pages 1149–1154, Menlo Park, CA, July 2006. AAAI Press.

Y. Huang and M. Valtorta. Pearl's calculus of interventions is complete. In R. Dechter and T.S. Richardson, editors, *Proceedings of the Twenty-Second Conference on Uncertainty in Artificial Intelligence*. AUAI Press, July 2006.

M. Kuroki and M. Miyakawa. Identifiability criteria for causal effects of joint interventions. *Journal of the Japan Statistical Society*, 29(2):105–117, 1999.

S. Lauritzen. Graphical models for causal inference. In O.E. Barndorff-Nielsen, D. Cox, and C. Kluppelberg, editors, *Complex Stochastic Systems*, chapter 2, pages 67–112. Chapman and Hall/CRC Press, London/Boca Raton, 2000.

J. Pearl and J.M. Robins. Probabilistic evaluation of sequential plans from causal models with hidden variables. In P. Besnard and S. Hanks, editors, *Uncertainty in Artificial Intelligence 11*, pages 444–453. Morgan Kaufmann, San Francisco, 1995.

J. Pearl. *Probabilistic Reasoning in Intelligent Systems*. Morgan Kaufmann, San Mateo, CA, 1988.

J. Pearl. Comment: Graphical models, causality, and intervention. *Statistical Science*, 8:266–269, 1993.

J. Pearl. Causal diagrams for empirical research. *Biometrika*, 82:669–710, December 1995.

J. Pearl. *Causality: Models, Reasoning, and Inference*. Cambridge University Press, NY, 2000.

James M. Robins, VanderWeele Tyler J., and Thomas S. Richardson. Comment on causal effects in the presence of non compliance: a latent variable interpretation by antonio forcina. *METRON*, LXIV(3):288–298, 2006.

J.M. Robins. A new approach to causal inference in mortality studies with a sustained exposure period – applications to control of the healthy workers survivor effect. *Mathematical Modeling*, 7:1393–1512, 1986.

J.M. Robins. A graphical approach to the identification and estimation of causal parameters in mortality studies with sustained exposure periods. *Journal of Chronic Diseases*, 40(Suppl 2):139S–161S, 1987.

J.M. Robins. Causal inference from complex longitudinal data. In *Latent Variable Modeling with Applications to Causality*, pages 69–117. Springer-Verlag, New York, 1997.

I. Shpitser and J. Pearl. Identification of conditional interventional distributions. In R. Dechter and T.S. Richardson, editors, *Proceedings of the Twenty-Second Conference on Uncertainty in Artificial Intelligence*, pages 437–444. AUAI Press, July 2006.

I. Shpitser and J. Pearl. Identification of joint interventional distributions in recursive semi-markovian causal models. In *Proceedings of the Twenty-First National Conference on Artificial Intelligence*, pages 1219–1226, Menlo Park, CA, July 2006. AAAI Press.

Ilya Shpitser and Judea Pearl. What counterfactuals can be tested. In *Twenty Third Conference on Uncertainty in Artificial Intelligence*. Morgan Kaufmann, 2007.

I. Shpitser and J. Pearl. Complete identification methods for the causal hierarchy. *Journal of Machine Learning Research*, 9:1941–1979, 2008.

Ilya Shpitser and Judea Pearl. Effects of treatment on the treated: Identification and generalization. In *Proceedings of the Conference on Uncertainty in Artificial Intelligence*, volume 25, 2009.

Ilya Shpitser, Thomas S. Richardson, and James M. Robins. Testing edges by truncations. In *International Joint Conference on Artificial Intelligence*, volume 21, pages 1957–1963, 2009.

P. Spirtes, C. Glymour, and R. Scheines. *Causation, Prediction, and Search (2nd Edition)*. MIT Press, Cambridge, MA, 2001.

J. Tian and J. Pearl. A general identification condition for causal effects. In *Proceedings of the Eighteenth National Conference on Artificial Intelligence (AAAI)*, pages 567–573, Menlo Park, CA, 2002. AAAI Press/The MIT Press.

J. Tian and J. Pearl. On the testable implications of causal models with hidden variables. In *Proceedings of the Conference on Uncertainty in Artificial Intelligence (UAI)*, 2002.

J. Tian and J. Pearl. On the identification of causal effects. Technical Report R-290-L, Department of Computer Science, University of California, Los Angeles, 2003.

J. Tian. Identifying conditional causal effects. In *Proceedings of the Conference on Uncertainty in Artificial Intelligence (UAI)*, 2004.

# Part IV: Reminiscences

# 31
# Questions and Answers

NILS J. NILSSON

Few people have contributed as much to artificial intelligence (AI) as has Judea Pearl. Among his several hundred publications, several stand out as among the historically most significant and influential in the theory and practice of AI. With my few pages in this celebratory volume, I join many of his colleagues and former students in showing our gratitude and respect for his inspiration and exemplary career. He is a towering figure in our field.

Certainly one key to Judea's many outstanding achievements (beyond dedication and hard work) is his keen ability to ask the right questions and follow them up with insightful intuitions and penetrating mathematical analyses. His overarching question, it seems to me, is "how is it that humans can do so much with simplistic, unreliable, and uncertain information?" The very name of his UCLA laboratory, the Cognitive Systems Laboratory, seems to proclaim his goal: understanding and automating the most cognitive of all systems, namely humans.

In this essay, I'll focus on the questions and inspirations that motivated his ground-breaking research in three major areas: heuristics, uncertain reasoning, and causality. He has collected and synthesized his work on each of these topics in three important books [Pearl 1984; Pearl 1988; Pearl 2000].

## 1 Heuristics

Pearl is explicit about what inspired his work on heuristics [Pearl 1984, p. xi]:

> The study of heuristics draws its inspiration from the ever-amazing observation of how much people can accomplish with that simplistic, unreliable information source known as *intuition*. We drive our cars with hardly any thought of how they function and only a vague mental picture of the road conditions ahead. We write complex computer programs while attending to only a fraction of the possibilities and interactions that may take place in the actual execution of these programs. Even more surprisingly, we maneuver our way successfully in intricate social situations having only a guesswork expectation of the behavior of other persons around and even less certainty of their expectations of us.

The question is "How do people do that?" The answer, according to Pearl, is that they use heuristics. He defines *heuristics* as "criteria, methods, or principles for deciding which among several alternative courses of action promises to be the most effective in order to achieve some goal." "For example," he writes, "a popular method for choosing [a] ripe cantaloupe involves pressing the spot on the candidate cantaloupe where it was attached to the plant, and then smelling the spot. If the spot smells like the inside of a cantaloupe, it is most probably ripe [Pearl 1984, p. 3]."

Although heuristics, in several forms, were used in AI before Pearl's book on the subject, no one had analyzed them as profitably and in as much detail as did Pearl. Besides focusing on several heuristic search procedures, including A*, his book beneficially tackles the question of how heuristics can be *discovered*. He proposes a method: consult "simplified models of the problem domain" particularly those "generated by *removing constraints* which forbid or penalize certain moves in the original problem [Pearl 1984, p. 115]."

## 2 Uncertain Reasoning

Pearl was puzzled by the contrast between, on the one hand, the ease with which humans reason and make inferences based on uncertain information and, on the other hand, the computational difficulties of duplicating those abilities using probability calculations. Again the question, "How do humans reason so effectively with uncertain information?" He was encouraged in his search for answers by the following observations [Pearl 1993]:

1. The consistent agreement between plausible reasoning and probability calculus could not be coincidental, but strongly suggests that human intuition invokes some crude form of probabilistic computation.

2. In light of the speed and effectiveness of human reasoning, the computational difficulties that plagued earlier probabilistic systems could not be very fundamental and should be overcome by making the right choice of simplifying assumptions.

Some ideas about how to proceed came to him in the late 1970s after reading a paper on reading comprehension by David Rumelhart [Rumelhart 1976]. In Pearl's words [Pearl 1988, p. 50]:

> In this paper, Rumelhart presented compelling evidence that text comprehension must be a distributed process that combines both top-down and bottom-up inferences. Strangely, this dual mode of inference, so characteristic of Bayesian analysis, did not match the capabilities of either the "certainty factors" calculus or the inference networks of PROSPECTOR – the two major contenders for uncertainty management in the 1970s. I thus began to explore the possibility of achieving distributed computation in a "pure" Bayesian framework, so as not to compromise its basic capacity to combine bi-directional inferences (i.e., predictive and abductive).

Previous work in probabilistic reasoning had used graphical structures to encode probabilistic information, and Pearl speculated that "it should be possible to use the links [in a graphical model] as message-passing channels, and [that] we could then update beliefs by parallel distributed computations, reminiscent of neural architectures [Pearl 1988, p. 51]." In the course of developing these ideas, Pearl says [Pearl 1988, p. 50]:

> it became clear that *conditional independence* is the most fundamental relation behind the organization of probabilistic knowledge and the most crucial factor facilitating distributed computations. I therefore decided to investigate systematically how directed and undirected graphs could be used as a language for encoding, decoding, and reasoning with such independencies.

Pearl's key insight was that beliefs about propositions and other quantities could often be regarded as "direct causes" of other beliefs and that these causal linkages could be used to construct the graphical structures he was interested in. Most importantly, this method of constructing them would automatically encode the key conditional independence assumptions among probabilities which he regarded as so important for simplifying probabilistic reasoning.

Out of these insights, and after much hard work by Pearl and others, we get one of the most important sets of inventions in all of AI – Bayesian networks and their progeny.

## 3 Causality

Pearl's work on causality was inspired by his notion that beliefs could be regarded as causes of other beliefs. He came to regard "causal relationships [as] the fundamental building blocks both of physical reality and of human understanding of that reality" and that "probabilistic relationships [were] but the surface phenomena of the causal machinery that underlies and propels our understanding of the world." [Pearl 2000, p. xiii]

In a Web page describing the genesis of his ideas about causality, Pearl writes [Pearl 2000]:

> I got my first hint of the dark world of causality during my junior year of high school.
>
> My science teacher, Dr. Feuchtwanger, introduced us to the study of logic by discussing the 19th century finding that more people died from smallpox inoculations than from smallpox itself. Some people used this information to argue that inoculation was harmful when, in fact, the data proved the opposite, that inoculation was saving lives by eradicating smallpox.

"And here is where logic comes in," concluded Dr. Feuchtwanger, "To protect us from cause-effect fallacies of this sort." We were all enchanted by the marvels of logic, even though Dr. Feuchtwanger never actually showed us how logic protects us from such fallacies.

It doesn't, I realized years later as an artificial intelligence researcher. Neither logic, nor any branch of mathematics had developed adequate tools for managing problems, such as the smallpox inoculations, involving cause-effect relationships.

So, the question is "How are we to understand causality?" Even though, as Pearl noted, most of his colleagues "considered causal vocabulary to be dangerous, avoidable, ill-defined, and nonscientific," he felt that his intuitions about causality should be "expressed, not suppressed." He writes that once he "got past a few mental blocks, I found causality to be smiling with clarity, bursting with new ideas and new possibilities." The key, again, was the use of graphical causal models.

Pearl's work on causality, the subject of his third book, has had major impacts even beyond the normal boundaries of AI. It has influenced work in philosophy, psychology, statistics, econometrics, epidemiology, and social science. Judging by citations and quotations from the literature, it is hard to identify another body of AI research that has been as influential on these related disciplines as has Pearl's work on causality.

One must be mathematically proficient to understand and to benefit from Pearl's work. Some have criticized him for "substituting mathematics for clarity." But, as Pearl points out [Pearl 1993, p. 51], "...it was precisely this conversion of networks and diagrams to mathematically defined objects that led to their current acceptance in practical reasoning systems." Indeed AI practitioners now acknowledge that successful applications depend increasingly on skillful use of AI's mathematically deep technology. Pearl, along with others in "modern AI," have made it so.

I'll close with a non-mathematical, but none-the-less important, topic. As we all know, Judea and Ruth Pearl's son, Danny, a *Wall Street Journal* reporter, was kidnapped and murdered by terrorists in Pakistan. In their grief, Judea and Ruth asked the question "How could people do this to to someone like Danny who 'exuded compassion and joy wherever he went'?" To help diffuse the hatred that led to this and other tragedies, Danny's family and friends formed the Daniel Pearl Foundation. Among the principles that the foundation hopes to promote are ones Judea himself has long exemplified: "uncompromised objectivity and integrity; insightful and unconventional perspective; tolerance and respect for people of all cultures; unshaken belief in the effectiveness of education and communication; and the love of music, humor, and friendship [Daniel Pearl Foundation]."

Shalom!

# References

Pearl, J. (1984). *Heuristics: Intelligent Search Strategies for Computer Problem Solving*, Reading, MA: Addison-Wesley Publishing Company.

Pearl, J. (1988). *Probabilistic Reasoning Systems: Networks of Plausible Inference*, San Francisco: Morgan Kaufmann Publishers.

Pearl, J. (2000). *Causality: Models, Reasoning, and Inference*, New York: Cambridge University Press (second edition, 2009).

Pearl, J. (1993). Belief networks revisited. *Artificial Intelligence 59*, 49–56.

Rumelhart, D. (1976). Toward an interactive model of reading. *Tech. Rept. #CHIP-56*. University of California at San Diego, La Jolla, CA.

Pearl, J. (2000). http://bayes.cs.ucla.edu/BOOK-2K/why.html.

Daniel Pearl Foundation. http://www.danielpearl.org/.

# 32
# Fond Memories From an Old Student
EDWARD T. PURCELL

I was very lucky to have been Professor Judea Pearl's first graduate student advisee in the UCLA Computer Science Department. Now I am further honored to be invited to contribute – in distinguished company – some fond memories of those early days studying under Professor Pearl.

In January 1972, after completing the core coursework for the M.S. degree, I took my first class in Artificial Intelligence from Professor Pearl. Thirty-eight calendar years seems like cyber centuries ago, such has been the incredible pace of growth of computer technologies and Computer Science and AI as academic disciplines.

The ARPAnet maps posted on the Boelter Hall corridor walls only showed a few dozen nodes, and AI was still considered an "ad hoc" major field of study, requiring additional administrative paperwork of prospective students. (Some jested, unfairly, this was because AI was one step ahead of AH — ad hoc.)

The UCLA Computer Science Department had become a separate Department in the School of Engineering only two and a half years earlier, in the Fall of 1969, at the same time it became the birthplace of the Internet with the deployment of the first ARPAnet Interface Message Processor node in room 3420 of Boelter Hall.

The computers available were "big and blue," IBM S/360 and S/370 mainframes of the Campus Computing Network, located on the fourth floor of the Mathematical Sciences Building, access tightly controlled. Some campus laboratories were fortunate to have their own DEC PDP minicomputers.

Programming was coded in languages like Assembly Language, Fortran, APL, PL/1, and Pascal, delimited by Job Control Language commands. Programs were communicated via decks of punched cards fed to card readers at the Campus Computing Network facility. A few hours later, the user could examine the program's output on print-out paper. LISP was not available at the Campus Computing Network. Time-sharing terminals and computers were just beginning to introduce a radical change in human-computer interaction: on screen programming, both input and output.

Professor Pearl's first "Introduction to AI" course was based on Nils Nilsson's *Problem-Solving Methods in AI,* a classic 1971 textbook focusing on the then two core (definitely non-ad-hoc) problem-solving methodologies in AI: search and logic. (As with the spectacular growth of computer technology, it is wondrous to regard how much Judea's research has extended and fortified these foundations of AI.) Supplemental study material included Edward Feigenbaum's 1963 compilation of articles on early AI systems, *Computers and Thought*, and a 1965 book by Nils Nilsson, *Learning Machines.*

In class I was immediately impressed and enchanted by Judea's knowledge, intelligence, brilliance, warmth and humor. His teaching style engaging, interactive, informative and fun. My interest in AI, dating back to pre-Computer Science undergraduate days, was much stimulated.

After enjoying this first AI class, I asked Professor Pearl if he would serve as my M.S. Advisor, and was very happy when he agreed.

Other textbooks Professor Pearl used in subsequent AI classes and seminars included Howard Raiffa's 1968 *Decision Analysis: Introductory Lectures on Choices under Uncertainty,* Duncan Luce and Howard Raiffa's 1957 *Games and Decisions*, and George Polya's *How to Solve it*, and the challenging 1971 three-volume *Foundations of Measurement,* by David Krantz, Duncan Luce, Patrick Suppes and Amos Tversky. The subtitles and chapter headings in this three-volume opus hint at Professor Pearl's future research on Bayesian networks: *Volume I: Additive and Polynomial Representations; Volume II: Geometrical, Threshold, and Probabilistic Representations; and Volume III: Representation, Axiomatization, and Invariance.*

It was always fun to visit Professor Pearl in his office. Along with the academic consultation, Judea had time to talk about assorted extra-curricular topics, and became like a family friend. One time, I found Judea strumming a guitar in his office, singing a South American folk song, *"Carnavalito,"* which I happend to know because of my U.S. diplomat's son upbringing in South America. I was happy to help with the pronunciation of the song's lyrics. It was nice to discover that we shared a love of music, Judea more in tune with classical music, myself more a jazz fan. Now and then I would see Judea and his wife Ruth at Royce Hall concerts, for example, a recital by the classical guitarist Narciso Yepes.

Judea's musical orientation (and humor) appeared in the title of a presentation a few years later at a Decision Analysis workshop, with the title acronym *"AIDA"* as Artificial Intelligence and Decision Analysis. The titles of other Pearl papers also revealed wry humor: *"How to Do with Probabilities What People Say You Can't,"* and *"Reverend Bayes on Inference Engines: a Distributed Hierarchical Approach."*

My M.S. thesis title was *"'A Game-Playing Procedure for a Game of Induction,"* and included results from a (PL/1) program for the induction game Patterns, a pattern sampling and guessing game introduced by Martin Gardner in his November 1969 *Scientific American* "Mathematical Games" column. (After sending Martin Gardner a copy of my M.S. thesis, I received a letter of appreciation from the game wizard himself.)

At a small public demonstration of the Patterns game-playing program in early 1973, a distinguished elderly scholar was very interested and asked many questions. After the presentation Professor Pearl asked if I knew who the inquisitive gentleman was. "No," I said. "That was Jacob Marschak," said Judea. Whenever I attend a Marschak Colloquium presentation at the UCLA Anderson School of Management, including several talks by Judea, I remember Professor Marschak's interest in my modest game-playing program.

Then, as now, seminars at Boelter Hall 3400 were an integral part of the UCLA Computer Science education. I remember several distinguished presentations there, for example, a seminar on coding theory given by Professor Andrew Viterbi, then still at UCLA, whom Professor Pearl engaged in an animated discussion, and another standing-room-only seminar on algorithms given by Donald Knuth, who listened attentively to Judea at a smaller, post-seminar gathering.

Soon enough, in June 1973, I was very proud and happy to receive my M.S. degree in Computer Science.

When I began my graduate studies in Computer Science at UCLA, I had only hoped to study for a Masters' degree. Though I was having a lot of fun studying AI and being mentored by Professor Pearl, I was not sure of my ability to pursue the doctorate degree. Encouraged and approved by Judea, I applied for and was accepted as a Ph.D. candidate, with Professor Pearl as my Advisor.

The early Ph.D. qualifying exams were challenging, because of the depth and breadth of topics covered, some topics beyond those covered in my classes. Thanks to Judea's guidance and support, I was able to overcome these challenges.

Professor Pearl's support extended beyond academic issues. On one lean occasion, I remember Judea lending me some funds to cover my registration fees. Fortunately, UCLA tuition fees were very modest in those days (unlike today's costs), and I was soon able to repay Judea's kind loan.

My classes were now mostly individual study seminars led by Professor Pearl. Despite a variety of readings and studies, I was stumped for a good dissertation topic. Judea suggested a very interesting topic: learning of heuristics for search algorithms.

I was immediately piqued by this topic, and soon formulated a perceptron-like learning-while-searching procedure for A*-like heuristic search algorithms. The unsupervised learning consisted of adjusting the weight vector **w** of a heuristic vector function **h**, trying to satisfy, on a local scale, necessary (but not sufficient) metric and order consistency properties of the perfect knowledge heuristic function h*. The learning samples derived from search observations of problem graph edge costs and node orderings, obtained as the search algorithm progressed.

The topic of learning heuristics for search algorithms was well received by the Ph.D. dissertation qualifying committee. I remember Professor Pearl telling me committee member Dr. Ken Colby of the UCLA School of Medicine expressed a favorable review of this topic and of my introductory overview of the topic.

I was able to complement and support my UCLA Computer Science studies with interesting part-time work, near campus and related to my studies. During 1974 and 1975 I worked part-time at Technology Service Corporation for William Meisel and Leo Breiman, and was invited to be co-author of a 1977 paper (*"Variable-Kernel Estimates of Multi-Variate Densities,"* **Technometrics**, vol. 19, no. 2, pp.

135-144, 1977), whose experimental results were based on my programming. (Many years later I learned this paper earned me an Erdős 4 number.)

In late 1976 and early 1977 I worked part-time for System Development Corporation, and was tasked by Drs. Jeff Barnett and Mort Bernstein with writing summaries of papers, reports and other documents on the emerging technology of knowledge-based systems, which contributed to a June 1977 System Development Corporation report (ADA044883), *"Knowledge-Based Systems: A Tutorial."*

Many of the early expert systems implemented the MYCIN - Prospector certainty factor calculus. Probabilities were dismissed because of the exponential number of joint probabilities presumed to be required. I remember Professor Pearl discussing the topic of uncertainty calculus with colleagues at a Workshop on Decision Analysis held at a hotel in Bel Air in the summer of 1977.

I thoroughly enjoyed those lean student days, commuting to campus on bicycle, studying Computer Science and AI under Professor Pearl. I remember many fun activities: a barbecue dinner hosted by Judea and Ruth Pearl for Donald Michie in May 1976, participating in experiments with Norman Dalkey's Delphi group decision-making system, attending Royce Hall concerts, playing perhaps too much soccer and rugby. (But I had good company in these sports activities: fellow UCLA Computer Science graduate student David Patterson was also a UCLA rugby teammate.)

The final hurdles on the doctoral track were more logistical and administrative rather than technical, and included scheduling (in pre-email days) five busy dissertation committee members to a common time and place, applying (in pre-PC days) for additional computer run time from Campus Computing Network, obtaining the approval of the UCLA School of Engineering bibliography checker, finding (in pre-TEXdays) a good typist, making copies of the dissertation, etc.

In June 1978, thanks to much encouragement, guidance and nurturing from Professor Pearl, I completed my Ph.D. dissertation, *"Machine Learning of Heuristics for Ordered-Search Algorithms."*

The fun memories associated with Professor Pearl continued after my graduation. During an AI conference in Miami in December 1984, a dinner with Judea at a restaurant in little Havana. Other AI conference dinners hosted by Professor Pearl for his graduate students. One day in 1985, when I visited Judea in his office enroute to a Computer Science Seminar, I remember him asking me which designation I liked better: "Bayes net" or "Bayesian network." I voted for the latter as more poetic. In November 1996 I was invited by Judea to attend his University of California Faculty Research Lecture at Schoenberg Auditorium. A capacity crowd listened attentively as Judea discussed *"The Art and Science of Cause and Effect."* Afterward, Judea and his family celebrated at a tea reception at the Chancellor's Residence. A special seminar for the publication of *"Causality"* in 2000. And the fond memories continue.

Many colleagues ask me, "Did you study under Judea Pearl?" "Yes!" I answer proudly. I am very proud to have been Professor Pearl's first student, even though

I was probably not worthy.

I cherish the memories of those student days in the UCLA Computer Science Department, studying under and learning from Professor Pearl.

With deep appreciation, I would like to thank you very much, Judea, for all your kindness, help, guidance and education through the years.

God bless you!

# 33

# Reverend Bayes and inference engines

DAVID SPIEGELHALTER

I first met Judea in 1986 at a conference in Paris on the "management of uncertainty in knowledge-based systems": this topic, which now sounds rather dated, was of consuming interest at the time and I was anxious about coming face-to-face with someone who might be considered a competitor in the field – what would he be like? I need not have worried.

This was an unusual research area for a statistician, but since the early 1980s I had been part of a group working on decision-support systems in medicine which used explicit probabilities for diagnosis and prognosis. There was a strong and (usually) good-natured rivalry between techniques based on formal probabilistic methods for so-called 'expert' or 'knowledge-based systems' and those arising in the computer science community that were more rooted in artificial intelligence and attempted to represent expert knowledge through a network of rules, with a separate 'inference engine' to control the reasoning process. The MYCIN team from Stanford were international leaders in this field with their system of 'certainty factors' to handle uncertainty in the rules. There was no rigid distinction between the approaches: everything was in the pot, from purely qualitative reasoning, through rather ad-hoc quantification, fuzzy logic, and attempts to use probabilities in a network of rules. It was a very exciting and competitive time, in which many disciplines were trying to establish authority.

A Royal Statistical Society discussion paper [Spiegelhalter and Knill-Jones, 1984] brought the controversy into the statistical world, but this was outdated even by the time of its publication. Struggling with the problem of handling probability models on directed graphs of arbitrary size and complexity (and, crucially, being totally unaware of Judea's work in this area), I realised that Wermuth and Lauritzen [1983] provided a connection between probabilistic models on directed graphs and unconnected graphs: the latter had been established for some time as an elegant way of representing conditional independence relations in contingency tables [Darroch, Lauritzen and Speed, 1980]. In April 1984 I met Steffen Lauritzen at the 150th Anniversary celebrations of the Royal Statistical Society and told him that his work on graphical models was of fundamental importance to artificial intelligence. He did not seem to believe me, but he was kind enough to take me seriously enough to invite me to visit.

So I took the long journey to Aalborg in the north of Jutland in Denmark. By a remarkable stroke of good fortune Steffen shared a department with a research team who were working on uncertainty in expert systems, and together we persuaded him that he should get involved. And he is still working in this area after 25 years.

As a bonus, the Danish group finally introduced me to Pearl [1982] and Kim and Pearl [1983]. These came as a shock: looking beneath the poor typography revealed fundamental and beautiful ideas on local computation that made me doubt we could contribute more. But Judea was working solely with directed graphs, and we felt the connection with undirected graphs was worth pursuing in the search for a general algorithm for probability propagation in arbitrary graphs.

I wrote to Judea who replied in a typically enthusiastic and encouraging way, and so at a 1985 workshop at Bell Labs I was able to try and put together his work with our current focus on triangulated graphs, clique separations, potential representations and so on [Spiegelhalter, 1986]. Then in July 1986 we finally met in Paris at the conference mentioned at the start of this article, where Judea was introducing the audience to d-separation. I have mentioned that I was nervous, but Judea was as embracing as ever. We ended up in a pavement café in the Latin quarter, with Judea drawing graphs on the paper napkin and loudly claiming that anyone could see that observations on a particular node rendered two others independent – grabbing a passer-by, Judea demanded to know whether this unfortunate Frenchman could recognise this obvious property, but the poor innocent man just muttered something and walked briskly away, pleased to have escaped these lunatics.

We continued to meet at conferences as he developed his propagation techniques based on directed graphs [Pearl, 1986] and we published our algorithm based on embedding the directed graph in a triangulated undirected graph that could be represented as a tree of cliques [Lauritzen and Spiegelhalter, 1988]. We even jointly presented a tutorial on probabilistic reasoning at the 1989 IJCAI meeting in Detroit, which I particularly remember as my bus got stuck in traffic and I was late arriving, but Judea had just carried on, extemporising from a massive pile of overhead slides from which he would apparently draw specimens at random.

Then I started on MCMC on graphical models, and he began on causality, which was too difficult for me. But I look back on that time in the mid 1980s as perhaps the most exciting and creative period of my working life, continually engaged in a certain amount of friendly rivalry with Judea, who always responded with characteristic generosity of spirit.

References

Darroch, J. N., Lauritzen, S. L. and Speed, T. P. (1980) Markov Helds and log-linear models for contingency tables. *Ann. Statist.*, 8, 522-539.

Kim, J. H. and Pearl, J. (1983) A computational model for causal and diagnostic reasoning in inference systems. In *Proc. 8th International Joint Conference on Artificial Intelligence*, Karlsruhe, pp. 190-193.

Lauritzen, S. L. and Spiegelhalter, D. J. (1988) Local Computations with Probabilities on Graphical Structures and Their Application to Expert Systems. *Journal of the Royal Statistical Society. Series B (Methodological)*, 50, 157-224.

Pearl, J. (1982) Reverend Bayes on inference engines: a distributed hierarchical approach. *Proc. AAAI National Conference on AI, Pittsburgh*, pp. 133-136.

Pearl. J. (1986) Fusion, propagation and structuring in belief networks. *Artificial Intelligence*, 29, 241-288.

Spiegelhalter, D. J. (1986) A statistical view of uncertainty in expert systems. In *Artificial Intelligence and Statistics (ed. W. Gale)*, pp. 17-56. Reading: Addison-Wesley.

# 34

# An old-fashioned scientist shaping a modern discipline

HECTOR GEFFNER

I took a course with Judea in September of 1984, while I was finishing my MS in Systems Science at UCLA. Right after, I switched to Computer Science, became his PhD student, and started working as a research assistant in his group. I finished my PhD five years later, a time during which I learned from him how science is done and how a scientist works, two things that were extremely valuable to me for at least two reasons. The first is that I was a pure science 'consumer', enthusiastic and well-informed but more inclined to read than to produce science. The second is that, unknown to me, AI was being redefined, with no one playing a larger role in the change than Judea.

While Judea published regularly in AI conferences from the late 70s on and the *Heuristics* book was about to be published, he still felt very much like an outsider in AI, even at UCLA, where the AI Lab, headed by former students of Roger Schank, used to get the spotligth, lavish funding, and most of the aspiring AI students. Judea, on the other hand, directed the Cognitive Systems Lab, which to my surprise was no more than a sign on the door of a secretary, whose main task, although not an easy one, was to input Judea's handwritings into the computer.

Judea's door was in front of the Lab with no sign revealing his name so that unwanted intrusions would be discouraged. Years later he added a sign, "Don't knock. Experiments in Progress" that remained there for more than 20 years. Judea liked to work at home early in the day, showing up by his office at 3pm, for meeting students and the secretary, for answering mail, and of course, for thinking, which is what he liked and needed the most. He kept administration to a minimum, and since the 80s at least, has not taught undergraduates (I still don't know how he got away with this). He also used to wear a pair of earplugs, and you could often discover that you said something interesting when you saw Judea taking them off.

What struck me first about Judea was not his research – I couldn't say much about it then – nor his classes, which I certainly liked but were not typical of the 'best teachers' (I still remember Judea correcting a slide in class with his finger, after dipping it into the coffee!), but his attitude toward students, toward science, and toward life in general. He was humble, fun, unassuming, respectful, intelligent, enthusiastic, full of life, very easy to get along with, and driven by a pure and uncorrupted *passion for understanding*. Judea doesn't just seek understanding, he needs it; it's something personal. I'm sure that this is the way scientists and philosophers like Hume, Newton, and Leibniz felt centuries ago, although I doubt that they were as much fun to be with.

In the late 80s, Judea had a small group of students, and we all used to meet weekly for the seminars. Judea got alone well with everyone, and had a lot of patience, in particular with me, who was a mix of rebel and dilettante, and couldn't get my research focused as Judea expected (and much less on the topics he was interested in, even if he was paying my research assistantship!). I remember telling him during the first couple of years that I didn't feel I was ready for research and preferred to learn more AI first. His answer was characteristic: "you do research now, you learn later — after your PhD". I told him also that I wanted to do something closer to the mainstream, something closer to Schankian AI for example, then in fashion. Judea wouldn't get offended at all. He would answer with calm "We will get there eventually", and he certainly meant it. Judea was probably a bit frustrated with me, but he never showed it; quite the opposite, he was sympathetic to my explorations, gave me full confidence and support, and eventually let me do my thesis in the area of non-monotonic reasoning using ideas from probability theory, something that actually attracted his interest at the time.

Since Judea was not an expert in this area (although, unsurprisingly, he quickly became one), I didn't get much technical guidance from him in my specific dissertation research. By that time, however, I had learned from him something much more important: I learned how science is done, and the passion and attitude that go into it. Well, may be I didn't learn this at all, and rather he managed to infect me with the 'virus'; in seminars, in conversations, by watching him work and ask questions, by osmosis. If so, by now, I'm a proud and grateful carrier. In any case, I have been extremely privileged and fortunate to have had the chance to benefit from Judea's generosity, passion, and wisdom, and from his example in both science and life. I know I wouldn't be the same person if I hadn't met him.

# 35
# Sticking With the Crowd of Four
Rina Dechter

I joined Judea's lab at UCLA at about the same time that Hector did, and his words echo my experience and impressions so very well. In particular, I know I wouldn't be the same person, scientist, and educator if I hadn't met Judea.

Interestingly, when I started this journey I was working in industry (with a company named Perceptronics). We had just come to the U.S. then, my husband Avi started his Ph.D. studies, and I was the breadwinner in our family. When I discussed my plans to go back to school for a PhD, I was given a warning by three former students of Judea who worked in that company (Chrolotte, Saleh, and Leal). They all said that working with Judea was fun, but not practical. "If you want a really good and lucrative career," they said, "you should work with Len Kleinrock." This was precisely what I did. I was a student of Kleinrock for three years (and even wrote a paper with him), and took AI only as a minor. During my 3rd year, I decided to ignore practical considerations and follow my interests. I switched to working with Judea.

At that time, Judea was giving talks about games and heuristic search to whoever was willing to listen. I remember one talk that he gave at UCLA where the audience consisted of me, Avi, and two professors from the math department. Judea spoke enthusiastically just like he was speaking in front of the Hollywood Bowl. Even the two math professors were mesmerized.

Kleinrock was a star already, and his students were getting lucrative positions in Internet companies. I congratulate myself for sticking with the crowd of four, fascinated by how machines can generate their own heuristics. Who could tell that those modest seminars would eventually give birth to the theories of heuristics, Bayesian networks, and causal reasoning?

Judea once told me that when he faces a really hard decision, a crossroad, he asks himself "What would Rabbi Akiva do?". Today, when I face a hard decision, I ask "What would Judea do?".

Thanks Judea for being such a wonderful (though quite a challenging) role model!

www.ingramcontent.com/pod-product-compliance
Lightning Source LLC
Chambersburg PA
CBHW082314230426
43667CB00034B/2700